A SWEETWATER RIVER PRESS
PUBLICATION

A FLORA OF
SAN DIEGO COUNTY, CALIFORNIA

A Flora of San Diego County, California

by R. Mitchel Beauchamp

National City, California

Sweetwater River Press

International Standard Book Number: 0-931950-01-5
0-931950-00-7 (pbk.)
Library of Congress Catalog Card Number: 80-51539

Printed in the United States of America
by Dover Litho Printing Company, Dover, DE
Typesetting by Central Graphics, San Diego, CA

With the heart-felt thanks of one native Californian
to another this work is

Dedicated to

"Harbie"

Charles Francis Harbison

Curator Emeritus of Entomology
San Diego Natural History Museum

Who shared with many his knowledge
and love of the natural history
wonders of Alta and Baja California.

TABLE OF CONTENTS

ACKNOWLEDGEMENTS . . . 1
INTRODUCTION . . . 3
SAN DIEGO COUNTY VEGETATION . . . 5
    Geographic Features of San Diego County . . . 5
    Climate . . . 5
    Plant Communities . . . 6
HISTORICAL CHANGES OF SAN DIEGO COUNTY VEGETATION . . . 16
INDEX TO VEGETATION MAP SECTORS . . . 17
VEGETATION MAP SECTORS . . . 18-25
FLORAL DIVERSITY AND ENDEMISM . . . 26
    Floral Districts . . . 26
    Edaphic Endemism . . . 28
FLORISTIC DIVERSITY COMPILATION . . . 29
CITED AND OTHER RELEVANT FLORISTIC LITERATURE . . . 30
CATALOG OF THE FLORA . . . 33
INDEX TO FAMILIES AND GENERA IN THE CATALOG OF THE FLORA . . 235

ACKNOWLEDGEMENTS

Revision of Ethel Bailey Higgins' 1949 annotated distributional floral checklist of San Diego County was proposed in 1970 as the topic of a Master's thesis. It soon became clear that the anticipated two years of course work were too short for the revision. The thesis topic became an analysis and discussion of the floral diversity of the county (Beauchamp 1972), but the checklist revision effort continued.

The present publication has involved a decade and a half of work, the majority being done at the New York Botanical Garden while a postgraduate student. There has been a tremendous amount of botanical investigation in San Diego County during the past two decades due to the requirements of environmental legislation. Employment as a botanical consultant, as well as Veterans Administration educational support while in school in New York and San Diego, allowed intermittent work on this floristic compilation.

Many botanists, naturalists and friends have given aid and moral support in this revision work. Primary among these is former Botany Curator, Reid Moran, who freely edited and criticized various editions of the manuscript. Long-time friend, Reed C. Pierce, aided in searching through various herbaria for new records and by adding to the checklist those common names he dredged up for the pacification of those who refuse to use botanical binomials. The mechanics of book production were shared with me by Bob Doherty.

Other botanical consultants, naturalists and friends who have greatly aided with this publication include Greg Abbott, Wayne P. Armstrong, Mark Bagley, Jeffrey Bohn, Morris H. Brinkman, Timothy L. Cass, Helen V. Chamlee, William D. Clemons, James Dice, Richard L. Ernenwein, Michael E. Evans, Michael U. Evans, Dylan P. Hannon, Jerilyn Hirshberg, Paul Johnson, Eric Jonsson, Jon E. & Sterling Carter Keeley, Steven Lacy, Bill and Bette Mackintosh, Eugene Mielcarek, John Oakley, Thomas A. Oberbauer, Clint Powell, Craig Reiser, John P. Rieger, Ann Marie Spolsky, Fred T. Sproul, Wayne Tyson, Susan Welker, and Bradford K. Williams.

Several other botanists have given constructive criticism on the manuscript as well as information on various taxonomic groups. These include my graduate advisor, H. Lee Wedberg, as well as Loren Anderson, Theodore Barkley and Eileen K. Scholfield, Rupert Barneby, David M. Bates, Dennis Clark, Arthur Cronquist, Barbara Ertter, Ray Fosberg, James Griffin, Ronald L. Hartman, Larry R. Heckard, James Hickman, Arthur, Noel and Patricia Holmgren, John Thomas Howell, Duane Isley, Daryl Koutnick, Arthur R. Kruckeberg, Stuart Lassetter, Elizabeth McClintock, Alan McCready, Nancy Morin, Philip A. Munz, Lorin Nevling, Marion Owenby, W. Robert Powell, Jack and James L. Reveal, Wayne Savage, Avi Schmida, Alan R. Smith, James P. Smith, jr., Otto T. Solbrig, G. Ledyard Stebbins, jr. , Henry J. Thompson, Robert F. Thorne, Ray E. Umber, Lowell E. Urbatsch, Henk van der Werk, David Verity and Grady L. Webster.

My greatest thanks go to my wife, Martha, who encouraged and supported me during the many years of research and who input the entire manuscript to the wordprocessor system.

15 December 1985
National City, California

INTRODUCTION

The diverse climatological and geological history of the region presently called San Diego County is reflected in its diverse flora. No other county within the continental United States appears to have a higher total of native plant taxa. Because of its location as a border region as well as the presence of a large, natural harbor, the region has been well investigated botanically since the visit of Archibald Menzies in 1793. Subsequent expeditions associated with survey of the United States/Mexican boundary, various railroad routes and an historic military campaign helped bring the flora of southwestern California to the attention of the scientific world. Botanists associated with these early expeditions in San Diego County include Coulter, Nuttall, Watson, Parry, Schott, Thurber and Vasey.

From about the 1880's several local botanists such as Spencer Stokes, Cleveland, Orcutt, Brandegee, Kimball, as well as Jones, Parish, Greene, Hall and Palmer, were actively investigating the foothill, mountain and desert areas of the region and sending their discoveries to botanists on the eastern seaboard. Establishment of strong botanical research centers at Berkeley, San Francisco, Stanford—and later Los Angeles and Claremont—heightened the intensity of botanical investigation in the San Diego region.

The flora of San Diego County has been the topic of several botanists' studies. LeRoy Abrams was involved in floristic studies in southern California, especially San Diego County, from 1905 to 1910 (Abrams 1910). Ira L. Wiggins prepared a flora of San Diego County as his PhD thesis in 1929 while at Occidental College. Philip A. Munz also covered San Diego County well in his 1932 Manual of Southern California Botany. Mrs. Ethel Bailey Higgins, former Curator of the herbarium of the San Diego Society of Natural History, produced a flora of the county in 1949. This last compilation has been out of print for over a decade. Also, because of many recent discoveries and revisions of various taxonomic groups based upon new diagnostic techniques, Mrs. Higgins' 1949 flora no longer reflects adequately the great floral diversity of San Diego County. Forest Shreve and Ira Wiggins *Flora of the Sonoran Desert* (1964) includes the eastern desert areas of San Diego County, but this work gives rather broad locality information relative to San Diego County. Earl Lathrop and Robert Thorne's (1985) Santa Rosa Plateau Flora includes the De Luz area of San Diego County and provides a few novelties to this present compilation.

The present work is arranged phylogenetically only so far as to the class level of the vascular flora. Below that level the arrangement is alphabetical. Although a phylogenetic arrangement would be more instructive, the precedent and ease of use set by the alphabetical arrangement of Munz (1974) make a phylogenetic arrangement relatively cumbersome.

The entries for the various taxa in this work are set in a rather terse format. The initial item is the scientific name of the taxon. Relevant synonyms follow, particularly if their types are from San Diego County. "Common" names are generally given, although no consistency is attempted in their derivation. Non-native taxa are indicated by an asterisk (*) and their presumed region of origin is stated.

An adjective which attempts to relate the frequency of the taxon in San Diego County is given. The set from which the values are chosen includes: Rare, Scarce, Uncommon, Infrequent, Occasional, Frequent, Common, Abundant, and Prevalent. The term "Local" has a special meaning denoting a generally limited distribution which is restricted to a few small areas. Clearly, the determination of the frequency value is a subjective estimate based upon 20 years of field work in San Diego County. The degree of frequency is not necessarily correlated with the number of collection localities given in a particular entry.

A short phrase gives the general habitat or plant community associated with the particular taxon. The classification of Munz & Keck (1949) is generally used in the species' treatments because of broader categories offered by their scheme; however, other categories are cited when appropriate.

Several terms are used to describe that portion of the county where the taxon occurs. These terms and their elevational limits are: montane—above 1200m; foothill—300 to 1200 m; coastal—below 300m on the westward coastal slope; desert—generally below 900m on the eastward slope; cismontane—westward draining areas, including westward-draining montane, foothill, and coastal; and transmontane—below 1900m, including desert and montane.

The collection localities are generally listed in a north to south and west to east manner. Unlike the Higgins flora, these localities do not have accession numbers cited. The majority of the localities are from vouchers at the San Diego Natural History Museum, Rancho Santa Ana Botanic Garden or Pomona College Herbarium (the latter two curated within the same facility). Additionally, several other major herbaria were investigated, especially for taxa which had few known collection sites. These additional herbaria are: Jepson Herbarium and University Herbarium, Berkeley; California Academy of Sciences and Dudley Herbarium, San Francisco; New York Botanical Garden, Bronx; Gray Herbarium, Cambridge; U.S. National Herbarium, Washington, D.C.; and Missouri Botanical Garden, St. Louis. Several small, local herbaria also were productive, such as that at San Diego State University, Anza-Borrego Desert State Park; Palomar Mountain State Park; Rancho Cuyamaca State Park, Camp Pendelton Natural Resource Office and the private herbarium of the late Darley F. Howe of Chula Vista. Finally, several locations are cited for which no voucher occurs. These are localities based upon field investigations by the author in conjunction with environmental consulting services. The localities cited include historic sites—no longer listed on current maps—as well as current sites. In locality listings, the use of e.g. indicates that several additional sites are known but not listed. Those listed, however, do give a general description of the taxon's county range.

The flowering or reproductive period of each taxon is cited. These periods are based upon dated herbarium material in flower or as stated in Munz (1974).

The type locality of the listed taxon is stated if it occurs within San Diego County. Many of these are cited in Higgins' *Type Localities of Vascular Plants in San Diego County* (1959).

Finally, somatic or gametic chromosome counts reported for the listed taxon are cited without reference.

Diagnostic keys are adapted from Munz (1974).

SAN DIEGO COUNTY VEGETATION

(Adapted from an analysis by Thomas A. Oberbauer of plant communities in San Diego County.)

GEOGRAPHIC FEATURES OF SAN DIEGO COUNTY

San Diego County is roughly trapezoidal shaped—containing about 2.7 million acres (11,870 hectares)—and is located in the extreme southwestern corner of the continental United States. San Diego County is bounded on the northwest by Orange County, on the north by Riverside County, on the east by Imperial County and on the south by the Mexican state of Baja California Norte. The Pacific Ocean is to the west.

The County lies in the Peninsular Range Geologic Province of Southern California. A mountainous zone with peaks up to 6533 feet (1965 meters) lies on a north-west to south-east trend through the central portion of the County. Various foothill peaks and valleys, the cismontane, extend west of this mountain region. West of the foothills, lies a coastal mesa region of the Tijuana, Otay, Sweetwater, San Diego, San Dieguito, Escondido, San Luis Rey and Santa Margarita River systems. Large, shallow water bays and lagoons exist where the major rivers empty into the ocean. In many areas, the mesas terminate at the ocean in an eroded bluff zone.

East of the mountain ridge, the transmontane drops abruptly in elevation to the desert floor—ultimately descending below sea level just to the east in Imperial County. Peaks protrude from the desert floor interspersed with large valleys and "badlands." San Diego County desert areas occupy the western portion of the Colorado Desert, a part of the Sonoran Desert of western North America. The occurrence of this arid region is attributable to the rain shadow effect of the northern Transverse (San Gabriel and San Bernardino) Mountain Ranges, as well as of the Peninsular (San Jacinto; Palomar, San Ysidro, Cuyamaca and Laguna) Mountain Ranges.

CLIMATE

Moving eastward through San Diego County, physiographic features range from seashore coastal mesas, coastal valleys, foothills, mountains and desert. Seasonal precipitation varies throughout the County associated with these major land forms and elevational differences. The coastal areas receive on the average between 10 and 12 inches (25-30cm) of rainfall annually with nearly all of it falling between October and May. The coastal mesas receive 2 to 4 inches (5-10cm) more rainfall than coastal valleys and up to twice as much as the beaches. Summer thunderstorms are occasional in the foothills and less frequent on the coast. The mountains are the wettest part of the County, with rainfall ranging from 20 to 45 inches (50-110cm) yearly including fairly intense, short summer thunderstorms. The driest part of the County, the desert, receives from 3 to 5 inches (7-13cm) yearly; up to half of that falling during the summer.

Temperatures also vary with elevation. Coastal areas are generally mild with occasional winter frost; a few days reach 100 degrees F (38 degrees C) in summer and fall. Yearly temperature variation increases inland. Coastal valleys have frequent winter frost—some weeks each summer have temperatures over 100 degrees F. Foothill areas are even cooler in winter and warmer in summer. Mountains have mild summers and cold winters with long daily periods of frost. Generally, snow cover in the mountains is of short duration. The desert temperatures are extremely variable.

Desert frosts are not rare in winter, and summer temperatures typically exceed 120 degrees F (49 degrees C) in the lower valleys. High temperatures and winds usually occur during the fall on coastal and foothill areas. These "Santa Ana" conditions are westward winds due to a high pressure cell in interior desert regions.

PLANT COMMUNITIES

Vegetation patterns are generally correlated with climatic and soil conditions of the County. Coast Sage Scrub vegetation was originally the dominant vegetation along the seashore, the southern coastal mesas and the coastal valleys. The major areas of Chaparral are found on the northern coastal mesas and up into the mountains. The drier-adapted Chamise Chaparral grows on the more exposed sites while Mixed Chaparral grows on the less xeric sites.

Oak Woodlands generally exist in two phases: a coastal canyon form that extends into the mountains, and the more open, form of foothill mesas and montane slopes. Coniferous Forest is found mostly in the mountains due to higher rainfall and cooler temperatures. Pinyon-Juniper Woodland vegetation exists on the high desert peaks, above 2500 feet (850 meters), in areas that receive more rainfall than the desert lowlands. Creosote Bush Scrub, the most widespread of vegetation communities in San Diego, exists in the summer-hot deserts. Riparian Woodland can be found in nearly all of the major geographic formations in San Diego County, growing in stream and riverbeds where soil moisture is close to the surface.

In many areas, however, vegetation communities are not distinct, but often blend in broad bands or ecotones at their borders. The discussions below cover the typical associations which represent the plant communities in San Diego County. The categories are derived from several classification schemes and are based upon floristic composition, i.e. Munz & Keck (1949, 1950), Thorne (1976), Cheatham & Haller (1975), Holstein (1980).

Coniferous Associations

Coniferous areas have pines, firs, Bigcone Douglas fir, or Incense cedar as the most conspicuous species. In many areas, they coexist with deciduous and evergreen species. Pinyon pine and juniper occurrences are referred to separately as Pinyon-Juniper Woodland. The tree species comprising Coniferous forest and woodland in San Diego County include the Coulter Pine (*Pinus coulteri*), Jeffrey Pine (*Pinus jeffreyi*), Ponderosa Pine (*Pinus ponderosa*), White Fir (*Abies concolor*), Incense Cedar (*Calocedrus decurrens*), Sugar Pine (*Pinus lambertiana*), Bigcone Douglas Fir (*Pseudotsuga macrocarpa*), Torrey Pine (*Pinus torreyana* ssp. *torreyana*), Cypress (*Cupressus arizonica* ssp. *stephensonii* and *C. guadalupensis* ssp. *forbesii*). Coniferous forests in San Diego County occur as Montane Coniferous Forest, Torrey Pine and Cypress Woodlands.

Montane Coniferous Forest

Montane Coniferous Forest is typically found above 3500-4000 feet (1050-1200 meters) in elevation, growing in the areas that receive greater than 18 to 20 inches (45-50cm) of annual precipitation; a portion of this falls as snow. Areas within San Diego County supporting typical coniferous forest include Mount Laguna, the Cuyamaca Mountains, Corte Madera Mountain, Volcan Mountain, Pine Valley, Pine Hills, Julian, Pine Mountain, Angel Mountain, Hot Springs and Bucksnort Mountains, the San Ysidro Mountains north of Ranchita, Palomar Mountain and Agua Tibia Mountains.

Montane Coniferous Forest differs in form and species composition between these areas. Forests on Cuyamaca and Palomar Mountains receive from 35 to 45 inches (87-112cm) of rainfall and are characterized by massive trees with relatively dense, heavily shaded understories. Coniferous Forest in these areas is composed of Ponderosa Pine, White Fir, Sugar Pine, on Cuyamaca and Hot Springs Mountain only; Bigcone Douglas Fir on Palomar and Volcan Mountains only; Jeffrey Pine and broad leaf trees such as Black Oak (*Quercus kelloggii*), Canyon Live Oak (*Quercus chrysolepis*) and Coastal Live Oak (*Quercus agrifolia*). In contrast, Coniferous Forest on the drier Mount Laguna is composed mostly of more open canopy, medium-sized Jeffrey Pine mixed with Black Oak and some Coulter Pine. Coniferous Forest in the dry floor of Pine Valley is composed almost entirely of Jeffrey Pine. The forest on rather xeric Corte Madera is solely Coulter Pine. Areas of Coniferous Forest are usually surrounded by Mixed and Chamise Chaparral vegetation. In a few places such as at Doane Valley and Laguna Meadow, Coniferous Forest comes into contact with meadow grassland vegetation.

Torrey Pine Woodland

On the coastal bluffs bearing its name, Torrey Pines (*Pinus torreyana*) associate with a unique coastal form of Mixed Chaparral. The presence of this woodland is suspected to be related to the occurrence of summer fogs from off-shore upwelling of cold water.

Cypress Woodland

The Cypress Woodland on Tecate, Otay and Guatay Mountains is composed of Tecate Cypress (*Cupressus guadalupensis* ssp. *forbesii*). The Cuyamaca Cypress (*Cupressus arizonica* ssp. *stephensonii*) occurs in stands on the south face of Cuyamaca Peak. Cypress Woodlands are adapted to infrequent fires as is the chaparral generally adjacent to them. Similar adaptation of Torrey Pine Woodland to fire is yet to be shown.

Oak Woodland

Oak Woodlands in San Diego County include both dense and sparse phases. Oak Woodland-dense phase or Canyon Oak Woodland are areas in which Coast Live Oak (*Quercus agrifolia*) and Canyon Live Oak (*Quercus chrysolepis*) predominate. These trees usually form a dense canopy in canyon bottoms—often called gallery forests—some north-facing slopes, and around the edges of small valleys. Associated species include Poison Oak (*Toxicodendron radicans* ssp. *diversilobum*), Currant and Gooseberry (*Ribes* ssp.), Elderberry (*Sambucus mexicana*, *S. caerulea*). The Oak Woodland-dense phase can be found through all cismontane climatic zones and to a minor extent, on the eastern slopes of the Peninsular Range crest. This type of Oak Woodland can still be found from San Clemente Canyon, Del Dios and De Luz eastward through Potrero, Descanso, Morena Valley, Alpine, Black Canyon, Boulevard, portions of Escondido, Fallbrook, Corte Madera, Guatay, the steep slopes and canyons of the mountains, Oak Grove and near Ranchita. The Coast Live Oak is more common in the woodlands in the lower elevations while the Canyon Live Oak occurs mainly in the areas above 4000 feet (1200 meters) and east of the mountain crest.

Oak Woodland-sparse phase or Southern Oak Woodland is mostly confined to the foothills and mountains. Estimation of the extent of the community by some (Kuckler 1977), appears to be unreasonably high based upon historic records and available habitat. The dominant tree in this community is the Engelmann Oak (*Quercus engelmannii*). It is sometimes associated with Coast Live Oaks and Black Oaks in the Mesa Grande area. This type of woodland is found in areas that receive more than 17 or 18 inches (43-45 cm.) of rainfall and have hot summers. This woodland often has the appearance of an oak savannah where cattle have long grazed; however, in other areas it is associated with White Sage (*Salvia apiana*). Infrequently, Chamise Chaparral is the associated vegetation community. Representative examples of oak woodland-sparse phase occur in the Paradise Mountain area east of Valley Center, Rancho Guajito region, and the hills east of Ramona including Witch Creek, Ballena, Santa Isabel, Mesa Grande, parts of Wynola.

Mixed Evergreen Forest

This community occurs in mesic areas, particularly along creeks and at springs in mountain areas, and in association with Mixed Chaparral or Montane Coniferous Forest. As it occurs in San Diego County, the community is very restricted and appears to be relictual due to its limited occurrence in sheltered areas. Plants of the community are Big Leaf Maple (*Acer macrophyllum*), Bay Laurel (*Umbellularia californica*), Flowering Dogwood (*Cornus nuttallii*), Western Azalea (*Rhododendron occidentale*) California Spikenard (*Aralia californica*), Madrone (*Arbutus menziesii*), Canyon Live Oak (*Quercus chrysolepis*) and Black Oak (*Q. kelloggii*).

Riparian Woodland

Riparian Woodland is composed mostly of winter-deciduous trees that require water near the soil surface. In this community Willow (*Salix* ssp.), White Alder (*Alnus rhombifolia*), California Sycamore (*Platanus racemosa*), Ash (*Fraxinus oregana* var. *velutina*) and Cottonwood (*Populus fremontii* and *P. trichocarpa*) form dense woodlands in moist canyons and drainage bottoms. Plants associated with these small to medium-sized trees include Mugwort (*Artemisia douglasiana*), False Indigo (*Amorpha fruticosa*), Mule-fat (*Baccharis glutinosa*), Stinging Nettle (*Urtica dioica* ssp. *holosericea*), and Wild Grape (*Vitis girdiana*).

Riparian Woodland is of great importance as a wildlife habitat, particularly for birds. Several animals apparently require it as habitat. Riparian Woodlands occur in relatively small canyons in the mountains, the eastern slopes of the mountains and local small cismontane creek beds but their primary development was originally in the large coastal river valleys, where they once covered large areas. The large tracts of Riparian Woodland that still remain in San Diego County occur in the Santa Margarita River and Sandia Creek area, north of Fallbrook; the San Luis Rey River; the San Diego River in portions west of Santee and east of Interstate 805, the Old Mission Dam area; the area upstream of Sweetwater Reservoir into Sloane Canyon on the Sweetwater River; portions of the Otay River Valley; the Tijuana River; and Cottonwood Creek areas south of Barrett Junction.

Shrub Community Types

The majority of Southern California natural vegetation was originally composed of woody shrubs. Summer drought and periodic fires are major factors influencing the structure, morphology and physiology of vegetation in coastal southern California. Desert shrub vegetation has resulted from little rainfall and intense summer heat and drought. The three major shrub community types in coastal San Diego County, i.e. Inland Sage Scrub, Chamise Chaparral and Mixed Chaparral, lie along a gradient of moisture conditions. Inland Sage Scrub is found in the driest areas with Chamise Chaparral and Mixed Chaparral, respectively, in moister areas. Where they come into contact, they often intergrade with one another. Creosote Bush Scrub grows in areas much drier than any of the coastal vegetation community types.

Inland Sage Scrub

Inland Sage Scrub consists primarily of summer drought-deciduous, aromatic shrubs and subshrubs. In most areas, this community is dominated by only a few species of shrubs. The more common are California Sagebrush (*Artemisia californica*), Flat-top Buckwheat (*Eriogonum fasciculatum*), White Sage (*Salvia apiana*) and occasionally Laurel Sumac (*Malosma laurina*). Other plants that are frequently associated with Inland Sage Scrub in San Diego are Broom (*Baccharis sarothroides*), San Diego Sunflower (*Viguiera laciniata*), Golden-yarrow (*Eriophyllum confertiflorum*) and Sawtooth Goldenbush (*Hazardia squarrosa*). These shrubs form a low, somewhat uniform vegetative cover.

Inland Sage Scrub grows in areas that receive between 9 and 15 inches (22-37cm) of rainfall each year. It is the vegetation community that has been impacted the greatest by urbanization. Only about 30 percent of its original extent remains within San Diego County.

The lack of a significant number of resprouting species in Inland Sage Scrub vegetation suggests that it is not adapted to fire as well as its closely-associated Chaparral counterpart.

Typical Inland Sage Scrub still exists in the hills about Poway, the Jamul-Dulzura area, the Jamacha-Singing Hills-Dehesa area, Santee and Lakeside, and on portions of Camp Pendleton.

Maritime Succulent Scrub

Coastal bluff areas contain Sage Scrub which includes many succulent species. This is best developed at Torrey Pines and Point Loma and is called here Maritime Succulent Scrub. Typical Inland Sage Scrub species are mixed with succulents and cacti including Velvet Cactus (*Bergerocactus emoryi*), Coast Barrel Cactus (*Ferocactus viridescens*), Bush Sunflower (*Encelia californica*), Prickly Pear Cactus (*Opuntia littoralis*), Cliff Spurge (*Euphorbia misera*), *Dudleya* spp. and Desert Thorn (*Lycium californicum*).

Chamise Chaparral

Chamise Chaparral grows in areas similar to that of Inland Sage Scrub but receives greater rainfall, or the rainfall is augmented by fog drip. Rainfall in areas of Chamise Chaparral ranges from about 12 to 25 inches (30-62cm). The predominant shrub in this vegetation community type is Chamise (*Adenostoma fasciculatum*). In some places, Chamise is practically the only shrub, forming a dense cover up to 10 feet (3 meters) tall. The plants most frequently associated with Chamise in this community type are Mission Manzanita (*Xylococcus bicolor*), Cleveland Sage (*Salvia clevelandii*), Black Sage (*S. mellifera*) and Coast Spice Bush (*Cneoridium dumosum*). Though the diversity of Chamise Chaparral may be low, these areas become carpeted with herbaceous wildflowers during the spring following a fire.

Chamise Chaparral is one of San Diego County's more widespread cismontane plant communities. It is found on the coastal mesas around Miramar, the hills north of Santee and Lakeside, large areas around Ramona, Alpine and Barrett Junction and on the eastern slopes of the mountains. Also included in this category are the areas dominated by Redshank (*Adenostoma sparsifolium*) such as Chihuahua Valley, Ranchita, and Boulevard. Redshank chaparral is taller (to 10 feet, or 3 meters) and often forms pure stands.

Mixed Chaparral

Mixed Chaparral is composed of shrubs with hard, broad leaves and stiff, woody stems. These shrubs form a dense canopy up to 13 feet (4 meters) high. Areas with this vegetation receive as little as 11 inches (27cm) of rain along the coast—often augmented with fog drip—and up to 35 inches (88 cm) on mountain slopes. The species composition of this community type varies in different parts of San Diego County.

The Mixed Chaparral along the coast, such as the remnants in the Encinitas area, is made up of Coast White-lilac (*Ceanothus verrucosus*), Ramona Lilac (*Ceanothus tomentosus*), the endemic Del Mar Manzanita (*Arctostaphylos glandulosa* ssp. *crassifolia*), Holly-leaf Redberry (*Rhamnus ilicifolia*), Smooth Mountain-Mahogany (*Cercocarpus minutiflorus*), Bush Poppy (*Dendromecon rigida*), Scrub Oak (*Quercus dumosa*) and Chamise.

In interior county areas, such as around Sutherland Reservoir, the most frequently encountered shrubs are White-stem Wild-lilac (*Ceanothus leucodermis*), Thickleaf Wild-lilac (*Ceanothus crassifolius*), Bigberry Manzanita (*Arctostaphylos glauca*), Our Lord's Candle (*Yucca whipplei*), Scrub Oak and again, Chamise.

The southern coastal peaks of Tecate, Otay, Jamul and San Miguel Mountains have an endemic-rich Mixed Chaparral including such localized areas of Cypress Woodland of Tecate Cypress (*Cupressus guadalupensis* ssp. *forbesii*), Southern Mountain Misery (*Chamaebatia australis*), San Miguel Savory (*Satureja chandleri*), Gander Pitcher Sage (*Lepechinia ganderi*), Otay Manzanita (*Arctostaphylos otayensis*), the more widespread Ramona Lilac (*Ceanothus tomentosus*) and Chamise. Mixed Chaparral in the Cuyamaca and Palomar Mountains consists of Mexican Manzanita (*Arctostaphylos pungens*), Coahuilla Manzanita (*Arctostaphylos pringlei*) and Palmer Ceanothus (*Ceanothus palmeri*). As seen above, Mixed Chaparral is a widespread vegetation community in San Diego County and also shows a great deal of floristic variability. Good stands of Mixed Chaparral can be seen in all of the locations mentioned above as well as Campo and a few areas in Rainbow and De Luz not yet planted to avocados.

Sagebrush Scrub

In San Diego County, this plant community is generally restricted to inland areas between 1500 feet and 4000 feet (600 to 1200 meters) in elevation. The only dominant of this community is Great Basin Sagebrush (*Artemisia tridentata*). In more northerly areas however, the community is more diverse. Precipitation in the areas of Sagebrush Scrub of San Diego County ranges from 12 to 20 inches (30 to 50cm) with periods of morning frost or snow cover during several weeks in winter. The commmunity is usually associated with Mottsville series soils and not infrequently occurs in association with Southern Oak Woodland. Areas of Sagebrush Scrub occur at Ranchita, Campo, Tierra del Sol, Pine Valley and Cottonwood Valley near Buckman Springs.

Desert Transition Chaparral

The category of Desert Transition Chaparral is applied to the vegetation on the lower portions of the eastern scarp of the mountains. It ranges from about 1500 to 3500 feet (460-1070 meters) in elevation. It contains plants from both the Pinyon-Juniper Woodland of higher elevations and the Creosote Bush Scrub of the lower elevations as well as species apparently restricted to it. In San Diego County, it is characterized, though not dominated by Turpentine Broom (*Thamnosma montana*), Desert Apricot (*Prunus fremontii*), Desert Jujube (*Ziziphus parryi*), Nolina (*Nolina bigelovii*), Blackbush (*Coleogyne ramosissima*) and scattered California Junipers (*Juniperus californica*) and Scrub Oak (*Quercus cornelius-mulleri*).

Desert Transition Chaparral is commonly found on rocky, boulder-strewn slopes. The vegetation form is that of scattered shrubs reaching up to about 6 feet (2 meters) in height. Desert transition vegetation occurs along nearly the entire eastern slope of the Cuyamaca, Volcan, Hot Springs and Laguna Mountain ranges. Typical examples can be seen in In-Ko-Pah Park, below Banner, on Montezuma Grade and much of San Felipe Valley.

Blackbush Scrub

This plant community is poorly developed within San Diego County. It occurs with greater diversity to the north, in the Mojave Desert area. The indicator of this community is Blackbush (*Coleogyne ramosissima*). This low scrub to 0.5 meters tall merges into Desert Transition Chaparral in the Earthquake Valley, Mountain Springs, and Montezuma Grade areas of the county.

Pinyon-Juniper Woodland

The main indicator plants of Pinyon-Juniper Woodland are Four-leaf Pinyon (*Pinus quadrifolia*) and One-leaf Pinyon (*Pinus monophylla*). Other associated shrubs include California Juniper (*Juniperus californica*), Scrub Oak (*Quercus dunnii, Q. cornelius-mulleri*) and rarely, Antelope Brush (*Purshia glandulosa*). These species grow in the form of a woodland of scattered, small trees, 15 to 30 feet (5-10 meters) tall, intermingled with small to medium-sized shrubs.

The San Diego County areas with Pinyon-Juniper Woodland are mainly located in the higher desert mountains. These include the Vallecito, parts of the San Ysidro, Santa Rosa, Sawtooth and eastern slopes of the Laguna Mountains. Continuous forests of Pinyon Pine can be found in the Sierra Juarez of northern Baja California directly south of Jacumba. Past fires have eliminated much of the former Pinyon-Juniper Woodland in southeastern San Diego County.

Creosote Bush Scrub

Creosote Bush Scrub is the vegetation community type covering the greatest area in San Diego County. Creosote Bush Scrub vegetation grows in the driest portions of San Diego County, from the lowest deserts with a 3-inch yearly rainfall such as Borrego Springs up to nearly 3000 feet (900 meters) and 7 inches (17cm) of rainfall near Jacumba. Creosote Bush Scrub is an open vegetation community in almost all locations; where it grows with shrubs, they are mostly lower than 6 feet (2 meters) in height.

Creosote Bush Scrub can be easily identified by the presence of Creosote Bush (*Larrea divaricata* ssp. *tridentata*) and Burroweed (*Ambrosia dumosa*). In most locations the diversity is low; in some places Creosote Bush is the only shrub.

During the wet years, the open spaces between shrubs may be inhabited by numerous showy annual plants including Desert Monkey-flower (*Mimulus bigelovii*), Desert Dandelion (*Malacothrix californica*), *Gilia*, *Phacelia*, Evening Primrose (*Oenothera deltoides*) and Suncup (*Camissonia* spp.).

Desert Succulent Scrub

On desert bajadas (alluvial fans) and steep, rocky slopes, the flora composition is more diverse than on flatter areas, with many stem-succulents and shrubs—particularly cacti. Precipitation ranges up to 7 inches (17 cm) annually but the more porous substrate influences the association toward a greater diversity than Creosote Bush Scrub.

Plants of this Desert Succulent Scrub plant community include Ocotillo (*Fouqueria splendens*), Brittlebush (*Encelia farinosa*), Chuparosa (*Justicia californica*), Chollas (*Opuntia bigelovii*, *O. fosbergii*), Barrel Cactus (*Ferocactus cylindraceus*), Dudleyas (*Dudleya arizonica*, *D. saxosa* ssp. *aloides*), Fish-hook Cactus (*Mammallaria dioica*, *M. tetrancistra*), Desert Senna (*Senna armata*) and Desert Thorn (*Lycium cooperi*).

Desert Microphyll Woodland

Desert Microphyll Woodland exists in desert arroyos and canyon bottoms. It consists of open clumps of large shrubs and small trees growing to heights of 12 to 15 feet (4 to 5 meters). Many of the plants in this community are phraeatophytic, having deep root systems that may reach into the water table. Thus, they are not directly dependent on the highly variable rainfall. The most common shrubs in this community are Catclaw (*Acacia greggii*), Smoke Tree (*Psorothamnus spinosus*), Desert Lavender (*Hyptis emoryi*), Cheesebush (*Hymenoclea salsola*), Ironwood (*Olneya tesota*), Mesquite (*Prosopis glandulosa*, *P. pubescens*), and Desert Willow (*Chilopsis linearis*). Fan Palms (*Washingtonia filifera*) grow in a few of the canyon oases on the eastern slope of the mountain.

Desert washes and canyons are occasionally inundated by flash floods during the late summer and early fall months. Extreme floods may completely clear washes of shrubs for a few years before seedlings are re-established. Desert wash areas exist in several parts of San Diego County such as Carrizo Wash, San Felipe Wash, Coyote Canyon and Borrego Palm Canyon.

Desert Dune

The reworking of erosional material by winds in the desert often results in sand dunes. Such systems occur in San Diego deserts only in the Borrego Valley-Ocotillo Wells area. Such dunes occur in association with Creosote Bush Scrub but contain a flora of herbs adapted to the shifting substrate. These include Dicoria (*Dicoria canescens*), Milk-vetches (*Astragalus* spp.) and Evening-primrose (*Oenothera deltoides*).

An interesting variation on the Desert Dune community is the Mesquite hummocks. These sand hills, usually 6 to 15 feet high (2-5 meters) accumulate about Mesquite shrubs (*Prosopis glandulosa* var. *torreyana*) through which the shrub continues to grow as the mound increases in height or moves downwind. These Mesquite hummocks are found in the eastern desert areas about Ocotillo Wells and Fish Creek Mine. They are best developed however, to the east in Imperial County.

Alkali Sink

Three desert sinks occur in eastern San Diego County. Two of these, i.e. Clark Dry Lake and Borrego Sink, contain a plant community adapted to the poorly drained alkaline conditions of these features. Plants of this associated community are Saltbush (*Atriplex polycarpa*), Iodinebush (*Allenrolfea occidentalis*), Alkali Isocoma (*Isocoma acradenia*) and Jackass-clover (*Wislizenia refracta*). At Borrego Sink the community has been invaded by the introduced Tamarisk (*Tamarix parvifolia*). The third sink, at Blair Valley, has alkali-associated *Lepidium felipense*.

Grasslands

Grassland is an extremely broad term. It includes areas dominated by herbaceous plants, grasses being most common. Grasslands occur in the mountains, foothills, and coastal regions. The historic extent of Grasslands in San Diego County has been greatly exaggerated by some (Dodge 1975) but the loss of perennial species in favor of annuals reflects the abusive overgrazing this plant community has experienced during the last two centuries.

Grassland species composition varies greatly. Grasslands in the coastal and foothill regions as they exist today, are made up mostly of disturbance-associated, weedy species introduced from the Mediterranean region. The most common species in these areas of San Diego County are Wild Oat (*Avena fatua*, *A. barbata*), Red Brome (*Bromus rubens*), Soft Chess (*Bromus mollis*), Ripgut (*Bromus diandrus*), Foxtail (*Hordeum murinum*) and Fescue (*Vulpia octoflora*, *V. megalura*). Beside these species, there are large numbers of herbaceous dicots including such ubiquitous non-natives as Filaree (*Erodium cicutarium*, *E. botrys*), Mustard (*Brassica nigra*, *B. rapa*, *B. geniculata*), Bur-clover (*Medicago polymorpha*), Star-thistle (*Centaurea melitensis*) and Hedypnois (*Hedypnois cretica*). Occasionally, native grasses (*Stipa*, *Elymus*, *Poa*, *Muhlenbergia*) occur in Grasslands in more upland sites, as at East Mesa in the Cuyamaca Mountains. The extensive grasslands about San Jose del Valle behind Lake Henshaw and,to a lesser degree,Chihuahua Valley have extensive stands of native perennial grasses including *Sporoblous cryptandrus, Stipa cernua* and native annual *Bouteloua aristioides*

Montane Meadow

The moist meadows of San Diego County mountains provide an excellent grassland habitat. Native perennial grasses (*Elymus, Muhlenbergia, Poa*), small rushes (*Juncus*) and sedges (*Carex*) are mixed with herbaceous dicots and annual grasses. Montane Meadows are located in the Laguna Mountains at Crouch and Laguna Meadows, around Cuyamaca Lake and several places on Palomar Mountain including Dyche, Mendenhall, French and Doane Valleys.

Coastal Strand

Coastal Strand vegetation is uncommonly scattered along the San Diego County coast on the narrow strip of sandy beach. It is best developed where sand dunes exist.

Coastal Strand vegetation is composed of sparse, low-growing plants, creeping over the sand. The most common plants on Coastal Strand are the Sea Rocket (*Cakile maritima*), Beach-Primrose (*Camissonia cheiranthifolia*), Bather's Delight or Beach-Bur (*Ambrosia chamissonis*), Beach Morning-Glory (*Calystegia soldanella*) and Milk-vetches (*Astragalus trichopodus*).

Most of the Coastal Strand habitat in San Diego County has been disturbed. The most extensive areas of Coastal Strand in San Diego County occur at Border Field State Park, Imperial Beach, the Silver Strand and the south end of Coronado. Small areas of Coastal Strand-like vegetation exist around the mouths of the larger lagoons.

Coastal Salt Marsh

Coastal Salt Marsh exists primarily in areas with tidal influence, although occasionally the community is found several miles upstream of such influence. Coastal Salt Marshes are inundated with water during very high tides and strong winter storms. Most of the plants occurring in this community type are low-growing, salt-tolerant succulents called halophytes. The common species are Pickle-weed (*Salicornia virginica*), *Frankenia salina*, *Jaumea carnosa*, Bigelow's Glasswort (*Salicornia bigelovii*), *Batis maritima*, Sea-lavender (*Limonium californicum*), Salt-grass (*Distichlis spicata*), Salt-cedar (*Monanthochloe littoralis*) and Dodder (*Cuscuta salina*). Most of these plants are confined to portions of the marsh standing above the water level. However, Cordgrass (*Spartina foliosa*) forms nearly pure stands in shallow levels of salt water.

Presently, large areas of Coastal Salt Marsh only occur in the Tijuana River Valley, the south end of San Diego Bay near the mouths of the Otay River, Sweetwater River and Paradise Creek; the San Diego River flood control channel; and in the northeastern part of Mission Bay. The mouth of Penasquitos Creek and the Santa Margarita River have Coastal Salt Marsh vegetation mixed in with salt flats. Numerous other small salt marshes are located around portions of the major coastal lagoons and estuaries on the northern coast of San Diego County.

Coastal Salt Marsh vegetation is very important for wildlife. Several rare and endangered species of birds as well as plants are dependent upon it for survival. Unfortunately, the Coastal Salt Marsh vegetation that remains is only a small remnant of what originally existed in San Diego County.

Marine Meadow

The calm waters of coastal inlets as well as the pounding surf areas of the coastline are the habitat of several submerged marine angiosperms, i.e. Eelgrass (*Zostera marina*) and Surfgrass (*Phyllospadix torreyi*). These plants, often referred to as marine grasses, although they are by no means true grasses, occur in colonies or beds which are exposed at very low tides. The Eelgrass serves as a food for wintering Brant in the San Diego area. Surfgrass beds are occasional along the coast, as at Point Loma and La Jolla, while Eelgrass occurs in both offshore areas, as at La Jolla, and in calmer inlets, such as Mission Bay, San Diego Bay and Agua Hedionda Lagoon.

Freshwater Marsh

Freshwater Marsh is emergent vegetation that grows in standing fresh water. It is dominated by plants such as Cattails (*Typha latifolia*, *T. domingensis*), Bulrushes

(*Scirpus americana*, *S. robusta*, *S. californicus*) and species of Smartweed (*Persicaria* spp.) and Dock (*Rumex* spp.).

Large areas of Freshwater Marsh are uncommon in San Diego County. They exist at San Elijo Lagoon, Buena Vista Lagoon and Guajome Marsh, the lower portion of the San Diego River and eastern part of Sweetwater Lake and at Carrizo Marsh in the Borrego Desert.

Vernal Pool

Vernal pools occur in naturally-formed depressions which, during the winter rainy season,fill with water. During the spring these pools dry out by evaporation or transpiration. These conditions restrict most plants from growing in vernal pools. A number of species have adapted specifically—and are thus restricted to vernal pools. Due to the restricted nature of vernal pool vegetation, several of the vernal pool plants are endemic.

Vernal pool vegetation is composed of small, prostrate herbaceous plants. Vernal pools generally occur in three areas: the coastal mesas and valleys, the foothill valley of Ramona, and—in the generic sense—Cuyamaca Lake. Both the vegetation in the pools and that surrounding them differ in these areas. In the coastal mesa and valley habitat, the most common species are the San Diego Mesa Mint (*Pogogyne abramsii*), Loma Alta Mint (*Pogogyne nudiuscula*), Popcorn Flower (*Plagiobothrys acanthocarpus*), Downingia (*Downingia cuspidata*), San Diego Coyote Thistle (*Eryngium aristulatum* ssp. *parishii*), Orcutt's Brodiaea (*Brodiaea orcuttii*), California Wiregrass (*Deschampsia danthonioides*) and Woolly Heads (*Psilocarpus tenellus*). Under natural conditions, the mesa pools such as at Miramar, are surrounded by Chamise Chaparral; those on Otay Mesa, by Inland Sage Scrub. Around Cuyamaca Lake, vernal pool-type species include Cuyamaca Lake Downingia (*Downingia concolor* ssp. *brevior*) and Parish Slender Meadowfoam (*Limnanthes gracilis* ssp. *parishii*). In the Ramona area *Downingia cuspidata*, *Myosurus minimus* and *Psilocarpus* are the indicators of remnant vernal pools.

HISTORICAL CHANGES OF SAN DIEGO COUNTY VEGETATION

Although large areas still show little or no effects of human influence, San Diego County vegetation is by no means pristine. Significant vegetation changes began with the first European settlers. Major vegetation changes have occurred in two ways. First, heavy grazing, vegetation clearing, and wood cutting have created grassy areas where they probably did not exist prior to European settlement (Oberbauer 1978). Second, modern development has eradicated large portions of San Diego County vegetation. In the last decade alone, avocado agriculture has taken an immense area of vegetation.

Another way in which vegetation is being changed now is through type conversion. Type conversion is intentionally producing grassland through continued burning or herbicide eradication of shrub species. Controlled burning of native chaparral and woodland understory has only recently come into practice. Almost a century of fire suppression may have changed the aspect of Chaparral by allowing accumulation of senescent vegetation—reducing productivity and threatening the natural fire cycle by too intense heating of dormant seed beds. Whether or not this is the case is still a point of academic investigation.

The impoundment and channelization of the several rivers which drain the west face of the Peninsular Ranges in San Diego County have reduced the dynamic flooding action which once allowed extensive Riparian Woodland to occur almost to the coast. The lack of scouring has resulted in a reduction in the age-class distribution of trees in these woodlands. Similarly, the removal of sand and gravel from these river beds, either by the dams themselves or mining, has eliminated the supply of sand to beaches, thus restricting or eliminating the Coastal Strand habitat.

Filling of coastal wetlands for harbor-oriented development, as well as the major coastal transportation corridors, has restricted the tidal exchange possible in the inlets, bays and estuaries where Coastal Salt Marsh vegetation occurs. This lack of tidal action often results in the filling of the mouths of these water features—lowering plant and animal diversity due to elevated salinity levels.

INDEX TO VEGETATION MAP SECTORS

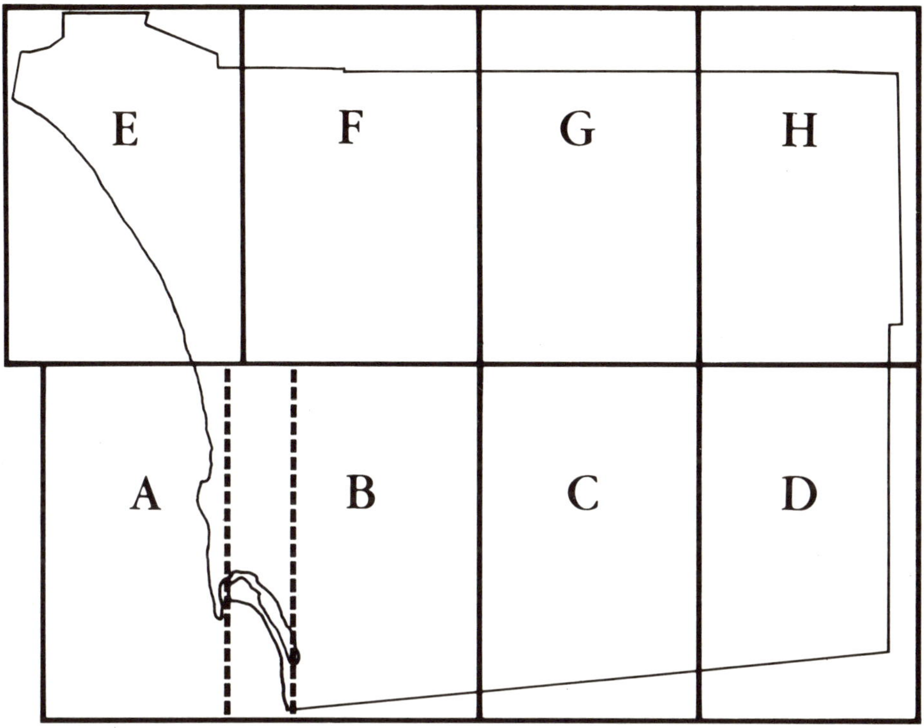

The generalized vegetation map of San Diego County, prepared by Thomas A. Oberbauer in 1977, is included here in a greatly reduced yet useful scale. The map is divided into 8 overlapping sectors, as shown above. Whole maps, at the 1:475,000 scale shown here, or at 1:125,000 scale are available as a separate black and white photograph print from the publisher or the 1:125,000 scale in blue-print form from the County of San Diego, Department of Public Works, Mapping Section.

VEGETATION MAP
Sector A

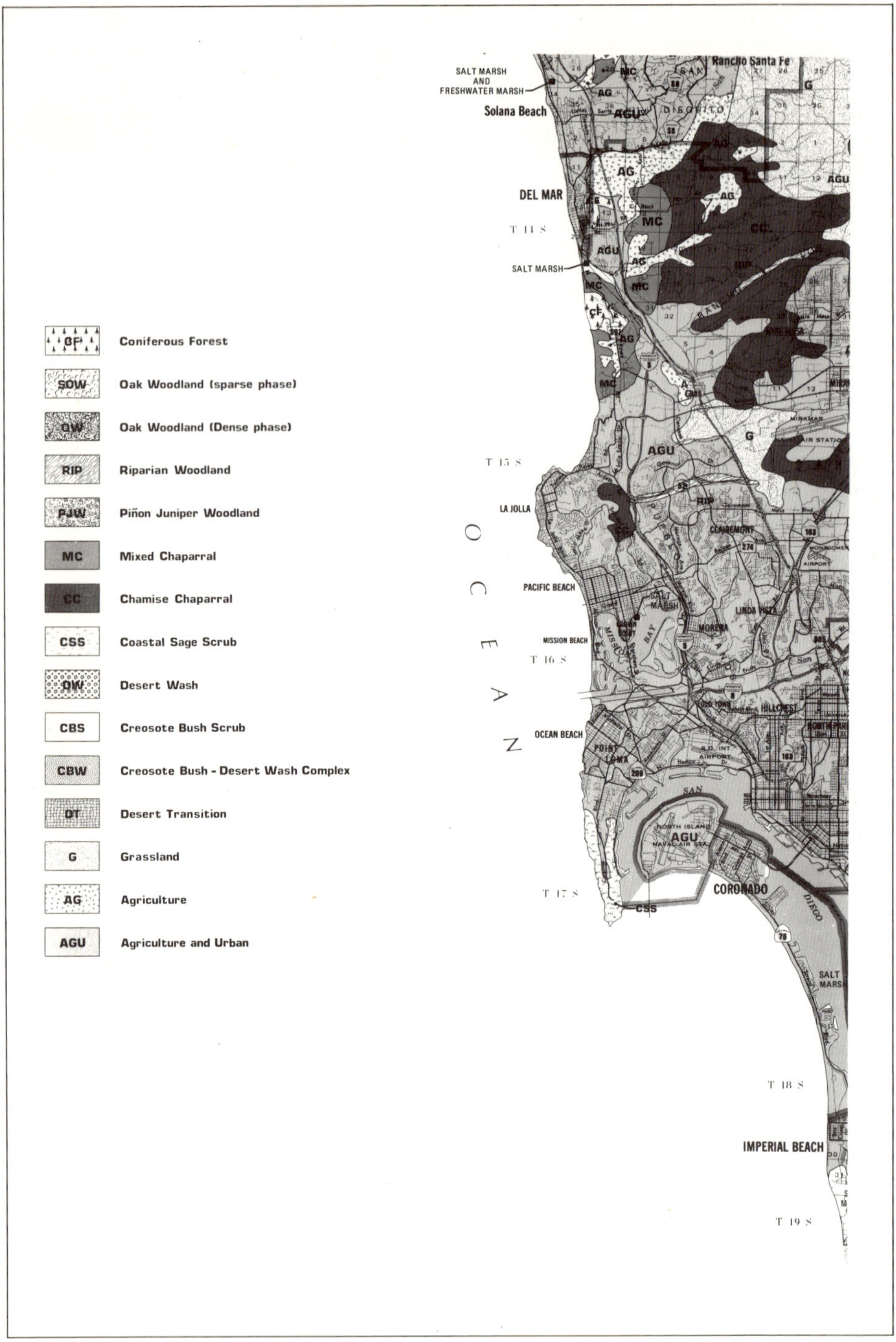

Sector B

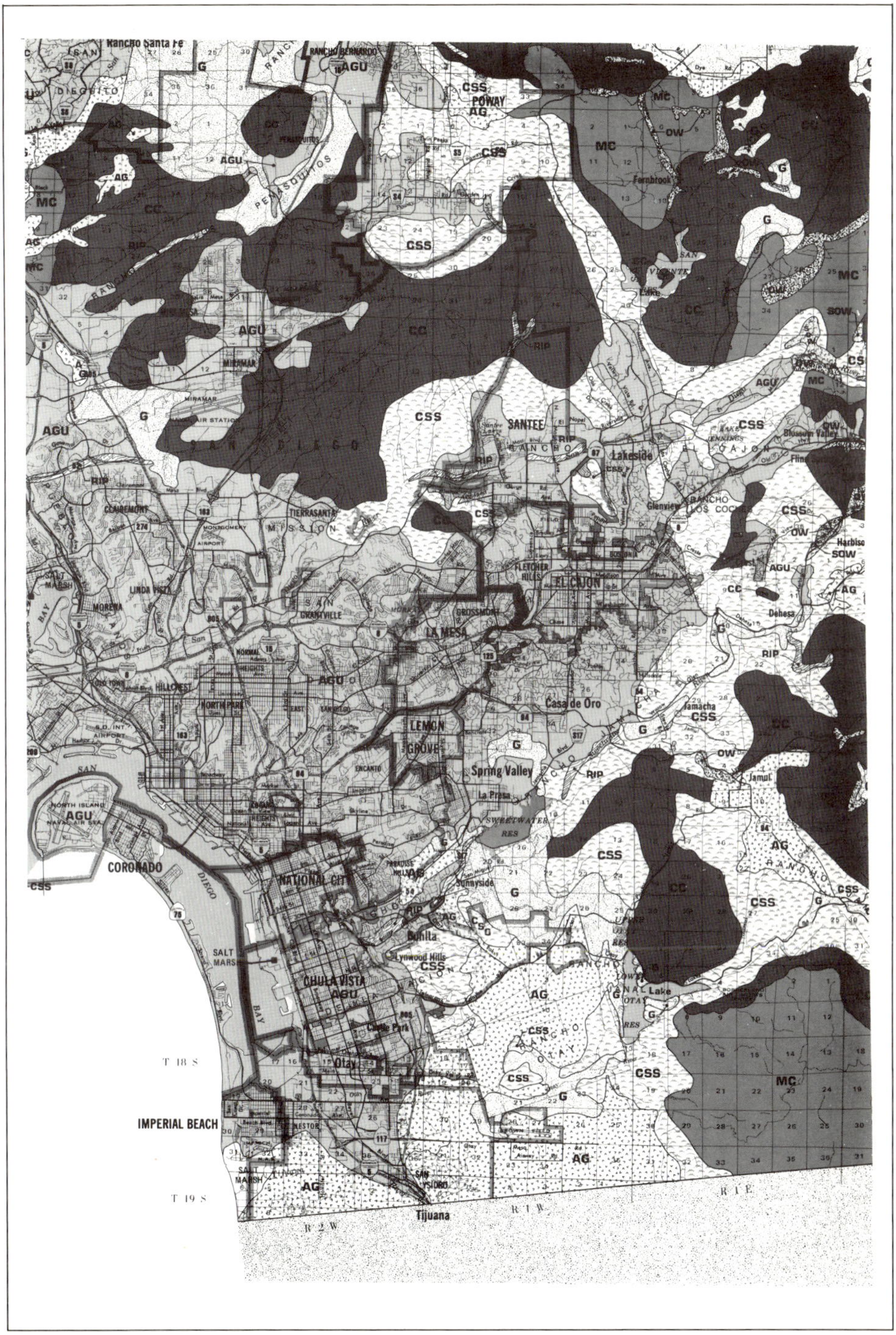

Rancho Santa Fe
RANCHO BERNARDO
POWAY
MIRAMAR
SANTEE
Lakeside
CLAIREMONT
TIERRASANTA
LINDA VISTA
MORENA
GRANTVILLE
LA MESA
EL CAJON
FLETCHER HILLS
GROSSMONT
HILLCREST
NORTH PARK
LEMON GROVE
Spring Valley
La Presa
Casa de Oro
Jamacha
Dehesa
Jamul
Glenview
Fernbrook
NATIONAL CITY
CORONADO
CHULA VISTA
Bonita
Sunnyside
Otay
IMPERIAL BEACH
Tijuana
SAN YSIDRO
NORTH ISLAND NAVAL AIR STA
SALT MARSH
T 18 S
T 19 S
R 2 W
R 1 W
R 1 E
CSS
CC
AGU
AG
MC
OW
G
RIP
SOW

VEGETATION MAP
Sector C

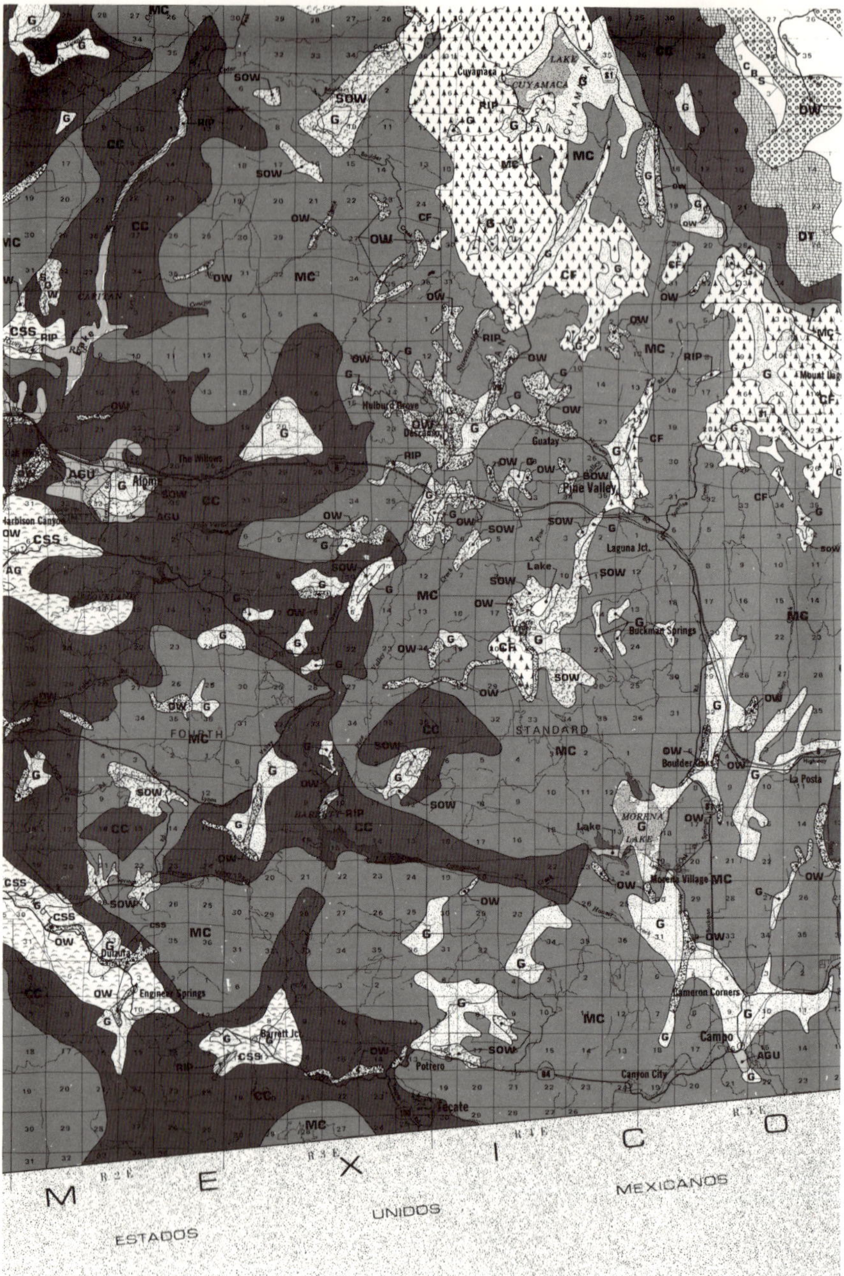

VEGETATION MAP
Sector D

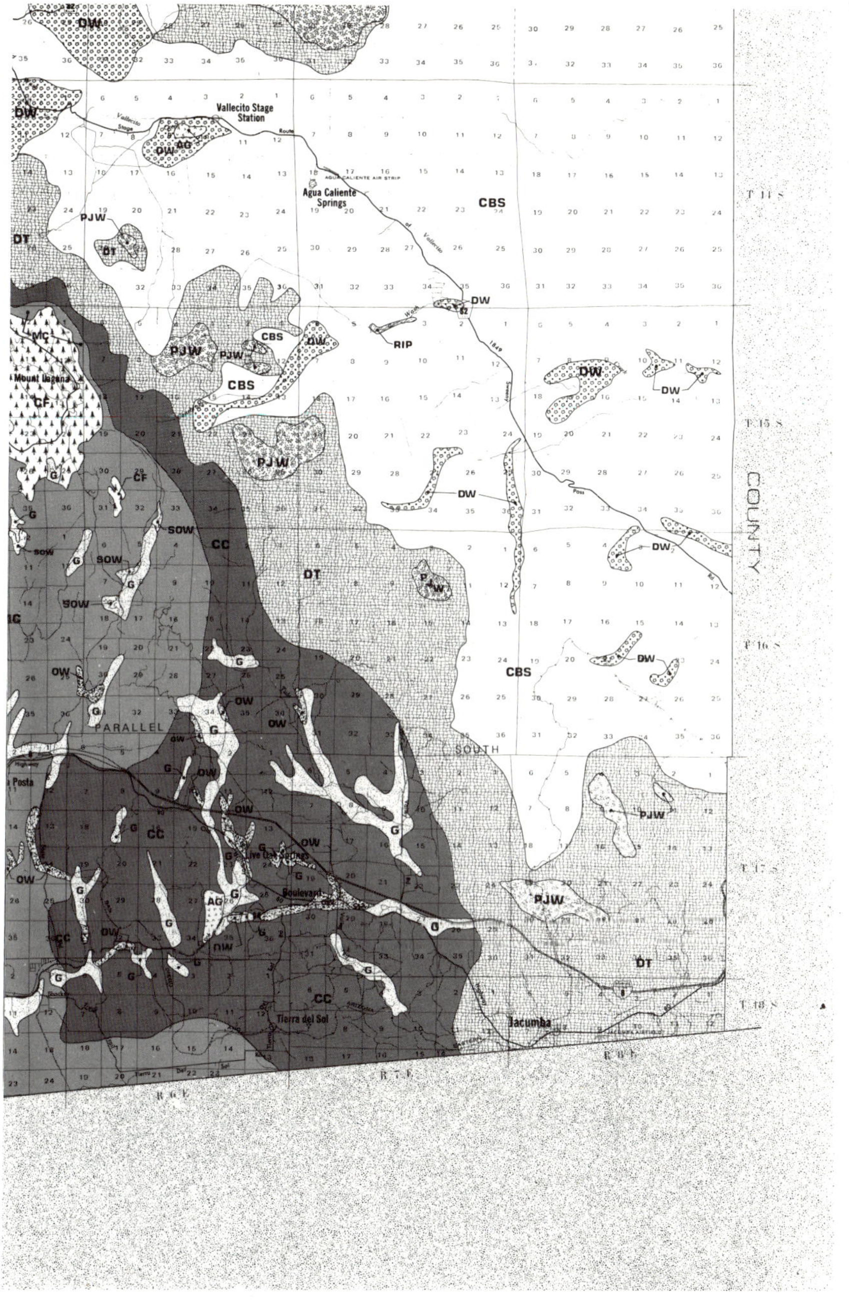

VEGETATION MAP
Sector E

VEGETATION MAP
Sector F

VEGETATION MAP
Sector G

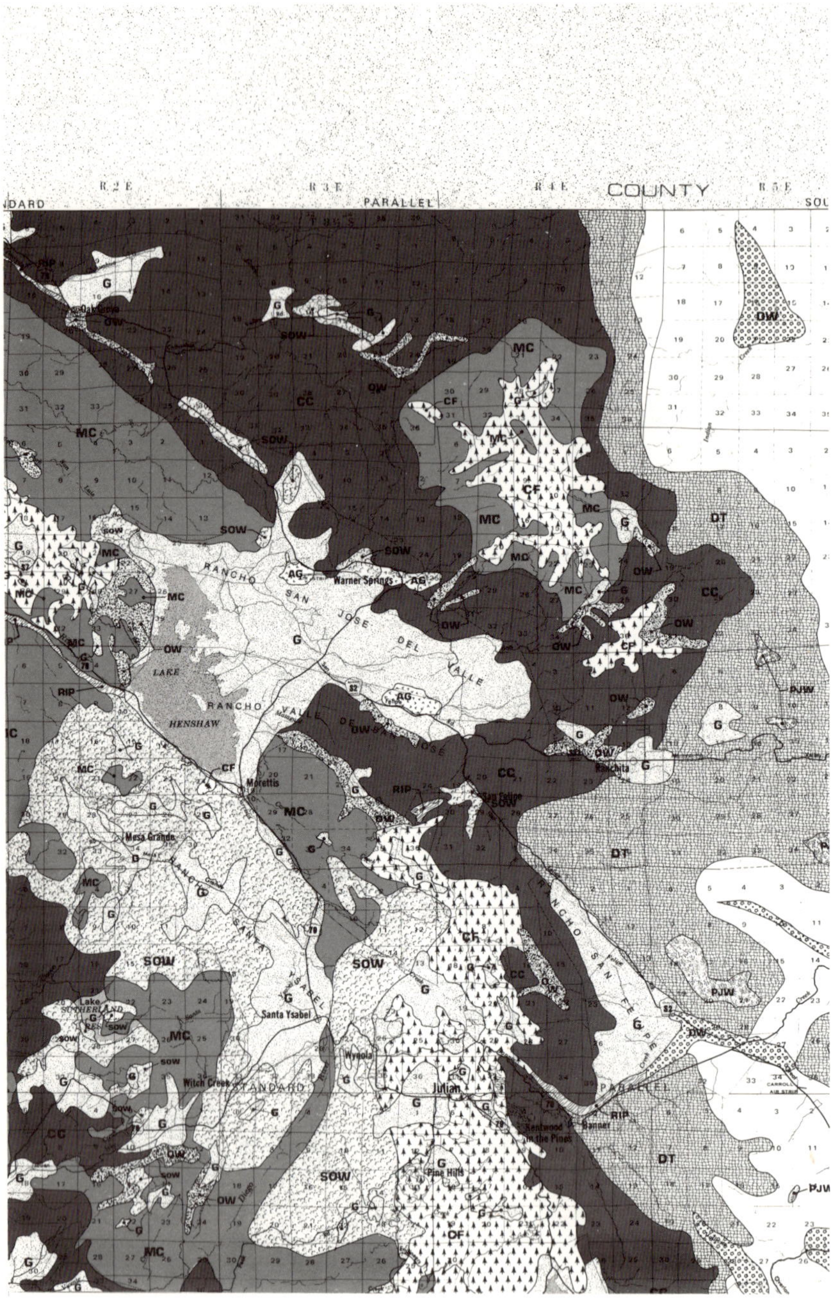

VEGETATION MAP
Sector H

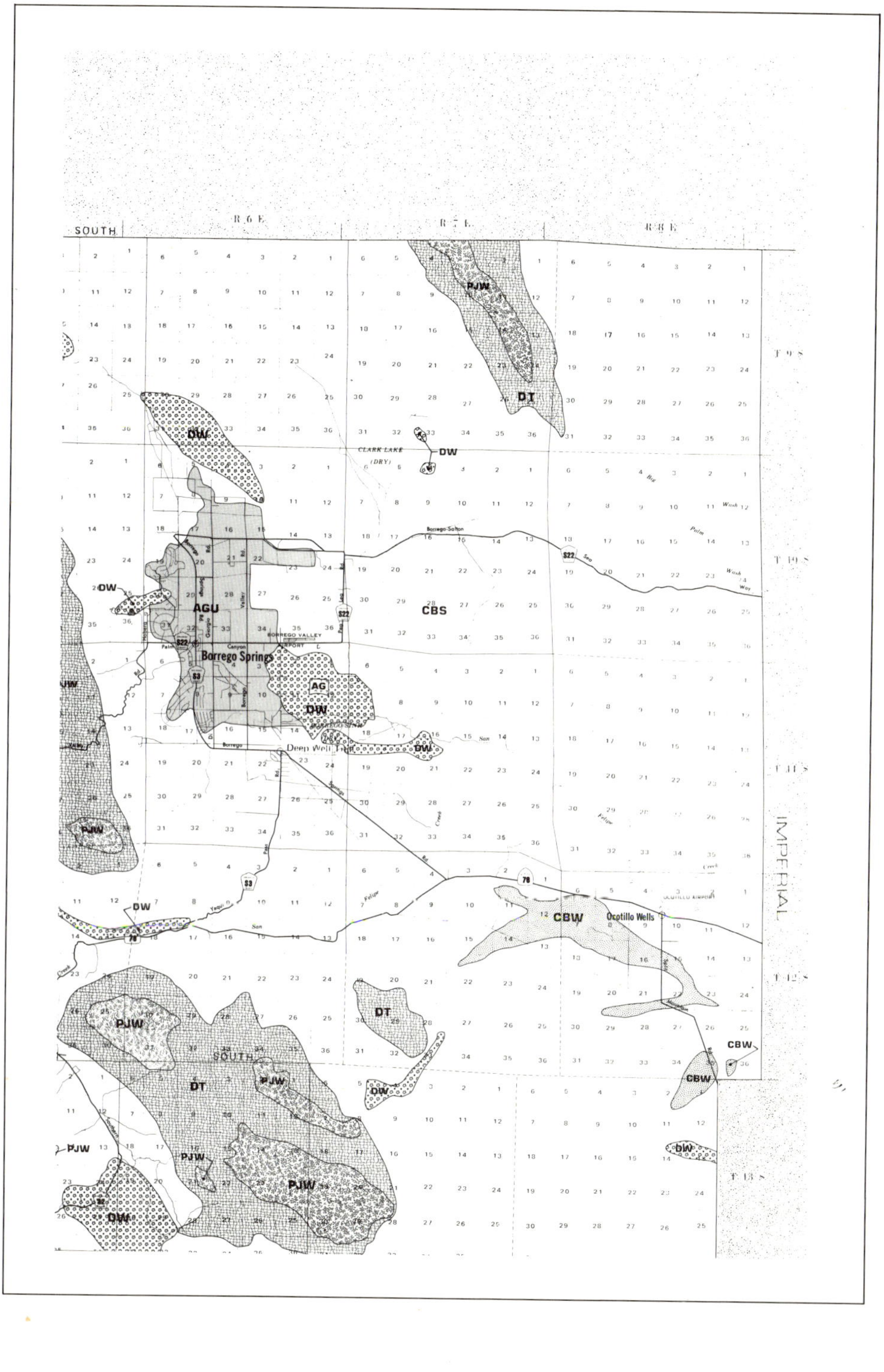

FLORAL DIVERSITY AND ENDEMISM

The various geological and edaphic (soil) situations and climates, as well as past geological events and paleoclimate (Raven and Axelrod 1978), are expressed in the diverse flora of San Diego County. Examination of the collective distribution patterns of various taxa tends to show certain floristic associations—some more strongly correlated than others. These associations, not to be confused with plant communities, are referred to as Floral Districts.

FLORAL DISTRICTS

Sierra Juarez Floral District

Many plants reach their northern distributional limits within San Diego County. In the Jacumba-Boulevard area, as well as westward to Tecate, several of these "northern limitaries" (Beauchamp 1972) occur. Most of these plants are generally more widespread in the Sierra Juarez and its foothills (Wiggins 1980). Taxa of this district include:

Astragalus douglasii var. **perstrictus**
Diplacus aridus
Fraxinus jonesii
Hemizonia floribunda
Ipomopsis tenuifolia
Linanthus bellus
Mahonia higginsiae
Potentilla saxosa
Astragalus coccineus
Ericameria brachylepis
Geraea viscida
Hulsea mexicana
Lathyrus splendens
Lupinus excubitus ssp. **medius**
Penstemon clevelandii
Ribes canthariforme

These elements of the district are best developed in the "high desert" portion of southeastern San Diego County, i.e. 800m to 1200m, but individually they range as far north as Ranchita (Bagley *et al.* 1980) and as far west as Dulzura. Curiously, some northern disjunction occurs in this area with Mohavean elements, such as **Astragalus coccineus** and **Potentilla saxosa**.

Ensenada Floral District

Several plant taxa reach their northern distributional limit on the coastal lowlands of San Diego County. Most of these taxa range as far north as the coastal mesas about San Diego and generally have more extensive ranges in northwestern Baja California Norte. Members assigned to this district include:

Acacia minuta
Adolphia californica
Ambrosia chenopodiifolia
Bergerocactus emoryi
Coreopsis maritima
Dudleya variegata
Ferocactus viridescens
Fremontodendron mexicanum
Lotus nuttallianus
Muilla clevelandii
Acanthomintha ilicifolia
Agave shawii
Ambrosia dumosa
Cordylanthus orcuttianus
Dudleya attenuata ssp. **orcuttii**
Euphorbia misera
Frankenia palmeri
Hazardia orcuttii
Mammillaria dioica
Opuntia parryi var. **serpentina**

Ornithostaphylos oppositifolia
Phacelia ixoides
Pogogyne nudiuscula
Salvia munzii
Selaginella cinerascens
Stipa diegoensis
Viguiera laciniata

Some plants of this district represent a southern extreme of the California Floristic Province, while some, such as **Mammillaria** represent northward excursions of desert species from around the southern end of the Peninsular Range, as represented by the Sierra San Pedro Martir. A large number of these plants occur in the Maritime Sage Scrub.

Otay Mountain Floral District

The summits and slopes of several south coastal peaks in San Diego County and adjacent Baja California Norte contain an assemblage of very restricted plants. Many of these plants reach their northern distributional limit in San Diego County, but, unlike the Ensenada Floral District elements, they occur higher, from 400 to 1000 meters on the peaks of San Miguel, Jamul, Otay and Tecate Mountains. These associated endemics include:

Arctostaphylos otayensis
Calochortus dunnii
Ceanothus otayensis
Chamaebatia australis
Cupressus guadalupensis ssp. **forbesii**
Lepechinia ganderi
Pickeringia montana ssp. **tomentosa**
Satureja chandleri
Solanum tenuilobatum

These plants occur on peaks of Cretaceous metavolcanic rock in an island-like fashion, with intervening areas covered by Grassland, Sage Scrub and Chamise Chaparral.

Cuyamaca Floral District

Above the 1100 meter level, particularly on the three peaks of the Cuyamaca Mountains, are the montane endemics and disjuncts of San Diego County. The pattern of disjunct distribution is found with several taxa in the Palomar, Volcan, Cuyamaca and Laguna Mountains. These taxa occur in the San Diego County mountains as the southern limit of their ranges. Their persistence here may be related to the high precipitation received by these mountain ranges. The plants have large areas of non-occurrence in intervening ranges; not growing particularly in the San Bernardino, San Gabriel and San Jacinto Mountains; and occur further north in the Sierra Nevada. These disjuncts of the Cuyamaca Floral District are:

Arnica cordifolia
Arenaria macrophylla
Arenaria pusilla ssp. **diffusa**
Ceanothus foliosus
Cercis occidentalis
Chimaphila menziesii
Euonymus occidentalis ssp. **parishii**
Malacothamnus aboriginum
Ranunculus aquatilis ssp. **hispidulus**
Rosa gymnocarpa
Salvia sonomensis
Sanguisorba occidentalis
Viola lobata

This disjunction pattern was the topic of a 1931 Master's thesis by Elizabeth Davenport Crow (now Norland), then a student of Philip A. Munz at Pomona College.

EDAPHIC ENDEMISM

The great diversity of geological substrates, and therefore, soils in San Diego County is responsible for an important segment of the floral diversity found there.

Gabbro Edaphic Association

The most stricking endemism is found on soils (Las Posas and Boomer series) derived from Basic Intrusive rocks, termed gabbro. This rock is similar to serpentine (Kruckeberg 1984) in that it possesses a high magnesium content. Gabbroic exposures in San Diego County occur as the Cuyamaca Mountains, and several more coast-ward peaks including Guatay, McGinty, Potrero, Viejas, Poser and Iron Mountains. The rock weathers more readily than the widespread and often adjacent granodiorite; gabbro peaks can be easily spotted by way of the reddish soil color and lack of large exposed boulders. Plants of this Edaphic Association include:

Acanthomintha ilicifolia
Calochortus dunnii
Caulanthus stenocarpus
Horkelia truncata
Salvia sonomensis
Tetracoccus dioicus
Calamagrostis densa
Caulanthus simulans
Delphinium hesperium ssp. **cuyamacae**
Nolina parryi
Senecio ganderi

The Otay Mountain Floral District frequently has taxa assigned to this Gabbro Edaphic Association. The relationship between gabbroic soils and those derived from the Cretaceous metavolcanics of the Otay Mountain Floral Association is in need of clarification.

Clay Edaphic Association

In several areas of coastal San Diego County, lenses of clay soils occur. Occasionally these are in association with gabbroic rocks, while at other times they are on marine sediments. The soils of this association include Diablo and Auld series. Due to the expansive nature of the clays, few shrubs do well in these soils. The Clay Edaphic Association includes:

Acanthomintha ilicifolia
Bloomeria crocea
Convolvulus simulans
Evax sparsiflora
Grindelia robusta
Muilla clevelandii
Adolphia californica
Brodiaea filifolia
Dudleya variegata
Fritillaria biflora
Hemizonia conjugens

FLORISTIC DIVERSITY COMPILATION

As is usual for a regional floristic treatment, an examination of the extent of diversity of the vascular flora is performed below. For the purposes of calculating the listed totals, those families and genera with at least one native taxon are assigned as native. When a single variety or subspecies is listed for a species, that entry is scored as a species. When two or more infra-specific taxa are listed for a species, those additional infra-specific taxa are scored as infra-specific taxa.

	Families		Genera		Species		Infra-specific taxa	
	Native	Non-native	Native	Non-native	Native	Non-native	Native	Non-native
Cryptogams	7	0	18	2	36	2	3	0
Gymnosperms	3	0	7	0	19	0	0	0
Angiosperms—Monocots	18	5	80	51	229	117	12	3
Angiosperms—Dicots	100	10	446	132	1232	345	210	2
	128	15	551	185	1516	464	225	5
	143		736		1980		230	

2210 Total Taxa

CITED AND OTHER RELEVANT FLORISTIC LITERATURE

Abrams LeRoy. 1910. A phytogeographic and taxonomic study of the southern California trees and shrubs. *Bull. N. Y. Bot. Gard.* 6:300-485.

____________. 1917. Flora of Los Angeles and Vicinity. Stanford University Bookstore, Stanford, California. 432pp.

____________. 1923-1960. Illustrated Flora of the Pacific States. Stanford University Press. 4 volumes

Armstrong, Wayne P. 1966. Ecological and Taxonomic Relationships of *Cupressus* in Southern California. MA thesis, Unpubl. MS. California State University, Los Angeles. 124 pp.

Axelrod, Daniel I. 1950 Studies in Late Tertiary Paleobotany. *Carnegie Inst. Wash. Pub.* 590. 306pp.

____________. 1958. Evolution of the Madro-Tertiary Geoflora. *Bot. Rev.* XXIV(7):433-509.

Bagley, Mark O., Richard L. Ernenwein & Joseph Liotta. 1980. Sensitive Plant Investigation of the Eastern San Diego County Planning Unit. U.S. Department of the Interior, Bureau of Land Management, El Centro Resource Area Office. 113pp.

Barbour, M. G. and J. Major, (editors). 1977. Terrestrial Vegetation of California. John Wiley & Sons, New York. 1002pp.

Beauchamp, R. Mitchel. 1972. Floral Diversity of San Diego County, California. M.Sc. thesis, Unpubl. MS. California State University, San Diego. 73pp.

Beauchamp, R. Mitchel and T. L. Cass. 1979. San Diego Vernal Pool Survey. California Department of Fish and Game, Non-Game Wildlife Investigation, Endangered Plant Program 145 Job I-1.0.

Benson, Lyman and Robert A. Darrow. 1981. Trees and Shrubs of the Southwestern Deserts, 3rd ed.. University of Arizona Press, Tucson. 416pp.

Cheatham, N. H. & J. Robert Haller. 1975. An Annotated List of California Habitat Types. University of California Natural Land & Water Reserve System. University of California Press, Berkeley.

Cooper, James Graham. 1874. The Botany of the Cuyamaca Mountains. *Am. Nat.* 8:90-98.

Davidson, Anstruther and George L. Moxley. 1923. Flora of Southern California. Times-Mirror Press, Los Angeles. 452pp.

Dodge, John Marvin. 1975. Vegetational Changes Associated with Land Use and Fire History in San Diego County. PhD Thesis, Unpubl. MS University of California, Riverside. 216pp.

Greene, Edward Lee. 1880 & 1881. Botanizing on the Colorado Desert. *Am. Nat.* 14:787-793; 15:24-32.

Higgins, Ethel Bailey. 1949. Annotated Distributional List of Ferns and Flowering Plants of San Diego County, California, *Occ. Pap. San Diego Soc. Nat. Hist.*, No. 8. 174pp.

____________. 1959. Type localities of vascular plants in San Diego County, California. *Trans. San Diego Soc. Nat. Hist.* 12(22):347-406.

Holstein, Glen. 1980. California Vegetation Cover Types. California Natural Diversity Data Base. California Department of Fish and Game, Sacramento.

Jepson, Willis Linn. 1923. The Trees of California. California School Book Depository, San Francisco. 240pp.

____________. 1925. A Manual of the Flowering Plants of California. Univerisity of California Press, Berkeley. 1238pp.

____________. 1909-1985. A Flora of California Vols. 1(1-7), 2, 3(1-3).

Kruckeberg, Arthur R. 1984. California Serpentines: Flora, Vegetation, Geology, Soils and Management. *Univ. Calif. Publ. Bot.* Vol. 78.

Kuckler, A. W. 1977. The Map of the Natural Vegetation of California. Appendix in Barbour & Major, 1977, pp. 909-938.

Lathrop, Earl W. and R.F. Thorne. 1978. A Flora of the Santa Ana Mountains, California. *Aliso* 9(2):197-278

____________. 1985. A Flora of the Santa Rosa Plateau, Southern California. Southern California Botanists, Special Publication No. 1. 39pp.

McMinn, Howard E. 1939. An Illustrated Manual of California Shrubs. University of California Press, Berkeley. 663pp.

Munz, Philip A. 1935. A Manual of Southern California Botany. Scripps Publishing Fund, Clairemont, California. 624pp.

____________. 1974. A Flora of Southern California. University of California Press, Berkeley. 1068pp.

Munz, P.A. and D.D. Keck. 1959. A California Flora. University of California Press, Berkeley. 1681pp.

____________. 1949. California Plant Communities. *Aliso* 2:87-105.

____________. 1950. California Plant Communities Supplement. *Aliso* 2:199-202.

Oberbauer, Thomas A. 1977. Generalized Vegetation Map of the County of San Diego. County of San Diego, Mapping Section.

____________. 1978. Distribution of Dynamics of San Diego County Grasslands. MSc Thesis, Unpubl. MS. San Diego State University. 120pp + map.

Orcutt, Charles Russell. 1885. Our Peninsular Flora. *West Am. Sci.* 1:25.

____________. 1886. Report on the Trees and Shrubs of San Diego County, California. *Bienn. Rep. Calif. State Board of Forestry* 1:45-52.

____________. 1887. Aquatic Plants of the Vicinity of San Diego. *West. Am. Sci.* 3:123.

Parish, Samuel Bonsall. 1903. A Sketch of the Flora of Southern California. *Bot. Gaz.* 36:203-222 and 259-279.

Purer, Edith A. 1939. Ecological Study of Vernal Pools, San Diego County. *Ecology* 20: 217-229.

____________. 1942. Plant Ecology of the Coastal Salt Marshlands of San Diego County, California. *Ecology Monograph* 12:81-111. January, 1942.

Raven, Peter H. and Daniel I. Axelrod. 1978. Origin and Relationships of the California Flora. *Univ. Calif. Publ. Bot.* 72. 134pp.

Robbins, W. W., Margaret R. Bellue & Walter S. Ball. ca. 1951. Weeds of California. California Department of Food and Agriculture. 547pp.

Scott, Carroll Dewilton. 1955. Story of the Wild Flowers of Southern California. Unpubl MS 78pp.

Shreve, Forest and I.L. Wiggins. 1964. Vegetation and Flora of the Sonoran Desert, Vols. 1 and 2. Stanford University Press, Stanford, California. 1740pp.

Spolsky, Ann Marie. 1978. A Report of the Status of the Rare and Endangered Plants of Anza-Borrego Desert State Park. Title VI CETA Project. 65pp.

Sudworth, George Bishop. 1888. Trees and Shrubs of San Diego County, California. *U.S. Dept. Agr., For. Div. Bull* 2:202-205.

Thorne, Robert F. 1976. The Vascular Plant Communities of California, pp. 1-31 in J. Latting (ed.) *Symposium Proceedings: Plant Communities of Southern California*. California Native Plant Society, Special Publication No. 2. Riverside, California.

Thurston, Carl. 1936. Wildflowers of Southern California. Esto Publishing Company, Pasadena, California. 412pp.

Warner, Richard E. & Kathleen M. Hendrix (eds.). 1984. California Riparian Systems: Ecology, Conservation, and Productive Management. University of California Press, Berkeley. 1035pp.

West, John and Mark Donnelly. ca. 1980. Botany of San Diego County, An Annotated Bibliography Compiled by John West, 3rd edition. San Diego State University Library. 176pp.

Wiggins, Ira Loren. 1929. Flora of San Diego County, California. PhD thesis, Stanford University. 888pp.

____________. 1980. Flora of Baja California. Stanford University Press, Stanford, California. 1025pp.

Zedler, P. H., T. A. Ebert, M. B. Balko, and R. M. Beauchamp. 1979. A Survey of Vernal Pools of Kearny Mesa, San Diego County, California. Spring 1979. San Diego District CalTrans Grant 06309. 152pp.

CATALOG OF THE FLORA

Division **TRACHEOPHYTA**

KEY TO MAJOR GROUPS

Plants without seeds or flowers, reproducing by 1-celled spores borne in sporangia
Plants not fern-like and not free-floating, the leaves mostly minute or somewhat scalelike (except in *Isoetes*), 1-veined; the fertile leaves with 1 sporangium on the adaxial surface
Stems not jointed; leaves not whorled and not forming a sheath at the nodes . . . . Subdivision *Lycopsida*
Stems of hollow joints; leaves whorled and forming a sheath at the solid nodes . . . . Subdivision *Sphenopsida*
Plants fern-like and with large leaves (fronds) or free-floating aquatics and with small or scalelike overlapping leaves; mostly elaborately veined; sporangia of spreading leaves usually abaxial . . . Subdivision *Pteropsida* Class *Filicae*
Plants with seeds usually produced by cones or flowers
Seeds not enclosed in ripened pistils, but naked and usually borne on the surface of a scale; the crowded or overlapping scales usually forming a cone or strobilus; plants not bearing typical flowers
Stems not jointed; leaves needlelike or linear, sometimes scalelike and then closely overlapping; plants mostly with resin; mostly trees or arborescent . . . . Class *Coniferae*
Stems jointed; leaves scalelike, in 2's or 3's in widely separated whorls; plants without resin; cones small, with thin scales; desert shrubs . . . . Class *Gnetae*
Seeds enclosed in ripened pistils; plants producing true flowers . . . . Class *Angiospermae*
Cotyledons 2; stems usually exogenous (increasing in diameter by cambial activity between the xylem and phloem); leaves mostly net-veined; flowers usually on the plan of 4 or 5 . . . . Subclass *Dicotyledones*
Cotyledons 1; stem mostly endogenous, or if with cambial activity, this adding whole vascular bundles; leaves mostly parallel-veined; flowers mostly on the plan of 3 . . . . Subclass *Monocotyledones*

Subdivision **LYCOPSIDA**

Class **LYCOPODIAE**

KEY TO ORDERS

Stems above ground, somewhat elongate and branched, covered with minute overlapping, scale-like leaves less than 1cm long. . . . . *Selaginellales*
Stems underground, cormlike, bearing tufted leaves 2.5-25cm long . . . . *Isoetales*

Order **SELAGINELLALES**

SELAGINELLACEAE. Spike-Moss Family

Selaginella Beauv. Spike-Moss, Mossfern

Stems erect or ascending, rooting only at or near the base . . . . *S. bigelovii*
Stems prostrate to decumbent, rooting at or near the apex
Leaves on the underside of the main stems not decurrent, lacking a terminal point or seta; stems forming an ashy carpet-like growth very near the ground . . . . *S. cinerascens*
Leaves on the underside of the main stems strongly decurrent
Leaves lacking a terminal seta at maturity; plants strongly dorsi-ventral; desert. . . . . *S. eremophila*
Leaves with a terminal seta; stems radially symmetrical . . . . *S. asprella*

S. asprella Maxon. Bluish Mossfern. Scarce, on rocky eastern slopes of the Laguna Mountains; montane, above 1650m: Desert View.

S. bigelovii Underw. Bigelow's Mossfern. Common, in rocky places; cismontane and desert: e.g. San Onofre Canyon; Rainbow Valley; Mission Gorge; Poway Grade; Del Mar; Dulzura; Morena Dam; Oriflamme Canyon.

S. cinerascens A. A. Eaton. [*S. bryoides* auct. non (Nutt.)Underw.] Mesa Mossfern. Frequent, on otherwise bare slopes and mesas; coastal, below 300m: e.g. Carlsbad; Poway Grade; Del Dios; Kearny Mesa; San Diego; Proctor Valley; Alpine; Woodwardia Canyon. Type from "bare ground, National City".

S. eremophila Maxon. [*S. parishii* Underw., in part] Desert Mossfern. Infrequent, in shaded canyons; desert, below 1000m: Hellhole Canyon; Sentenac Canyon; Oriflamme Canyon; Campbell Ranch; Mountain Palm Springs.

Order **ISOETALES**

ISOETACEAE. Quillwort Family

Isoetes L. Quillwort

Corms usually 2-lobed, velum or basal flap completely covering the sporangium . . . . *I. howellii*
Corms usually 3-lobed; velum usually nearly or quite covering the sporangium . . . . *I. orcuttii*

I. howellii Engelm. [*I. h.* var. *minima* (A. A. Eat.)Pfeif.] Howell's Quillwort. Rare, in vernal pools; coastal: Kearny Mesa, 150m.

I. orcuttii A. A. Eaton. [*I. nuttallii* var. *o.* Clute] Orcutt's Quillwort. Scarce, in vernal pools and other ephemeral ponds; coastal and foothill, below 400m: Rainbow Valley; San Marcos; Poway Junction; Kearny Mesa; Bottle Peak, Lake Wohlford. Type from San Diego from "low depressions on top of the mesas".

Subdivision **SPHENOPSIDA**

Class **EQUISETAE**

Order **EQUISETALES**

EQUISETACEAE. Horsetail Family

Equisetum L. Horsetail; Scouring-Rush

Aerial stems all green, not of two kinds, stomates in regular rows
  Sheaths on stem much longer than broad, somewhat funnel-shaped, green, normally with a narrow black band at the top . . . . . *E. laevigatum*
  Sheaths nearly or quite as broad as long, cylindrical, usually ashy with 2 black bands . . . . . *E. hyemale*
Aerial stems of 2 kinds, the fertile ones not green or branched, the sterile green and with many slender branches; stomates scattered
  Sterile stems stout, generally over 3mm thick and 6-12dm tall; sheaths with 20-30 teeth . . . . . *E. telmateia*
  Sterile stems slender, generally less than 3mm thick and less than 6dm tall; sheaths with 8-12 teeth. . . . . *E. arvense*

E. arvense L. Common Horsetail. Local; in moist places in lower Doane Valley, 1370m.

E. hyemale L. ssp. **affine** (Engelm.)Calder & Taylor. [*E. h.* var. *robustum* (A. Br.)A. A. Eaton; *E. h.* var. *a.* (Engelm.)A.A.Eaton] Tall Scouring-Rush. Infrequent, in moist areas; montane, above 1200m: Palomar and Cuyamaca Mountains. Occasionally escaping *ex hort* as at San Marcos.

E. laevigatum A. Br. [*E. funstonii* A. A. Eaton, *E. kansanum* Schaffn.] Smooth Scouring-Rush. Frequent, in moist places; cismontane: e.g. Fallbrook; Henshaw Dam; Doane Valley; Pine Hills; Shaw Canyon; Buckman Spring; Tecate Peak.

E. telmateia Ehrh. var. **braunii** Milde. [*E. maximum* Lam.] Braun's Horsetail. Uncommon, in moist areas; cismontane, below 1500m: De Luz Creek; La Jolla Indian Reservation; below Cuyamaca Dam.

Subdivision **PTEROPSIDA**

Class **FILICAE** Ferns

KEY TO ORDERS

Plants fernlike, terrestrial or on mud, not floating . . . . . *Filicales*
Plants not particularly fern-like, i.e. with dissected laminar portions
  Spores produced on a small, erect stalk issuing from the base of an entire green blade up to 4cm tall *Ophioglossales*
  Spores borne on structures on the underside of the branches or at the base of the leaves
    Fronds or leaves uncoiling at emergence . . . . . *Marsileales*
    Fronds or leaves not uncoiling, sessile, mostly floating on water . . . . . *Salviniales*

Order **FILICALES**

POLYPODIACEAE. Fern Family

[Segregate families are shown parenthetically]

KEY TO GENERA

Sporangia borne near or at the margins of the leaves, i.e. near the apex of the veins or decurrent on the veins or completely covering them
  Sporangia following the veins throughout, hence sporangia on the back of the fronds
    Fronds deltoid-pentagonal to deltoid-ovate; stipes about twice as long as the blade . . . . . *Pityrogramma*
    Fronds elongated, stipes not as long
      Pinnae broad, the ultimate segments at least 1cm wide, somewhat fan-shaped, ovate, or reniform-orbicular, each with a slender, stipe-like stalk . . . . . *Adiantum*
      Pinnae narrower, usually somewhat elongate and mostly sessile or even decurrent
        Fronds large, to 1.5m long from running underground rhizomes; blades 3-pinnate to 1m long, pubescent beneath . . . . . *Pteridium*
        Fronds mostly less than 5dm long, or tufted on a short rhizome
          Fronds 1-3-pinnate, the segments fairly large, more than 2mm long, somewhat articulate, oval to narrowly elliptic with long, entire margins, or sometimes with 1 or 2 large smooth lobes; rachis usually dark, shining, glabrous . . . . . *Pellaea*
          Fronds 2-4-pinnate, the segments small, bead-like or with lead-like lobes usually less than 2mm long . . . . . *Cheilanthes*
  Sporangia not marginal, but borne in separate sori away from the margins
    Sori round to oval
      Indusia, i.e. the covering of the sporangia, lacking, the sori naked . . . . . *Polypodium*
      Indusia present although sometimes soon concealed by the maturing sporangia; stipes not jointed to the rhizome
        Indusium round, centrally attached; leaves once pinnate, the basal pinnae sometimes pinnate; texture coriaceous
          Veins free. . . . . *Polystichum*
          Veins anastomosing. . . . . *Cyrtomium*
        Indusium not centrally attached, leaves at least bipinnate throughout; texture not coriaceous
          Stipes slender, less than 1.5mm in diameter; blade of frond 1-1.5dm long and 5-8cm wide . . . . . *Cystopteris*
          Stipes coarser, 2-4mm in diameter; blade of frond 2.5-5dm or more long and 1-2dm wide
            Blades with distinctly toothed segments; veins 2-several times forked. . . . . *Dryopteris*
            Blades with entire or subentire segments; veins simple or once forked . . . . . *Thelypteris*
    Sori oblong, linear to lunate, or horseshoe-shaped
      Fronds 6-20dm tall; indusium often more or less curved. . . . . *Woodwardia*
      Fronds 0.3-2.5dm tall, pinnate; indusium straight . . . . . *Asplenium*

Adiantum L. Maidenhair
[Adiantaceae]

Blade segments fan-shaped, deeply and somewhat irregularly lobed; stipes slender, nearly black, about as long as the blades . . . *A. capillus-veneris*
Blade segments rounded or somewhat reniform, scarcely lobed . . . *A. jordani*

A. capillus-veneris L. Southern Maidenhair. Occasional, on shaded, moist ledges; coastal and foothill, below 1000m: Murray Dam; Barrett Dam. Also upper Borrego Palm Canyon; Indian Canyon. *n*=30.

A. jordani K. Muell. California Maidenhair. Common, in shaded areas; coastal and foothill, below 800m: e.g. De Luz; Valley Center; Olivenhain; El Cajon Mountain; Pine Creek; Tecate Peak; Point Loma. 2*n*=30 pairs.

Asplenium L. Spleenwort
[Aspleniaceae]

A. vespertinum Maxon. Western Spleenwort. Occasional, in moist rocky areas; foothill, below 1000m: Fallbrook; Pala; Witch Creek; Ramona; Barona Valley; Foster; Alpine; Morena Dam; Potrero Grade; Tecate Peak; Jamul; Otay Mountain; Lyons Peak; Mission Gorge; Poway. Type from San Miguel Mountain.

Bommeria Fourn. in Baillon
[Gymnogrammaceae]

B. hispida (Kuhn)Underw. An erroneous citation cited for an area in or near San Diego County (*Biotropica* 8:14 & *Journ. Arnold Arb.* 60(4):470). Otherwise known from Arizona.

Cheilanthes Sw. Lipfern, Cottonfern
[Adiantaceae]

Leaves with glabrous pinnae
  Leaf blades viscid, 2-3-pinnate, lower pinnae distant; stipes slender, dark brown, viscid glandular; rachis glandular and hairy . . . *C. viscida*
  Leaf blades not viscid, rachis not glandular and hairy
    Leaf blades 4-pinnate, segments bright green, decurrent 2-3mm long; stipes dark brown, wiry, 0.5-2.5dm long; fronds to 3.5dm long . . . *C. californica*
    Leaf blades 2- or 3-pinnate; plants tufted, less than 15cm high . . . *C. deserti*
Leaves with hairy or scaly pinnae
  Blades villous above and beneath with long flexuous hairs, and with elongate brown scales on rachis beneath; blades 3-4-pinnate with gray-green segments; rhizomes short with tufted scales 3-4mm long; fronds 0.5-1.5dm long with dark brown fibrillose-paleaceous stipes . . . *C. ×parishii*
  Blades not villous and scaly; blades mostly 2- or 3-pinnate (but ranging 1-4 pinnate)
    Pinnae more or less densely scaly beneath, often glabrous above
      Leaf scales not ciliate, except possibly at the very base; scales imbricate, white to brownish, somewhat translucent; pinnae beadlike; stipes brown to purplish, with narrower scales; rhizomes short, with appressed scales; fronds 0.5-3dm high . . . *C. covillei*
      Leaf scales conspicuously ciliate, brown or light brown; not imbricate . . . *C. clevelandii*
    Pinnae woolly, tomentose, hairy or fibrillose, but not scaly
      Pinnae in about 10 pairs, the lowers distant; pinnule segments about 1mm long . . . *C. newberryi*
      Pinnae in 5-9 pairs; pinnule segments about 2-3mm long . . . *C. parryi*

C. californica (Hook.)Mett. [*Aspidotis c.* Nutt. ex Copl.] California Lacefern. Occasional, on shaded slopes; foothill, below 1200m: e.g. upper San Onofre Canyon; Pala; Mount Woodson; San Pasqual; Pine Valley; Harbison Canyon; Lyons Peak; Tecate Peak; Otay Mountain. 2*n*=30 pairs.

C. clevelandii D.C. Eaton. Cleveland's Lipfern. Occasional, in dry, rocky areas; foothill, below 1500m: e.g. Red Mountain; Oak Grove; Witch Creek; Pine Valley; Morena Dam; Otay Mountain; San Miguel Mountain; Mount Woodson. Type from "a mountain about forty miles from San Diego, California, at an elevation of about 2500 feet".

C. covillei Maxon. Coville's Lipfern. Infrequent, in rocky places; cismontane and desert, below 1500m: Hot Springs Mountain; upper Borrego Palm Canyon; Angelina Spring; upper Blair Valley; Sheep Canyon; Shaw Canyon; Oriflamme Canyon; Vallecito Stage Station; southeast corner of county. Also Cedar Creek Bridge; Spencer Valley; Ramona. 2*n*=30 pairs.

C. deserti Mickel. [*Notholaena californica* D.C. Eaton; *N. candida* var. *accessita* Jeps., the type from upper Vallecito; *Aleuritopteris cretacea* auct. non (Liebm.)Fourn.] California Cloakfern. Infrequent, in rocky areas; coastal and desert, below 750m: Mission Gorge; Spring Valley; Borrego Palm Canyon; Yaqui Well; Sentenac Canyon; lower Box Canyon; Vallecito Stage Station; Bow Willow Canyon; InKoPah Gorge. Type from Spring Valley.

C. newberryi (D.C. Eaton)Domin. [*Notholaena n.* D.C. Eaton.] California Cottonfern. Occasional, in rocky areas; foothill, below 1200m: e.g. upper San Onofre Canyon; Rainbow; Barona Valley; Ramona; Barrett Reservoir; Tecate Peak; Dulzura; Mission Gorge; Adobe Falls. Type from San Diego.

C. × parishii Davenp. (*C. covillei* × *C. parryi*) Parish's Lipfern. Rare, among rocks in Sentenac Canyon, 630m. *n*=60.

C. parryi (D.C. Eaton)Domin. [*Notholaena p.* D.C. Eaton] Parry's Cloakfern. Occasional, in rocky areas; desert, below 800m: e.g. northeast corner of county; Coyote Canyon; Borrego Palm Canyon; upper Blair Valley; Sentenac Canyon; Vallecito Stage Station; Mountain Palm Springs; Canebrake Canyon. 2*n*=30 pairs.

C. viscida Davenp. Viscid Lipfern. Uncommon, in rocky areas; desert, below 800m: Borrego Palm Canyon; Blair Valley; InKoPah County Park; Sentenac Canyon; La Puerta.

Cyrtomium Presl
[Aspidiaceae]

***C. falcatum** (L. f.)Presl. [*Polypodium f.* L. f.] House Hollyfern. Native of eastern Asia. Local, naturalized at La Jolla, 80m. *n*=82.

Cystopteris Bernh.
[Aspidiaceae]

C. fragilis (L.)Bernh. Brittle Bladderfern. Infrequent, in moist areas; montane, above 1050m: Julian; Mount Palomar; Pine Valley; Corte Madera; Spencer Valley. *n*=84.

Dryopteris Adans. Woodfern
[Aspidiaceae]

D. arguta (Kaulf.)Watt. Coastal Woodfern. Frequent, in shaded areas; cismontane, below 1500m: e.g. San Onofre Canyon; Palomar Mountain; Moosa Canyon; La Costa; Banner Grade; Spencer Valley; Olivenhain; Morena Dam; Campo; Otay Mountain. *n*=41.

Pellaea Link. Cliff-Brake
[Adiantaceae]

Pinnules mucronulate, usually less than 4mm long . . . . . *P. mucronata*
Pinnules not mucronulate, 4-8mm wide . . . . . *P. andromedifolia*

P. andromedifolia (Kaulf.)Fee. [*P. a.* var. *pubescens* D.C. Eaton] Coffee Fern. Frequent, in dry, rocky areas; coastal and foothill, below 1200m: e.g. mouth of Santa Margarita River; Vista; Pala; Del Mar; Ramona; Banner; Point Loma; Morena Dam; San Ysidro. *n*=29, 87, 32, 64.

P. mucronata (D.C. Eaton)D.C. Eaton. Bird's Foot Cliff-Brake. Frequent, in dry, rocky areas, below 1450m: e.g. De Luz; Escondido; Mount Palomar; Borrego Palm Canyon; Pinyon Mountain; San Felipe; southeast corner of county; Spencer Valley; Otay.

Pityrogramma Link.
[Adiantaceae]

Upper surface of fronds sparsely yellowish-glandular . . . . . *P. triangularis* var. *maxonii*
Upper surface of fronds glabrous, green
  Under surface of fronds white-powdery . . . . . *P. triangularis* var. *viscosa*
  Under surface of fronds yellow-powdery . . . . . *P. triangularis* var. *t.*

P. triangularis (Kaulf.)Maxon var. **maxonii** Weath. Desert Goldenback Fern. Uncommon, on rocky slopes; desert: Jacumba, 900m.

P. triangularis var. **triangularis**. California Goldenback Fern. Common, in rocky regions; foothill and desert, below 1500m: e.g. De Luz; Pala; Lakeside; Borrego Palm Canyon; Banner Grade; Spencer Valley; Pine Valley; Shaw Canyon; Tecate Junction; Lyons Peak. *n*=30, 45, 60.

P. triangularis var. **viscosa** (Nutt. ex D. C. Eaton) Weath. Silverback Fern. Occasional, in shaded areas; coastal, below 500m: e.g. Poway Grade; Rainbow; Mission Gorge; Point Loma; Monument Beach; Torrey Pines State Reserve. Also Dulzura; Morena Dam. Type from San Diego. *n*=30.

Polypodium L. Polypody
[Polypodiaceae]

P. californicum Kaulf. California Polypody. Frequent, in shaded, rocky areas; coastal and foothill, below 1200m: e.g. upper San Onofre Canyon; Henshaw Dam; Spencer Valley; San Marcos Creek; Point Loma; Chula Vista; Tecate Peak. *n*=37, 74.

Polystichum Roth. Swordfern
[Aspidiaceae]

Fronds 3-5dm long . . . . . *P. imbricans* ssp. *imbricans*
Fronds 4-8dm long . . . . . *P. imbricans* ssp. *curtum*

P. imbricans (D.C. Eaton)Wagner ssp. **curtum** (Ewan)Wagner. [*P. munitum* (Kaulf.)Presl ssp. *curtum* Ewan] Canyon Swordfern. Uncommon, in shaded areas of Montane Coniferous Forest; montane: Cuyamaca Mountains, 1500m.

P. imbricans ssp. **imbricans**. [*P. munitum* ssp. *imbricans* (D.C. Eaton)Munz, *P. m.* var. *i.* (D.C. Eaton)Maxon] Imbricate Sword-Fern. Occasional, in damp areas of Montane Coniferous Forest; montane, above 1500m: Palomar, Hot Springs, Laguna, and Cuyamaca Mountains. Also San Felipe.

Pteridium Scop. Bracken
[Dennstaedtiaceae]

P. aquilinum (L.)Kuhn var. **pubescens** Underw. [*P. a.* var. *lanuginosum* auct. non (Bong.)Fern.] Western Bracken. Frequent, in fields and glens; foothill and montane, above 200m: e.g. De Luz; Palomar, Hot Springs, Volcan, Cuyamaca, and Laguna Mountains; Dehesa; Mussey Grade. *n*=52.

Thelypteris Schott
[Thelypteridaceae]

***T. acuminata** (Houtt.)Morton. [*Polypodium a.* Houtt.; *Cyclosorus a.* (Houtt.) Nakai] Acuminate Woodfern. Native of east Asia. Local, once naturalized at Rosecroft Gardens, Point Loma, 95m.

Woodwardia Sm. Chainfern
[Blechnaceae]

W. fimbriata Sm. in Rees. [*W. chamissoi* Brack.] Giant Chain-Fern. Uncommon, in moist, shaded areas; cismontane: San Onofre; Rainbow Creek; San Marcos, Palomar, Hot Springs, Cuyamaca, and Laguna Mountains; San Felipe Canyon; Dehesa; Tecate Peak; Woodwardia Canyon; Sloane Ranch. n=34.

Order **MARSILEALES**

MARSILEACEAE. Marsilea Family

KEY TO GENERA

Fronds or leaves like a 4-leaf clover, plants hairy . . . . *Marsilea*
Fronds or leaves grasslike, plants glabrous . . . . *Pilularia*

Marsilea L. Cloverfern

M. vestita Hook. & Grev. [*M. mucronata* A. Br., sensu lato] Hairy Cloverfern. Occasional, on muddy banks of ponds and reservoirs; cismontane: Stuart Mesa; Lake Wohlford; Cuyamaca Reservoir; Laguna Lakes; Sweetwater Lake; Rickey Dam; Lake Murray; Ramona; Otay Mesa; Otay Lake.

Pilularia L. Pillwort

P. americana A. Br. American Pillwort. Uncommon, in vernal pools and other ephemeral ponds; coastal and foothill, below 800m: Ysidora; San Marcos; Poway Junction; Ramona; Kearny Mesa; Fletcher Hills; East San Diego.

Order **OPHIOGLOSSALES**

OPHIOGLOSSACEAE Adder's-Tongue Family

Ophioglossum L. Adder's-Tongue

O. californicum Prantl. [*O. lusitanicum* ssp. *c.* (Prantl)Clausen] California Adder's-Tongue. Infrequent, on grassy slopes and around vernal pools; coastal and foothill, below 900m: Escondido; Olivenhain; Valley Center; Encinitas; Grossmont College; Lake Jennings; Mount Soledad; Kearny Mesa; Balboa Park; East San Diego; El Cajon Mountain; Lee Valley; Point Loma. Type from San Diego.

Order **SALVINIALES**

SALVINIACEAE. Salvinia Family

Azolla Lam. Mosquitofern

A. filiculoides Lam. Pacific Mosquitofern. Frequent, in bodies of fresh water and slow streams; coastal and foothill, below 1200m: Buena; Lake Hodges; Rancho Santa Fe; Echo Valley; San Diego; Willow Tree Dam.

Subdivision **PTEROPSIDA** Seed-Bearing Plants

Class **CONIFERAE** Cone-Bearing Plants

Order **CONIFERALES**

KEY TO FAMILIES

Leaves scale-like, mostly decurrent, thickly clothing the branches . . . . *Cupressaceae*
Leaves needlelike or linear, not scalelike . . . . *Pinaceae*

CUPRESSACEAE. Cypress Family

KEY TO GENERA

Fruit a woody cone; leaves scale-like, decussate
  Cones and their 6 scales oblong, imbricated or valvate, maturing the first year, seeds 2 to each scale . . . . *Calocedrus*
  Cones subglobose, their scales shield- or wedge-shaped, maturing in 1 or 2 years; seeds many on each scale *Cupressus*
Fruit berrylike, formed by the fusion of the scales; ovules 1-2; leaves in 2's or 3's . . . . *Juniperus*

Calocedrus Kurz.

C. decurrens (Torr.)Florin. [*Libocedrus d.* Torr., *Heyderia d.* (Torr.)K. Koch] California Incensecedar. Occasional, on sheltered slopes of Montane Coniferous Forest; montane, above 1050m: Palomar, Hot Springs, Cuyamaca, and Laguna Mountains. Apr.-May.

Cupressus L. Cypress

Dorsal surface of mature leaves usually lacking an active gland or pit; foliage green, not grayish . . . . *C. guadalupensis*
Dorsal surface of mature leaves usually with an active gland or pit; foliage blue-gray or gray-green . . . . . . *C. arizonica*

C. guadalupensis ssp. **forbesii** (Jeps.)Beauchamp. [*C. guadalupensis* var. *f.* (Jeps.)Little] Tecate Cypress. Local, in Chaparral; foothill, 150-1100m: Otay, Tecate, and Guatay Mountains. Also naturalized in Peutz Valley, near Alpine where it may have been originally planted. Apr.-June. Type from Cedar Canyon, between El Nido and Dulzura.

C. arizonica Greene ssp. **stephensonii** (C.B. Wolf)Beauchamp. [*C. s.* Wolf] Cuyamaca Cypress. Local, in Chaparral on south slope of Cuyamaca Peak, 1200-1250m. Apr.-June. Type from "upper limits of King Creek, a tributary to the south fork of the San Diego River, Cuyamaca Mountains...at about 4000 feet elevation on a dry ridge near the creek".

Juniperus L. Juniper

J. californica Carr. California Juniper. Frequent, as an indicator of Pinyon-Juniper Woodland, on flats and slopes; desert, 300-1500m: e.g. Yucca Valley; Montezuma Valley; San Felipe Valley; Banner; Oriflamme Canyon; Shaw Canyon; Earthquake Valley; Pinyon Mountain; Jacumba. Jan.-Mar.

PINACEAE. Pine Family

KEY TO GENERA

Cones erect on branches, scales falling separately at maturity; leaves solitary, blunt, flattened; terminal buds blunt and enveloped in resin . . . . . *Abies*
Cones not erect, scales persistent; leaves solitary or in fascicles, acute to sharp at apex; terminal buds mostly pointed, not enveloped in resin
  Leaves fascicled, needle-like, in axillary 1- to 5-leaved clusters, enclosed basally by a membranous sheath . . . . *Pinus*
  Leaves scattered, linear, not enclosed basally . . . . . *Pseudotsuga*

Abies Mill. Fir

A. concolor (Gord. & Glend.)Lindl. ssp. **concolor**. Rocky Mountain White Fir. Frequent, in Montane Coniferous Forest; montane, above 1500m: Palomar, Hot Springs, Bucksnort, Volcan, and Cuyamaca Mountains. May-June. n=12.

Pinus L. Pine

Sheaths of leaf-clusters deciduous in the 1st year; base of bracts of leaf-bearing spurs not decurrent; leaves with 1 central vascular strand (Subg. *Haploxylon*)
  Leaves 5 in a cluster . . . . . *P. lambertiana*
  Leaves 1-5 in a cluster, if 5, then stout, not slender
    Leaves commonly in 1's or 2's, more than 1mm wide, somewhat greyish-green, not very glaucous. . *P. monophylla*
    Leaves commonly in 4's, sometimes 5, glaucous, scarcely 1mm wide . . . . . *P. quadrifolia*
Sheaths of leaf-clusters persistent, base of spur-bracts decurrent; leaves with 2 vascular strands (Subg. *Diploxylon*)
  Leaves 5 in a cluster, grey-green. 20-30cm long; cones 10-15cm long; coastal . . . . . *P. torreyana*
  Leaves mostly 3 in a cluster
    Cones subterminal; cone-scale tips not incurved
      Leaves dull grey-green; branches glaucous; cones 15-35cm long . . . . . *P. jeffreyi*
      Leaves yellow-green; branches not glaucous; cones 7-15cm long; cone-scale tips straight. . . . . *P. ponderosa*
    Cones lateral; cone-scale tips stout, incurved, flattened . . . . . *P. coulteri*

P. coulteri D. Don. Coulter Pine. Common, on Montane Coniferous Forest slopes; montane, above 1200m: Palomar, Bucksnort, Hot Springs, Pine, Volcan, Angel, San Ysidro, Cuyamaca, Laguna, Corte Madera, and Los Pinos Mountains. May-June. Reforestation with hybrid and introgressed seedlings of *P. coulteri* parentage makes identification of this taxon as well as that of *P. ponderosa* and *P. jeffreyi* arbitrary at times.

P. jeffreyi Grev. & Balf. in A. Murr. [*P. ponderosa* var. *j.* (Murr.)Balf. ex Vasey] Jeffrey Pine. Prevalent, on Montane Coniferous Forest slopes; montane, above 1200m: Cuyamaca and Laguna Mountains; less prevalent on Palomar and Hot Springs Mountains. A rather disjunct, low elevation (1080m) population occurs in Pine Valley. Apr.-June. n=12.

P. lambertiana Dougl. Sugar Pine. Occasional, in Montane Coniferous Forest; montane, above 1350m: Cuyamaca and Hot Springs Mountains. May-June. 2n=24.

P. monophylla Torr. & Frem. [*P. cembroides* var. *m.* (Torr. & Frem.)Voss] One-Leaf Pinyon. Uncommon, as an element of Pinyon-Juniper Woodland, but also in Chaparral, on slopes of the main cordillera; desert and montane, above 900m: Villager Peak area; McCain Valley; southeast corner of county; InKoPah County Park; Laguna and Pinyon Mountains. May.

P. ponderosa Dougl. ex P.C. Lawson Ponderosa Pine. Frequent, on sheltered slopes of Montane Coniferous Forest; montane, above 1200m: Palomar, Hot Springs, and Cuyamaca Mountains. May-June. n=12.

P. quadrifolia Parl. ex Sudw. [*P. juarezensis* Lanner, *P. cembroides* var. *parryana* (Engelm.)Voss] Parry Pinyon. Infrequent as an indicator of Pinyon-Juniper Woodland, on slopes of the main cordillera; desert, above 900m: McCain Valley; Desert View Point; Monument Peak; Mountain Springs. Apr.-June. Type from the mountains east of San Diego.

P. torreyana Parry ex Carr ssp. **torreyana**. Torrey Pine. Local, about coastal headlands, below 160m: Torrey Pines State Reserve; Torrey Pines Terrace; Del Mar; San Dieguito Valley. Jan.-Mar. Type from "bluffs near the mouth of Soledad Creek, 10 miles north of San Diego".

Pseudotsuga Carr.

P. macrocarpa (Vasey)Mayr. Big-Cone Douglas-Fir. Occasional, on slopes and canyons of Montane Coniferous Forest; montane, above 1000m: Palomar and Volcan Mountains; Julian; east slope of Mesa Grande. Apr.-June. Type from "mountains near San Felipe". n=12.

Class **GNETAE**

Order **EPHEDRALES**

EPHEDRACEAE. Ephedra Family

Ephedra L. Mormon-Tea

Leaves and bracts in 3's; cones sessile
- Leaves persisting, becoming shreded and grey in age, 5-13mm long; tips of twigs spinose . . . . *E. trifurca*
- Leaves deciduous, 2-6mm long; tips of twigs not spinose . . . . *E. californica*

Leaves and bracts in 2's; cones often pedunculate
- Seeds solitary, or if more, light in color
  - Seeds smooth, brown; leaf bases brown and persistent; stems usually rough . . . . *E. aspera*
  - Seeds furrowed or scabrous, light brown to grey-green; leaf bases grey and deciduous; stems usually smooth . . . . *E. fasciculata*
- Seeds paired, brown to almost black
  - Leaf bases brown, seeds slightly furrowed longitudinally; branchlets green, not glaucous, 1-1.5mm in diameter, numerous and tending to be erect and parallel . . . . *E. viridis*
  - Leaf bases grey; seeds smooth; branchlets grey-green, more or less glaucous, ca. 2mm in diameter, fewer and spreading . . . . *E. nevadensis*

E. aspera Engelm. ex Wats. Boundary Ephedra. Infrequent, on slopes and mesas; desert, below 1200m: Fish Creek; The Narrows; Sentenac Canyon; upper Blair Valley; Vallecito Stage Station; Mountain Springs. Mar.-Apr.

E. californica Wats. California Ephedra. Infrequent, on slopes and mesas; coastal, foothill, and desert, below 1000m: Lone Palm; San Felipe; The Narrows; Vallecito Stage Station; InKoPah County Park; Jacumba; Campo; Tecate; Barrett Junction; Jamul; Boundary Monument #258; Silver Strand; North Island; Point Loma. Also San Luis Rey. Mar.-Apr. Type from "promontory near San Diego (Point Loma?) and Jamul Valley".

E. fasciculata A. Nels. var. **clokeyi** (Cutler)Clokey. [*E. c.* Cutler] Clokey's Ephedra. Uncommon, on slopes and flats; desert, below 1000m: Coyote Canyon; Borrego Palm Canyon; Sentenac Canyon; Box Canyon; Mountain Palm Springs; Mountain Springs. Mar.-Apr.

E. nevadensis Wats. Nevada Ephedra. Rare, in sandy areas; desert, below 1000m: Lone Palm; between Jacumba and Mountain Springs. Apr.-May.

E. trifurca Torr. in Emory. Three-Fork Ephedra. Scarce, in sandy areas; desert, below 1000m: Yaqui Well; Agua Caliente Hot Springs; Mountain Springs; Mason Valley. Jan.-Apr.

E. viridis Cov. var. **viridis**. Green Ephedra. Rare, on rocky slopes; desert, below 1000m: Jacumba; Mountain Springs. July. n=14.

Class **ANGIOSPERMAE** Flowering Plants

Subclass **MONOCOTYLEDONES**

[Segregate families are shown parenthetically]

KEY TO FAMILIES

Plants small, floating on water or stranded on mud, not differentiated into stem and leaf, sometimes with 1 to few simple roots . . . . *Lemnaceae*

Plants with stem and leaves, the latter sometimes scalelike
- Perianth wanting or reduced, its parts often bristles or mere scales, not petallike in color and texture
  - Flowers in axils of chaffy or husklike scales and somewhat concealed by them, or stamens and style protruding; flowers in spikes, spikelets or heads
    - Leaf-sheaths split lengthwise on the side opposite the blade; leaves usually 2-ranked; stems mostly hollow and terete; anthers versatile . . . . *Poaceae*
    - Leaf-sheaths continuous around the stem or ruptured in age; leaves mostly 3-ranked; stems often triangular in cross-section, usually with a pith; anthers basifixed . . . . *Cyperaceae*
  - Flowers not in the axils of chaffy bracts, or, if subtended by bracts, exceeding or equaling them and not concealed
    - Plants floating or submerged; flowers floating or submerged, or barely raised above the surface of the water
      - Marine plants; flowers on 1 side of a flattened axis . . . . *Zosteraceae*
      - Freshwater plants or of brackish water
        - Flowers in spikes or heads
          - Flowers perfect, in peduncled axillary spikes; sepals 4 . . . . *Potamogetonaceae*
          - Flowers unisexual, in globose heads, the lower pistillate, the upper staminate . . . . *Sparganiaceae*
        - Flowers axillary and solitary or very few in groups
          - Leaves alternate; fruits in umbelliform clusters . . . . *Potamogetonaceae*
          - Leaves opposite
            - Carpels 2 or more; leaves; mostly 3-10cm long . . . . *Zannichelliaceae*
            - Carpels 1; leaves mostly 1-3cm long . . . . *Najadaceae*
    - Plants of land or shallow water; usually leaves and flowers well emersed
      - Flowers on a spadix, surrounded by a yellow or white spathe . . . . *Araceae*

Flowers not on a spadix
Inflorescence a dense, elongate spike
Plants 1-4dm high; staminate and pistillate flowers intermingled or flowers perfect . . . . . *Juncaginaceae*
Plants 10-20dm high; stamens usually 3, the staminate ones in a separate, upper portion of the spike, the pistillate below . . . . . *Typhaceae*
Inflorescence of subglobose heads, racemes or open clusters
Flowers unisexual, the lower heads pistillate . . . . . *Sparganiaceae*
Flowers bisexual; perianth-segments distinguishable
Ovaries 3 or 6, separating at maturity . . . . . *Juncaginaceae*
Ovary 1, forming a capsule . . . . . *Juncaceae*
Perianth well developed, at least the inner segments petaloid in color and texture
Carpels more or less free, 1-loculed, mostly 1-ovuled, maturing into a bunch or whorl of akenes . . . *Alismataceae*
Carpels united into a mostly 3-12-loculed ovary maturing into a capsule or berry
Leaves large, plicately folded in bud; plant arborescent, unbranched; flowers in a large panicle with a spathaceous bract . . . . . *Arecaceae*
Leaves not as above and the plant branched if arborescent
Plants quite woody, with long stiff swordlike leaves . . . . . *Agavaceae*
Plants herbaceous, or if woody, climbing vines
Ovary superior
Inflorescence with a spathe, submersed or floating aquatics with perianth-segments partly connate
Flowers blue . . . . . *Pontederiaceae*
Flowers white . . . . . *Aponogetonaceae*
Inflorescence without a spath; plants not floating
Sepals green; petals colored; flowers in umbels; prostrate introduced plants . . . . . *Commelinaceae*
Sepals and petals concolored, or if unlike, the flowers not in umbels
Flowers in a scapose umbel subtended by membranous spathe-like bracts . . . . . *Alliaceae*
Flowers not in a scapose umbel with spathe-like bracts . . . . . *Liliaceae*
Ovary inferior
Flowers in a scapose umbel subtended by a membranous, spathe-like bract . . . . . *Amaryllidaceae*
Flowers not scapose
Aquatic plants with mostly submersed leaves . . . . . *Hydrocharitaceae*
Land plants, the leaves not submersed
Flowers regular; ovary 3-loculed; stamens 3; leaves equitant . . . . . *Iridaceae*
Flowers irregular; ovary 1-loculed; stamens 1-2; leaves not equitant . . . . . *Orchidaceae*

AGAVACEAE. Agave Family

KEY TO GENERA

Ovary inferior; flowers yellow to greenish; short-stemmed or stemless plants with broad fleshy leaves . . . . . *Agave*
Ovary superior; flowers white to purplish; plants stemless to arborescent; leaves narrow, scarcely fleshy
Perianth less than 1cm long; flowers both perfect and imperfect . . . . . *Nolina*
Perianth 3-13cm long; flowers all perfect . . . . . *Yucca*

Agave L. Century Plant
[Agavaceae]

Leaves mostly less than 6dm long, less than 10cm wide
Stems below ground, leaves glaucous; marginal teeth pale . . . . . *A. deserti*
Stems rising above ground and decumbent in age; leaves deep green; marginal teeth red to blackish . . . . . *A. shawii*
Leaves 15-25dm long, 10-25cm wide, glaucous . . . . . *A. americana*

***A. americana** L. American Agave. Native of Mexico. Local, persisting about abandoned dwellings; coastal, below 200m: e.g. Olivenhain; Sweetwater Valley. Feb.-Oct.

A. deserti Engelm. [*A. consociata* Trel., the type from San Felipe.] Desert Agave. Prevalent, on outwash fans and sandy flats; desert, below 1000m: Borrego Palm Canyon; The Narrows; San Felipe Valley; Vallecito Valley; Carrizo Creek; Dos Cabezas; Bow Willow; Jacumba; Mountain Springs. May-July. Type from "eastern base of the Southern California mountains and in the adjoining deserts". n=30 (5 long, 25 short).

A. shawii Engelm. Coastal Agave. Rare, on bluffs and slopes; coastal, 10-40m: Torrey Pines State Reserve; Point Loma; Boundary Monument #258. Sep.-May. Type from "the arid hills which overlook the sandy strand of the Pacific in the southwest corner of California, where the boundary is marked by the initial monument". n=30 (5 long, 25 short). The Torrey Pines location is doubtfully native.

Nolina Michx.
[Dracaenaceae]

Leaves mostly 15-30mm wide above expanded base
Leaves smooth on margins and with shredding fibers, glaucous; perianth 2-2.5mm long . . . . . *N. bigelovii*
Leaves hispid-serrulate on margins, not shredding fibrous, green; perianth 5mm long . . . . . *N. wolfii*
Leaves mostly 8-15mm wide above expanded base . . . . . *N. parryi*

N. bigelovii (Torr.)Wats. Bigelow's Bear-Grass. Scarce, on slopes of the main cordillera; desert and montane, 400-1300m: southeast corner of county; InKoPah County Park; Rockhouse Canyon; Montezuma View Point. May-June. n=19.

N. parryi (Wats.)L. Benson. [*N. interrata* Gentry, type from "slope west of Dehesa".] Parry's Bear-Grass. Uncommon, on slopes of the main cordillera and desert ranges; cismontane, below 1900m: Pala; McGee Truck Trail; Palomar Mountain; Viejas Mountain; Japatul Valley; Dehesa; McGinty Mountain; Sequan Peak; near Wood Valley. . Apr.-June; n=19.

N. wolfii (Munz)Munz. [*N. parryi* ssp. *w.* Munz.] Wolf's Bear-Grass. Uncommon, in Chaparral and Pinyon-Juniper Woodland, 1000-1600m: Villager Peak; near Lost Valley; Pinyon Mountains; Whale Peak, Vallecitos Mountains; desert slopes of San Ysidro Mountains, 1400m. June.

Yucca L. Yucca
[Agavaceae]

Plants caulescent; leaves green, with fibrous-shredding margins . . . . . . *Y. schidigera*
Plants acaulescent; leaves bluish-grey green with serrulate margins . . . . . . *Y. whipplei*

Y. schidigera Roezl ex Ortgies. Mohave Yucca. Frequent, in Sage Scrub and Chaparral as well as in western Creosote Bush Scrub areas; coastal, foothill, and desert, below 1500m: e.g. Fallbrook; Torrey Pines State Reserve; Point Loma; Tijuana River Valley; Descanso; San Felipe; Oak Grove; upper Box Canyon; Montezuma View Point; Jacumba; southeast corner of county. Apr.-May. Type from near San Diego.

Y. whipplei Torr. ssp. **whipplei**. Our Lord's Candle. Frequent, on exposed slopes; cismontane, below 1900m: e.g. Encinitas, De Luz; Rainbow; Palomar Mountain; Agua Tibia Mountains; Montezuma View Point; Rattlesnake Valley; Descanso; Cuyamaca Mountains; San Felipe Valley; Hipass; Otay Mountain. Apr.-May. Type from "rocks near San Pasqual". n=30 (5 long, 25 short).

ALISMATACEAE. Water-Plantain Family

KEY TO GENERA

Fruits in a ring; stamens 6 . . . . . . *Alisma*
Fruits in a dense head; stamens 6 to 30 . . . . . . *Echinodorus*

Alisma L. Water-Plantain

A. triviale Pursh. [*A. plantago-aquatica* L. var. *americanum* Schultes & Schultes; *A. p-a.* auct. non L.] American Water-Plantain. Scarce, in standing and sluggish fresh water: Shrine Camp; Laguna Mountains; Lake Hodges; Alvarado Canyon. June-July. 2n=14, 28.

Echinodorus Rich. Burhead

E. berteroi (Spreng.)Fassett. [*E. cordifolius* auct. non Griseb.] Upright Burhead. Infrequent, along streams and ponds; coastal, below 300m: Lake Hodges; Carroll Canyon; Lindo Lake; Lake Murray; Galloway Valley; Sweetwater Valley; Otay Lake. May-July. n=11.

ALLIACEAE. Onion Family

KEY TO GENERA

Pedicels jointed at summit
  Perianth segments distinct or nearly so; flowers yellow . . . . . . *Bloomeria*
  Perianth segments united below; flowers purplish
    Leaves keeled below; stigma with 3 small lobes which run into wings toward apex of the style; seeds sharply angled . . . . . . *Dichelostemma*
    Leaves not keeled below; stigma 3-parted with elongate, recurved spreading lobes; seeds obtusely angled *Brodiaea*
Pedicels not jointed at summit
  Plants with onion-like odor and taste; rootstock bulbous . . . . . . *Allium*
  Plants lacking such odor and taste
    Rootstock bulbous, garden escape . . . . . . *Nothoscordum*
    Rootstock cormous, native . . . . . . *Muilla*

Allium L. Onion

Leaves mostly 2-several
  Bulbs oblong, long-necked, clustered on a more or less well-developed rhizome, stamens included *A. haematochiton*
  Bulbs round or ovoid, mostly solitary or enclosed by a common bulb-coat
    Ovary prominently 6-crested; bulb-coat reticulations within sinous walls . . . . . . *A. campanulatum*
    Ovary crestless or with less prominent crests, bulb-coat reticulations thick-walled, prominent under magnification
      Reticulations reticulate-polygonal, more or less isodiametric . . . . . . *A. lacunosum*
      Reticulations transverse elongate, commonly in regular serrate rows
        Ovary without crests or with very obscure crests; reticulations of bulb-coats more or less irregular: *A. praecox*
        Ovary with more or less evident crests; reticulations of bulb-coats very regularly serrate and in vertical rows
          Perianth 6-7mm long, becoming papery; flowers in subglobose umbel; ovary with 6 lateral crests . . . . . . *A. amplectens*
          Perianth usually 8-14mm long, not papery in age; flowers in open umbels; ovary with 6 central crests . . . . . . *A.peninsulare*
Leaves solitary and terete; ovary prominantly 6-crested
  Perianth purple to rose . . . . . . *A.fimbriatum* var. *f.*
  Perianth mostly pale pink to white . . . . . . *A. fimbriatum* var. *parryi*

A. amplectens Torr. in Whipple. Paper Onion. Local, in Montane Coniferous Forest and Chaparral; montane, above 1300m: Chimney Creek; slopes around Cuyamaca Lake; Guatay Peak. Mar.-June. 2n=14?, 21, 28.

A. campanulatum Wats. [*A. bisceptrum* auct. non Wats.; *A. tenellum* Davids., in part, the type from Julian; *A. bullardiae* Davids., the type from near Julian.] Sierra Onion. Infrequent, in Montane Coniferous Forest and Chaparral; montane, above 1200m: Palomar Mountain; Julian; Cuyamaca Lake; Cuyamaca Mountains; Vallecito Canyon; Laguna Camp. May-July. n=7, 14.

A. fimbriatum Wats. var. **fimbriatum**. Desert Onion. Occasional, on rocky slopes; desert, below 1000m: e.g. San Felipe; Banner; Blair Valley; Mason Valley; Box Canyon; Vallecito Stage Station; Mountain Springs; Jacumba. Apr.-May. *n*=7.

A. fimbriatum var. **parryi** (Wats.)Owenby & Aase. [*A. p.* Wats.; *A. tenellum* Davids., in part, the type from Julian; *A. sanbornii* auct. non Wood] Parry's Onion. Uncommon, in Montane Coniferous Forest and Chaparral; montane, above 1200m: Julian; Cuyamaca Lake; Vallecito Canyon. June. *n*=7.

A. haematochiton Wats. Red-Skin Onion. Frequent, on open slopes; coastal and foothill, below 1000m: e.g. upper San Onofre Canyon; Palomar Mountain; Solana Beach; La Jolla; Miramar; south fork of Featherstone Creek; Barona Indian Reservation; Morena Butte; Pine Valley; Laguna Junction; Corte Madera Lake; Lyons Peak; Potrero; Otay Mountain. Also Borrego Palm Canyon. Apr.-May. *n*=7.

A. lacunosum Wats. var. **davisiae** (M.E. Jones)McNeal & Owenby. Jones' Onion. Local, in Valle de San Jose, 930m: Warner Hot Springs; Warner Ranch. May. *n*=7.

A. peninsulare Lemmon. Red-Flower Onion. Frequent, on open slopes; coastal and foothill, below 1000m: e.g. Moosa Canyon; Linda Vista; Mission Hills; Santa Isabel; Witch Creek; Barona Valley; Clevenger Canyon; Ramona; Barrett Dam; Harbison Canyon; Lakeside; Tecate Peak; Otay Mountain. Apr.-June. *n*=7. Middle Peak, also.

A. praecox Bdg. Early Onion. Prevalent, on open slopes and mesas; coastal, below 150m: e.g. Oceanside; Bernardo; Poway; San Marcos Mountain; Point Loma; San Diego; Encanto; Chula Vista; El Cajon; Otay Mesa. Feb.-May. Type is given as "common about San Diego, on northern slopes of canyons, especially near the coast." *n*=14.

Bloomeria Kell. Golden Stars

B. crocea (Torr.)Cov. ssp. **crocea**. Common Golden-Stars. Frequent, on slopes and mesas; cismontane, below 1800m: e.g. O'Neill Lake; Fallbrook; Pauma Valley; Doane Valley; Barona Valley; Escondido; Rancho Santa Fe; San Marcos; Otay Ranch; Cuyamaca Lake; Espinosa Trail; Laguna Mountains; Mountain Springs. Apr.-June. Type from "summit of the mountains east of San Diego". *n*=9.

Brodiaea Sm. Brodiaea

Perianth-tube firm in fruit and not splitting; staminodia obtuse or acute
- Filaments 2-5mm long, dilated at base only if at all; staminodia not incurved at apex . . . . *B.terrestris*
- Filaments 1mm long, broadly triangular; staminodia incurved at apex . . . . *B.jolonensis*

Perianth-tube thin-membranous and splitting in fruit; staminodia acute or lacking
- Filaments 1mm long; staminodia present . . . . *B.filifolia*
- Filaments 4-6mm long; staminodia lacking . . . . *B. orcuttii*

B. filifolia Wats. Thread-Leaf Brodiaea. Rare, in grasslands and vernal pools, 100m: Alta Loma Creek; Vista at Melrose Lane; Thibodo Ranch; Buena; San Marcos Hills; Agua Hedionda. May. *n*=12.

B. jolonensis Eastw. Mesa Brodiaea. Occasional, on grass-covered mesas; coastal, below 600m: Pauma Valley; Rancho Santa Fe; San Marcos; Linda Vista; Del Cerro; Kearny Mesa; Bonita; Proctor Valley; Alpine; Otay Mesa. Apr.-June. 2*n*= 12, 36.

B. orcuttii (Greene)Baker. [*Hookera multipedunculata* Abrams, the type from heavy soils near Cuyamaca Lake.] Orcutt's Brodiaea. Infrequent, in grasslands and near vernal streams and pools; coastal and foothills, below 1500m: Santa Margarita Mountains; San Marcos; Poway; Henshaw Dam; Santa Ysabel; Cuyamaca Mountains, Cuyamaca Lake; Miramar Naval Air Station; Linda Vista; Ramona; Buschalaugh Canyon; Japatul Valley; Woodwardia Canyon; Lower Otay Reservoir. Apr.-July. Type from "San Diego county, near the city of that name, and also thirty miles to the northward". *n*=12, 2*n*=24.

B. terrestris Kell. ssp. **kernensis** (Hoover)Niehaus. [*B. t.* auct. non Kell.] Dwarf Brodiaea. Uncommon, in grasslands and on slopes; montane, above 1000m: Palomar Mountain; Pine Hills, Julian; Cuyamaca Lake; Laguna Mountains. May-July. *n*=24, 2*n*=36.

Dichelostemma Kunth.

D. pulchellum (Salisb.)Heller. [*Brodiaea p.* (Salisb.)Greene; *B. capitata* Benth.; *D. p.* var. *pauciflorum* (Torr.)Mort.] Blue Dicks. Abundant, in grassy areas; cismontane; rare in desert, below 1500m: e.g. Las Flores; Palomar, Hot Springs, Cuyamaca, Laguna Mountains; Borrego Palm Canyon; Mountain Springs; Barona Valley; Jamul; Campo; Otay Mountain; San Diego. Feb.-May. *n*=9, 18, 36; 2*n*=18, 36, 45, 54, 72.

Muilla Wats. Muilla

Perianth bright yellow; filaments with appendages 2-3mm long . . . . *M. clevelandii*
Perianth greenish-white and brown; filaments dilated at base, without appendages . . . . *M. maritima*

M. clevelandii (Wats.)Hoover. [*Bloomeria c.* Wats.] Cleveland's Golden-Stars. Infrequent, on mesas and slopes; coastal, below 150m: Rancho Santa Fe; Lopez Canyon; Miramar; Mission Hills; Linda Vista; Montezuma Mesa; Murray Dam; Grossmont; Foster; San Diego; San Miguel Mountain, Otay Mesa, Otay Mountain; Mar.-May. Type from "mesas near San Diego". 2*n*=14.

M. maritima (Torr.)Wats. [*M. serotina* Greene] Rough Muilla. Occasional, on slopes and mesas; cismontane and desert, below 900m: San Marcos; Escondido; Ramona; Barona Valley; Proctor Valley; East San Diego; Kearny Mesa; Harbison Canyon; El Cajon Mountain; San Felipe; Banner; Box Canyon; Vallecito Canyon. Mar.-May. 2*n*=14, 16, 20.

Nothoscordum Kunth.

***N. bivalve** (L.)Britt. [*Ornithogalum b.* L.] Yellow False-garlic. Native of southeastern United States. Local, weed in gardens: Fallbrook, 200m. May.

AMARYLLIDACEAE. Amaryllis Family

Narcissus L.

***N. papyraceus** Ker-Gawl. ssp. **papyraceus**. Paperwhite Narcissus. Native of the Old World. Escaped or persisting about abandoned dwellings: Barrett Junction; Bonita.

APONOGETONACEAE.

Aponogeton L. f. Pondweed

***A. distachyon** L. f. Cape Pondweed. Native of South Africa. Escaped aquarium plant in ponds or ditches. Miramar. March.

ARACEAE. Arum Family

Zantedeschia Spreng. Calla

***A. aethiopica** (L.)Spreng. [*Calla a.* L.] White Calla. Native of South Africa. Local, once escaping from cultivation, 1.5km north of Salt Works, Chula Vista; San Luis Rey River, Oceanside. Mar.-May. $2n=32$.

ARECACEAE. Palm Family
(Palmae)

KEY TO GENERA

Leaves with veins radiating from a common locus . . . . *Washingtonia*
Leaves with veins issuing from an elongated rachis . . . . *Phoenix*

Brahea Mart.

B. armata Wats. [*Erythea a.* (Wats.)Wats.] Baja Blue-Palm. Unconfirmed reports of this distinctive palm in desert canyons of south San Diego County may be based upon the rumored scattering of Blue Palm seed by Laurence M. Huey, early Curator of Mammals at the San Diego Natural History Museum. The natural range of the palm is otherwise known as close as 30 miles to the U.S. - Mexican border at Cantu Palms, Baja California Norte.

Phoenix L. Date Palm

Plants single-stemmed; leaves more or less green, not glaucous, arching . . . . *P. canariensis*
Plants suckering prominently; leaves grey-green and glaucous, more or less erect and straight . . . . *P. dactylifera*

***P. canariensis** Chaub. Canary Island Date Palm. Native of the Canary Islands. Local, persisting around old dwellings and frequently escaping from cultivation, especially in riparian areas; coastal, below 1000m: San Mateo Creek; Cuca Grade; Encinitas; Bonita; San Ysidro. May-Oct.

***P. dactylifera** L. Edible Date Palm. Native of northern Africa. Local, persisting around old dwellings or sometimes escaping from cultivation; coastal and desert, below 200m: National City; Travertine Palms. May-Oct.

Washingtonia Wendl. Fan Palm

W. filifera (Linden)Wendl. California Fan Palm. Occasional, in seepage areas; desert, below 800m: e.g. Middle Willows; Coyote Creek; Sheep Canyon; Seventeen Palms; Borrego Palm Canyon; Hellhole Canyon; Palm Spring; Indian Gorge; Bow Willow Canyon, Rockhouse Canyon; Carrizo Gorge; Mortero Palms; Mountain Springs. June. Type "Des bords du Colorado dans l'Arizona", Roezl, 1869; or specimen material may have come from trees cultivated in San Diego -Parish. $2n=26$.

COMMELINACEAE. Spiderwort Family

Tradescantia L.

***T. fluminensis** Vell. Wandering Jew. Native of South America. Local, escaping in moist, shaded areas and dumps; coastal, below 200m: Balboa Park; National City. Apr.-Sep. $2n=60$.

CYPERACEAE. Sedge Family

KEY TO GENERA

Flowers perfect, or perfect and staminate; akenes naked
- Scales or spikelet spirally imbricated
  - Perianth bristles present but included within the scales
    - Style-base deciduous from summit of akene; 1 or more involucral leaves present . . . . *Scirpus*
    - Style-base persistent as a tubercle at apex of akene; involucral leaves none . . . . *Eleocharis*
  - Perianth bristles absent
    - Hyaline perianth-member present between flower and rachis of spikelet . . . . *Hemicarpha*
    - Hyaline perianth-member absent
      - Style-base not swollen, plants 4-20cm high . . . . *Scirpus*
      - Style-base swollen; plants 20-70cm high . . . . *Fimbristylis*
- Scales of spikelet 2-ranked
  - Spikelets elongated, many-flowered, in more or less loose umbels or heads . . . . *Cyperus*
  - Spikelets minute, with 1 perfect flower, in dense glomerate heads . . . . *Kyllinga*

Flowers all unisexual; akene surrounded by sac-like bractlet (perigynium) . . . . *Carex*

Carex L. Sedge

Perigyna pubescent; styles 3
  Spikes with few to ca. 25 perigynia
    Perigynia strongly 2-keeled with subglobose body; bracts of the non-basal pistillate spikes sheathless or nearly so . *C. globosa*
    Perigynia sharply triangular, obovoid; bracts of the non-basal pistillate spikes sheathing. . . . . *C. triquetra*
  Spikes with 25-50 perigynia. . . . . *C. lanuginosa*
Perigyna glabrous
  Styles 3; akenes triangular
    Spikes one on each culm . . . . *C. multicaulis*
    Spikes more than one on each culm. . . . . *C. spissa*
  Styles 2; akenes lenticular
    Terminal spikes staminate and lateral spikes elongated, or if terminal, spike gynaecandrous (pistillate above staminate), lateral spikes peduncled
      Lowest bract long-sheathing; pistillate spikes rather few-flowered; perigynia beakless. . . . . *C. hassei*
      Lowest bract sheathless, or if sheathing, then pistillate spikes with very numerous flowers; perigynia short-beaked
        Leaf-sheaths not becoming filamentous when breaking; beak of perigynium 0.5-1mm long . . . *C.nebrascensis*
        Leaf-sheaths breaking and becoming filamentose
          Perigynium-beak 0.5mm long, bidentate, hsipidulous between the teeth . . . . *C. barbarae*
          Peryginium-beak ca. 0.25mm long, emarginate or entire
            Leaf-sheaths strongly carinate dorsally; blades 6-12mm broad; culms 10-15dm tall . . . . *C. schottii*
            Leaf-sheaths slightly carinate or rounded dorsally; blades 3-5mm broad; culms 3-10dm tall. . . . . *C. senta*
    Terminal spike partly pistillate, or if staminate, the lateral spikes short and sessile
      Rootstocks long-creeping, the culms arising singly or few together
        Culms generally smooth and less than 2dm long; leaves folded-involute near the apex; perigynia finely nerved ventrally. . . . . *C. douglasii*
        Culms generally roughened above and more than 2dm long; leaves flat or canaliculate; perigynia nerveless ventrally. . . . . *C. praegracilis*
      Rootstocks short-creeping; plants cespitose
        Perigynia narrowly to broadly wing-margined
          Lower bracts well developed, conspicuous or at least quite evident . . . . *C. teneraeformis*
          Lower bracts inconspicuous or wanting
            The sheaths strongly prolonged upward at the mouth opposite the blades . . . . *C. fracta*
            The upper sheaths concave or truncate at mouth opposite the blades
              Leaves 3-6mm wide; spikes about 10; perigynia nerved on both faces. . . . . *C. multicostata*
              Leaves 1.5-3mm wide; spikes 4-8; perigynia nerved on inner face
                Culms slender; leaf-blade 1-2.5mm wide; perigynia spreading, ascending . . . . *C. teneraeformis*
                Culms stiffish; leaf blade 2-3.5mm wide; perigynia appressed. *C. subfusca*
        Perigynia not evidently wing-margined
          Perigynia 3.5mm long, lightly nerved on both faces . . . . *C. alma*
          Perigynia 4-4.5mm long, strongly nerved on both faces . . . . *C. densa*

C. alma L. H. Bailey. Sturdy Sedge. Frequent, in wet habitats; montane, above 1000m: e.g. Doane Valley; Volcan Mountain; Spencer Valley; Julian; Scissors Crossing; Pine Hills; Cuyamaca Lake. May.

C. barbarae Dewey. Barbara Sedge. Uncommon, in wet habitats; cismontane, below 1000m: De Luz Canyon. Apr.

C. densa (L. H. Bailey)Bailey. Dense Sedge. Rare, in meadows; montane, above 1400m: Volcan and Cuyamaca Mountains. May-July.

C. douglasii F. Boott in Hook. Douglas' Sedge. Rare, known in the county only at Doane Valley, 1350m. May.

C. fracta Mkze. Fragile-Sheath Sedge. Infrequent, in meadows and on moist slopes; montane, above 1300m: Chimney Creek; Hot Springs and Cuyamaca Mountains; Cuyamaca Lake. May-Aug.

C. globosa F. Boott. Round-Fruit Sedge. Uncommon, on dry slopes; cismontane, below 1500m: San Diego, Cuyamaca Lake. Apr.-July.

C. hassei L. H. Bailey. Hasse's Sedge. Uncommon, along streams; montane, above 1200m: Palomar Mountains. May- Aug.

C. lanuginosa Michx. Woolly Sedge. Infrequent, in wet meadows and marshes; cismontane, above 100m: Fallbrook. Mar.-June. n=39.

C. multicaulis L. H. Bailey. Many-Stemmed Sedge. Occasional, on dry slopes; foothill and montane, above 800m: e.g. De Luz; Doane Valley; Fry Creek; Hot Springs Mountain; Cuyamaca Lake; Middle Peak. May-July.

C. multicostata Mkze. Many-Ribbed Sedge. Uncommon, in meadows; montane, above 1350m: Cuyamaca Lake. May.

C. nebrascensis Dewey. Nebraska Sedge. Infrequent, in meadows and marshes; cismontane, above 400m: Rainbow; French Valley. Aug. n=33.

C. praegracilis W. Boott. Cluster Field-Sedge. Frequent, in wet places; cismontane: e.g. Dyche Valley; Lake Henshaw; Cuyamaca Lake; Laguna Lakes; Middle Peak; Morena Dam; Pine Valley; Campo; Canyon City; Ramona; Murray Dam; Felicitas Park. Apr.-June. Type from San Diego.

C. schottii Dewey. Schott's Sedge. Uncommon, along streams; cismontane, below 1500m: Richman Ranch near De Luz. Apr.-June.

C. senta F. Boott. Rough Sedge. Occasional, along streams; montane, above 1100m: e.g. Doane Valley; Chimney Creek; Eagles Nest; Cedar Creek Falls; meadows below Cuyamaca Lake; Laguna Mountains; Glen Cliff Campground. May-Aug.

C. spissa L.H. Bailey. San Diego Sedge. Common, along streams; cismontane, below 900m: e.g. San Onofre Bluffs; Lone Palm Spring; De Luz Creek; Hormiguero Ranch; Clevenger Canyon; Felicitas Park; Grapevine Canyon; El Capitan Mountain; Lakeside; Adobe Falls; Sweetwater Valley; Viejas Grade; Barona Valley; Harbison Canyon; Tecate Creek; Tecate Peak; Jamul; Spring Valley; Otay Mountain. May-July. Type from San Diego.

C. subfusca W. Boott in Wats. Rusty Sedge. Occasional, in meadows; montane, above 1000m: Doane Valley; Henshaw Grade; Cuyamaca Lake; Cuyamaca Mountains; Laguna Public Camp; Shrine Camp. May-July.

C. teneraeformis Mkze. Sierra Slender-Sedge. Occasional, in meadows; montane, above 1300m: Doane Valley; Pine Hills; Cuyamaca Lake; Harper Ranch; Laguna Public Camp. June-Aug.

C. triquetra F. Boott. [*C. monticola* Dewey in Torrey, the type from mountains east of San Diego.] Triangular-Fruit Sedge. Common, on shaded slopes; coastal and foothill, below 800m: e.g. De Luz; Bonsall; Lilac; Olivenhain; Poway; Foster; Barona Valley; Japatul Valley; Harbison Canyon; Alpine; Barrett; San Miguel Mountain; Point Loma; North Island; Balboa Park; Adolphia Canyon; Kearny Mesa. Apr.-June.

Cyperus L. Umbrella-Sedge

Inflorescence with 100-200 rays, stem 2-5m tall . . . *C. papyrus*
Inflorescence with 0-25 rays, stem to 2m tall
  Style 2-fid; akenes lenticular
    Akene laterally flattened . . . *C. niger*
    Akene dorsally flattened . . . *C. laevigatus*
  Style 3-flowered; akenes triangular
    Spikelets persistent on spike; scales deciduous
      Rachis not winged
        Perennials with short rhizome
          Spikelets 10-20mm long
            Culms bluntly triangular, smooth . . . *C. eragrostis*
            Culms sharply triangular, scaberulous . . . *C. virens*
          Spikelets 4-7mm long . . . *C. alternifolius*
        Annual, with fibrous roots
          Scales acuminate or awned, the tip curved . . . *C. squarrosus*
          Scales obtuse . . . *C. difformis*
      Rachis winged with a pair of inner hyaline appendages at each node
        Plants perennial; akenes obtuse or mucronulate
          Stoloniferous with tubers; leaves about as long as culms
            Scales dull brown or yellow-brown, scarcely keeled . . . *C. esculentus*
            Scales shining red-brown, keeled . . . *C. rotundus*
          Lacking tubers
            Tufted perennial; leaves equaling culms . . . *C. strigosus*
            Short-rhizomatous; leaves much shorter than culms . . . *C. parishii*
        Robust annuals; akenes distinctly mucronate . . . *C. erythrorhizos*
    Spikelets disarticulating above basal pair of scales, or breaking up into 1-fruited joints; scales persistent
      Spikelets disarticulating above sterile basal pair of scales; rachis-wings thin and hyaline . . . *C. parishii*
      Spikelets breaking up into 1-fruited joints; rachis-wings firm and brown . . . *C. odoratus*

***C. alternifolius** L. African Umbrella-Plant. Native of Africa. Infrequent, escaping from cultivation in aquatic habitats; coastal, below 800m: De Luz Creek Crossing; Encinitas. July-Sep. $2n$=32.

***C. difformis** L. [*C. lateriflorus* Torr.] Rice Flatsedge. Native of Asia. Infrequent, weed in wet places; coastal, below 200m: Balboa Park. Oct.

C. eragrostis Lam. [*C. vegetus* Willd.] Tall Flatsedge. Frequent, in wet places; coastal, below 200m: e.g. De Luz Creek; Cottonwood Creek; Rolando; Alvarado Canyon. May-Nov.

C. erythrorhizos Muhl. Red-Root Flatsedge. Occasional, around reservoirs; coastal, below 400m: De Luz Creek; Cottonwood Creek; Lake Hodges; Lake Murray; Lower Otay Reservoir. May-Sep. n=48.

***C. esculentus** L. Yellow Nutsedge. Native of tropical regions. Common, weed in disturbed and moist habitats; coastal, below 400m: e.g. San Onofre; San Pasqual; Poway Junction; Foster; San Clemente Canyon; Balboa Park; Padre Dam; Murray Dam; Lindo Lake; Jamacha. June-Oct. n=9, ca. 54.

C. laevigatus L. Smooth Flatsedge. Uncommon, in moist habitats; coastal and desert, below 300m: Oceanside; Padre Dam; Palm Spring; Bow Willow Canyon. July-Dec.

***C. ligularis** L. Native of tropical regions. Rare, weed in disturbed, moist areas; coastal, below 200m: Balboa Park. July.

C. niger R. & P. var. **capitatus** (Britt.)O'Neill. [*C. n.* var. *castaneus* (Wats.) Kuekenth] Brown Umbrellasedge. Frequent, in moist places; coastal, below 350m: e.g. Santa Margarita River Ford; Carlsbad; San Pasqual; Murray Dam. July-Nov.

C. odoratus L. [*C. ferax* Rich.] Fragrant Flatsedge. Infrequent, in disturbed and wet habitats; coastal, below 200m: Carlsbad Slough; Mussey Grade; Alvarado Canyon; Murray Dam. July-Oct.

***C. papyrus** L. [*P. antiquorum* Willd.] Egyptian Papyrus. Native of Africa and Asia Minor. Infrequent, escaping in aquatic habitats; coastal, below 100m: San Diego; Balboa Park. July-Sep.

C. parishii Britt. Parish's Flatsedge. Uncommon, in wet areas; foothill and montane, above 1000m: Laguna Mountains; Pine Valley; Hipass; White Star. June-Oct.

***C. rotundus** L. Purple Nut-Grass. Native of Eurasia. Rare, weed in cultivated areas: Escondido. Nov. n=54.

C. squarrosus L. [*C. aristatus* auct. non Rottb.] Beard Flatsedge. Frequent, in wet areas; cismontane: e.g. Doane Valley; Olivenhain; San Pasqual Grade; San Clemente Canyon; Padre Dam; Barona Valley; Mussey Grade; Jamacha; Japatul Valley; Pine Valley; Laguna Public Camp. May-Sep.

***C. strigosus** L. Strawcolor Flatsedge. Native of northern California and elsewhere. Rare, in disturbed areas; Escondido. Nov. n=54.

***C. virens** Michx. Green Flatsedge. Native of Neotropics. Rare, adventive in moist areas; coastal, below 150m: San Diego River at Old Town. Sep.

Eleocharis R. Br. Spike-Sedge

Akenes plump or triangular; style trifid
  Akenes with several longitudinal ridges and many fine horizontal lines
    Dwarf annual with fibrous roots; montane marshes . . . . *E. bella*
    Perennials with rhizomes or stolons; lowland marshes . . . . *E. acicularis*
  Akenes not longitudinally ribbed
    Spikelets 2-7-flowered . . . . *E. quinqueflora*
    Spikelets 10 to many flowered
      Tubercle conic or pyramidal, constricted at base
        Spikelets ovoid, blunt; bristles 2-6; akenes pitted-reticulate under magnification . . . *E. montevidensis* var. *m.*
        Spikelets linear-lanceolate, acute; bristles 6-7; akenes smooth or finely pitted under magnification . . . . *E. montevidensis* var. *parishii*
      Tubercule long-subulate, continuous with apex of akene . . . . *E. rostellata*
Akenes lenticular; style bifid
  Akene yellowish-brown; plant perennial with long-creeping rhizomes . . . . *E. macrostachya*
  Akene shining black; plants annual with cespitose culms . . . . *E. geniculata*

E. acicularis (L.)R. & S. [*E. a.* var. *occidentalis* Svens.] Needle Spike-Sedge. Uncommon, in moist habitats; cismontane: Lake Hodges; Cuyamaca Lake; Palomar Mountain; Kearny Mesa. May-Aug. n=10, 15-19, 25-29.

E. bella (Piper)Svens. [*E. acicularis* var. *b.* Piper] Delicate Spike-Sedge. Occasional, around aquatic habitats: e.g. Rainbow; Doane Valley; Cuyamaca Lake; Cuyamaca Mountains; Ramona; San Pasqual; Laguna Public Camp. May-Aug.

E. geniculata (L.)R. & S. Capitate Spike-Sedge. Uncommon, in marshy areas; desert: Mountain Palm Springs, 400m. Dec.

E. macrostachya Britt. in Small. [*E. palustris* auct. non (L.)R. & S.] Pale Spike-Sedge. Abundant, in wet places: e.g. Rainbow; Doane Valley; Lake Henshaw; Pine Hills; Cuyamaca Lake; Harper Ranch; San Pasqual; Ramona; Poway; Kearny Mesa; Point Loma; San Diego Bay; San Ysidro; National City; Rolando; Lake Murray; Corte Madera; Agua Caliente; Vallecito Stage Station. Apr.-Sep. n=5, 8, 16, 19, 21, 23.

E. montevidensis Kunth var. **montevidensis**. [*E. montana* auct.] Dombey's Spike-Sedge. Common, in wet places; cismontane, below 1200m: e.g. De Luz; upper San Onofre Canyon; O'Neill Lake; Rainbow; Valley Center; Lake Henshaw; Eagles Nest; Ballena; Cuyamaca Lake; San Pasqual; La Mesa; Adobe Falls; Tijuana River Valley; Sears Ranch; Jamacha; Jamul Mountain; Sloane Ranch; Barrett; Clover Flat; Hipass; Pine Valley. May-Sep.

E. montevidensis var. **parishii** (Britt.)V. Grant. [*E. parishii* Britt.] Parish's Spike-Sedge. Frequent, in wet places; cismontane, above 300m: e.g. Doane Valley; Palomar Mountain; Eagles Nest; Borrego Palm Canyon; Banner; Scissors Crossing; Oriflamme Canyon; Cuyamaca Lake; El Capitan; Barona Valley; San Luis Rey; San Clemente Canyon; Poway Junction; Pine Valley; Glen Cliff Campground. May-Sep. (The type, Parish 1569, is cited from Agua Caliente, Colorado Desert, San Diego County, which is now referred to as Palm Springs, Riverside Co.)

E. quinqueflora (Hartm.)O. Schwarz. [*E. pauciflora* (Lightf.)Link] Few-Flower Spike-Sedge. Rare, in wet places; montane, above 1200m: Palomar Mountain. May-Aug.

E. rostellata (Torr.)Torr. Beak Spike-Sedge. Rare, in alkaline marshes; desert: Vallecito Stage Station, 480m. May-Aug.

Fimbristylis Vahl.

***F. vahlii** (Lam.)Link. Vahl's Fimbristylis. Native of southeastern United States and Neotropics. Rare, adventive in San Clemente Canyon at Highway I-15, 120m. July-Nov.

Hemicarpha Nees & Arn.

***H. micrantha** (Vahl)Britt. var. **minor** (Schrad.)Friedl. Common Hemicarpha. Native of northern California and elsewhere. Rare, adventive in San Clemente Canyon, 90m. July.

Kyllinga Rottb.

***K. brevifolia** Rottb. [*Cyperus b.* (Rottb.)Hassk.] Short-leaf Flatsedge. Native of Neotropics. Infrequent, weed in moist areas; coastal, below 200m: San Diego; Balboa Park; San Diego State University. July-Sep. 2*n*=120.

Scirpus L. Bulrush

Involucral leaves 2-5, usually exceeding the inflorescence; culms leafy
  Spikelets 3-6mm long; inflorescence a loose, spreading compound umbel . . . . *S. microcarpus*
  Spikelets 10-40mm long; inflorescence capitate or with 1-several elongated rays . . . . *S. robustus*
Involucral leaf solitary, often appearing as a continuation of the culm
  Culm leafy, triangular or subterete; spikelets 1-12
    Plants perennial with rhizomes; culms not filiform; bristles present
      Leaf blades convolute; involucral leaf 3-10cm long; plant to ca. 1m tall. . . . . *S. pungens*
      Leaf blades flat; involucral leaf 1-4cm long; plant to over 2m tall . . . . *S. americanus*
    Plants annual with fibrous roots; culms filiform; bristles absent . . . . *S. cernuus*
  Culm with the leaves reduced to basal sheaths with short blades and not more than 16cm long; culm stout and terete; spikelets numerous, in umbels
    Bristles 2-4, broad, ciliate or plumose, but never barbed. . . . . *S. californicus*
    Bristles usually 6, filiform, downwardly barbed. . . . . *S. acutus*

S. acutus Muhl. ex Bigel. Viscid Bulrush. Common, around pools and reservoirs; coastal and foothill, below 1400m: e.g. Fallbrook; Lake Henshaw; Eagles Nest; Witch Creek; Cedar Creek Falls; Valley Center; Encinitas; San Pasqual; Poway Junction; Barona Valley; National City; Edgemoor; Lakeside; Alvarado Canyon; Otay; Otay Reservoir; El Capitan Reservoir; Pine Valley; Boulevard. May-Aug. *n*=20, 21.

S. americanus Pers. [*S. olneyi* Gray] Olney's Bulrush. Occasional, in wet places; coastal, foothill, and desert, below 700m: Santa Margarita River ford; O'Neill Lake; Poway Junction; Lake Henshaw; Alvarado Canyon; Padre Dam; Harbison Canyon; Spring Valley; Barrett; Scissors Crossing; Carrizo Stage Station; mouth of Shaw Canyon. June-Aug. *n*=39.

S. californicus (C. A. Mey.)Steudel. California Bulrush. Occasional, along marshes and lakes; coastal, below 200m: e.g. San Onofre; De Luz Creek Crossing; O'Neill Lake; Carlsbad; Scripps Ranch; Lake Murray; Sweetwater Reservoir; Lower Otay Reservoir. June-Sep.

S. cernuus Vahl ssp. **californicus** (Torr.)Thorne. California Club-Rush. Scarce, in moist areas: Poway Junction; Escondido; lower Vallecito Canyon. May-Aug.

S. microcarpus Presl. Small-Fruit Bulrush. Frequent, in wet places; cismontane, above 800m: e.g. De Luz Creek Crossing; Fallbrook; Pala; Doane Valley; Chimney Flats; Eagles Nest; Mesa Grande; San Felipe Valley; Banner; Pine Hills; Oasis; Paso Picacho; Ramona. May-Aug.

S. pungens Vahl. [*S. americanus* auct.] Three-Square. Infrequent, in wet areas; coastal and foothill, below 200m: Carlsbad; Escondido; San Pasqual; Padre Dam. May-Aug. *n*=38, 39, 40.

S. robustus Pursh. Prairie Bulrush, Bull Tule. Frequent, in alkaline or brackish marshes; coastal, below 300m: e.g. Oceanside; Carlsbad; San Marcos; Poway Junction; Penasquitos Slough; Padre Dam; Edgemoor; Lindo Lake; National City; Sweetwater Lake; Chula Vista. Apr.-Aug. *n*=53, 55, 57.

HYDROCHARITACEAE Frog's-Bit Family

Hydrilla L. Rich.

***H. verticillata** (L. f.)Royle. Hydrilla. Native of Old World. Local introduction in reservoirs; Lake Murray.

IRIDACEAE. Iris Family

KEY TO GENERA

Inflorescence not umbellate; perianth-segments more then 2cm long . . . . *Iris*
Inflorescence umbellate; perianth-segments all alike, not more than 2cm long . . . . *Sisyrinchium*

Iris L.

I. missouriensis Nutt. Rocky Mountain Blue-Flag. Rare, along ponds and lakes; montane, above 1400m: below Fern Flat Spring, Cuyamaca Rancho State Park; Cuyamaca Lake; Big Laguna Lake. May-June. 2*n*=38, 48.

Sisyrinchium L.

S. bellum Wats. [*S. eastwoodiae* Bickn.; *S. hesperium* Bickn.] California Blue-Eyed-Grass. Common, on open slopes; cismontane, below 1550m: e.g. Santa Margarita Mountain; Oceanside; Point Loma; Otay; Escondido; Viejas Mountain; Ramona; Palomar Mountain; Cuyamaca Lake; Hipass. Apr.-July. 2*n*=32.

JUNCACEAE. Rush Family

KEY TO GENERA

Leaf sheaths open; stems usually with spongy pith. . . . . *Juncus*
Leaf sheaths closed; stems hollow . . . . *Luzula*

Juncus L. Rush

Plants annual
  Plants mostly 10-25cm high with short, flat, sometimes involute leaves
    Capsules oblong, 3-4.5mm long; perianth 4-6mm long
      Flowers borne singly on branch. . . . . *J. bufonius* var *b.*
      Flowers congested . . . . *J. bufonius* var. *halophilus*

Capsules subglobose, 2-3mm long; perianth 3-4mm long . . . *J. sphaerocarpus*
Plants mostly 0.5-10cm high with more or less setaceous leaf blades
Heads often 2-several-flowered; bracts 2, subequal, 1-1.5mm long
Anthers longer than filaments; plants 3-10cm high; style 1.5-3mm long . . . *J. triformis*
Anthers shorter than filaments; plants 0.5-4cm high; style 0-1mm long . . . *J. tiehmii* & *J. luciensis*
Heads 1-flowered; bract 1 or 0, or if 2 then unequal and shorter . . . *J. bryoides*
Plants perennial
Inflorescence seemingly lateral, lowest bract terete and exactly like a continuation of the stem
Flowers inserted in capitate clusters arranged in a panicle; bracts but not bractlets present
Perianth segments 2-3mm long, the outer obtuse; capsules longer than perianth . . . *J. acutus*
Perianth segments 4-5mm long, the outer acute; capsules ca. as long as the perianth . . . *J. cooperi*
Flowers inserted singly on racemose branches of the panicle, each with a pair of bractlets at their base
Leaf blades usually well-developed on upper basal leaf-sheaths; stems compressed . . . *J. mexicanus*
Leaf blades wanting or rudimentary; stems terete
Perianth 2-2.5mm long; anthers not longer than filaments
Stamens 3; capsules ovoid, not apiculate . . . *J. effusus*
Stamens 6; capsules subglobose, apiculate . . . *J. patens*
Perianth 3.5-5mm long
Stems smooth or irregularly ridged when dry; capsules acute . . . *J. balticus*
Stems finely and evenly ridged; capsules obtuse . . . *J. textilis*
Inflorescence usually seemingly terminal, lowest bract unlike a continuation of the stem, or if so then channeled along inner side
Leaf blades transversely flattened (inserted with flat surface facing stem), not septate
Flowers inserted singly, each subtended by 2 bractlets; leaves with scarious, elongate auricles; perianth 3-4mm long . . . *J. tenuis*
Flowers inserted in heads, these subtended by bracts; flowers without bractlets . . . *J. macrophyllus*
Leaf blades terete or equitant, leaves with transverse ribbing due to internal septae
Leaf blades subterete or terete, septa usually complete
Epidermis of stems and leaves conspicuously transversely rugose under a hand lens . . . *J. rugulosus*
Epidermis smooth
Anthers shorter than filaments; perianth 4-5mm long; capsules subulate; heads many-flowered . . *J. torreyi*
Anthers equal to or longer than filaments . . . *J. dubius*
Leaf blades flattened with edge towards stem, equitant; septa incomplete
Perianth 4.5-5mm long; capsules acute below the beak and almost as long as the perianth . . *J. phaeocephalus*
Perianth 3-4mm long
Style at least as long as ovary at anthesis; anthers mostly longer than filaments . . . *J. oxymeris*
Style shorter; anthers much shorter than filments . . . *J. xiphioides*

J. acutus L. ssp. **leopoldii** Snogerup. [*J. a.* var. *sphaerocarpus* Engelm.] Southwestern Spiny Rush. Common, in moist, saline, or alkalne areas; coastal, foothill, and desert, below 900m: e.g. Agua Hedionda; Encinitas; Rancho Santa Fe; Mission Bay; Old Town; Silver Strand; Tijuana River Valley; Alvarado Canyon: Edgemoor; Sweetwater Valley; Salt Works; Jamul Mountain; Barrett Junction; Flynn Springs; Mountain Springs; Mountain Palm Springs. May-June.

J. balticus Willd. Wire Rush. Uncommon, in moist areas; coastal and foothill, below 700m: Santa Margarita River ford; Barona Valley; San Diego. May-Aug. $2n$=40.

J. bryoides F. J. Hermann. One-Flower Dwarf Rush. Uncommon, in meadows; montane, above 1500m: Laguna Camp. May-Aug.

J. bufonius L. var. **bufonius**. Common Toad-Rush. Abundant, in wet habitats; cismontane: e.g. San Onofre; O'Neill Lake; De Luz; Fallbrook; Rainbow; Doane Valley: San Marcos; Valley Center; Lake Henshaw; Ramona; Poway Junction; Barona Valley; Santa Ysabel; San Pasqual; Escondido; Kearny Mesa; El Cajon Mountain; Jamul; Jamacha; Lee Valley; Barrett; Clover Flat. Also Canebrake Canyon. Apr.-Sep. $2n$= ca. 60, 120.

J. bufonius var. **halophilus** Buch. & Fern. [*J. b.* var. *congestus* Wahl] Fasciculate Toad-Rush. Occasional, in brackish water; cismontane, below 1000m: Santa Margarita River Ford; Old Town; Kearny Mesa; Balboa Park; Upper Otay Reservoir; Pine Valley; Agua Caliente. May-Aug.

J. cooperi Engelm. Cooper's Rush. Uncommon, in alkaline areas; desert, below 300m: Carrizo Stage Station. May.

J. dubius Engelm. Mariposa Rush. Frequent, in wet places; cismontane, below 1100m: e.g. Valley Center; San Marcos; Cedar Creek; Barona Valley; Santa Ysabel; Mesa Grande; Hatfield Creek; Mount Woodson; Mussey Grade; San Clemente Canyon; Poway Grade; Point Loma; Otay Mesa; Lakeside; Jamul Mountain; Ramona; Barrett; Campo; Japatul Valley; Espinosa Trail; White Star; Dubber. Apr.-Aug.

J. effusus L. var. **pacificus** Fern. & Wieg. [*J. e.* ssp. *p.* (Fern.& Wieg.)Lint var. *australis* Lint, nom. nud.] Pacific Rush. Common,in wet habitats; montane, above 900m: e.g. Palomar Mountain; Eagles Nest; Cuyamaca and Laguna Mountains; Mesa Grande; Julian; Buckman Springs; San Felipe Creek. June-Aug.

J. luciensis Ertter in edit. Lucia Rush. Infrequent, in moist places; montane, above 1300m: Cuyamaca Lake.

J. macrophyllus Cov. Long-Leaf Rush. Frequent, in wet meadows; foothill and montane, above 600m: e.g. Wyss Ranch; Deer Flat; Harper Ranch; Witch Creek; Barona Valley; Descanso; Cuyamaca and Laguna Mountains; Laguna Lakes. May-Aug.

J. mexicanus Willd. ex Roem. & Schult. f. [*J. arcticus* Willd. var. *m.* (Willd.)Balslev; *J. balticus* Willd. var. *m.* Kuntz; *J. a.* ssp. *m.* (Willd.)Lint, nom. nud.] Mexican Rush. Common, in moist, usually alkaline areas; cismontane: e.g. Santa Margarita River; Fallbrook; Palomar Mountain; Hot Springs Mountain; San

Felipe; Sentenac Canyon; Cuyamaca and Laguna Mountains; Pine Valley; San Ysidro. Also Jacumba and Boulevard. May-Aug.

J. oxymeris Engelm. Pointed Rush. Infrequent, in moist habitats; foothill and montane, above 400m: Doane Valley; Cuyamaca Lake; Ramona. June-Aug.

J. patens E. Mey. Spreading Rush. Rare, known in the county only from a collection from Cuyamaca Peak, 1900m. June-July.

J. phaeocephalus Engelm. var. **paniculatus** Engelm. Brown-Head Rush. Occasional, in wet places; cismontane, below 1400m: Fallbrook; Valley Center; Julian; San Pasqual; Sweetwater Valley; Pine Hills; Cuyamaca Lake. May-Aug.

J. rugulosus Engelm. Wrinkled Rush. Frequent, along streams; cismontane: e.g. De Luz; San Marcos; Doane Valley; North Peak; Cuyamaca Peak; Arroyo Seco; Laguna Public Camp; Terrys Lodge. May-Sep.

***J. sphaerocarpus** Nees in Funk. Round-Fruit Toad-Rush. Native of Old World. Uncommon, in wet areas; cismontane: Palomar Mountain; Santa Ysabel; Julian; Cuyamaca Lake; Laguna Campground; Imperial Beach. Apr.-Aug. $2n=36$.

J. tenuis Willd. Western Rush. Infrequent, in damp, grassy places; montane, 1200-1600m: Wyss Ranch; Pine Hills. June-Aug. $2n=30$.

J. textilis Buchen. Basket Rush. Occasional, in shaded, moist places; foothill, 200-1000m: e.g. Mesa Grande; Bear Valley; El Cajon Mountain. May-June.

J. tiehmii Ertter in edit. [*J. kelloggii* auct. non Engelm.; *J. triformis* var. *uniflorus* Engelm.] Kellogg's Dwarf Rush. Uncommon, in moist areas; foothill and montane, above 800m: lower Doane Valley; Morgan Hill; Harper Ranch; Ramona; Campo. May-July.

J. torreyi Cov. Torrey's Rush. Uncommon, in wet places; foothill and montane, above 1000m: Buena Vista Creek near Lake Henshaw; Laguna Mountains. July-Sep.

J. triformis Engelm. Yosemite Dwarf-Rush. Occasional in vernal pools and other ephemeral ponds; cismontane: e.g. Rainbow; Ramona; El Cajon Mountain; Morena Dam; Campo; Barona Valley; Kearny Mesa. May-July.

J. xiphioides E. Mey. Iris-Leaf Rush. Common, in wet places: e.g. De Luz Creek Crossing; Lone Palm Spring; Lake Henshaw; upper Borrego Palm Canyon; Montezuma Valley; Indian Canyon; Canebrake Canyon; Cedar Creek; Banner; Chariot Creek; Agua Hedionda; La Costa; Lake Hodges; Escondido; Miramar Naval Air Station; Sweetwater Valley; Jamul; Laguna Meadows; Buckman Spring; Summers Ranch; Mountain Springs. May-Oct.

Luzula DC. Wood-Rush

L. comosa E. Mey. [*L. campestris* var. *comosa* (E. Mey.)Fern. & Wieg.] Common Wood-Rush. Occasional, in moist meadows; montane, above 1500m: e.g. Doane Valley; Dyche Valley; Hot Springs Mountain. May-Aug. $2n=24$.

JUNCAGINACEAE. Arrowgrass Family

KEY TO GENERA

Perianth present; stamens 3 or 6 . . . . *Triglochin*
Perianth absent; stamen 1 . . . . *Lilaea*

Lilaea Humb. & Bonpl.

L. scilloides (Poir.)Haum. [*L. subulata* Humb. & Bonpl.] Flowering Quillwort. Occasional, in shallow ponds, slow streams, and vernal pools; cismontane: Cuyamaca Lake; Laguna Camp; Pine Hills; Ramona; San Pasqual; Miramar; Kearny Mesa; Falls of the San Diego River; Otay Mesa. Feb.-May. $2n=12$.

Triglochin L. Arrowgrass

Ligules 2-parted to base; blade nearly terete . . . . *T. concinna*
Ligules simple, entire or emarginate; blade obcompressed . . . . *T. maritima*

T. concinna Davy var. **concinna**. Round-Leaf Arrowgrass. Infrequent, in Coastal Salt Marsh; below 5m: Imperial Beach; Silver Strand; Coronado. Mar.-Aug. $2n=48$.

T. maritima L. Saltmarsh Arrowgrass. Occasional, in tidal flats of Coastal Salt Marsh; below 5m: Mission Bay; Silver Strand; Chula Vista Bayfront; Salt Works; Imperial Beach; Monument Beach. May-Aug. $2n=12$, 48, 144.

LEMNACEAE. Duckweed Family
Contributed by Wayne P. Armstrong

KEY TO GENERA

Fronds flattened with two lateral (marginal) budding pouches; stamens 2, pistil 1, enclosed in membranous spathe within budding pouch; roots 1-5
  Roots 2-5 . . . . *Spirodela*
  Roots solitary . . . . *Lemna*
Fronds globoid-ellipsoid (generally 1mm long or less) with single budding pouch at basal end; stamen 1, pistil 1, spathe 0, produced in floral cavity on dorsal side; roots absent . . . . *Wolffia*

Lemna L. Duckweed

Nerve (vein) 1, sometimes obscure
  Fronds oval, usually symmetrical at base, 1-2.5mm long, generally thickened in middle with thin margin, often 2 attached; nerve extending less than 2/3 distance from node (point of root attachment) to apex; variable—changes into larger growth from 2-4mm long with thin transparent-green fronds in 4's. . . . . *L. minuscula*
  Fronds oval-oblong, often asymmetrical (oblique) at base, 2-4mm long, thin transparent-green with uniform thickness and texture, often 4-8 attached; nerve extending 3/4 distance from node to apex; variable—may grow in tangled masses under other aquatic plants . . . . *L. valdiviana*
Nerves 3-5
  Frond oval-orbicular with asymmetrical apex, often gibbous (swollen) on ventral side with enlarged air spaces; dorsal surface with slight median ridge, often yellowish-green mottled with reddish anthocyanin; variable—nongibbous forms very difficult to identify. . . . . *L. gibba*
  Frond oval and symmetrical, flattened on ventral side; dorsal surface smooth or with row of minute papules along midline, often shiny green
    Fronds producing rootless dark green or brown starch-filled turions in budding pouches during fall; dorsal side often with row of minute papules along midline; lower side developing reddish blotches of anthocyanin . . . . *L. turionifera*
    Fronds not producing overwintering turions; dorsal surface smooth or with obscure row of papules; lower side without reddish anthocyanin. . . . . *L. minor*

L. gibba L. Wind-Bags. Gibbous Duckweed. Common, in standing water: Carrizo Creek; Lakeside; southwest of Lake Hodges; Ramona. Mar.-Sep. $2n=40$.

L. minor L. Water Lentil, Lesser Duckweed. Infrequent, in stagnant ponds and quiet waters; coastal and foothill, below 1500m: Santa Margarita River estuary; Pamo Valley; San Pasqual Valley; Padre Dam; Descanso; El Monte; White Star. Mar.-Sep. $2n=40$

L. minuscula Herter. [*L. minuta* H.B.K.; *L. minima* Humb. ex Phil.; *L. cyclostasa* auct. non Thomps.] Least Duckweed. Frequent, in quiet waters; cismontane: Lake Henshaw; Spencer Valley; Cold Spring; Ramona; Descanso; Buckman Spring; Jacumba; Pine Valley; McCain Ranch; Mission Valley; Rolando; Lake Hodges. Mar.-Sep.

L. turionifera Landolt. [*L. minor* pro. parte auct. non L.] Turion Duckweed. Uncommon, in calm waters; cismontane: French Creek, Palomar Mountain; San Pasqual Valley; Moosa Canyon; Woods Valley.

L. valdiviana Phil. [*L. cyclostasa* (Ell.)Thomps.] Chile Duckweed. Infrequent, in stagnant ponds and quiet waters: Las Flores Creek estuary; Palomar Mountain; Warner Hot Springs; Vallecito Canyon; Lakeside. Mar.-Sep.

Spirodela Schleiden Greater Duckweed

S. punctata (G.F.W.Meyer)Thompson. [*S. oligorrhiza* (Kurz)Hegelm.] Rare, in ponds; cismontane: north Twin Oaks Valley.

Wolffia Horkel ex Schleid.

Fronds ovoid to ellipsoid with flattened dorsal surface; apex somewhat pointed (acute) and slightly upturned; floating with entire dorsal surface above water; greatest width of frond near water surface; brown pigment cells present . . . . *W. borealis*
Fronds ovoid or globoid (often ellipsoid or cylindrical in *W. globosa*); apex of fronds rounded; floating with only the central portion of dorsal surface above water; brown pigment cells lacking
  Fronds 0.8-1mm long; dorsal surface rounded or slightly flattened, minutely roughened by enlarged cells . . . . *W. columbiana*
  Fronds 0.4-0.9mm long; dorsal surface flattened and minutely roughened by a few enlarged cells . . . . . . *W. globosa*

W. borealis (Engelm. ex Hegelm.)Landolt. [*W. punctata* auct. non Griseb.] Rare, known in the county only from a collection at Lake Hodges.

W. columbiana Karst. Rare, known from pond 1.5km below Lake Hodges; San Dieguito River; Lake Hodges.

***W. globosa** (Roxb.)Hartog & Plas. [*W. cylindracea* Hegelm.] Native of Asia. Rare, in calm water: pond near Lake Hodges.

LILIACEAE. Lily Family
[Segregate families are shown parenthetically]

KEY TO GENERA

Plants not bulbous; rootstock tuberous
  Foliage consisting of needlelike or flattened branches borne in axils of scale-like. . . . . *Asparagus*
  Foliage of true leaves, not borne in axils of scale leaf
    Leaves all basal, terete, fleshy . . . . *Asphodelus*
    Leaves cauline, broad
      Plants 1-2m tall; leaves 2.5-4dm long; fruit a capsule . . . . *Veratrum*
      Plants 3-6dm tall; leaves 5-15cm long; fruit a berry . . . . *Smilacina*
Plants bulbous or cormous
  Stems leafy, sometimes branched above
    Styles 3 and distinct; perianth segments with a prominent gland . . . . *Zigadenus*
    Style 1, more or less lobed at summit
      Perianth segments dissimilar (outer sepaloid, inner petaloid) . . . . *Calochortus*
      Perianth segments alike, all petaloid
        Perianth segments 15-30mm long, purplish or brownish . . . . *Fritillaria*
        Perianth segments 60-90mm long, orange or yellow. . . . . *Lilium*

Stems scape-like, leaves mostly basal
  Perianth funnelform, 4-6cm long; deserts . . . . . *Hesperocallis*
  Perianth more or less rotate, to 3cm long; coastal . . . . . *Chlorogalum*

Asparagus L. Asparagus
[Asparagaceae]

Branchlets leathery, ovate to 5cm long . . . . . *A. asparagoides*
Branchlets filiform
  Branchlets up to 10 in a cluster . . . . . *A. officinalis*
  Branchlets up to 20 in a cluster, forming a flat, leaf-like spray . . . . . *A. setaceus*

***A. asparagoides** Wight. Florist's-Smilax, Smilax Asparagus. Native of South Africa. Local, escaping from culivation; coastal: Rancho Santa Fe; Eden Gardens; North Park; National City; Casa de Oro. June.

***A. officinalis** L. Garden Asparagus. Native of Europe. Infrequent, persisting from cultivation or escaping in moist areas; coastal, below 200m: San Pasqual Valley; San Elijo Lagoon; East San Diego; Murray Dam; National City. May-June. *n*=10.

***A. setaceus** (Kunth)Jessop. [*A. plumosus* Baker] Asparagus-Fern. Plume Asparagus. Native of South Africa. Local, escaping in moist areas; coastal, below 100m: San Diego. June.

Asphodelus L. Asphodel
[Asphodelaceae]

***A. fistulosus** L. Hollow-Stem Asphodel. Native of Mediterranean region. Infrequent, escaping in cultivated areas and along roadsides; coastal: Camp Pendleton, I-5; Vista; San Marcos; San Diego; National City. Feb.-May.

Calochortus Pursh Mariposa-Lily.
[Liliaceae]

Flowers globose to closed-campanulate, pendant; capsules nodding
  Flowers white . . . . . *C. albus*
  Flowers yellow . . . . . *C. amabilis*
Flowers more open than above, upright
  Surface of gland naked or nearly so (bordering membrane may be fringed and extend over gland); petals fimbriate . . . . . *C. weedii*
  Surface of gland densely hairy, often with peculiar thickened hairs
    Glands not depressed or only slightly so, without surrounding membrane
      Inflorescence with distinct axis, internodes short
        Glands covered with linear hairs or processes
          Processes on gland dark; capsule oblong, obtuse . . . . . *C. catalinae*
          Processes on gland yellow; capsule linear, acute . . . . . *C. dunnii*
        Glands with clavate or fungoid processes; anthers purplish or blue . . . . . *C. splendens*
      Inflorescence subumbellate
        Gland quadrate, each petal with dark red blotch . . . . . *C. venustus*
        Gland not quadrate, shaped like an inverted "V" . . . . . *C.superbus*
    Glands depressed, surrounded by a membrane
      Flowers white to purplish . . . . . *C. invenustus*
      Flowers yellow . . . . . *C. concolor*

C. albus Dougl. ex Benth. var. **albus**. White Fairy-Lantern. Occasional, on shaded slopes; foothill and montane, above 700m. e.g. Palomar, Cuyamaca, and Laguna Mountains; Hutcut Valley; Mesa Grande; Pine Hills; Guatay; Potrero Grade; Tecate Peak. Also De Luz. Apr.-June. *n*=10.

C. amabilis Purdy. Reported by Clint Powell from Volcan Mountain. July. *n*=10.

C. catalinae Wats. Catalina Mariposa. Rare, on grassy slopes; foothill, below 700m: Lake Hodges; Ramona. Apr.-June. *n*=7.

C. concolor (Baker)Purdy. Golden-Bowl Mariposa. Occasional, on open slopes; foothill, montane, and transmontane, above 750m: Mesa Grande; Witch Creek; Doane Valley; Warner Hot Springs; San Felipe Hills; Pine Hills; Cuyamaca Lake; Vallecito Valley; Canebrake Canyon; Laguna Mountains; Laguna Junction; Jacumba; Campo; Descanso. May-June. Type from "Laguna, on the edge of the desert, San Diego County". 2*n*=ca. 14.

C. dunnii Purdy. Dunn's Mariposa. Rare, on mountain slopes; foothill and montane, above 1000m: lower Doane Valley; Julian; Inspiration Point; Cuyamaca Mountains; East Mesa; Descanso; Guatay Mountain; San Miguel Mountain; Otay Mountain. June-July. Type from "near Julian". *n*=7.

C. invenustus Greene. [*C. nuttallii* var. *australis* Munz.] Shy Mariposa. Scarce, on granite outcrops of Montane Coniferous Forest; montane, above 1700m: Cuyamaca Mountains; Laguna Camp; Garnet Peak; Laguna Observatories. May-June. 2*n*=7 pairs.

C. splendens Dougl. ex Benth. [*C. davidsonianus* Abrams, the type from grassy slopes between Onofre Mountains and the sea.] Splendid Mariposa. Common, on grassy slopes; cismontane, below 1500m: e.g. Palomar, Cuyamaca, and Laguna Mountains; Banner Grade; Moosa Canyon; Camp Pendleton; Fallbrook; Pala; Chihuahua Valley; Mesa Grande; Santa Ysabel; Hellhole Canyon; Warner Hot Springs; Pine Valley; Descanso; Tecate Peak; Otay Lake; Mission Hills; Oceanside; Mount Soledad; Jacumba. Apr.-July. *n*=7, 14.

C. superbus Purdy ex J.T. Howell. Superb Mariposa. Rare, known in the county only from a collection at Doane Valley, Palomar Mountain, 1500m. June. *n*=6, 7.

C. venustus Dougl. ex Benth. Square Mariposa. Scarce, on Palomar Mountain, above 1200m: lower Doane Valley; Observatory Meadow. June. *n*=7.

C. weedii Wood var. **weedii**. Weed's Mariposa. Common, on open slopes; cismontane, below 1300m: e.g. Rainbow; Palomar Mountain; Rincon Grade; San Marcos; Pine Hills; Santa Ysabel; Ballena; Alpine; Flynn Springs; Potrero Grade; Jacumba; Laguna Junction; Buckman Spring; Morena; Jamul; Torrey Pines State Reserve; San Diego; Point Loma. May-July. Type from San Diego. *n*=9.

Chlorogalum Kunth. Soap-Plant
[Hyacinthaceae]

Flowers diurnal; style exceeding perianth; segments 5-8mm long . . . . *C. parviflorum*
Flowers vespertine; style not exceeding perianth; segments 8-30mm long . . . . *C. pomeridianum*

C. parviflorum Wats. Small-Flower Soap-Plant. Frequent, in open areas; coastal and foothill, below 700m: e.g. De Luz; Oceanside; Moosa Canyon; Barona Indian Reservation; Alpine; Ballena; Jacumba; Otay Mountain; Lee Valley; San Diego; Del Mar. May-June. Type from Cajon Valley. *n*=30.

C. pomeridianum (DC.)Kunth. Wavy-Leaf Soap-Plant. Infrequent, in open areas; northern coastal and foothill, below 800m: Upper San Onofre Canyon; Twin Oaks Valley; Lake San Marcos; Reidy Canyon; San Marcos Slough. May-June. *n*=15, 18.

Fritillaria L. Fritillary
[Liliaceae]

F. biflora Lindl. California Chocolate Lily. Occasional, in clay soils; on open slopes; coastal, below 700m: e.g. upper San Onofre Canyon; San Marcos; Rancho Santa Fe; Bernardo; Del Mar; San Diego; Point Loma; Dehesa; McGinty Mountain; Santee; San Miguel Mountain; Jamul; Bonita; San Ysidro. Jan.-Apr. *n*=12.

Hesperocallis Gray. Desert Lily
[Funkiaceae]

H. undulata Gray. Desert Lily. Occasional, in sandy areas; desert, below 300m: e.g. Rockhouse Canyon; Borrego Valley; Benson Dunes; Ocotillo Wells; Vallecito Stage Station; Carrizo Creek; Sweeney Pass. Feb.-Apr. 2*n*=48.

Lilium L. Lily
[Liliaceae]

Leaves mostly in conspicuous whorls
  Flowers more or less orange; plants of xeric places
    Flowers orange-yellow; stems 9-15dm high. . . . . *L. humboldtii* var. *bloomerianum*
    Flowers orange-red; stems to 3.5m high. . . . . *L. humboldtii* var. *ocellatum*
  Flowers yellow; plants of wet places . . . . *L. pardalinum*
Leaves scattered or only the lowest whorled; flowers yellow. . . . . *L. parryi*

L. humboldtii Roezl & Leichtl. var. **bloomerianum** (Kell.)Jeps. [*L. fairchildii* M.E. Jones, the type from pine woods in deep shade, Palomar Mountain.] San Diego Tiger-Lily. Infrequent, in shaded canyons and slopes; cismontane, above 300m: e.g. De Luz Creek; Castro Canyon; Palomar, Hot Springs, Cuyamaca, and Laguna Mountains; Lake Henshaw; San Felipe Canyon; Japatul Valley; Featherstone Creek; Woodwardia Canyon. June-Aug.

L. humboldtii var. **ocellatum** (Kell.)Elwes. Red-Eye Tiger-Lily. Infrequent, in shaded canyons; northern montane, above 800m: Palomar Mountains; Tenaja Canyon. July.

L. pardalinum Kell. [*L. californicum* Lindl.] Leopard Lily. Rare, in shaded, springy areas of Montane Coniferous Forest; montane, above 1300m: Cuyamaca and Laguna Mountains; Azalea Creek below Fern Flat Spring; Noble Canyon Hiking Trail. May-July. *n*=12.

L. parryi Wats. Lemon Lily. Rare, in Montane Coniferous Forest; montane, above 1200m: Palomar Mountain; Cuyamaca Mountains; Cuyamaca Lake. July-Aug. *n*=12.

Smilacina Desf. Solomon's-Seal
[Asparagaceae]

S. stellata (L.)Desf. Star Solomon's-Seal. Uncommon, in Montane Coniferous Forest; montane, above 1300m: below Cuyamaca Dam; near Milk Ranch in the Cuyamaca Mountains; below Fern Flat Spring. Apr.-July. *n*=18.

Veratrum L. Corn-Lily
[Melanthiaceae]

V. californicum Durand. California Corn-Lily. Local, on Palomar Mountain; montane, above 1200m: Observatory Meadow; Wyss Ranch; Doane Valley. Aug. *n*=16.

Zigadenus Michx. Zygadene
[Melanthiaceae]

Stamens longer than perianth in mature flowers; segments 3-4mm long . . . . *Z. venenosus*
Stamens not exceeding perianth; segments 10-12mm long
  Stems 3-10dm tall . . . . *Z. fremontii* var. *f.*
  Stems 1-2dm tall . . . . *Z. fremontii* var. *minor*

Z. fremontii (Torr.)Torr. ex Wats. var. **fremontii**. [*Toxicoscordion f.* (Torr.)Rydb.] Fremont's Camas. Frequent, in open areas; coastal and foothill, below 700m: e.g. Pala; Barona Valley; Torrey Pines State Reserve; Point Loma; Otay Mountain; Potrero Grade. Apr.-May. n=11.

Z. fremontii var. **minor** (H.& A.)Jeps. [*Toxicoscordion f* var. *m.* (H. & A.)Jeps. Few-Flowered Camas. Occasional, in open areas; south coastal, below 400m: San Marcos; Mount Soledad; Rancho Santa Fe; Poway; Mission Gorge; Fletcher Hills; San Diego; San Ysidro. Feb.-Mar. n=11.

Z. venenosus Wats. [*Toxicoscordion v.* (S.Wats.)Rydb.; *Z. diegoensis* Davids., the type from Palomar Mountains.] Death Camas. Occasional, on open slopes; coastal and montane: Calaveras Lake; La Jolla; Mount Soledad; Point Loma; San Diego; Olivenhain; Middle Peak; Cuyamaca Lake; Laguna Mountains. May-July.

NAJADACEAE. Nymph Family

Najas L.

Leaves coarsely toothed; plants dioecious . . . . . *N. marina*
Leaves almost entire; plants monoecious. . . . . *N. guadalupensis*

N. guadalupensis (Spreng.)Morong. Southern Naias. Rare, in freshwater ponds and lakes: San Diego. July-Aug. $2n$=12, 36, 42, 48, 54, 60.

N. marina L. [*N. m.* var. *californica* Rendle] Large Naias. Rare, in freshwater ponds and lakes: De Luz Creek; San Diego. July-Aug. n=6.

ORCHIDACEAE. Orchid Family

KEY TO GENERA

Leaves reduced and scale-like; plants saprophytic, not green . . . . . *Corallorhiza*
Plants not as above, foliage green
  Flowers with distinct, slender spur 4-12mm long . . . . . *Habenaria*
  Flowers spurless
    Perianth segments 10-18mm long, greenish or purplish-reddish . . . . . *Epipactis*
    Perianth segments 2-5mm long, whitish . . . . . *Spiranthes*

Corallorhiza Chat. Coralroot

C. maculata (Raf.)Raf. Spotted Coralroot. Occasional, on shaded slopes of Montane Coniferous Forest; montane, above 1100m: Palomar, Volcan, Hot Springs, and Cuyamaca Mountains. June-Aug. $2n$=42.

Epipactis Sw.

E. gigantea Dougl. ex Hook. Stream Helleborine. Occasional, along perennial streams: Fallbrook; Lake Henshaw; Cuyamaca Mountains, northwest slope of Cuyamaca Peak; Noble Mine; Oasis Spring; Hot Springs Mountain; Hellhole Canyon; Mason Valley; Barret Dam; Padre Dam. Apr.-Aug. n=20.

Habenaria Willd. Rein-Orchid

Sepals 1-nerved; leaves usually near base of stem and withering before anthesis
  Spur at least twice as long as lip; raceme densely-flowered . . . . . *H. elegans*
  Spur not much longer than lip; raceme loosely-flowered . . . . . *H. unalascensis*
Sepals 3-nerved; leaves usually scattered and green at anthesis
  Flowers white; lip abruptly dilated at base. . . . . *H. leucostachys*
  Flowers green; lip not prominently dilated at base . . . . . *H. sparsiflora*

H. elegans (Lindl.)Boland. [*Piperia e.* (Lindl.)Rydb.; *P. leptopetala* Rydb., the type from mountains east of San Diego.] Elegant Habenaria. Occasional, on shaded slopes of Montane Coniferous Forest; montane, above 1100m: e.g. Palomar, Volcan, Cuyamaca, and Laguna Mountains; Pine Hills; Boulder Creek; Pine Valley; Corte Madera. May-July. $2n$=21 pairs.

H. leucostachys (Lindl.)Wats. [*H. dilatata* var. *l.* (Lindl.)Ames.; *Platanthera l.* Lindl.] Sierra Nevada Bog-orchid. Infrequent, in seepage areas in meadows; montane, above 1200m: upper Doane Valley; Palomar Mountain Country Club; meadows below Cuyamaca Dam. June-Aug. n=21.

H. sparsiflora Wats. var. **sparsiflora**. [*Limnorchis s.* (S.Wats.)Rydb.] Sparseflower Bog-orchid. Rare, known in the county only from the Wyss Ranch, Palomar Mountain, 1800m. June-July.

H. unalascensis(Spreng.)Wats. [*Piperia u.* (Spreng.)Rydb.; *H. cooperi* S. Wats., the type from clay hills near San Diego.] Slenderspire Piperia. Occasional, on grassy slopes; cismontane: e.g. Fallbrook; Palomar, Hot Springs, Cuyamaca, and Laguna Mountains; Volcan Mountain; Julian; Descanso; Barrett Lake; Campo; Paradise Valley; Point Loma; La Jolla; Carlsbad; Rice Canyon. Mar.-July.

Spiranthes Rich. Ladies'-Tresses

S. romanzoffiana Cham. var. **porrifolia** (Lindl.)Ames & Correll. [*S. porrifolia* Lindl.] Western Ladies'-Tresses. Rare, montane seepage areas; above 1000m: below Cuyamaca Dam; Pine Valley; northwest slope of Cuyamaca Peak. July-Aug. $2n$=60.

POACEAE. Grass Family
(Gramineae)
[Tribes shown parenthetically]

KEY TO DIAGNOSTIC GROUPS OR GENERA

Spikelets with glumes persistent, rachilla articulated above them, 1-many-flowered, often prolonged beyond upper lemma; upper lemmas frequently empty
- Spikelets borne in an open or spikelike raceme or panicle, usually upon distinct pedicels
  - Spikelets 1-flowered
    - Spikelets with 2 sterile or staminate lemmas below fertile lemma; palea 1-nerved . . . . . *Phalaris*
    - Spikelets without sterile lemmas below fertile lemma; palea 2-nerved . . . . . *GROUP 1*
  - Spikelets 2-many-flowered
    - Lemma usually shorter than empty glumes; awn dorsal, usually bent . . . . . *GROUP 2*
    - Lemma usually longer than empty glumes; awn terminal and straight or none (rarely dorsal in *Bromus*) *GROUP 3*
- Spikelets borne in 2 rows, sessile or nearly so
  - Spikelets on one side of the continuing axis, forming 1-sided spikes; spikes usually more than 1 . . . . . *GROUP 4*
  - Spikelets alternate on opposite sides of the often articulated axis; spike terminal, single . . . . . *GROUP 5*

Spikelets falling from the pedicels entire, articulate below the glumes, naked or enclosed in bristles or bur-like involucres, 1-flowered, or if 2-flowered, then the lower flower staminate; no lemmas empty; rachilla not extending beyond upper lemma
- Plants monoecious, staminate and pistillate spikes in separate inflorescences . . . . . *Zea*
- Plants not monoecious
  - Lemma and palea hyaline, thin, much more delicate in texture than the glumes
    - Spikelets in pairs, 1 sessile, other pedicellate . . . . . *GROUP 6*
    - Spikelets not in pairs . . . . . *GROUP 1*
  - Lemma, at least that of the perfect flower, similar in texture to glumes, or thicker and firmer, never hyaline and thin
    - Lemma and palea membranous, 1st glume usually the larger . . . . . *Hilaria*
    - Lemma and palea chartaceous to coriaceous, very different in color and appearance from the glumes *GROUP 7*

KEY TO GENERA

GROUP 1

Rachilla articulate below glumes, these falling with spikelet
- Glumes long-awned . . . . . *Polypogon*
- Glumes awnless
  - Glumes united towards base, ciliate on keel; inflorescence not capitate and bracteate . . . . . *Alopecurus*
  - Glumes not united, glabrous; inflorescence capitate, in axils of broad bracts . . . . . *Crypsis*

Rachilla articulate above the glumes
- Lemma indurate when mature and very closely embracing grain; callus usually well-developed and bearded
  - Awn 3-branched, the lateral sometimes obsolete and then no line of demarcation between lemma and awn *Aristida*
  - Awn simple, with line of demarcation between awn and lemma
    - Awn twisted and bent, persistent, several to many times longer than the fruit . . . . . *Stipa*
    - Awn not twisted, deciduous, rarely more than 3-4 times as long as the plump fruit . . . . . *Oryzopsis*
- Lemma usually hyaline or membranaceous at maturity; callus not well-developed
  - Glumes usually longer than lemma (sometimes not much longer in *Agrostis*)
    - Glumes compressed-carinate, stiff-ciliate on keel; panicle dense, cylindric or ellipsoid . . . . . *Phleum*
    - Glumes not compressed-carinate, not ciliate
      - Glumes saccate at base; lemma long-awned; panicle contracted, shining. . . . . *Gastridium*
      - Glumes not saccate at base; lemma awned or not; panicle open or contracted
        - Floret bearing a tuft of hairs at the base from the short callus; palea well-developed, rachilla prolonged behind palea as a hairy bristle . . . . . *Calamagrostis*
        - Floret without hairs at base or with short hairs; palea usually small or obsolete . . . . . *Agrostis*
  - Glumes not longer than the lemma, usually shorter (except sometime for the awn-tips)
    - Lemma awned from the tip or mucronate, usually 3-5-nerved . . . . . *Muhlenbergia*
    - Lemma awnless or awned from the back
      - Fruit at maturity falling from the lemma and palea; seed loose in the pericarp; lemma 1-nerved; panicle open and contracted. . . . . *Sporobolus*
      - Fruit not falling from lemma and palea, but permanently enclosed by them; seed adnate to the pericarp; panicle spikelike
        - Floret without hairs at base
          - Panicle short, partly enclosed in the sheath; annual . . . . . *Crypsis*
          - Panicle elongate; perennial . . . . . *Muhlenbergia*
        - Floret with tuft of hairs at base from the short callus; lemma and palea chartaceous, awnless . . *Ammophila*

GROUP 2

Florets 2, the lower perfect, awnless; upper staminate and awned. . . . . *Holcus*

Florets 2 or more, all alike except the reduced upper ones
- Lemmas bifid at apex, awned or mucronate between the lobes; spikelets several-flowered
  - Awns conspicuous, flat, bent; spikelets 1cm or more long . . . . . *Danthonia*
  - Awns minute to obsolete; spikelets not more than 5mm long . . . . . *Schismus*
- Lemmas toothed but not bifid and awned, nor mucronate between the lobes
  - Glumes 2-3.5cm long, 7-9-nerved; spikelets 2-flowered, or with a rudimentary 3rd one, pendulous; annual *Avena*
  - Glumes not more than 1cm long, 1-5-nerved; spikelets not pendulous; mostly perennial
    - Lemmas keeled and awnless, or awn straight, from toothed apex . . . . . *Koeleria*
    - Lemmas convex, awned from below the middle
      - Rachilla prolonged from behind the upper floret; lemmas truncate and erose-dentate at summit *Deschampsia*
      - Rachilla not prolonged; lemmas tapering into 2 slender teeth . . . . . *Aira*

GROUP 3

Lemmas divided at top into 5 or more awns or awn-like teeth or lobes . . . . *Orcuttia*
Lemma awnless or 1-3-awned
  Plants tall stout reeds with large, plumose panicles; lemmas or rachilla with silky hairs as long as lemma
    Leaves crowded at base of culms . . . . *Cortaderia*
    Leaves distributed along the culms
      Lemmas naked; rachilla hairy . . . . *Phragmites*
      Lemmas hairy; rachilla naked . . . . *Arundo*
  Plants low or to ca. 1.5m high, not reed-like
    Plants perennial, dioecious
      Spikelets obscure, scarcely differentiated from the short and rigid, crowded leaves . . . . *Monanthochloe*
      Spikelets in narrow, simple, exserted panicles . . . . *Distichlis*
    Plants not dioecious (except some *Poa* spp., these with villous lemmas)
      Spikelets of 2 forms, sterile and fertile intermixed; panicle dense, more or less secund
        Fertile spikelets 2-3-flowered; sterile spikelets with numerous, rigid, awn-shaped lemmas; panicle dense, spikelike . . . . *Cynosurus*
        Fertile spikelets with 1 perfect floret, long-awned; sterile spikelets with many, obtuse, sterile lemmas; panicle branches short, nodding . . . . *Lamarckia*
      Spikelets all alike in the same inflorescence
        Lemmas 3-nerved, nerves prominent
          Inflorescence a few-flowered, woolly, capitate panicle overtopped by the leaves; lemmas toothed or cleft: deserts . . . . *Erioneuron*
          Inflorescence an exserted, open or spikelike panicle . . . . *Eragrostis*
        Lemmas 5-many-nerved, nerves sometimes obscure
          Lemmas as broad as long, with outspread margins; florets closely imbricate, horizontally spreading. . *Briza*
          Lemmas longer than broad, margins clasping the palea; florets not horizontally spreading
            Lemmas keeled on back (sometimes rounded in *Poa*)
              Spikelets strongly compressed, crowded in 1-sided clusters at ends of stiff, naked panicle-branches . . . . *Dactylis*
              Spikelets not strongly compressed, not in 1-sided clusters
                Lemmas awned from minutely bifid apex or nearly awnless; spikelets large . . . . *Bromus*
                Lemmas awnless; spikelets small . . . . *Poa*
            Lemmas rounded on back
              Glumes papery; lemmas firm, strongly nerved, scarious-margined; upper florets sterile, often reduced to rudiments infolded by the broad upper lemmas; spikelets tawny or purplish . . . . *Melica*
              Glumes not papery; upper florets like the others
                Nerves of lemma parallel, not converging above or only slightly so . . . . *Puccinellia*
                Nerves of lemma converging toward the summit, lemmas narrowed at apex
                  Lemmas awned or awn-tipped from minutely bifid apex (rarely awnless); palea adhering to caryopsis
                    Spikelets in open to contacted panicles; stigmas at the sides at summit of ovary . . . . . . *Bromus*
                    Spikelets nearly sessile in strict racemes; stigmas terminal on the ovary. . . . . . . *Brachypodium*
                  Lemmas entire, pointed, awnless or awned from the tip
                    Spikelets awned except in a few perennial spp.; lemmas pointed
                      Plants perennial . . . . *Festuca*
                      Plants annual . . . . *Vulpia*
                    Spikelets awnless; second glumes 1-3-nerved; spikelets smaller; perennials. . . . . *Poa*

GROUP 4

Spikelets with more than 1 perfect floret
  Spikelets many, slender, racemose or on elongate axis . . . . *Diplachne*
  Spikelets few, digitate or nearly so
    Rachis of spike not prolonged beyond spikelets . . . . *Eleusine*
    Rachis of spike extending beyond spikelets. . . . . *Dactyloctenium*
Spikelets with only 1 perfect floret, sometimes with imperfect florets above or below
  Spikelets without additional modified florets, rachilla sometimes prolonged
    Rachilla articulate below glumes, spikelets falling entire; spikes arranged on elongate axis, not digitate . . *Spartina*
    Rachilla jointed above glumes; spikes digitate . . . . *Cynodon*
  Spikelets with 1 or more modified florets above the perfect one
    Spikes digitate or nearly so . . . . *Chloris*
    Spikes racemose along main axis . . . . *Bouteloua*

GROUP 5

Spikelets solitary at each node of rachis
  Spikelets 1-flowered, sunken in hollows in rachis; spikes slender, cylindric; plants low annuals
    First glume wanting; spikes 1-2dm long . . . . *Monerma*
    First glume present; spikes 0.7-1dm long . . . . *Parapholis*
  Spikelets 2-several-flowered, not sunken in rachis
    Spikelets placed edgewise to rachis, lateral ones with single glume . . . . *Lolium*
    Spikelets placed flatwise to rachis
      Plants perennial . . . . *Agropyron*
      Plants annual
        Glumes ovate, 3-nerved . . . . *Triticum*
        Glumes subulate, 1-nerved . . . . *Secale*
Spikelets usually more than 1 at each node or rachis
  Spikelets 3 at each rachis-node, 1-flowered, lateral pair pedicellate, usually reduced to awns . . . . *Hordeum*
  Spikelets 2 or more (sometimes 1 in *Elymus*) at each rachis-node, alike, 2-6-flowered
    Rachis continuous, rarely disarticulating and then only tardily; glumes not greatly elongate . . . . *Elymus*
    Rachis disarticulating into joints at maturity, glumes usually setaceous and greatly elongate . . . . *Sitanion* and × *Elysitanion*

GROUP 6
[Andropogoneae]

Spikelets unisexual, dissimilar . . . . . . *Zea*
Spikelets ordinarily bisexual, but if some are staminate or neuter then these borne in the same raceme or spike
  Racemes of several joints
    Fertile spikelet with hairy-pointed callus, formed of attached supporting rachis joint or pedicel; awns strong, brown . . . . . . *Heteropogon*
    Fertile spikelet without callus, rachis disarticulating below spikelet; awns slender
      Pedicels and rachis internodes without central groove or membranous area . . . . . . *Schizachyrium*
      Pedicels and upper rachis usually with central groove or membranous area . . . . . . *Bothriochloa*
  Racemes reduced to 1 or few joints, these in a compound panicle . . . . . . *Sorghum*

GROUP 7
[Paniceae]

Spikelets sunken in cavities of the flattened, corky rachis; coarse, creeping grass (escaping from cultivation) . . . . . . *Stenotaphrum*
Spikelets not sunken in rachis
  Spikelets subtended or surrounded by many, distinct or connate bristles forming an involucre
    Bristles persistent, spikelets deciduous . . . . . . *Setaria*
    Bristles falling with spikelets at maturity
      Bristles not united at base, slender, often plumose . . . . . . *Pennisetum*
      Bristles united into burlike involucre, retrorsely barbed . . . . . . *Cenchrus*
  Spikelets not subtended by bristles
    Glumes awned or mucronate; apex of palea not enclosed by lemma
      Inflorescence paniculate; spikelets silky . . . . . . *Rhynchelytrum*
      Inflorescence of unilateral racemes along common axis; spikelets not silky . . . . . . *Echinochloa*
    Glumes awnless; apex of palea usually enclosed by lemma
      Spikelets in panicles . . . . . . *Panicum*
        Second glume and sterile lemma glabrous or somewhat scaberulous only on the nerves toward the apex . . . . . . *Panicum*
        Second glume and sterile lemma short pilose all over . . . . . . *Dichanthelium*
      Spikelets in 1-sided, spikelike racemes
        Racemes slender, 3-12 in number . . . . . . *Digitaria*
        Racemes stout, in pairs . . . . . . *Paspalum*

Agropyron Gaertn. Wheatgrass
[Triticeae]

Lemmas awned; plants typically without creeping rhizomes . . . . . . *A. parishii*
Lemmas awnless or nearly so; plants rhizomatous or densely tufted
  Lemma not keeled toward apex . . . . . . *A. parishii*
  Lemma keeled toward apex . . . . . . *A. intermedium*

***A. intermedium** (Host)Beauv. Intermediate Wheatgrass. Native of Eurasia. Local, frequently planted on burns or fuel breaks or type conversions: Hot Springs, Cuyamaca, and Laguna Mountains. $2n$=42.

A. parishii Scribn. & Sm. var. **laeve** Scribn. & Sm. California Wheatgrass. Infrequent, on slopes; Montane Coniferous Forest, above 1300m: Palomar Mountain; Doane Valley; Cuyamaca and Laguna Mountains. June-July. Type from Fowley's, Cuimaca [Cuyamaca] Mountains.

Agrostis L. Bent
[Aveneae]

Plants annual, of ephemeral ponds
  Lemma 4-toothed at apex, the awn 3.5-4mm long . . . . . . *A. microphylla*
  Lemma deeply bifid, the awn 5-6mm long . . . . . . *A. tandilensis*
Plants perennial
  Palea present, 1mm or more long; 2-nerved, at least half as long as lemma
    Panicle branches in dense verticils; the panicle contracted and densely flowered; glumes 2mm or less long; plants ascending with creeping stolons . . . . . . *A. semiverticillatus*
    Panicle branches not in dense verticils; panicle contracted but usually not densely flowered; glumes, at least some, over 2mm long; plants usually with both stolons and rhizomes . . . . . . *A. stolonifera*
  Palea obsolete or a minute nerveless scale sometimes up to 0.5mm long
    Plants with long, slender creeping rhizomes; anthers 1.1-1.6mm long . . . . . . *A. diegoensis*
    Plants without rhizomes, stolons sometimes developed; anthers 0.3-0.6mm long
      Panicle dense to loose, the branches crowded and densely flowered to base; lemma awnless or awned *A. exarata*
      Panicle very diffuse, the branches capillary, not flexuous, the spikelets arranged only above the middle *A. scabra*

A. diegoensis Vasey. Leafy Bent. Occasional, in meadows and on open slopes; cismontane, above 150m: Palomar Mountain, Cuyamaca Lake; Laguna Mountains; Harbison Canyon; Jamul Mountain; Proctor Valley. July-Aug. Type from either Chollas Valley or Palomar Mountain, San Diego. $2n$=42.

A. exarata Trin. var. **minor** Hook. Spike Redtop. Occasional, on interior slopes; cismontane, above 500m; e.g. Palomar Mountain; Cuyamaca Lake; Laguna Mountains; Barona Valley. $2n$=28, 42, 56.

A. microphylla Steud. Small-Leaf Bent. Infrequent, in vernal pools; coastal, below 200m: Kearny Mesa; San Diego. May-July. n=28.

A. scabra Willd. [*A. hiemalis* auct. non (Walt.)B.S.P.] Rough Ticklegrass. Infrequent, in moist meadows and on slopes; montane, above 1300m: Doane Valley; Cuyamaca Mountains. July-Sep. $2n$=28, 42, 43.

***A. semiverticillatus** (Forsk.)C. Chr. [*Polypogon s.* (Forsk.)Hylander] Water Bent. Native of Europe. Commonly naturalized in moist, disturbed places; below 1000m: e.g. Santa Margarita River ford; San Pasqual; Sentenac Canyon; Buckman Springs; Morena Village; Jacumba; Campo. May-June. $2n$=14,28.

***A. stolonifera** L. var. **palustris** (Hudson)Farwell. [*A. alba* auct. non L.] European Redtop. Native of Europe. Occasionally adventive in moist areas: Palomar Mountain; Dehr Creek; Pine Hills; Paso Picacho; Cuyamaca Mountains; La Mesa; Balboa Park. June-Sep. $2n$=28, 46, 56.

A. tandilensis (Mtze.)Parodi. [*A. kennedyana* Beetle, the type from San Diego.] San Diego Bent. Rare, about vernal pools; cismontane; La Jolla; San Diego; Laguna Meadow. Apr.-June.

Aira L. Hairgrass
[Avenaeae]

***A. caryophyllea** L. Silver Hairgrass. Native of Europe. Uncommon in open areas; cismontane below 1200m: Lake Henshaw Resort; San Miguel Mountain.

Alopecurus L. Foxtail
[Avenaeae]

Plants perennial; spikelets 2-2.5mm long . . . . . *A. aequalis*
Plants annual
  Spikelets 3-3.5mm long . . . . . *A. howellii*
  Spikelets 2-2.5mm long . . . . . *A. carolinianus*

A. aequalis Sobol. Short-Awn Foxtail. Rare, in moist places; montane, above 1400m: Laguna Mountains. May-July. $2n$=7 pairs.

A. carolinianus Walt. Carolina Foxtail. Rare, in moist, open ground: Cuyamaca Lake. May-June.

A. howellii Vasey. [*A. californicus* Vasey, in part.] Howell's Foxtail. Rare, about drying mud flats; coastal, below 150m: Kearny Mesa; San Diego. Mar-June.

Ammophila Host.
[Avenaeae]

**A. arenaria* (L.)Link. European Beachgrass. Native of Europe. Once introduced at Spanish Bight, Coronado, but not recently collected. May-Aug. n=7, 14.

Aristida L.
[Aristideae]

Neck of fruit jointed at base, disarticulating with slight pressure at maturity . . . . . *A. californica*
Neck of fruit not jointed, persistent
  Plants annual
    Awn segments less than 2cm long . . . . . *A. adscensionis*
    Awn segments 4-7cm long, widely divergent . . . . . *A. oligantha*
  Plants perennial
    Glumes about equal
      Panicle open with spreading branches naked at the base
        Summit of lemma narrowed into a twisted neck 2-5mm long . . . . . *A. divaricata*
        Summit of lemma somewhat narrowed but not twisted . . . . . *A. hamulosa*
      Panicle narrow, the branches not horizontally spreading . . . . . *A. parishii*
    Glumes definitely unequal
      Panicle rather slender often nodding branches; purple. . . . . *A. purpurea* var. *p.*
      Panicle erect, stiff with mostly appressed reddish branches
        Panicle more than 1-2.5cm long, the branches several-flowered; awns about 10-30mm long . . . . . *A. purpurea* var. *glauca*
        Panicle mostly less than 5-11cm long, the branches few-flowered; awns 2-several centimeters long
          Culms 2-3dm tall; lemma gradually narrowed above, scaberulous on upper half; leaves mostly in the short curly cluster at base of plant . . . . . *A. purpurea* var. *longiseta*
          Culms 3-5dm tall; lemma scarcely narrowed above, scaberulous only at tip; leaves not conspicuously basal . . . . . *A. purpurea* var. *robusta*

A. adscensionis L. [*A. bromoides* H.B.K., *A. a.* var. *abortiva* Beetle, *A. a* var. *modesta* Hack. in Stuckert] Six-Weeks Three-Awn. Frequent, in open, dry areas; coastal and desert, below 400m: e.g. Santa Margarita Ranch; San Luis Rey; San Pasqual; northeast corner of county; Fish Creek; Vallecito Stage Station; Mountain Palm Springs; Box Canyon; Dulzura; Dehesa; Alvarado Canyon. Feb.-June. n=11.

A. californica Thurb. California Three-Awn. Scarce, in dry, sandy areas; desert, below 300m: Borrego Springs. Apr.-May.

A. divaricata Humb. & Bonpl. in Willd. Poverty Three-Awn. Rare, on dry slopes; cismontane, below 1000m: Pala. May-July. n=11.

A. hamulosa Henr. Mesa Three-Awn. Uncommon, on dry slopes: Lemon Grove; Rolando. May-Nov. n=22.

***A. oligantha** Michx. Prairie Three-Awn. Native of eastern United States. Reported by Crampton to occur in San Diego County, generally below 1000m. July-Sept.

A. parishii Hitchc. Parish's Three-Awn. Uncommon, in dry, rocky places; desert, below 1200m: San Felipe; Agua Caliente Hot Springs. Apr.-June.

A. purpurea Nutt. var. **glauca** (Nees)A. & N. Holmgren. [*A. reverchonii* Vasey; *A. wrightii* Nash] Blue

Three-Awn. Scarce, in rocky canyons; desert, below 800m: Borrego Palm Canyon; Box Canyon; San Felipe; Vallecito Stage Station. Also Jacumba. Mar.-May. $2n$=44. =A.p.g.

A. purpurea var. **longiseta** (Steudel)Vasey. [*A. fendleriana* Steud.] Fendler's Three-Awn. Rare, known in the county only at the head of Box Canyon, Vallecitos area, southern Borrego Desert. May.

A. purpurea var. **purpurea**. Purple Three-Awn. Occasional, on dry slopes; cismontane, below 1000m: Hormiguero Ranch; El Cajon Mountain; Harbison Canyon. May-July. n=11. =A.p.p.

A. purpurea var. **robusta** (Merr.)A. & N. Holmgren. [*A. longiseta* Steud. var. *robusta* Merr.] Robust Three-Awn. Rare, on dry, rocky slopes; desert, below 800m: Box Canyon. Mar.-May.

Arundo L.
[Arundineae]

***A. donax** L. Giant Reed. Native of Old World. Frequently persisting in moist areas; below 400m: e.g. Sorrento; Foster; Grapevine Canyon; Borrego Palm Canyon; National City; Chula Vista; San Ysidro. Mar.-Sep. n=55.

Avena L. Oats
[Aveneae]

Teeth of lemmas awned or setaceous; pedicels capillary . . . . *A. barbata*
Teeth of lemmas acute, not setaceous; pedicels stoutish
  Lemmas pubescent with long brown hairs; spikelets usually 3-flowered . . . . *A. fatua*
  Lemmas subglabrous; spikelets mostly 2-flowered . . . . *A. sativa*

***A. barbata** Brot. Slender Oat. Native of Old World. Common weed in disturbed and overgrazed areas: e.g. San Onofre; Palomar Mountain; Hot Springs Mountain; Cuyamaca Mountains; Live Oak Springs; Tecate. Mar.-June. n=14.

***A. fatua** L. Wild Oat. Native of Europe. Common weed in disturbed and overgrazed areas: e.g. San Onofre Bluffs State Beach; San Luis Rey; Solana Beach; Pacific Beach; Otay Mesa. Apr.-June. n=21.

***A. sativa** L. Cultivated Oat. Native of Europe. Occasionally escaping from cultivation: San Luis Rey; Anahuac School; Lakeside. Apr.-June. n=21.

Bothriochloa Ktze.
[Andropogoneae]

B. barbinodis (Lag.)Herter. [*Andropogon b.* Lag.; *A. saccharoides* auct. non Swartz.] Plumed Beardgrass. Locally common, on dry slopes, below 400m: O'Neill Lake; Agua Hedionda; Harrison Grade; Borrego Palm Canyon; San Diego; Dehesa. Feb.-Sep. n=90.

Bouteloua Lag.
[Chlorideae]

Plants perennial . . . . *B. curtipendula*
Plants annual
  Spikelets pectinately arranged at maturity, the spikes persistent, the florets falling from the perisistent glumes . . . . *B. barbata*
  Spikelets not pectinately arranged, the spikes falling entire at maturity . . . . *B. aristidoides*

B. aristidoides (H.B.K.)Griseb. Needle Grama. Uncommon, in dry, sandy areas; desert, below 800m: Sentenac Canyon; Vallecitos; Egg Mountain; Carrizo Badlands Overlook. Apr.-Sep.

B. barbata Lag. Six-weeks Grama. Infrequent, in rocky areas; desert, below 1000m: San Felipe; Blair Valley; Vallecito Stage Station; Canebrake Canyon; Mountain Palm Springs; Canyon Sin Nombre; south Dolomite Mine. July-Dec. $2n$=20, 40.

B. curtipendula (Michx.)Torr. in Marcy var. **caespitosa** Gould & Kapadia. Side-oats Grama. Rare, known in the county only at Jacumba, 900m.

Brachypodium Beauv.
[Poeae]

***B. distachyon** (L.)Beauv. [*Bromus distachyos* L.] Purple Falsebrome. Native of Eurasia. Uncommon, in disturbed areas; cismontane: Burnt Mountain, Escondido; Palomar Airport; Penasquitos Canyon; Mission Bay; Torrey Pines Mesa. Apr.-May.

Briza L.
[Poeae]

***B. minor** L. Little Quaking-Grass. Native of Europe. Infrequently naturalized in disturbed areas; below 1200m: Miramar Naval Air Station; Alvarado Canyon; Corte Madera Lake. Apr.-July. n=5, 7.

Bromus L.
[Poeae]

Plants annual
  Spikelets strongly flattened, the lemmas compressed-keeled and with terminal teeth not more than 0.5mm long
    Lemmas awnless or the awn less than 3mm long
      Lemmas usually with 13 vein . . . . *B. willdenovii*
      Lemmas usually with 9 veins . . . . *B. unioloides*

Lemmas with awns 7-15mm long
Spikelets 5-7 flowered; second glume almost equal to the lowest lemma . . . . *B. arizonicus*
Spikelets 6-10 flowered; second glume shorter than the lowest lemma . . . . *B. carinatus*
Spikelets terete or somewhat flattened, the lemmas not compressed-keeled, their terminal teeth mostly 0.6-5mm long
Awn usually geniculate, twisted at the base; lemma-teeth aristate . . . . *B. trinii*
Awn straight or spreading, sometimes minute, not twisted at base
Lemmas broad, rounded apically, not acuminate, the teeth mostly less than 1mm long; first lemma 3-5 nerved
Panicles compact, the spikelets mostly longer than the pedicels; lemmas membranaceous or scarious, prominently nerved unequal, the lower longer than the upper . . . . *B. mollis*
Panicles more open, the spikelets mostly not longer than the pedicels; lemmas somewhat coriaceous, not prominently veined . . . . *B. arenarius*
Lemmas narrow, elongate, tapering at the tip, the teeth 2-5mm long; first glume 1-nerved
Panicle erect, contracted; awn 1-2cm long
Culms pubescent below the dense panicle; sheaths pubescent . . . . *B. rubens*
Culms glabrous below the slightly open panicle; sheaths mostly smooth . . . . *B. madritensis*
Panicle open, with spreading or drooping branches
Awn 12-14mm long; lemmas about 1cm long; spikelets 1-2cm long . . . . *B. tectorum*
Awn 35-50mm long; lemmas 2.5-3cm long; spikelets 2.5-4cm long . . . . *B. diandrus*
Plants perennial
First glume 3-5-nerved
Spikelets strongly flattened, the lemmas compressed-keeled
Blades canescent, densely short-pilose, 2-5mm wide, often involute; blades 1-3mm wide . . . . *B. breviaristatus*
Blades not canescent, glabrous to puberulent or sparsely pilose; blades mostly 4-12 mm wide
Awns 7-15mm long; spikelets 2-3cm long . . . . *B. carinatus*
Awns 4-7mm long; spikelets 2.5-3.5cm long . . . . *B. marginatus*
Spikelets terete before anthesis or somewhat flattened but the lemmas not compressed-keeled
Second glume 5-nerved . . . . *B. pseudolaevipes*
Second glume 3-nerved
Panicle 1-1.5cm long; erect or pyramidal; awn 5-8mm long
Blades glabrous . . . . *B. orcuttianus* var. *o.*
Blades soft-pubescent . . . . *B. orcuttianus* var. *hallii*
Panicle 1.5-2dm long, with slender drooping branches; awn 3-6mm long . . . . *B. grandis*
First glume 1-nerved . . . . *B. inermis*

***B. arenarius** Labill. Australian Chess. Native of Australia. Occasional weed in disturbed areas: e.g. Lake Henshaw; Hot Springs Mountain; Foster; Banner; Cuyamaca Mountains; Laguna Mountains. Apr.-July.

B. arizonicus (Shear.)Stebb. Arizona Chess. Infrequent in dry, open places; cismontane, elevation 100m: Encinitas; Ramona; Alpine; La Jolla; Otay Mesa; Lakeside. Mar.-May. $2n=42$.

B. breviaristatus Buckl. [*B. carinatus* var. *linearis* Shear] Narrowleaf Brome. Occasional, in dry, open areas; montane, above 900m: Palomar, Hot Springs, and Laguna Mountains; Pine Hills; Pine Valley. Also La Mesa. May-July. $n=28$

B. carinatus H. & A. [*B. c.* var. *californicus* Shear; *B. c.* var. *hookerianus* (Thurb.)Shear] California Brome. Frequent, in dry, open places; cismontane: e.g. San Onofre; Fallbrook; Palomar Mountain; Encinitas; Witch Creek; Buckman Springs; Eckener Pass; Point Loma. Apr.-Aug. $n=28$.

***B. diandrus** Roth. [*B. rigidus* auct. non Roth; *B. diandrus* var. *gussonei* (Parl.)Coss & Durieu] Common Ripgut-Grass. Native of Europe. Common in disturbed areas; cismontane: e.g. San Onofre; Agua Tibia Mountians; Hot Springs Mountain; Cuyamaca Mountains; La Posta; Tecate; Silver Strand. Apr.-June. $n=21$, $2n=28$

B. grandis (Shear)Hitchc. in Jeps. [*B. orcuttianus* var. *g.* Shear] Infrequent, on open slopes; cismontane, below 800m: Richman Ranch, De Luz; Castro Canyon, Agua Tibia; El Monte. Apr.-July. Type from "La Maite, San Diego" = El Monte. $2n=14$.

***B. inermis** Leysser. Smooth Brome. Native of Eurasia. Occasionally introduced onto burns: North Peak, Cuyamaca Mountains. May-Aug. $2n=28$, 42, 49, 54-58, 70.

***B. madritensis** L. Compact Chess. Native of Europe. Infrequent in disturbed areas; cismontane below 1000 m. Ramona; Poway; Otay.

B. marginatus Nees. [*B. m.* var. *seminudus* Shear; *B. m.* var. *latier* Shear] Mountain Brome. Infrequent, on open slopes; cismontane, below 800m: San Diego. Apr.-July.

***B. mollis** L. [*B. hordeaceus* auct. non L., *B. molliformis* Lloyd] Soft Chess. Native of Europe. Prevalent weed in disturbed and undisturbed areas; cismontane: e.g. San Onofre; Oceanside; Palomar Mountain; Henshaw Dam; Hot Springs Mountain; Cuyamaca Mountains; Descanso Junction; Barona Valley; Lyons Valley; Miramar Naval Air Station; San Diego. Apr.-July. $2n=14$, 28.

B. orcuttianus Vasey var. **hallii** Hitchc. in Jeps. Hall's Brome. Scarce, in meadows and on protected slopes; montane, above 900m: Palomar Mountain; Banner Grade. June-Sep.

B. orcuttianus var. **orcuttianus**. Orcutt's Brome. Occasional, on open slopes; cismontane: Chimney Creek, Palomar Mountain; Pine Hills; Cuyamaca Mountains; Alvarado Canyon. June-Sep. Type from mountains near San Diego (Palomar Mountain).

B. pseudolaevipes Wagnon. Woodland Brome. Rare, in shaded places; coastal, below 700m: Santa Margarita River ford. Apr.-June. $2n=14$.

***B. rubens** L. Foxtail Chess. Native of Europe. Abundant weed of disturbed and undisturbed areas: e.g. San Onofre Bluffs; De Luz; San Luis Rey; Palomar Mountain; Lake Henshaw; Hot Springs Mountain; Laguna Mountains; Live Oak Springs; Tecate Peak; Miramar Naval Air Station; Point Loma. Also Borrego Palm Canyon; Tubb Canyon; Loki Canyon. Mar.-June. $2n=28$.

***B. tectorum** L. [*B. t.* var. *glabratus* Spenner; *B. t.* var. *nudus* Klett. & Richt.] Cheat-Grass Brome. Native of Europe. Occasional weed; montane, above 750m: Palomar Mountain; Lake Henshaw; Boulder Creek Road, 3 miles north of Descanso; Cuyamaca Rancho Headquarters; Monument Peak; Mason Valley; La Posta Creek; Crouch Meadow, Laguna Mountains. May-June. n=7.

B. trinii Desv. in Gay. Chile Brome. Infrequent, on open slopes; cismontane and desert, below 1200m: Sheep Canyon, Cottonwood Valley. Feb.-May. 2n=42.

***B. unioloides** H.B.K. [*B. haenkeanus* (Presl.)Kunth] Rescue Grass. Native of South America. Infrequent in cultivated and otherwise disturbed areas; cismontane below 500m: Otay Mesa.

***B. willdenovii** Kunth. [*B. catharticus* Vahl, nom. confus.] Prairie Brome. Native of South America. Local introduction at Chula Vista and San Diego. Apr.-Nov. 2n=28, 42, 56.

Calamagrostis Adans.
[Aveneae]

C. densa Vasey. [*C. vilfaeformis* Kearney; *C. koelerioides* Vasey, the type from Julian.] San Diego Reedgrass. Infrequent, in open areas and disturbed sites; foothill and montane, above 400m: Julian; Inspiration Point; Barona Valley along south fork of Featherstone Creek; Los Terranitos; Otay Mountain. June-July. Type from near Julian. n=14.

Cenchrus L.
[Paniceae]

Burs of one whorl of united flattened spines subtended by one to several whorls of smaller, finer bristles with a ring of slender bristles at base; spikelets usually 4 in a bur . . . . *C. echinatus*
Burs of several whorls of united flattened spines, the spines enlarged at irregular intervals through the bur with no ring of bristles at base; spikelets usually 2 in each bur . . . . *C. incertus*

***C. echinatus** L. Hedgehog Sandbur. Native of eastern North America and elsewhere. Local weed; cismontane, below 300m: San Diego; Balboa Park; Bostonia. Oct. n=17; 2n=68.

C. incertus M.A. Curtis. [*C. pauciflorus* Benth.] Mat Sandbur. Scarce, in open areas; cismontane, below 500m: San Luis Rey Mission; Rincon; Coronado; Nestor; Tijuana River Valley. July-Sep. n=18; 2n=34.

Chloris Sw.
[Chlorideae]

Plants annual . . . . *C. virgata*
Plants perennial
  Lemmas firm, dark brown, awnless or mucronate . . . . *C. distichophylla*
  Lemmas pale or fuscous, distinctly awned . . . . *C. gayana*

***C. distichophylla** Lag. Finger Grass. Native of South America. Local escape at Balboa Park and San Ysidro. June-Sep. n=10; 2n=40.

***C. gayana** Kunth. Rhodes Grass. Native of Africa. Local escape at La Mesa, San Diego, and Otay. Aug.-Sep. n=10; 2n=40.

***C. virgata** Sw. Showy Chloris, Feather Fingergrass. Native of eastern United States and Tropics. Infrequent, in disturbed, moist areas, below 1000m: San Diego; Vista. Apr.-Sep. n=10

Cortaderia Stapf
[Arundineae]

Inflorescence elevated above the foliage by a bare peduncle . . . . *C. dioica*
Inflorescence emerging through the foliage with peduncle not evident . . . . *C. jubata*

***C. dioica** (Spreng.)Speg. [*C. selloana* (Schult.)Asch.] Selloa Pampas-Grass. Native of Argentina. Persisting from cultivation in coastal region, below 300m: Lomas Santa Fe; Mission Valley. Sep.-Dec. 2n=72.

***C. jubata** Stapf. [*C. atacamensis* (Phil)Pilger] Atacama Pampas Grass. Native of Chile. Local escape in coastal region, below 300m: Balboa Park; Mission Valley; Pontos; Agua Hedionda Lagoon; Mission Bay. Aug.-Oct. n=38; 2n=108.

Crypsis Ait. Pricklegrass
[Eragrosteae]

Collar and margins of sheaths glabrous; glume shorter than lemma; anther 0.9mm long . . . . *C. schoenoides*
Collar and margins of sheaths pilose; glume equalling or exceeding lemma; anthers 0.6-0.7mm long . . . *C. vaginiflora*

***C. schoenoides** (L.)Lam. [*Heleochloa s.* (L.)Host.] Swamp Timothy. Native of Europe. Naturalized at Pilgrim Creek, Oceanside. 2n=36.

***C. vaginiflora** (Forsk.)Opiz. [*C. niliaca* Fig. & DeNot. *C. aculeata* Calif. auct. non (L.)Ait.] Sharp-Leaf Crypsis. Native of Mediterranean region. Locally naturalized at Cuyamaca Lake. June-Sep.

Cynodon Rich. in Pers.
[Chlorideae]

***C. dactylon** (L.)Pers. [*Panicum d.* L.] Common Bermuda-Grass. Native of Old World. Abundant weed in disturbed areas; below 1500m: e.g. De Luz; San Marcos; Warner Hot Springs; Santa Ysabel; Bow Willow Canyon; Jacumba; Dulzura; National City; Point Loma. June-Aug. 2n=18, 27, 30, 36, 40, 54.

Cynosurus L.
[Poeae]

***C. echinatus** L. Hedgehog Dogtail. Native of Europe. Infrequent weed in cismontane: Pine Hills. May-July. *n*=7.

Dactylis L.
[Poeae]

***D. glomerata** L. Orchardgrass. Native of Europe. Uncommon weed in disturbed areas; below 1000m: Balboa Park; Presidio Park; San Diego; Aguanga Ridge; Harbison Canyon; San Marcos. May-Aug. 2*n*=14, 21, 27-31, 4-2.

Dactyloctenium Willd.
[Eragrosteae]

***D. aegyptium** (L.)Beauv. Durban Crowfoot-Grass. Native of Paleotropics. Local weed at Pala; Bonsall and Miramar Naval Air Station. June-Sep. *n*=18.

Danthonia Lam. & DC. Oatgrass
[Danthonieae]

Panicle of 2-5 spikelets . . . . *D. californica*
Panicle usually of a single spikelet. . . . . *D. unispicata*

D. californica Bol. var. **americana** (Scribn.)Hitchc. Western Oatgrass. Uncommon, on dry, rocky slopes; montane, above 1400m: Cuyamaca Mountains. May-July.

D. unispicata (Thurb.)Munro ex Macoun. One-Spike Oatgrass. Uncommon, on dry, rocky slopes; montane, above 1400m: Cuyamaca Lake. May-July. Type from San Diego to San Francisco. *n*=18.

Deschampsia Beauv.
[Aveneae]

Plants annual
  Glumes 6-8mm long . . . . *D. danthonioides* var. *d.*
  Glumes 4mm long . . . . *D. danthonioides* var. *gracilis*
Plants perennial . . . . *D. elongata*

D. danthonioides (Trin.)Munro in Benth. var. **danthonioides**. Annual Hairgrass. Occasional, in moist areas; cismontane: Palomar Mountain; Ramona; Poway Junction; Cuyamaca Lake; Cuyamaca and Laguna Mountains; Chollas Heights; San Diego; Kearny Mesa. Mar.-Aug. *n*=26.

D. danthoniodes var. **gracilis** (Vasey)Munz. Graceful Hairgrass. Uncommon, in mud flats and vernal pools; coastal, below 300m: Ramona; San Marcos; Kearny Mesa. Mar.-June. Type from mesas about San Diego.

D. elongata (Hook.)Munro in Benth. Slender Hairgrass. Uncommon, in moist places; montane, above 1300m: Doane Valley; Palomar Mountain; Cold Spring, Cuyamaca Mountains. May-Aug. 2*n*=26.

Dichanthelium (Hitchc. & Chase)Gould Dichotomous Panicums
[Paniceae]

D. lanuginosum (Elliott)Gould var. **fasciculatum** (Torr.)Spellenberg. [*Panicum occidentale* Scribn.; *P. pacificum* Hitchc. & Chase] Western Panicgrass. Occasional, in moist sandy places; cismontane: Cedar Creek Falls; El Cajon Mountain; Barona Valley; San Diego. Apr. July. *n*=9.

Digitaria Heister Crabgrass
[Paniceae]

Sheaths pilose or villous; spikelets 2.5-3.5mm long; fertile lemma pale . . . . *D. ciliaris*
Sheaths glabrous; spikelets 2mm long; fertile lemma brown . . . . *D. ischaemum*

***D. ciliaris** (Retz)Koel. [*D. adscendens* (H.B.K.)Henr. *D. sanguinalis* auct. non Scop.] Large Crabgrass. Native of Old World. Occasional weed of disturbed, moist areas: e.g. San Luis Rey; Montezuma Valley; Mussey Grade; Morena; San Diego; Padre Dam; Sweetwater. June-Sep. 2*n*=36, 54, 60, 70, 72.

***D. ischaemum** (Schreb.)Schreb. ex Muhl. Smooth Crabgrass. Native of Eurasia. Infrequent weed in disturbed areas: Foster. Sep.-Nov. 2*n*=36, 45.

Diplachne Beauv.
[Eragrosteae]

D. uninervia (Presl)Parodi. [*Leptochloa uninervia* (Presl)Hitchc. & Chase. *L. imbricata* Thurb. in Wats.] Dense-Flower Sprangle-Top. Infrequent, in disturbed or alkaline spots; cismontane, below 700m: Encinitas; Escondido; San Diego; Balboa Park; Otay Mesa; Jacumba. Mar.-Dec. *n*=10.

Distichlis Raf.
[Aeluropodeae]

Panicle congested . . . . *D. spicata* ssp. *spicata*
Panicle scarcely congested, with pedicels easily visible . . . . *D. spicata* ssp. *stricta*

D. spicata (L.)Greene ssp. **spicata**. [*D. spicata* var. *stolonifera* Beetle; *D. spicata* var. *nana* Beetle] Coastal Salt-Grass. Common, in Coastal Salt Marsh and occasionally further inland; coastal and foothill, below 1000m: e.g. San Onofre Bluffs; Carlsbad; Oak Grove; Henshaw Dam; Imperial Beach. Apr.-Sep. *n*=20.

D. spicata ssp. **stricta** (Torr.)Thorne. [*D. spicata* var. *stricta* (Torr.)Beetle.] Desert Salt-Grass. Infrequent, in alkaline interior areas; foothill and desert, below 1000m: Dubber; Carrizo Stage Station; Vallecito Stage Station. May-Aug. *n*=20.

Echinochloa Beauv.
[Paniceae]

Spikelets with at least some awns 16mm or more long . . . . . *E. crus-galli*
Spikelets awnless or with awns 15mm or less long
  Primary inflorescence branches simple, usually 2cm or less long, with small (2.5-3mm long), awnless spikelets in 4 regular rows; palea of lower floret well developed; hairs of the inflorescence branches and spikelets not papilla-based . . . . . *E. colonum*
  Primary inflorescence branches at least the lower, usually rebranched, the lower branches commonly more than 2cm long; hairs of the inflorescence and spikelets papilla-based . . . . . *E. crus-pavonis*

***E. colonum** (L.)Link. Jungle-Rice. Native of Old World. Infrequent weed in disturbed places; cismontane, below 1000m: Morena; San Diego. July-Sep. *n*=36.

***E. crus-galli** (L.)Beauv. Common Barnyard-Grass. Native of Tropics. Infrequent weed in moist, disturbed areas; cismontane, below 1000m: De Luz; San Pasqual; Balboa Park; La Punta. July-Oct. *n*=18.

E. crus-pavonis (H.B.K.)Schult. [*E.c.* var. *macera* (Wiegand)Gould]. Mexican Barnyard-Grass. Infrequent, in moist places; cismontane, below 1100m: Lake Hodges; East San Diego; Clover Flats. July-Oct.

Eleusine Gaertn.
[Eragrosteae]

***E. indica** (L.)Gaertn. Goose Grass. Native of Europe. Uncommon weed as at Old Town State Park. *n*=9, 18.

Elymus L.
[Triticeae]

Spikelets disarticulating at the nodes, glumes with awns 2-5cm long
  Glumes enitre or 2-cleft . . . . . *E. longifolius*
  Glumes cleft into at least 3 divisions . . . . . *E. multisetus*
Spikelets not so, glumes awnless or with awn to 1cm long
  Lemmas awned, the awns mostly 10-30mm long
    Awns of lemmas curving outward at maturity . . . . . *E. glaucus* ssp. *jepsonii*
    Awns of lemmas more or less straight, not curving outward . . . . . *E. glaucus* ssp. *g.*
  Lemmas awnless or the awns not more than 6mm long
    Glumes broadly lanceolate, strongly 3-9 nerved . . . . . *E. glaucus* ssp. *virescens*
    Glumes subulate, or if lanceolate, inconspicuously nerved
      Spikelets 3 to several at a node; culms mostly 6-10mm in diameter; blades 15-35mm wide . . . . . *E. condensatus*
      Spikelets mostly 1-2 at a node; culms mostly less than 6mm in diameter; blades 3-15mm . . . . . *E. triticoides*

E. condensatus Presl. Giant Wild Rye. Occasional, on open slopes; cismontane, below 1200m: e.g. San Onofre Bluffs; Encinitas; Vista; Chihuahua Valley; Palomar Mountain; San Felipe. June-Aug. *n*=14, 28.

E. glaucus Buckl. ssp. **glaucus**. Blue Wild Rye. Frequent, on open slopes; cismontane: e.g. San Onofre Canyon; Palomar Mountain; Hot Springs Mountain; Spencer Valley; Cuyamaca Mountains; San Diego. June-Aug. *n*=14.

E. glaucus ssp. **jepsonii** (Davy)Gould. Jepson's Wild Rye. Infrequent, in Montane Coniferous Forest; montane, above 1000m: Pine Hills; Cuyamaca Lake; Oriflamme Canyon; Rattlesnake Valley. June-Aug.

E. glaucus ssp. **virescens** (Piper)Gould. Pacific Wild Rye. Local, on slopes; foothill, 150-500m: Harbison Canyon; Campo. May-Aug. *n*=14.

E. longifolius (J.G. Sm.)Gould. [*Sitanion l.* J.G. Sm.] Longbristle Squirreltail. Uncommon, on open slopes; foothill and montane, above 500m: Julian; Cuyamaca Lake; Live Oak Springs; Corte Madera Ranch; Campo. May-Aug.

E. multisetus (J.G.Sm.)Davy. [*Sitanion jubatum* J.G.Sm.] Big Squirreltail. Infrequent, on brushy slopes; montane, above 800m: Cuyamaca Lake; Guatay. May-July. *n*=14.

E. triticoides Buckl. [*Leymus t.* (Buckl.)Pilger; *E. orcuttianus* Vasey] Beardless Wild Ryegrass. Occasional, in wet places; cismontane: e.g. Palomar Mountain; Cuyamaca Mountains; Pine Valley; Jamacha; Mission Dam; Tijuana River Valley. June-July. *n*=14, 21.

× Elysitanion Bowden
[Triticeae]

× Elysitanion hansenii. [*Sitanion* × *hansenii* (Scribn.)J.G.Sm. (*Elymus glaucus* × *Sitanion* ssp.) Hansen's Squirreltail. Uncommon, on dry slopes; foothill, above 400m: Ballena; Crouch Meadow. June-Aug. *n*=14.

Eragrostis Beauv. Lovegrass
[Eragrosteae]

Plants perennial; spikelets gray-green, 7-11 flowered . . . . . *E. curvula*
Plants annual
  Plants glandular or warty along the leaf-margins, keel of lemmas, or on the panicle branches . . . . . *E. cilianensis*

Plants not glandular on lemmas, panicle-branches, or leaf-margins, sometimes so on sheaths and near nodes
Spikelets about 1mm wide, slender, linear
Spikelets 3-5mm long; lemmas 1-1.5mm long; pedicels mostly longer than the spikelets; surface of grain smooth . . . . . *E. pilosa*
Spikelets 5-7mm long; lemmas about 2mm long; pedicels mostly shorter than the spikelets; surface of grain reticulate . . . . . *E. mexicana* ssp. *virescens*
Spikelets about 1.5mm wide or wider, linear to ovoid
Some or all of the spikelets on pedicels 1-5mm long; spikelets linear at maturity, appressed along primary panicle-branches
Primary branches of panicle simple or the lower with a branchlet bearing 2-3 spikelets; culms slender, mostly less than 3dm high . . . . . *E. pectinacea*
Primary branches of panicle usually bearing branchlets with few to several spikelets; culms rather robust, over 3dm tall . . . . . *E. diffusa*
Some or all of the spikelets on pedicels 5-8mm long; spikelets ovate to linear, not appressed along the primary panicle branches . . . . . *E. mexicana* ssp. *m.*

E. cilianensis (All.)Mosher. [*E. megastachya* (Koel.)Link] Common Stink-Grass. Native of Europe. Uncommon weed in disturbed places: San Luis Rey; Montezuma Valley; San Felipe; San Diego; Bonita. June-Oct. n=10, 20.

E. curvula (Schrad.)Nees. Weeping Lovegrass. Native of Africa. Rare weed in disturbed places: Escondido; Doane Valley.

E. diffusa Buckl. Spreading Lovegrass. Infrequent, in disturbed places; cismontane, below 300m: San Luis Rey; San Diego. June-Sep. $2n$=60.

E. mexicana (Hornem.)Link ssp. **mexicana**. [*E. neomexicana* Vasey] Mexican Lovegrass. Scarce, in open disturbed areas; cismontane: San Luis Rey; Montezuma Valley; Castle Park. July-Oct. n=28, 30.

E. mexicana ssp. **virescens** (Presl)Koch & Sanchez. [*E. orcuttiana* Vasey.] Orcutt's Lovegrass. Rare, in disturbed, moist places; cismontane, below 300m: Florida Canyon; San Diego; Dehesa. May-Oct. Type from San Diego.

E. pectinacea (Michx.)Nees. Uncommon, in disturbed area; cismontane: San Diego River flood channel. $2n$=20,30.

E. pilosa (L.)Beauv. India Lovegrass. Native of Europe. Locally naturalized in Balboa Park. July-Sep.

Erioneuron Nash
[Eragrosteae]

E. pulchellum (H.B.K.)Tateoka. [*Tridens p.* (H.B.K.)Hitchc.; *Triodia p.* H.B.K.] Low Fluffgrass. Occasional, on rocky slopes; desert, below 900m: northeast corner of county; Yaqui Well; Desert Ironwoods; Fish Creek; Vallecito Stage Station; Mountain Palm Springs. Feb.-May. $2n$=16.

Festuca L.
[Poeae]

Awns scarcely if at all evident, less than 1mm long
Panicle 1.5-3.5dm long; lemmas 7-10mm long . . . . . *F. arundinacea*
Panicle 1-2dm long; lemmas 5-7mm long . . . . . *F. pratensis*
Awns well developed, 2-many millimeters long . . . . . *F. rubra*

F. arundinacea Schreb. Reed Fescue. Native of Europe. Uncommon weed as at San Vicente Ranch, Los Terranitos, Descanso. May-June. $2n$=28, 42, 56, 63, 70.

F. pratensis Huds. [*F. elatior* auct. non L.] Meadow Fescue. Native of Europe. Locally escaping from cultivation: Rancho Santa Fe. May-July. $2n$=14, 28, 42, 70.

F. rubra L. Red Fescue. Native of more northerly regions. Uncommon, adventive in disturbed places; below 300m: Escondido; Balboa Park. May-July. n=7; $2n$=14, 21, 28, 42, 46, 49, 53, 56, 64, 70.

Gastridium Beauv.
[Aveneae]

G. ventricosum (Gouan)Schinz & Thell. Nitgrass. Native of Europe. Occasional weed of open or disturbed areas; coastal, below 400m: e.g. O'Neill Lake; Rainbow Valley; Carlsbad; Escondido; San Pasqual; El Cajon Mountain; Dulzura; Potrero; San Diego. May-Sep. n=7.

Heteropogon Pers.
[Andropogoneae]

H. contortus (L.)Beauv. ex Roem. & Schult. Common Tanglehead. Native of tropics. Apparent introduction at East San Diego and Dehesa. Mar.-Apr.

Hilaria H.B.K.
[Chlorideae]

H. rigida (Thurb.)Benth. ex Scribn. Big Galletagrass. Common, in sandy areas; desert, below 1100m: e.g. northeast corner of county; Fish Creek; Borrego Valley; Borrego Palm Canyon; San Felipe; The Narrows; La Puerta; Carrizo Stage Station; Dubber; Mountain Springs. Feb.-June. $2n$=18.

Holcus L. Velvetgrass
[Aveneae]

***H. lanatus** L. Common Velvetgrass. Native of Europe. Occasionally adventive in moist places: Palomar Mountain; Julian; Pine Hills; Cuyamaca Lake; San Diego. June-Aug. *n*=7.

Hordeum L. Barley
[Triticeae]

Plants perennial
  Glumes and awns 1.8-2.8cm long . . . . . *H. jubatum*
  Glumes and awns less than 1.8cm long
    Blades pubescent, 1.5-5mm wide; anthers usually 1.5-3mm long; glumes of central spikelet 1.5-2.5 times as long as palea. . . . . *H. californicum*
    Blades glabrous, 3-9mm wide; anthers mostly 1-1.5mm long; glumes of central spikelet often scarcely longer than palea . . . . . *H. brachyantherum*
Plants annual
  Glumes of cental spikelet and the inner ones of the lateral spikelets with ciliate margins
    Spike with 6-8 spikelets to each centimeter of the rachis; stamens of central florets included at anthesis . . . . . *H. murinum* ssp. *glaucum*
    Spike with 3-5 spikelets to each centimeter of rachis; stamens of central florets exserted at anthesis . . . . . *H. murinum* ssp. *leporinum*
  Glumes not ciliate
    Auricles long; rachis continuous; all 3 spikelets sessile, fertile . . . . . *H. vulgare*
    Auricles obsolete; rachis articulate; lateral spikelets pedicellate, usually not fertile
      Inner glumes of lateral spikelets strongly broadened, 0.6-1.8mm wide . . . . . *H. intercedens*
      Inner glumes and outer linear-subulate, less than 0.6mm wide
        Spike ovate to ovate-oblong, usually less than 5cm long; awns and glumes strongly spreading at maturity; bases of glumes of lateral spikelets prominent above the pedicel . . . . . *H. geniculatum*
        Spike linear-oblong, usually over 5cm long; awns and glumes suberect; bases of glumes of lateral spikelets not prominent above the pedicel . . . . . *H. depressum*

H. brachyantherum Nevski. Meadow Barley. Uncommon, in moist places; cismontane: Cuyamaca Lake. June. *n*=14.

H. californicum Covas & Steb. [*H. nodosum* auct. non L.] California Barley. Infrequent, on open slopes; cismontane: Palomar Mountain; North Peak, Cuyamaca Mountains; Kearny Mesa. Apr.-Aug. *n*=7.

H. depressum (Scribn. & Sm.)Rydb. Alkali Barley. Scarce, in moist, alkaline places: Julian; Cuyamaca Lake; San Diego; Jamul; Otay. Apr.-May. *n*=14.

***H. geniculatum** Allioni. [*H. hystrix* Roth; *H. gussoneanum* Parl.] Mediterranean Barley. Native of Europe. Occasional weed in disturbed places; cismontane: e.g. De Luz; San Marcos; Palomar Mountain; Lake Henshaw; Pine Hills; Cuyamaca Mountains; Cuyamaca Lake; Rattlesnake Valley. Apr.-June. *n*=7.

H. jubatum L. Foxtail Barley. Frequent in disturbed areas, coastal, below 1000m: e.g. Fallbrook; Encinitas; Escondido; San Diego; National City. Apr-June.

H. intercedens Nevski. [*H. pusillum* Nutt.] Little Barley. Uncommon, in alkaline spots; cismontane, below 1000m: Witch Creek; San Diego. Apr.-May. *n*=7; 2*n*=14, 28.

***H. murinum** L. ssp. **glaucum** (Steud.)Tzvel. [*H. stebbinsii* Covas; *H. g.* Steud.] Glaucous Barley. Native of Europe. Occasional weed in disturbed areas: San Onofre Bluffs; De Luz; San Luis Rey; Henshaw Dam; Linda Vista; East San Diego; Pine Valley; Jamul; Imperial Beach; Otay Mesa. Apr.-July. *n*=7.

***H. murinum** ssp. **leporinum** (Link)Arcang. [*H. murinum* auct. non L.; *H. l.* Link] Hare Barley. Native of Europe. Common weed in disturbed places; cismontane: e.g. San Onofre; Banner; Cuyamaca Mountains; East San Diego; Roseville; Point Loma. Apr.-July. *n*=7, 14.

***H. vulgare** L. Cultivated Barley. Native of Old World. Infrequent escape from cultivation: San Onofre; Lake Henshaw; Ramona; San Diego; Kearny Mesa; Silver Strand; Jacumba. Apr.-July. *n*=7; 2*n*=14, 28.

Koeleria Pers. Junegrass
[Aveneae]

K. pyramidata (Lam.)Beauv. [*K. nitida* Nutt. *K. cristata* auct. non (L.)Pers.; *K. macrantha* (Ledeb)Spreg.; *K. n.* var. *californica* subvar. *transiens* Domin., the type from San Diego; *K. n.* var. *c.* subvar. *vestita* Domin, the type from San Diego; *K. pseudocristata* var. *californica* Domin, the type from San Diego.] Junegrass. Occasional, on dry slopes; cismontane: Santa Margarita Ranch; Escondido; Palomar Mountain; Pine Hills; Cuyamaca Lake; Laguna Mountains; Pine Valley; Encinitas; Mount Soledad; Harbison Canyon; El Cajon Mountain; San Diego. May-July. *n*=4; 2*n*=14, 28.

Lamarckia Moench
[Poeae]

***L. aurea** (L.)Moench. Golden-Top. Native of Mediterranean region. Common weed in moist, disturbed places; cismontane: e.g. San Onofre; Cuca Indian Reservation; Ramona; San Diego; Glen Lonely. Feb.-May. *n*=7.

Lolium L. Ryegrass
[Poeae]

Glumes ordinarily shorter than the spikelet
  Lemmas, or at least some of them, awned . . . . . *L. multiflorum*

Lemmas all awnless . . . . *L. perenne*
Glumes as long or longer than the spikelet. . . . . *L. temulentum*

***L. multiflorum** Lam. [*L. perenne* ssp. *m.* (Lam.)Husnot.] Italian Ryegrass. Native of Europe. Common weed; cismontane, below 700m: e.g. San Onofre Bluffs; Vista; Carlsbad; Encinitas; Linda Vista; Mission Valley; San Diego. June-Aug. n=7.

***L. perenne** L. English Ryegrass. Native of Europe. Occasionally escaping from cultivation or naturalized: Mount Woodson; Cuyamaca Mountains; Mission Hills; south San Diego. May-Sep. n=7; 2n=14,28

***L. temulentum** L. Darnel. Native of Europe. Frequent weed; cismontane; e.g. De Luz; Encinitas; Escondido; El Cajon Mountain; Murray Dam; Potrero. Apr.-June. n=7.

Melica L. Melic
[Meliceae]

Spikelets 8-15mm long; fertile florets 2 to several . . . . *M. frutescens*
Spikelets 4-6mm long; fertile florets 1 or rarely 2. . . . . *M. imperfecta*

M. frutescens Scribn. Tall Melic. Occasional, on dry slopes; coastal, foothill and desert, below 1000m: e.g. De Luz; San Felipe; Borrego Palm Canyon; Montezuma Grade; Sentenac Canyon; Box Canyon; Vallecito Stage Station; Barrett Dam; Bernardo; San Diego. Mar.-May. n=9.

M. imperfecta Trin. [*M. i.* var. *flexulosa* Bol; *M. i.* var. *minor* Scribn.; *M. i.* var. *refracta* Thurb. in Wats.; *M. poaeoides* Nutt., the type from San Diego?] Coast Range Melic. Common, on dry slopes; cismontane: e.g. Santa Margarita Ranch; Lilac; Bonsall; Oak Grove; Lake Henshaw; Hot Springs Mountain; Encinitas; San Diego; Otay; Tecate; Canyon City; Hipass; Jacumba. Apr.-May. n=9.

Monanthochloe Engelm.
[Aeluropodeae]

M. littoralis Engelm. Salt-cedar. Prevalent, in Coastal Salt Marsh; coastal, below 20m: e.g. Oceanside; Cardiff; Del Mar; Chula Vista; Imperial Beach; Tijuana River Estuary. May-June.

Monerma Beauv.
[Monermeae]

***M. cylindrica** (Willd.)Coss. & Dur. [*Lepturus c.* (Willd.)Trin.] Thintail. Native of Old World. Frequent, in Coastal Salt Marsh and other alkaline spots; coastal, below 300m: Oceanside; La Jolla; Spring Valley; San Diego; Alpine; Jamul Valley; Otay Mesa. May-July. n=13.

Muhlenbergia Schreb. ex Gmel. Muhly
[Eragrosteae]

Plants annual; lemma awn 1cm or more long. . . . . *M. microsperma*
Plants perennial; lemma awnless or very short-awned
  Rhizomes present, usually prominent, scaly, creeping, often branching
    Panicle open, the pedicels capillary, 2-15mm long. . . . . *M. asperifolia*
    Panicle narrow, condensed, the pedicels stouter, appressed, shorter . . . . *M. richardsonis*
  Rhizomes not developed, the culms tuffed
    Panicle 5-10cm long, open with spreading branches . . . . *M. porteri*
    Panicle 25-50cm long, slender, mostly spicate . . . . *M. rigens*

M. asperifolia (Nees & Mey.)Parodi. [*Sporobolus a.* Nees & Mey.] Common Scratchgrass. Local, in moist spots; cismontane: Encanto; Campo; Pine Valley; Laguna Public Camp; White Star; Mountain Springs. July-Oct. 2n=20,22,28.

M. microsperma (DC.)Kunth. Littleseed Muhly. Common, on open, often rocky slopes; cismontane, below 1200m: e.g. Moosa Canyon; Del Mar; San Diego; El Cajon Mountain; Steele Canyon; Cottonwood; Laguna Mountains. Mar.-May.

M. porteri Scribn. ex Beal. Bush Muhly. Uncommon, on rocky slopes; desert, below 1000m: San Felipe; Sentenac Canyon. June-Oct. 2n=20,40.

M. richardsonis (Trin.)Rydb. Mat Muhly. Known in the county only in a spring east of Wooded Hill, Laguna Mountain Recreation Area, 1600m. Aug. 2n=40.

M. rigens (Benth.)Hitchc. [*Epicampes r.* Benth.] California Deergrass. Frequent, in damp places; cismontane and desert: e.g. De Luz; Gopher Canyon; Deer Flat, Palomar Mountain; Borrego Palm Canyon; San Felipe; Cuyamaca Mountains; Pine Valley; Dulzura; San Diego. June-Sep. n=20.

Orcuttia Vasey
[Orcuttieae]

O. californica Vasey. Southern Orcuttgrass Rare, from vernal pools and slump ponds of the coastal mesas, below 200m: Miramar Naval Air Station, at runway; south of San Clemente Canyon; Otay Mesa.

Oryzopsis Michx. Ricegrass
[Stipeae]

Lemmas smooth, rarely pubescent; glumes 3mm long. . . . . *O. miliacea*
Lemmas pubescent; glumes 6-7mm long . . . . *O. hymenoides*

O. hymenoides (R. & S.)Ricker. Indian Ricegrass. Scarce, in sandy areas; desert, below 1000m: Las Arenas; Jacumba. Apr.-July. 2n=28,48,65,130.

***O. miliacea** (L.)Benth. & Hook. ex Ascher & Schweinf. Millett Ricegrass. Native of Mediterranean regions. Infrequent, in disturbed places; cismontane: San Diego; Spring Valley; Bostonia; Harbison Canyon; Cuyamaca Mountains; Jacumba. Apr.-Sep. *n*=12.

Panicum L.
[Paniceae]

P. capillare L. var. **occidentale** Rydb. Western Witchgrass. Occasional, in moist places; cismontane, below 1600m: e.g. Lake Hodges; Doane Valley; Cuyamaca Lake; Pine Valley; Japatul Valley; Alvarado Canyon; Santee; Jamacha. July-Sep. 2*n*=18.

Parapholis C.E. Hubb
[Monermeae]

***P. incurva** (L.)C.E.Hubb. European Sicklegrass. Native of Europe. Infrequent, in Coastal Salt Marsh; coastal, below 20m: Del Mar; San Diego Bay; Nestor; Imperial Beach. Apr.-June. *n*=18,21.

Paspalum L.
[Paniceae]

Racemes a pair, at the summit of the culms . . . . . *P. paspalodes*
Racemes several to many, forming a panicle . . . . . *P. dilatatum*

***P. dilatatum** Poir. Dallisgrass. Native of South America. Occasional weed in disturbed places; cismontane, below 300m: e.g. Coronado; San Diego; Balboa Park; Lemon Grove. May-Nov. *n*=20; 2*n*=50.

***P. paspalodes** (Michx.)Scribn. [*P. distichum* auct. non L.] Common Knotgrass. Occasional, in moist spots; cismontane; Santa Margarita River ford; Palomar Mountain; Warner Hot Springs; Pine Hills; Alvarado Canyon. June-Oct. *n*=20; 2*n*=60.

Pennisetum Rich.
[Paniceae]

Culms extensively creeping; spikelets few, hidden in the upper sheath . . . . . *P. clandestinum*
Culms not creeping; panicle exserted
  Plants with woody culms . . . . . *P. nervosum*
  Plants herbaceous
    Panicle oval, tawny, 3-10cm long . . . . . *P. villosum*
    Panicle elongate, purple or rosey, 15-35cm long . . . . . *P. setaceum*

***P. clandestinum** Hochst. ex Chiov. Kikuyugrass. Native of Africa. Locally adventive weed of lawns and orchards: Mission Bay Park; Balboa Park. 2*n*=36.

***P. nervosum** (Nees)Trin. Native of South America. Adventive along drainage courses; cismontane, below 100m: Carlsbad; Bonita. Sep.-Jan.

***P. setaceum** (Forsk.)Chiov. [*P. ruppelii* Steud.] African Fountain Grass. Native of Africa and western Asia. Occasional escape along roads; coastal, below 600m: e.g. San Diego; Grossmont; Murphy Canyon; Bernardo; Wildcat Canyon. July-Oct. 2*n*=27.

***P. villosum** R. Br. African Feathertop. Native of northeast Africa and Arabia. Occasional escape from cultivation: San Diego; National City. June-Aug. 2*n*=45.

Phalaris L.
[Aveneae]

Spikelets in groups of 6-7, 1 fertile surrounded by 6 sterile, the group falling entire . . . . . *P. paradoxa*
Spikelets all alike, not falling entire in groups
  Plants perennial . . . . . *P. aquatica*
  Plants annual
    Glumes broadly winged; panicle ovate to short-oblong
      Sterile lemma solitary; fertile lemma 3mm long . . . . . *P. minor*
      Sterile lemmas two; fertile lemmas 5-6mm long . . . . . *P. canariensis*
    Glumes wingless or nearly so; panicles linear or oblong
      Glumes wingless, acuminate; fertile lemma turgid, the apex smooth, acuminate . . . . . *P. lemmonii*
      Glumes narrowly winged toward the summit, abruptly pointed or acute; fertile lemmas less turgid, villous to the acute apex
        Panicle subcylindric, mostly 6-15cm long . . . . . *P. angusta*
        Panicle tapering to each end, mostly 2-6cm long . . . . . *P. caroliniana*

P. angusta Nees ex Trin. [*P. intermedia* var. *angustata* Beal, the type from San Diego] Timothy Canary Grass. Uncommon, in moist places; cismontane, below 400m: San Diego; Fallbrook; Encinitas; Harbison Canyon. May-June. *n*=7; 2*n*=14.

***P. aquatica** L. [*P. tuberosa* var. *stenoptera* Hitchc.] Harding Grass. Native of Europe. Occasional, introduced; in moist places: Cuyamaca and Laguna Mountains. May-Sept. *n*=14

***P. canariensis** L. Mediterranean Canary Grass. Native of Mediterranean regions. Local, escaping from cultivation in moist areas: East San Diego. Apr.-July. 2*n*=12,28.

***P. caroliniana** Walt. Carolina Canary Grass. Uncommon, in open areas; cismontane: Witch Creek. Apr.-May. 2*n*=14.

P. lemmonii Vasey. Lemmon's Canary Grass. Occasional, in mud flats and vernal pools; cismontane, below 950m: San Marcos; Ramona; Witch Creek; Murray Dam; Linda Vista; Torrey Pines Mesa; San Diego. Apr.-June. *n*=7.

***P. minor** Retz. Littleseed Canary Grass. Native of Mediterranean region. Frequent, in disturbed areas: Rainbow Valley; Pine Valley; The Narrows; Dulzura; San Diego. Apr.-July. *n*=14.

***P. paradoxa** L. var. **praemorsa** (Lam.)Coss. & Dur. Paradox Canary Grass. Native of Mediterranean region. Local introduction at San Marcos; San Diego and Spring Valley. May-Aug. *n*=7.

Phleum L.
[Aveneae]

***P pratense** L. Hay Timothy. Native of Eurasia. Occasional escape from cultivation: e.g. Palomar Mountain; Dehr Creek; Pine Hills; below Cuyamaca Dam; La Mesa; San Diego. May-June. 2*n*=14,21,28-84.

Phragmites Trinus.
[Arundineae]

P. australis (Cav.)Trin. ex Steud. [*P. communis* var. *berlandieri* (Fourn.)Fern.] Common Reed, Carrizo. Frequent, in moist areas; desert, below 900m: e.g. Carrizo Gorge; Agua Caliente Hot Springs; Canebrake Canyon. Also Lake Henshaw; San Dieguito River estuary. July-Nov. 2*n*=36,48,54,72,84,ca. 96.

Poa L.
[Poeae]

Plants annual
  Lemmas pubescent, but not webbed at base . . . . . *P. annua*
  Lemmas glabrous or with a tuft of cobwebby hairs at base
    Panicle at length open, the elongated branches not floriferous to base . . . . . *P. bolanderi*
    Panicle narrow, the short appressed branches floriferous to base . . . . . *P. bigelovii*
Plants perennial
  Creeping rhizomes present
    Panicle open, pyramidal, its elongated lower branches lax and floriferous only in the outer half
      Lemmas glabrous or pubescent on nerves but not webbed at base; florets all pistillate . . . . . *P. nervosa*
      Lemmas with a tuft of cobwebby hairs at base; florets perfect . . . . . *P. pratensis*
    Panicle more contracted, oblong, its short lower branches not lax; lemmas not webbed at base . . *P. atropurpurea*
  Creeping rhizomes lacking . . . . . *P. secunda*

***P. annua** L. Annual Bluegrass. Native of Europe. Common weed in damp places; e.g. Palomar Mountain; Cardiff; Cuyamaca Moutains; San Felipe; Clover Flats; San Diego. Jan.-July. 2*n*=14,24-26,28.

P. atropurpurea Scribn. San Bernardino Bluegrass. Rare, in montane meadows, above 1500m: Crouch Meadow and Laguna Meadow, Laguna Mountains. May-June.

P. bigelovii Vasey & Scribn. in Vasey. Bigelow's Bluegrass. Rare, in canyons; desert, below 1000m: Coyote Canyon; southeast corner of county. Mar.-May.

P. bolanderi Vasey ssp. **howellii** (Vasey & Scribn.)Keck. Howell's Bluegrass. Infrequent, on shaded slopes; montane, above 1200m: Palomar Mountain; Hot Springs Mountain. Apr.-June.

P. nervosa (Hook.)Vasey var. **wheeleri** (Vasey)C.L. Hitchc. Wheeler's Bluegrass. Rare, known in the county only at Cuyamaca Lake. June-Aug. 2*n*=28,29,56,61-91.

***P. pratensis** L. Kentucky Bluegrass. Infrequent, in moist meadows; montane, above 900m: Palomar Mountain; Hot Springs Mountain; Banner; Cuyamaca Lake. May-Aug. 2*n*=21-147.

P. secunda Presl. [*P. sandbergii* Vasey; *P. scabrella* (Thurb.)Benth. ex Vasey; *P. capillaris* Scribn.; *P. orcuttiana* Vasey, type from "near San Diego"] Malpais Bluegrass. Common, on open slopes; cismontane: e.g. De Luz; Palomar Mountain; Montezuma View Point; Cuyamaca Mountains; Pine Valley; Laguna Mountains; southeast corner of county, Descanso; Cowles Mountain; Otay. Feb.-June. 2*n*=44,63,66,74,81,82,84,86,87,91,104.

Polypogon Desf.
[Aveneae]

Awns of glumes 6-8mm long; plants annual . . . . . *P. monspeliensis*
Awns of glumes 3-5mm long; plants perennial . . . . . *P. interruptus*

***P. interruptus** H.B.K. [*P. lutosus* auct. non (Poir.)Hitchc.] Ditch Beardgrass. Native of Europe. Infrequent, in moist, disturbed places; cismontane, below 600m: Lake Hodges; La Jolla; Sears Ranch; Santa Ysabel; Buckman Spring; Campo; Grapevine Canyon. May-Aug. 2*n*=28,42.

***P. monspeliensis** (L.)Desf. Rabbitfoot Beardgrass. Native of Europe. Prevalent weed in disturbed, moist places: e.g. O'Neill Lake; Encinitas; Henshaw Dam; Sentenac Canyon; Vallecito Stage Station; desert slopes of Laguna Mountains; Buckman Spring; Point Loma. Apr.-Aug. 2*n*=14,28.

Puccinellia Parl.
[Poeae]

***P. distans** (L.) Parl. European Alkaligrass. Native of Europe. Uncommon, in alkaline places: San Diego. June-July. *n*=7,21.

Rhynchelytrum Nees
[Paniceae]

***R. roseum** (Nees) Stapf & Hubb. [*Tricholaena r.* Nees] Natal Grass. Native of South Africa. Local escape along roadways; coastal, below 400m: San Diego; La Mesa; Allied Gardens; Casa de Oro; Dehesa. June-Sep.

Schismus Beauv.
[Danthonieae]

***S. barbatus** (L.)Thell. Mediterranean Schismus. Native of Eurasia. Widespread in coastal and desert areas, below 1000m: Carlsbad; San Diego; Otay; Chula Vista; Borrego Springs; Yaqui Well; Vallecito Valley; Mountain Springs. Mar.-Apr. *n*=6.

Schizachyrium Nees
[Andropogoneae]

**S. cirratum* (Hack)Woot. & Standl. [*Andropogon cirratus*] Texas Bluestem. Attributed to the Jamacha region by Canby. A voucher does exist at US National Herbarium, but no additional collections have been made in that area which is out of the otherwise-reported range of the plant.

Secale L.
[Triticeae]

***S. cereale** L. Common Rye. Native of southwest Asia. Local escape from cultivation, as at Santa Ysabel and Jacumba. May-Aug. 2n=14.

Setaria Beauv.
[Paniceae]

Bristles below each spikelet at least more than 5
  Plants annual; spikelets 3mm long; lower floret staminate, the palea well developed. . . . . *S. lutescens*
  Plants perennial; spikelets 2-3mm long . . . . *S. geniculata*
Bristles below each spikelet 1 or 3 . . . . *S. viridis*

***S. geniculata** (Lam.)Beauv. Knotroot Bristlegrass. Native of Tropics. Infrequent, on open slopes; coastal and foothill, below 300m: Murray Dam; Presidio Park; Chollas Creek estuary; San Diego. May-Sep. n=36; 2n=36, 72.

***S. lutescens** (Weigel)Hubb. [*S. glauca* auct. non (L.)Beauv.] Yellow Foxtail. Native of Old World. Occasional, in disturbed places: San Diego; Lakeside. June-Oct. *n*=18.

***S. viridis** (L.)Beauv. Green Bristlegrass. Native of Europe. Occasionally adventive in irrigated places; cismontane: Oceanside; San Luis Rey; Del Mar; National City; Jacumba. June-Aug. 2*n*=18, 35, 36.

Sitanion Raf.
[Triticeae]

S. hystrix (Nutt.)J.G. Sm. [*S. minus* J.G. Sm.] Bottlebrush Squirreltail. Common, on open slopes; foothill and montane, above 300m: e.g. Lone Palm Springs; Hot Springs Mountain; Cuyamaca Lake; Shrine Camp: Eckener Pass; San Felipe; Jacumba. July-Aug. *n*=14.e

Sorghum Moench
[Andropogoneae]

Plants annual . . . . *S. bicolor*
Plants perennial . . . . *S. halepense*

***S. bicolor** (L.)Moench. Milo. Native of Old World. Infrequent, weed in moist places: Bonita; Rancho Santa Fe.

***S. halepense** (L.)Pers. [*Holcus h.* L.] Johnson Grass. Native of the Mediterranean region. Frequent weed in moist places; cismontane, below 1000m: e.g. San Luis Rey; Encinitas; Ramona; Santa Ysabel; Mesa Grande; Mission Valley; Jamacha. May-Aug. *n*=10, 20.

Spartina Schreb.
[Chlorideae]

S. foliosa Trin. California Cord-Grass. Abundant, in Coastal Salt Marsh; coastal, below 20m: e.g. San Mateo Creek; Mission Bay; Ocean Beach; La Punta; Tijuana River estuary. July-Nov. 2*n*=56.

Sporobolus R. Br. Dropseed
[Eragrosteae]

Glumes subequal, much shorter than lemma . . . . *S. poiretii*
Glumes unequal, or if equal, as long as lemma
  Sheath with tuft of hairs at throat; glumes scabrous on keel
    Panicle included in the sheath, the branches ascending . . . . *S. cryptandrus*
    Panicle exserted, the branches spreading. . . . . *S. flexuosus*
  Sheath naked or sparingly ciliate at the throat; glumes glabrous . . . . *S. airoides*

S. airoides (Torr.)Torr. var. **wrightii** (Munro ex Scribner)Gould. [*S. altissimus* Vasey] Alkali Sacaton, Alkali Dropseed. Local, in alkaline places: Oceanside; Jamul Mountain; La Punta; Potrero; Cameron Valley; Sweetwater Dam; Tijuana Valley; Thing Valley; Jacumba. Apr.-Oct. 2*n*=80, 90, 108, 126. Type from San Diego.

S. cryptandrus (Torr.)Gray. Sheathed Dropseed. Local in grassland at Chihuahua Valley and San Jose del Valle. 800-1100m. May-Aug. *n*=18, 19, 36; 2*n*=36.

S. flexuosus (Thurb.)Rydb. Mesa Dropseed. Rare, in sandy areas; desert: San Felipe Valley. Sep.-Oct. 2*n*=38.

***S. poiretii** (R. & S.)Hitchc. Rattail Smutgrass. Native of eastern United States and South America. Local adventive at Pine Valley and Chula Vista. July-Sep. *n*=18.

Stenotaphrum Trin.
[Paniceae]

***S. secundatum** (Walt.) Ktze. St. Augustine Grass. Native of southeast United States and Neotropics. Local escape from cultivation: San Diego. July-Sep. *n*=10.

Stipa L.
[Stipeae]

Panicle broad, open, loose, with spreading branches; lemma at least 4 times as long as palea
  Lemma less than 7mm long; awn up to 4cm long . . . . *S. lepida*
  Lemma more than 7mm long; awn much more than 4cm long
    Lemma slender, cylindrical; middle culm leaves 1.2-2.4mm broad; foliage usually glaucous; awn slender, flexuous beyond the 2nd bend, mostly 9-12 times as long as lemma. . . . . *S. cernua*
    Lemma fusiform; middle culm leaves 2.4-6mm broad; foliage usually green; awn stout, stiff, mostly 7-9 times as long as lemma. . . . . *S. pulchra*
Panicle narrow, slender or contracted; lemma less than 4 times as long as palea
  Lemma densely appressed-villous, with hairs 3-4mm long and rising above the summit like a pappus-crown
    Culms 1-2dm tall. . . . . *S. coronata* var. *c.*
    Culms 3-6dm tall . . . . *S. coronata* var. *depauperata*
  Lemma sparsely strigose to subglabrous, never long-villous
    Awn pubescent, most commonly plumose . . . . *S. speciosa*
    Awn scabrous to subglabrous, rarely appressed-hispid, not plumose
      Lemma broad; first glume 5-nerved . . . . *S. lemmonii*
      Lemma narrow; first glume 3-nerved . . . . *S. diegoensis*

S. cernua Steb. & Love. Nodding Stipa. Infrequent on dry slopes, cismontane, below 1500m: San Miguel Mountain; San Diego; Henshaw Dam. Apr.

S. coronata Thurb. in Wats. var. **coronata**. Giant Stipa. Frequent, on dry slopes; cismontane: e.g. De Luz; Encinitas; Palomar Mountain; Hot Springs Mountain; Julian; desert slopes of Laguna Mountains; Hipass. Type from hills near Julian City. Apr.-June. *n*=20.

S. coronata var. **depauperata** (Jones)Hitchc. [*S. parishii* Vasey] Parish's Stipa. Infrequent, on dry slopes, below 1200m: Bernardo; Hipass; San Felipe; Jacumba. May-Aug.

S. diegoensis Swall. San Diego Stipa. Local, along vernal streams and on clay slopes; southwest cismontane, 300-700m: Proctor Valley; Jamul Mountain; McGinty Mountain; near Lee Valley; Otay Mountain. Type from along vernal stream in Chaparral, Proctor Valley, near Jamul. May-June.

S. lemmonii (Vasey)Scribn. Lemmon's Stipa. Rare, in open montane areas, above 1600m: Laguna Meadow; Harper Ranch; Rattlesnake Valley, Laguna Mountains. May-July. *n*=18.

S. lepida Hitchc. [*S. l.* var. *andersonii* (Vasey)Hitchc., *S. emiens* auct.; *S. l.* var. *hitchcockii* Vasey] Foothill Stipa. Frequent, on dry slopes, coastal, cismontane below 1200m: e. g. De Luz; Rainbow Valley; San Pasqual; Barona Valley; Point Loma; San Diego. Also Jacumba. Mar.-May. *n*=17.

S. pulchra Hitchc. Purple Stipa. Common, on dry slopes; southwest cismontane, below 900m: e.g. Encinitas; Point Loma; Foster; Proctor Valley. Also Jacumba. Mar.-May *n*=32.

S. speciosa Trin. & Rupr. Desert Stipa. Occasional, on dry, rocky slopes; desert and foothill, below 1000m: e.g. Montezuma View Point; Vallecito Stage Station; southeast corner of county; Canyon City; Box Canyon. Apr.-June. 2*n*=60, 64.

Triticum L. Wheat
[Triticeae]

***T. aestivum** L. Cereal Wheat. Native of Old World. Infrequent escape from cultivation: San Diego; Lemon Grove. Apr.-July. 2*n*=42.

Vulpia K.C. Gmelin
[Poeae]

Spikelets densely 5-12-flowered; lemmas without scarious margin
  Lemmas hirtellous or pubescent. . . . . *V. octoflora* var. *hirtella*
  Lemmas glabrous or scabrous . . . . *V. octoflora* var. *o.*
Spikelets loosely 1-5-flowered; lemmas with narrow scarious margin
  Inflorescence narrow, branches ascending or appressed-ascending
    Lemmas conspicuously long-ciliate toward apex . . . . *V. myuros* var. *hirsuta*
    Lemmas not ciliate
      Lower glume 2/3 to 3/4 as long as the second . . . . *V. bromoides*
      Lower glume less than half as long as the second . . . . *V. myuros* var. *m.*
  Inflorescence broader, the principal branches spreading
    Spikelets glabrous or somewhat scabrous, but not with long well developed hairs *V. microstachys* var. *pauciflora*
    Spikelets pubescent, with long well developed hairs
      Lemmas glabrous; glumes pubescent . . . . *V. microstachys* var. *confusa*
      Lemmas hirsute; glumes glabrous or pubescent . . . . *V. microstachys* var. *ciliata*

*V. **bromoides** (L.)S.F. Gray. [*Festuca dertonensis* (All.)Asch. & Graebn.] Six-Weeks Fescue. Native of Europe. Rare adventive, at Carrizo Stage Station. Apr.-June. $2n$=14, 42.

V. **microstachys** (Nutt.)Munro ex Benth. var. **ciliata** (Beal)Lonard & Gould. [*Festuca grayi* (Abrams)Piper] Gray's Fescue. Rare, in interior valleys as at Cañada de Buena Vista, Warners Pass, and Cuyamaca Mountains. Mar.-June. 2n=42.

V. **microstachys** var. **confusa** (Piper)Lonard. [*Festuca c.* Piper.] Hairy-Leaf Fescue. Scarce, on dry slopes; montane, above 1000m: Hot Springs Mountain; Pine Valley. Mar.-June.

V. **microstachys** var. **pauciflora** (Scribn. ex Beal)Lonard & Gould. [*Festuca pacifica* Piper; *Festuca reflexa* Buckl.] Pacific Fescue. Occasional, on dry slopes and burns; cismontane: Ysidora; Fallbrook; Lilac; Palomar Mountain; Montezuma View Point; Cuyamaca Mountains; Foster; Hidden Glen; Kitchen Creek; Campo; Otay Mountain. Mar.-June.

*V. **myuros** (L.)K.C.Gmelin var. **hirsuta** Hack. [*Festuca megalura* Nutt.] Foxtail Fescue. Native of Europe. Common, on dry, open places; cismontane: e.g. Encinitas; Palomar Mountain; Lake Henshaw; Hot Springs Mountain; Cuyamaca Mountains; Rattlesnake Valley; Boulevard; La Posta; Tierra Santa; San Diego. Apr.-June.

*V. **myuros** var. **myuros**. [*Festuca m.* L.] Rattail Fescue. Native of Europe. Infrequent adventive in disturbed areas: Moro Hill; San Luis Rey; Clevenger Canyon; Wild Rose Spring; Kitchen Creek; Point Loma. Mar.-May. n=7; $2n$=42.

V. **octoflora** (Walt.)Rydb. var. **hirtella** (Piper)Henr. [*Festuca o.* ssp. *h.* Piper.] Tufted Fescue. Occasional, on open slopes and burns, below 1300m: e.g. Lilac; Pala; San Pasqual; Montezuma View Point; northeast corner of county; Borrego Palm Canyon; Hipass; Campo; Mahogany Canyon. Apr.-June. $2n$=14.

V. **octoflora** var. **octoflora**. [*Festuca o.* Walt.] Slender Fescue. Infrequent, on open slopes and burns; cismontane: e.g. Palomar Mountain; Lake Henshaw; Poway Grade; Rincon; Cottonwood; Lilac. Apr.-June.. $2n$=14.

Zea L. Maize
[Andropogoneae]

*Z. **mays** L. Cultivated Corn. Native to subtropical areas. Escaping or persisting after cultivation; cismontane, below 100m: Chula Vista, Rancho Santa Fe.

PONTEDERIACEAE. Pickerel-Weed Family

Eichhornia Kunth Water-Hyacinth

*E. **crassipes** (Mart.)Solms. Common Water-Hyacinth. Native of Neotropics. Local, adventive in slow-running streams or permanent ponds; cismontane: northwest of Ramona; Tijuana River Valley. June-Oct. n=16; $2n$=32.

POTAMOGETONACEAE. Pondweed Family

KEY TO GENERA

Sepals 4, distinct; stamens 4; carpels sessile . . . . *Potamogeton*
Sepals none; stamens 2; carpels stipitate in fruit . . . . *Ruppia*

Potamogeton L. Pondweed

Leaves all submerged and similar
  Leaves linear
    Stipules fused with the base of leaf to form a sheath at least 1cm long . . . . *P. pectinatus*
    Stipules free from the leaves or nearly so
      Leaves without basal glands; nerves 3-5 . . . . *P. foliosus*
      Leaves with a pair of basal glands; nerves mostly 3 . . . . *P. pusillus*
  Leaves broad, never linear . . . . *P. illinoensis*
Leaves of two types, floating (broad and coriaceous) and submerged (broad or narrow, but thin) . . . . *P. nodosus*

P. **foliosus** Raf. var. **foliosus**. [*P. pauciflorus* var. *californicus* Morong, the type from San Diego County.] Leafy Pondweed. Occasional, in brackish ponds: Ramona; Mission Valley; Jamul Creek; Campo; Pine Valley; Canebrake Canyon. Apr.-Nov. $2n$=28.

P. **illinoensis** Morong. [*P. lucens* auct. non L.] Shining Pondweed. Infrequent, in ponds and lakes: upper Mason Valley; Otay Lake; Rickey Dam; Mission Valley; Penasquitos Creek. June-Sep. $2n$=52

P. **nodosus** Poir. [*P. americanus* Cham. & Schlecht.] Long-Leaf Pondweed. Occasional, in lakes and ponds; montane, above 1000m: Pine Hills; Laguna Lakes; Corte Madera Lake. May-Aug. n=26.

P. **pectinatus** L. Fennel-Leaf Pondweed. Occasional, in ponds and quiet streams: San Mateo Creek; Mission Valley; Cuyamaca Lake; Big Laguna Lake; Otay Lake; Sweetwater Lake. May-July. $2n$=42, 78.

P. **pusillus** L. var. **pusillus**. [*P. panormitanus* Biv.] Small Pondweed. Rare, known in the county only from a collection at Cuyamaca Lake, 1400m. May-June. $2n$=26.

Ruppia L. Ditch-Grass

R. **maritima** L. Beakfruit Sea-Tassle. Occasional, in brackish waters of Coastal Salt Marsh: Santa Margarita River estuary; Carlsbad; Agua Hedionda Lagoon; San Elijo Lagoon; National City; Otay River estuary. Apr.-July. n=8; $2n$=16, 20, 40.

SPARGANIACEAE. Bur-Reed Family

Sparganium L. Bur-Reed

S. eurycarpum Engelm. Broad-Fruit Bur-Weed. Infrequent, in brackish waters: San Mateo Creek estuary; Las Flores Creek estuary. Apr.-Aug.

TYPHACEAE. Cat-Tail Family

Typha L. Cat-Tail

Leaves mostly 5mm wide, strongly convex on back, dark green; pistillate spikes much overtopped by leaves and with bracts very dark brown, opaque and firm . . . . . *T. angustifolia*
Leaves 6-15mm wide; pistillate spikes not or moderately overtopped by leaves and with bracts light brown and translucent or lacking
  Leaf-blades moderately convex on back, light yellow-green; bracts of staminate spikes cuneate, laciniate, brown; bracts of pistillate spikes ovate and apiculate; interval between staminate and pistillate spikes 1-4cm . . . . . *T. domingensis*
  Leaf-blades flat, light green; bracts of staminate spikes simple, hairlike, white; bracts of pistillate spikes none; no interval between spikes . . . . . *T. latifolia*

T. angustifolia L. Narrowleaf Cattail. Occasional, in ponds, lakes, and seepage areas: Agua Hedionda Lagoon; Lake Murray; Sweetwater Lake; Pine Hlls; Famosa Slough. June-July. n=15.

T. domingensis Pers. Tule Cattail. Occasional, in ponds, lakes, and seepage areas: San Mateo Creek; Las Flores Creek; Santa Margarita River estuary; Lemon Grove. June-July. 2n=30.

T. latifolia L. Soft Flag. Common in ponds, lakes, and seepage areas: San Mateo Creek; Las Flores Creek; Santa Margarita Ranch; Encinitas; Cuyamaca Lake; Shrine Camp; Campo; Lakeside; Otay Mountain; Telegraph Creek estuary; Sweetwater Dam; Lake Murray; Alvarado Canyon; Rolando; Poway Junction; La Jolla. June-July. 2n=30.

ZANNICHELLIACEAE. Horned-Pondweed Family

Zannichellia L. Horned-Pondweed

Z. palustris L. Occasional, in ponds and slow streams; cismontane and desert, below 1000m: San Mateo Creek estuary; Las Flores Creek estuary; San Pasqual; Famosa Slough; Sears Ranch; Jamul Creek; Sweetwater Dam; Descanso; Vallecito Stage Station. Mar.-Nov. n=12; 2n=12, 24, 28, 32, 36.

ZOSTERACEAE. Eel-Grass Family

KEY TO GENERA

Plants dioecious, fruit cordate at base; leaves 1.5-4mm wide . . . . . *Phyllospadix*
Plants monoecious; fruit ovoid; leaves mostly 6-12mm wide . . . . . *Zostera*

Phyllospadix Hook. Surf-Grass

Pistillate spadix usually 1, on flowering stems 1-2dm long . . . . . *P. scouleri*
Pistillate spadices 2-6, borne on flowering stems 3-10dm long . . . . . *P. torreyi*

P. scouleri Hook. Scouler's Surf-Grass. Abundant, in surf-pounded, rocky shorelines of the coast: La Jolla; Bird Rock. May-Aug.

P. torreyi Wats. Torrey's Surf-Grass. Abundant, on rocky coastline and in calmer waters: Oceanside; La Jolla; Pacific Beach; Mission Bay; Coronado. May-July.

Zostera L. Eel-Grass

Z. marina L. [*Z. marina* var. *latifolia* Morong.] Common Eel-Grass. Abundant, in sublittoral zone of coastline and in calm waters of bays and estuaries: Oceanside; La Jolla; Pacific Beach Mission Bay; Coronado; San Diego Bay. May-Sep. n=6.

Class ANGIOSPERMAE Flowering Plants

Subclass DICOTYLEDONES

KEY TO FAMILIES

Petals lacking or not evident (sometimes simulated by the petal-like calyx, as in *Fremontodendron* and *Nyctaginaceae*)
  Plants definitely woody; well developed trees and shrubs . . . . . **Group 1**
  Plants herbaceous, sometimes slightly woody at the base only . . . . . **Group 2**
Petals present, evident
  Corolla of separate petals
    Stamens numerous, more than twice as many as the petals . . . . . **Group 3**
    Stamens few, not more than twice as many as the petals . . . . . **Group 4**
  Corolla of mostly connate petals, these often markedly grown together . . . . . **Group 5**

Group 1. Woody plants without petals, or petals not evident

Leaves compound, the blade divided into separate parts
Plants woody vines climbing by twisting petioles; leaves opposite; stamens and pistils many . . . *Ranunculaceae* (*Clematis*)
Plants trees or shrubs, not climbing
Leaves alternate; fruit a nut . . . *Juglandaceae*
Leaves opposite; fruit a samara
Leaflets 3; samaras pubescent . . . *Aceraceae* (*Acer negundo*)
Leaflets 3-7; samaras glabrous . . . *Oleaceae* (*Fraxinus*)
Leaves simple, the blade entire, toothed or lobed
Plants parasitic on branches of trees and shrubs and not connected with the ground . . . *Viscaceae*
Plants growing normally on the ground
Flowers (at least the staminate) in catkins
Leaves alternate; plants often monoecious
Calyx present in staminate flowers
Fruit a nut enclosed in a leafy involucre, or fruit cone-like; plants deciduous . . . *Betulaceae*
Fruit an acorn with a basal cup; plants mostly evergreen . . . *Fagaceae*
Calyx none, the staminate flowers subtended by bracts . . . *Salicaceae*
Leaves opposite; plants dioecious
Catkins erect; prostrate plant with succulent, linear-oblanceolate leaves . . . *Bataceae*
Catkins drooping; upright shrub with broader leaves . . . *Garryaceae*
Flowers not in catkins
Plants normally trees with a well defined central trunk
Leaves entire, lanceolate to ovate, strongly aromatic when crushed
Ovary superior, maturing into a drupe; stamens in 3 series of 3 each, not highly colored, native *Lauraceae*
Ovary inferior, maturing into a woody capsule; stamens many, white to yellow or red, introduced . . . *Myrtaceae* (*Eucalyptus*)
Leaves toothed to lobed, ovate to roundish, not strongly aromatic
Leaves 12-25cm broad, deeply 5-lobed; flowers in round heads . . . *Platanaceae*
Leaves 2-5cm broad, somewhat serrate; flowers in small clusters or solitary . . . *Ulmaceae*
Plants normally shrubs with several stems from near base
Flowers yellowish, 2.5-6cm broad, the wide sepals or calyx-lobes petaloid *Sterculiaceae* (*Fremontodendron*)
Flowers smaller
Ovary 3-4-loculed; styles 3-4
Fruit a lobed capsule; leaves often alternate . . . *Euphorbiaceae*
Fruit a smooth cylindrical capsule; leaves opposite . . . *Simmondsiaceae*
Ovary 1-celled (or if 2-3-celled, the style single); styles 1- 3
Fruit an akene or utricle
Styles 2-3; ovary not surrounded by a floral tube
Flowers subtended by an involucre of united bracts; akene triangular or lenticular . . . *Polygonaceae*
Flowers without an involucre of bracts; fruit mostly a utricle or depressed-globose akene . . . *Chenopodiaceae*
Style 1, plumose in fruit; ovary surrounded by floral tube . . . *Rosaceae* (*Cercocarpus*, *Coleogyne*)
Fruit a samara, drupe or berry
Leaves opposite; plants glabrous; fruit a samara or blackish drupe . . . *Oleaceae*
Leaves alternate; fruit a drupe or berrylike . . . *Rhamnaceae*

Group 2. Herbaceous plants without petals, or petals not evident

Aquatic plants growing somewhat submerged, or, as water dries up, on wet mud
Leaves largely dissected, whorled
Ovary superior, fruit an akene; leaves 3-times forked . . . *Ceratophyllaceae*
Ovary inferior; fruit a nutlet or splitting into 4 nutlets; leaves somewhat pinnatifid . . . *Haloragaceae*
Leaves entire, often opposite and often crowded into terminal rosettes
Stamen 1; ovary superior, splitting into 4 parts when ripe; leaves 3-8mm long . . . *Callitrichaceae*
Stamens 4-5; ovary inferior, forming a capsule . . . *Onagraceae* (*Ludwigia*)
Land plants, sometimes growing in wet places
Ovary superior, free from the calyx, although sometimes surrounded by it
Perianth lacking entirely
Flowers white, perfect, borne in a bracted spike which is subtended by a conspicuous involucre . . *Saururaceae*
Flowers imperfect, either staminate or pistillate
Flowers borne in clusters often surrounded by an involucre resembling a perianth; capsules 3-lobed; sap often milky; plants monoecious . . . *Euphorbiaceae*
Flowers borne in catkins disposed in terminal spikes; plants dioecious; seashore . . . *Bataceae*
Perianth present as a calyx (often petaloid)
Pistils more than 1; stamens 10-many . . . *Ranunculaceae*
Pistil 1
Flowers perigynous, the ovary enclosed in or seated in a floral tube
Stipules present; leaves alternate; fruit an akene . . . *Rosaceae*
Stipules none, leaves opposite
Stamens many; fruit circumscissile . . . *Aizoaceae* (*Sesuvium*)
Stamens 3-5; fruit an akene . . . *Nyctaginaceae*
Flowers hypogynous; the ovary not so enclosed
Style and stigma single
Calyx not tubular
Leaves subulate, squarrose-spreading, 3-6mm long . . . *Caryophyllaceae* (*Loeflingia*)
Leaves ovate, the blades 10-20mm long
Calyx 5-6 parted; stamens 6-7 . . . *Euphorbiaceae* (*Eremocarpus*)
Calyx largely 4-parted; stamens 4 . . . *Urticaceae*
Calyx tubular, corolla like, subtended by bracts often forming a calyx-like involucre . . . . *Nyctaginaceae*

Styles and stigmas more than 1
Leaves mostly deeply palmately 5-7-lobed or with 5-7 leaflets; flowers imperfect. . . . . . . *Cannabaceae*
Leaves mostly entire or shallowly lobed; flowers perfect or imperfect
Ovary 1-loculed (except sometimes in *Silene* which has a tubular gamosepalous calyx)
Ovary or seed solitary; fruit mostly an akene or urticle; leaves often alternate
Flowers borne in a tubular to campanulate involucre . . . . *Polygonaceae*
Flowers not borne in an involucre
Leaves with evident stipular sheath above each node . . . . *Polygonaceae*
Leaves lacking such stipular sheath
Calyx mostly 6-cleft; stamens 3, 6, or 9 . . . . *Polygonaceae*
Calyx-lobes or sepals 1, 4, or 5
Bracts subtending flowers not scarious; plants mostly mealy, scurfy or fleshy . . . . *Chenopodiaceae*
Bracts subtending flowers scarious; plants not mealy, scurfy or fleshy. . . . *Amaranthaceae*
Ovules or seeds more than 1; fruit a capsule; leaves opposite
Leaves of each pair very unequal . . . . *Aizoaceae*
Leaves of each pair normally ca. equal . . . . *Caryophyllaceae*
Ovary more than 1-loculed; fruit a berry. . . . . *Phytolaccaceae*

Ovary generally inferior, partly or wholly adnate to the floral tube
Flowers imperfect; stamens 8-many; fruit a 1-loculed capsule
Leaves divided; large green herb. . . . . *Datiscaceae*
Leaves reduced to minute bracts; very small parasite on *Psorothamnus*. *Rafflesiaceae*
Flowers perfect
Fruit a bony nut, indehiscent; leaves rhombic-ovate, entire . . . . *Aizoaceae* (*Tetragonia*)
Fruit a capsule . . . . *Saxifragaceae*

Group 3. Plants with separate petals; stamens numerous, more than twice as many as petals

Ovary superior or appearing so
Plants woody, i.e. shrubs or trees
Plants with evident stipules
Plants with many exserted stamens; leaves pinnate or the pinnules suppressed and the petiole flattened into a leaf-like vertical structure; pistil 1, forming a legume . . . . *Fabaceae* (*Acacia*)
Plants not as above
Stamens united into a tube around the pistil; leaves rounded, often stellate-pubescent . . . . *Malvaceae*
Stamens not united into a tube; leaves various, not stellate- pubescent . . . . *Rosaceae*
Plants without stipules
Filaments connate into a cylindric staminate column. . . . . *Meliaceae*
Filaments free to their bases
Pistil 1; sepals 2, rarely 3 or 4, often caducous . . . . *Papaveraceae* (*Dendromecon*)
Follicles 2-9, separate; calyx of 5 persistent lobes . . . . *Crossosomataceae*
Plants herbaceous, or woody at base only; not definite shrubs
Sepals 2, rarely 3 or 4 (joined into a pointed cap in *Eschscholzia*)
Plants fleshy; sepals persistent. . . . . *Portulacaceae*
Plants not fleshy; sepals caducous at anthesis . . . . *Papaveraceae* (*Romneya*)
Sepals normally more than 2
Large, peltate-leaved aquatic plants. . . . . *Nelumbonaceae*
Leaves not so
Stamens united into a tube around the pistil . . . . *Malvaceae*
Stamens not so united
Maturing ovary at summit exposing the contained seeds; flowers irregular . . . . *Resedaceae*
Maturing ovary closed at summit; flowers mostly regular
Stamens hypogynous
Leaves opposite
Leaves punctate with pellucid dots; stamens usually in 3 or 5 bundles . . . . *Hypericaceae*
Leaves not so punctate; stamens separate . . . . *Cistaceae*
Leaves alternate
Ovary 3-5-celled
Styles 3, bilobed; stamens united in ranks of 5. . . . . *Euphorbiaceae* (*Argythamnia*)
Style 1, with 3 stigmas; stamens not so united . . . . *Cistaceae* (*Helianthemum*)
Ovary or ovaries 1-celled; stamens spirally arranged
Sepals persistent; anthers maturing in centrifugal sequence. *Paeoniaceae*
Sepals deciduous; anthers maturing centripetally. . . . . *Ranunculaceae*
Stamens perigynous; stipules usually well developed . . . . *Rosaceae*

Ovary at least partly inferior
Ovary only partly inferior
Fruit a dry capsule . . . . *Saxifragaceae*
Fruit a fleshy pome . . . . *Rosaceae*
Ovary wholly inferior
Petals numerous (technically may be modified stamens); plants succulent
Plants not spiny; leaves well developed . . . . *Aizoaceae*
Plants spiny, with leaves much reduced and caducous, or wanting . . . . *Cactaceae*
Petals few; plants not succulent
Trees or shrubs; fruit fleshy . . . . *Rosaceae*
Herbs; fruit a capsule . . . . *Loasaceae*

Group 4. Plants with separate petals: stamens not more than twice as many as petals

Pistils more than 1, nearly or quite separate
Plants fleshy, at least the leaves so; leaves simple ... *Crassulaceae*
Plants not succulent
Insertion of the stamens on the floral tube; flowers perigynous
Stipules usually absent; pistils mostly fewer than sepals ... *Saxifragaceae*
Stipules usually present; pistils mostly as many as sepals ... *Rosaceae*
Insertion of stamens hypogynous; small herbs with long tail-like spike of pistils ... *Ranunculaceae* (*Myosurus*)

Pistil 1, of 1 or more generally united carpels
Plants climbing by means of tendrils; leaves palmately veined
Fruit a berry; fruits borne in clusters, grapes ... *Vitaceae*
Fruit a pepo, the fruits borne singly, gourds, melons, etc. ... *Cucurbitaceae*
Plants not climbing by means of tendrils
Styles 2-5, separate to near base
Plants definitely woody, well developed shrubs or trees
Leaves small, scalelike, appressed; flowers minute, in large clusters ... *Tamaricaceae*
Leaves well developed
Ovary superior
Fruit a samara; leaves palmately veined, opposite ... *Aceraceae*
Fruit not a samara; leaves pinnately veined, alternate ... *Anacardiacae*
Ovary appearing mostly inferior
Fruit a pome, the dry pistils technically superior and surrounded by a fleshy floral tube; leaves mostly oblong, pinnately veined ... *Rosaceae* (*Heteromeles*)
Fruit a berry, inferior; leaves mostly palmately lobed ... *Grossulariaceae* (*Ribes*)
Plants herbaceous
Plants submerged aquatics or on wet mud flats and rooting at nodes
Leaves mostly finely dissected ... *Haloragaceae*
Leaves entire, opposite ... *Elatinaceae*
Plants normal terrestrial
Ovary superior
Leaves compound with 3 leaflets; stamens 10 ... *Oxalidaceae*
Leaves simple
Leaves largely basal
Plants fleshy, not generally maritime ... *Portulacaceae*
Plants not fleshy, maritime ... *Plumbaginaceae*
Leaves mostly cauline
Leaves for the most part alternate
Sepals 2, or if more, the leaves fleshy
Sepals 2; plants not large vines; stamens 1-3 ... *Portulacaceae*
Sepals 5; plants large vines; stamens 5 ... *Basellaceae*
Sepals 3 several; plants not fleshy
Stamens 10; plants generally gray-pubescent ... *Euphorbiaceae* (*Argythamnia*)
Stamens 3-8; plants not gray-pubescent
Flowers regular, 5-merous; capsules closed until mature
Ovary 1-celled ... *Caryophyllaceae*
Ovary 2-5 celled ... *Linaceae*
Flowers irregular, 2-7-merous; capsules open at top before maturity ... *Resedaceae*
Leaves opposite
Stamens 5 or 10; capsules 3-10-valved or -toothed; placenta largely central ... *Caryophyllaceae*
Stamens 4-7; capsules 2-4-valved; placenta basal but parietal ... *Frankeniaceae*
Ovary mostly inferior
Ovules several in each cavity of the ovary
Fruit a capsule ... *Saxifragaceae*
Fruit a many-seeded berry ... *Grossulariaceae* (*Ribes*)
Ovules solitary in each cavity of the ovary; flowers in umbels
Fruit a berry; leaves 3-15dm long ... *Araliaceae*
Fruit dry, splitting into 2 carpels; leaves mostly smaller ... *Apiaceae*

Style 1, sometimes divided toward the apex
Ovary inferior
Plant scabrous with short barbed hairs; fruit dry ... *Loasaceae* (*Petalonyx*)
Plant not scabrous with short barbed hairs
Plants well developed shrubs or arborescent; fruit fleshy
Flowers racemose or solitary; leaves alternate; stamens usually 5 ... *Grossulariaceae* (*Ribes*)
Flowers in heads, cymes or umbels; leaves mostly opposite; stamens 4 ... *Cornaceae*
Plants herbaceous, sometimes suffrutescent; fruit dry; flowers mostly 4-merous ... *Onagraceae*
Ovary superior, but sometimes appearing inferior because enclosed in but not adnate to floral tube
Plants well developed shrubs or trees
Flowers irregular, the petals not all alike
Petals 3 (2 forming a pair, the 3rd hooded above); sepals 5, the lateral petaloid ... *Polygalaceae*
Petals 4-5; calyx not petaloid
Flowers papilionaceous, fruit a legume; leaves mostly compound; stamens more than 4 ... *Fabaceae*
Flowers not papilionaceous; fruit a spiny 1-seeded pod; leaves simple; stamens 4 ... *Krameriaceae*
Flowers regular, the petals essentially alike
Leaves compound, consisting of 2 or more leaflets (separately deciduous)
Fruit a samara; wing of fruit terminal; leaflets 3-9; leaves opposite ... *Oleaceae (Fraxinus)*
Fruit not a samara
Fruit fleshy; petals 3 plus 3; leaflets mostly more than 3 ... *Berberidaceae*

Fruit dry; petals in 1 series
Ovary 1-loculed, becoming a legume; stamens mostly 10 . . . *Fabaceae*
Ovary more than 1-loculed
Leaflets 13-25; flowers 5-merous . . . *Burseraceae*
Leaflets 2-3
Sepals and petals 5; fruit not inflated . . . *Zygophyllaceae*
Sepals and petals 4; fruit inflated . . . *Capparaceae* (*Cleome*)

Leaves simple, but sometimes divided, looking like leaflets, but the parts not separately deciduous
Fruit stipitate, 2-locular; flowers yellow . . . *Brassicaceae* (*Stanleya*)
Fruit sessile
Fruit a legume, oblong, flat; stamens 10; flowers red-purple . . . *Fabaceae* (*Cercis*)
Fruit not a legume
Stamens connate into a tube or cup . . . *Sterculiaceae* (*Ayenia*)
Stamens free and distinct
Leaves pellucid-punctate (gland-dotted); petals 4, white; leaves linear; stamens 8 . . . *Rutaceae* (*Cneoridium*)
Leaves not as above; stamens not 8; petals not 4
Trees with large palmately lobed leaves; flowers imperfect, reduced, in spherical clusters . . . *Platanaceae*
Trees or shrubs, the leaves not large and palmately lobed
Stamens 4-5
Stamens opposite the petals . . . *Rhamnaceae*
Stamens alternate with the petals or more numerous
Ovules 1-2 in each ovary-cell . . . *Celastraceae*
Ovules several to many in each ovary-cell . . . *Pittosporaceae*
Stamens 10, cismontane Chaparral . . . *Rosaceae* (*Adenostoma*)

Plants herbaceous
Sepals 2 or 3
Plants fleshy, succulent; sepals 2, persistent . . . *Portulacaceae*
Plants not succulent
Sepals caducous, 2 or 3; stamens 6-12; leaves entire . . . *Papaveraceae*
Sepals not early deciduous; leaves dissected; stamens in 2 sets of 3 each; petals 4, in 2 dissimilar pairs . . . *Fumariaceae*
Sepals 4 or 5, sometimes more
Flowers nearly or quite regular
Leaves compound, the leaflets separately deciduous
Fruit 1-loculed, forming a legume; leaves pinnate . . . *Fabaceae* (*Senna*)
Fruit 2-5-loculed; fruit not a legume
Leaflets 3; leaves alternate
Stamens 10; foliage sour to taste, not ill-smelling . . . *Oxalidaceae*
Stamens 2-6; foliage acrid, ill-smelling . . . *Capparaceae*
Leaflets 10-16; leaves opposite . . . *Zygophyllaceae*
Leaves simple and entire or divided, but not truly compound
Ovary appearing inferior, actually superior and free in the floral tube; flowers 4-6merous . *Lythraceae*
Ovary appearing superior
Sepals 4; filaments 6, 4, rarely 2, distinct . . . *Brassicaceae*
Sepals mostly 5; filaments 5, 8, or 10, often united at base
Carpels in mature fruit separating from each other as 1-seeded structures
Stipules scarious; carpels tailed in mature fruit; lower leaves usually opposite . . . *Geraniaceae*
Stipules none; carpels not tailed; leaves alternate . . . *Limnanthaceae*
Carpels when mature forming a many-seeded capsule . . . *Ericaceae*
Flowers quite definitely irregular, the petals unlike
Flowers papilionaceous; fruit a legume . . . *Fabaceae*
Flowers not papilionaceous; fruit not a legume
Leaves not peltate; fruit a 1-loculed, 3-valved capsule . . . *Violaceae*
Leaves peltate; fruit 3-lobed, 3-loculed, each locule 1-seeded . . . *Tropaeolaceae*

Group 5. Plants with petals generally connate

Ovary superior
Stamens more than 5
Corolla not markedly sympetalous (not urn-shaped or tubular) the petals united only near the base
Pistils 4-5; stamens 10; plants succulent . . . *Crassulaceae*
Pistil 1
Corolla regular
Flowers minute, in dense heads or spikes; stamens 10 to many; fruit a legume; shrubs or trees . . . *Fabaceae*
Flowers not as above; fruit not a legume
Stamens many, united in a tube about the pistil; leaves simple, palmately veined or lobed . . . *Malvaceae*
Stamens 10, united at very base only; leaves 3-foliolate . . . *Oxalidaceae*
Corolla irregular, the petals not all alike
Petals 4, in 2 unlike pairs; sepals 2; leaves dissected . . . *Fumariaceae*
Petals 3, appearing to be 5 because of 2 petaloid sepals; other 3 sepals not petaloid; leaves entire . . . *Polygalaceae*
Corolla urn-shaped or tubular or with petals markedly united
Stamens nearly or quite free from the corolla
Style 1; stigma 1; plants not thorny . . . *Ericaceae*
Styles 3, united to middle; anthers opening lengthwise; plants thorny virgate, desert shrubs *Fouquieriaceae*
Stamens inserted in corolla-throat; root parasites . . . *Lennoaceae*

Stamens not more than 5
Plants parasitic, lacking chlorophyll
Twining or trailing vines with small haustoria and minute flowers; no connection with ground when mature . . . . *Convolvulaceae* (*Cuscuta*)
Root-parasites
Corolla regular; stamens 5. . . . . *Lennoaceae*
Corolla 2-lipped; stamens 4 . . . . *Orobanchaceae*
Plants not or only partially parasitic and having chlorophyll
Corolla regular
Pistils 2 (ovaries distinct, but styles or stigmas united); plants with milky juice
Styles united; stamens connivent around the stigma . . . . *Apocynaceae*
Styles distinct below; stamens monadelphous and adnate to the stylar column. . . . . *Asclepiadaceae*
Pistil 1
Ovary 1-loculed, 1-ovuled; style and stigma 1; fruit dry, hard; the seemingly colored corolla really a calyx inside a 1-flowered involucre . . . . *Nyctaginaceae*
Ovary and fruit not as above
Stamens as many as the corolla-lobes and opposite them, or with an additional series of staminoidia; placentae central or basal
Fruit a several-seeded capsule; style 1 . . . . *Primulaceae*
Fruit 1-seeded; styles 5 . . . . *Plumbaginaceae*
Stamens as many as or fewer than the corolla-lobes and alternate with them
Foliage punctate with translucent dots . . . . *Myoporaceae*
Foliage not punctate with translucent dots
Corolla small, dry-scarious, veinless; stamens 2 or 4 . . . . *Plantaginaceae*
Corolla not dry-scarious, veiny; capsules usually not as above.
Ovary 4-celled, commonly 4-lobed, each lobe forming a nutlet when mature (although some may abort); inflorescence usually a scorpioid cyme . . . . *Boraginaceae*
Ovary 1-, 2-, or 3-celled
Stamens 2, opposite each other; capsule circumscissile . . . . *Oleaceae* (*Menodora*)
Stamens 4-7; fruit not opening by a lid
Anthers opening by terminal pores; flowers large, white or yellowish . . . . *Ericaceae* (*Rhododendron*)
Anthers opening by longitudinal slits
Style 3-cleft; ovary 3-loculed; capsule 3-valved . . . . *Polemoniaceae*
Style not 3-cleft; ovary 1-2-celled
Calyx 4-5-toothed or -cleft; style 1, entire
Ovary 1-celled; stigmas 2; leaves opposite . . . . *Gentianaceae*
Ovary 2-celled; stigma usually 1; leaves alternate . . . . *Solanaceae*
Calyx of 5 distinct sepals, or sepals united only at very base; styles 2 or 1, usualy partly divided
Plants twining or trailing, inflorescence not coiled; corolla plaited in bud . . . . *Convolvulaceae*
Plants erect or diffuse; inflorescence cymose, often coiled; corolla not plaited in bud . . . . *Hydrophyllaceae*

Corolla mostly irregular
Fruit of 2-4 nutlets; leaves opposite
Ovary not lobed; style apical, entire . . . . *Verbenaceae*
Ovary 4-lobed; style arising between the ovary lobes, cleft at apex . . . . *Lamiaceae*
Fruit a capsule
Ovary 1-celled or incompletely 2- or 4-celled
Plants shrubby or arborescent. . . . . *Avicenniaceae*
Plants herbaceous. . . . . *Martyniaceae*
Ovary 2-celled
Seeds not winged; placentae axile
Calyx with a pair of bractlets at base; stamens 2; anther cells oblique . . . . *Acanthaceae*
Calyx without bractlets; stamens 5, 4, or rarely 2; anther cells opposite. . . . . *Scrophulariaceae*
Seeds winged; placentae parietal; capsules linear, 15-25cm long . . . . *Bignoniaceae*

Ovary inferior or partly so
Stamens more than 5
Anthers opening by terminal pores or chinks; fruit fleshy . . . . *Ericaceae*
Anthers opening by longitudinal slits; fruit dry . . . . *Styracaceae*
Stamens 5 or fewer
Stamens distinct
Leaves alternate; flowers regular (somewhat irregular in *Nemacladus* and *Parishella*); stamens 5 . . . . *Campanulaceae*
Leaves opposite or whorled
Stamens 1-3; flowers irregular; fruit 1-seeded . . . . *Valerianaceae*
Stamens 4-5, rarely 2; flowers often regular
Ovary 1-celled; flowers in short spikes or involucral heads; fruit an akene; introduced herbs. . *Dipsacaceae*
Ovary 2-5-celled; flowers not so arranged; fruit usually not a solitary akene
Leaves opposite and evidently stipulate, or whorled and not stipulate . . . . *Rubiaceae*
Leaves opposite or perfoliate, rarely stipulate and when so, the stipules minute
Leaves pinnately-compound. . . . . *Adoxaceae*
Leaves simple . . . . *Caprifoliaceae*
Stamens united by the anthers
Plants bearing tendrils; leaves palmate; stamens 5, with 2 pairs united and appearing to be 3 . . . *Cucurbitaceae*
Plants not tendril-bearing
Flowers not in involucrate heads; stamens free from the corolla . . . . *Campanulaceae*
Flowers in involucrate heads; stamens adnate to the corolla . . . . *Asteraceae*

ACANTHACEAE. Acanthus Family

Justicia L.

J. californica (Benth.) D. Gibson. [*Beloperone c.* Benth.] Chuparosa. Common on mesas and in washes; desert, below 800m: e.g. Coyote Creek; Borrego Valley; Borrego Palm Canyon; The Narrows; Fish Creek; Vallecito Valley; Canebrake Canyon; Mountain Palm Springs; Carrizo Gorge. Also occurs disjunctly on the north slope by the San Diego River, east of Wildcat Canyon. The yellow-flowered form occurs in Collins Valley and at the San Diego River site. Mar.-June $2n$=28.

ACERACEAE. Maple Family

Acer L. Maple

Leaves simple, palmately lobed; petals present . . . . *A. macrophyllum*
Leaves pinnate with 3 leaflets; petals absent . . . . *A. negundo*

A. macrophyllum Pursh. Big-leaf Maple. Rare, in canyons; montane, above 450m: Castro Canyon; Morgan Hill; Oasis Spring. Apr.-May. $2n$=26.

A. negundo L. ssp. **californicum** (T.& G.)Wesmael. California Box-Elder. Scarce, along streams; foothill and montane, above 450m: Doane Valley; Valley Center; Panawatt Springs; La Posta Reservation. Also along Montezuma Road at Fairmont Avenue, East San Diego. Mar.-Apr.

ADOXACEAE. Adoxus Family

Sambucus L. Elderberry

Leaflets mostly 5-9, oblong-lanceolate, 6-15cm long, gradually short-acuminate, sharply and often rather deeply serrate; inflorescence 0.5-2dm across, montane . . . . *S. caerulea*

Leaflets mostly 3-5, roundish to ovate or oblong-lanceolate, rather abruptly acuminate, mostly 1.5-6cm long, finely serrate; inflorescence mostly 0.3-1dm across . . . . *S. mexicana*

S. caerulea Raf. Blue Elderberry. Frequent, in open areas; montane and foothill, above 900m; e.g. Palomar Mountain; Lost Valley; Warner Springs; Vallecitos Canyon. June-Sep.

S. mexicana Presl ex DC. [*S. velutina* Dur. & Hilg.] Desert Elderberry. Common, in open areas; coastal and foothill, below 1000m; e.g. Castro Canyon; San Luis Rey; San Marcos; Descanso; La Jolla; Lakeside; Jacumba; Campo; Otay Mesa. Mar.-Sep. $2n$=ca. 36.

AIZOACEAE. Carpet-weed Family

KEY TO GENERA

Petals present, numerous; ovary inferior
  Leaves largely alternate . . . . *Mesembryanthemum*
  Leaves largely opposite; perennials; herbage often smoothish
    Leaf-blades flat, expanded; flowers red, 1cm wide . . . . *Aptenia*
    Leaf-blades semiterete or 3-sided, linear
      Leaves 3-angled; stigmas and ovary-locules 8-20 . . . . *Carpobrotus*
      Leaves mostly semiterete; stigmas and ovary-locules 4-6
        Flowers pale pink-purple, less than 2cm in diameter . . . . *Drosanthemum*
        Flowers more intensely colored, larger
          The flowers orange-red, to 5cm diameter; leaves glaucous . . . . *Malephora*
          The flowers intense red, to 4cm diameter . . . . *Lampranthus*
Petals none; calyx sometimes colored within and petaloid
  Leaves alternate; flower-tube united to ovary; leaves triangular-ovate . . . . *Tetragonia*
  Leaves opposite or whorled; flower-tube free from ovary
    The leaves whorled; flower-tube nearly or quite wanting; capsule loculicidal
      Plants glabrous; flowers with a filiform pedicel; sepals distince to base . . . . *Mollugo*
      Plants tomentulose; flowers essentially sessile; calyx cleft only to middle . . . . *Glinus*
    The leaves opposite
      Leaves not stipulate; ovary three- to five-locular . . . . *Sesuvium*
      Leaves stipulate; ovary one- to two-locular . . . . *Cypselea*

Aptenia N.E. Br.

***A. cordifolia** (L.f.)Schwant. [*Mesembryanthemum c.* L.f.] Shrubby Dewplant. Native of South Africa. Local, escaping from cultivation; coastal, below 200m: La Jolla; San Diego; National City; Point Loma. Apr.-May. $2n$=18.

Carpobrotus N.E.Br.

Leaves straight, not serrate, 3-5cm long; flowers rose-magenta, 3-5cm broad . . . . *C. aequilaterus*
Leaves curved, serrate on lower angle, becoming 7-10cm long; flowers yellow or purple, 8-10cm broad . . . . . *C. edulis*

C. aequilaterus (Haw.)N.E.Br. [*Mesembryanthemum chilense* Molina] Sea-Fig. Common, on sandy beaches and banks; coastal, below 200m: e.g. Solana Beach; Del Mar; Silver Strand; Boundary Monument. Often planted for erosion control. Apr.-Sep.

***C. edulis** (L.)Bolus. [*Mesembryanthemum e.* L.] Hottentot-Fig. Native of South Africa. Common, naturalized on sandy beaches and disturbed areas; coastal, below 200m: e.g. Oceanside; Carlsbad; Del Mar. Often planted for erosion control. Feb.-Oct.

Cypselea Turp.

***C. humifusa** Turp. Native of Caribbean area. Rare, collected at Corte Madera Lake. July.

Drosanthemum Schwant.

***D. hispidum** (L.)Schwant. Rosea Iceplant. Native of South Africa. Infrequent, naturalized on banks and dumps; coastal, below 150m: Del Mar; San Diego; National City. Dec.-Feb.

Glinus L.

Seeds reddish-brown. . . . . *G. radiatus*
Seeds blackish-brown . . . . *G. lotoides*

***G. lotoides** L. [*Tidestromia lanuginosa* Calif. auct.] Native of Europe. Stream beds and pond shores, coastal below 300m: Highway 163—Carrol Canyon; Lower Otay Lake. June-Nov.

***G. radiatus** (R. & P.)Rohrb. Native of eastern North America. Local on the shores of ponds and reservoirs, coastal, below 500m: Sweetwater Lake; Rickey Dam; Upper and Lower Otay Lakes. July-Oct.

Lampranthus N.E.Br.

***L. coccineus** (Haw.)N.E.Br. [*Mesembryanthemum c.* Haw.] Native of South Africa. Local, naturalized on banks and dumps; coastal, below 150m: San Diego. Nov.-Mar.

Malephora N.E. Br.

***M. crocea** (Jacq.)Schwant. var. **purpureo-crocea** (Haw.)Jacobs & Schwant. [*Hymenocyclus p.* (Haw.)L. Bolus] Croceum Ice Plant. Native of South Africa. Local, naturalized on dumps and in disturbed areas, below 150m: Encinitas; Del Mar; Lomas Santa Fe; San Diego; National City. Nov.-Sep.

Mesembryanthemum L. Ice Plant

Leaves ovate to broadly spatulate, flat . . . . *M. crystallinum*
Leaves linear, semi-terete . . . . *M. nodiflorum*

***M. crystallinum** L. [*Cryophytum c.* (L.)N.E.Br.; *Gasoul c.* (l.)Rothm.] Crystal Ice Plant. Native of South Africa. Frequent, naturalized in disturbed areas; coastal, below 150m: e.g. Del Mar; Point Loma; Silver Strand; National City; Otay Mesa; Boundary Monument. Mar.-Oct. $2n=18$.

***M. nodiflorum** L. [*Cryophytum n.* (L.)L.Bolus; *Gasoul n.* (L.)Rothm.] Little Ice Plant. Native of South Africa. Frequent, naturalized in disturbed areas; coastal, below 150m: e.g. Del Mar; Mission Beach; Silver Strand; Otay Mesa; Boundary Monument. Apr.-Nov. $n=18$.

Mollugo L.

Leaves linear; plant erect or ascending . . . . *M. cerviana*
Leaves spatulate; plant prostrate . . . . *M. verticillata*

***M. cerviana** (L.)Ser. Native of Old World. Common, in sandy areas; after summer rains; desert, below 1000m: Scissors Crossing; Blair Valley; Vallecitos; Mountain Palm Spring; Carrizo Badland Overlook. Sept.

***M. verticillata** L. India Carpetweed. Native of Neotropics. Local, naturalized about Sweetwater Reservoir, 75m. May-Nov. $2n=64$.

Sesuvium L. Sea-Purslane

S. verrucosum Raf. Western Sea-Purslane. Occasional, in disturbed and alkaline places; cismontane and desert, below 300m: e.g. Oceanside; Carlsbad; San Diego; Lindo Lake; Otay Mesa; Sweetwater Reservoir; Rickey Dam; Carrizo Stage Station. Apr.-Nov.

Tetragonia L.

***T. tetragonioides** (Pallas)Kuntze. [*T. expansa* Murr.] New Zealand Spinach. Native of Australia. Infrequent, naturalized in alkaline and disturbed places; coastal, below 150m: Oceanside; Carlsbad; Solana Beach; San Elijo Lagoon; Imperial Beach. Apr.-Sep. $2n=32$.

AMARANTHACEAE. Amaranth Family

KEY TO GENERA

Leaves alternate; anthers 4-celled; plants nearly or quite glabrous . . . . *Amaranthus*
Leaves largely opposite; anthers 2-celled; plants white stellate-woolly or villous
  Flowers glomerate, with an involucre of upper leaves. . . . . *Tidestromia*
  Flowers in axillary headlike spikes, without involucre . . . . *Alternanthera*

Alternanthera Forsk.

***A. peploides** (H. & B.)Urban. [*A. repens* auct. non L.; *Achyranthes r.* auct. non L.] Mexican Alternanthera. Native of Mexico. Local, escaping from cultivation; coastal, below 200m: Oceanside. Aug.-Dec.

Amaranthus L. Pigweed

Sepals of pistillate flowers lacking; flowers dioecious; plants erect, stout, to 1.5m tall . . . . *A. rudis*
Sepals present on staminate flowers and perfect flowers
  Sepals of pistillate flowers spatulate, the claw narrow, the calyx somewhat urceolate
    Plants dioecious . . . . *A. palmeri*
    Plants monoecious . . . . *A. fimbriatus*
  Sepals of pistillate flowers oblong or obovate, not with a narrow claw, the calyx not urceolate; plants monoecious
    Utricle indehiscent, fleshy; spikes terminal; prostrate plant with slender stems . . . . *A. deflexus*
    Utricle dehiscent, the top falling away like a lid
      Inflorescence of terminal or axillary spikes; sepals 5
        Style-branches recurved; lateral branches of inflorescence few to none . . . . *A. powellii*
        Style-branches erect, lateral spikes of inflorescence numerous, crowded
          Sepals oblong, acute, straight; midrib of bract long, excurrent; style branches with slender bases *A. hybridus*
          Sepals narrowly obovate, emarginate, recurved; midrib of bracts barely excurrent; style-branches with moderately stout bases . . . . *A. retroflexus*
      Inflorescence wholly of axillary glomerules
        Sepals 4-5 in both staminate and pistillate flowers; stems prostrate; seeds 1.5mm broad . . . . *A. blitoides*
          Sepals 1-3; seeds less than 1mm broad
            Sepals of pistillate flowers 2-3, but only 1 well developed; stems slender, prostrate . . . . *A. californicus*
            Sepals of pistillate flowers 3, subequal; stems ascending . . . . *A. albus*

***A. albus** L. [*A. graecizans* auct. non L.] White Tumbleweed. Native of Neotropics. Frequent, weed in disturbed and cultivated places; cismontane: e.g. Oak Grove; Doane Valley; Encinitas; Murphy Canyon; Balboa Park; Poway; Bear Valley; Ballena; Point Loma; Boulevard; Glenn Cliff Campground; Jamul. June-Oct. $2n$=32.

A. blitoides Wats. [*A. graecizans* auct. non L.] Prostrate Amaranth. Occasional, in sandy areas; cismontane, below 700m: e.g. Oak Grove; Ponto; Lake Hodges; San Felipe Wash; Santa Ysabel; Ramona; Boulevard; Glenview; Cottonwood; Jamacha; Spring Valley. July-Nov. $2n$=32.

A. californicus (Moq.)Wats. California Amaranth. Occasional, in moist places; montane, above 800m: Doane Valley; Lake Henshaw; Pine Valley; Little Laguna Lake. July-Oct.

***A. deflexus** L. Low Amaranth. Native of Old World. Occasional, weed in disturbed places; coastal, below 300m: Escondido; San Diego; Balboa Park; Point Loma. May-Nov. $2n$=34.

A. fimbriatus (Torr.)Benth. ex Wats. Fringe Amaranth. Infrequent, in sandy areas; desert, below 950m: Rockhouse Canyon; San Felipe Valley; Banner; Blair Valley; Jacumba. Aug.-Nov.

***A. hybridus** L. Slender Pigweed. Native of Neotropics. Infrequent, in disturbed or cultivated places; cismontane: Vista; San Diego. June-Nov. $2n$=32.

A. palmeri Wats. Palmer's Amaranth. Occasional, in waste places and fields; cismontane: e.g. Warner Springs; Moretti Junction; Jacumba; San Luis Rey School; San Felipe Valley; Lakeside. Aug.-Nov. Type from Larkin's Station (Jacumba). $2n$=34, 45.

A. powellii Wats. Powell's Amaranth. Infrequent, in waste places; cismontane: Ballena; San Diego. July-Oct. $2n$=34.

***A. retroflexus** L. Rough Pigweed. Native of Neotropics. Frequent, in disturbed areas; cismontane: e.g. Oak Grove; Hormiguero Ranch; San Luis Rey; Witch Creek; Julian; White Star; Foster; Chollas Valley. June-Nov. $2n$=ca. 32.

***A. rudis** Sauer. [*A. tamariscinus* Nutt.] Western Water-Hemp. Native of central United States. Local, introduced at Pala, 140m. July-Sep.

Tidestromia Standl.

T. oblongifolia (Wats.)Standl. ssp. **cryptantha** (Wats.)Wiggins. Salton Sea Honeysweet. Infrequent, on sandy mesas and slopes; desert, below 1000m: Mountain Palm Springs; near Arroyo Salado; Palm Spring; Carrizo Stage Station. Apr.-Dec.

ANACARDIACEAE. Sumac Family

KEY TO GENERA

Stamens 5; native shrubs, mostly with entire or 3-foliolate leaves
  Leaflets 3- (rarely 5-) foliolate
    Flowers whitish, in loose axillary panicles; fruit glabrous; whitish; leaves subglabrous, shining; branches stiff, erect or climbing . . . . *Toxicodendron*
    Flowers yellowish in sessile spikes; fruit hairy, red; leaves usually pubescent; branches tending to be arched *Rhus*
  Leaflets simple, persistent, coriaceous . . . . *Malosma*
Stamens 10; introduced trees mostly with pendent pinnate leaves . . . . *Schinus*

Malosma (Nutt.)Raf.

M. laurina (Nutt. in T. & G.)Nutt. ex Abrams. [*Rhus l.* Nutt. in T. & G.] Laurel Sumac. Common, on slopes and in canyons; cismontane, below 1000m: e.g. San Mateo Creek; De Luz; Rainbow; Escondido; Mount Woodson; Santa Maria Valley; Tule Springs; Olivenhain; Descanso; Barrett Junction; Potrero Valley; Lyons Peak; Dulzura; Imperial Beach; San Diego. May-July.

Rhus L.

Leaves 3- (rarely 5-) foliolate, deciduous
New branchlets densely pubescent; leaves tomentose at least beneath . . . . *R. trilobata* var. *pilosissima*
New branchlets and leaves mostly glabrate, at least not densely pubescent
Terminal leaflet not deeply lobed; leaflets mostly glabrate . . . . *R. trilobata* var. *ansiophylla*
Terminal leaflet deeply lobed; leaflets mostly somehwhat pubescent . . . . *R. trilobata* var. *quinata*
Leaves simple, persistent, coriaceous
Leaflets oblong-elliptic . . . . *R. integrifolia*
Leaflets ovate, somewhat folded along midrib . . . . *R. ovata*

R. integrifolia (Nutt. in T. & G.)Brew. & Wats. [*Styphonia serrata* Nutt. in T. & G., type from "the margins of cliffs, &c. near the sea, around St. Diego & St. Barbara."] Lemonade-Berry. Common, on slopes and in canyons; coastal, below 300m: e.g. San Onofre; Ysidora; Oceanside; Live Oak Park; Olivenhain; La Jolla; Mount Soledad; Spring Valley; Imperial Beach; Monument Beach; San Diego; Point Loma. Feb.-May. Type from "the margins of cliffs, &c. near the sea, around St. Diego & St. Barbara; common."

R. ovata Wats. Sugar Bush. Common, on open slopes; foothill, 150-1600m: e.g. Rainbow; Pala; Oak Grove; upper Borrego Palm Canyon; Banner; Barona Valley; Descanso Junction; Buckman Spring; Jaucmba; Campo; Potrero Valley; Lyons Peak; Otay Mountain; Poway; Del Dios. Mar.-May. Type from "San Diego Mtns.".

R. trilobata Nutt. ex T. & G. var. **ansiophylla** (Greene)Jeps. Desert Basketbush. Local, in open areas of Chaparral; foothill, desert and montane, above 950m: Oak Grove; Hot Springs Mountain; Pinyon Mountain. Apr.-May.

R. trilobata var. **pilosissima** Engler in DC. [*R. t.* var. *malacophylla* (Greene)Munz] Pubescent Basketbush. Occasional, in open areas and on Chaparral slopes; foothill and montane, above 150m: e.g. De Luz; Fallbrook; Rainbow; Fry Creek; Oak Grove; Eagles Nest; Witch Creek; Garnet Peak; Laguna Junction; Pine Valley; Jacumba; Potrero; Dulzura; Barona Valley; Bear Valley. Mar.-May.

R. trilobata var. **quinata** (Greene)Jeps. [*Schmaltzia cruciata* Greene, the type from "Hot Springs in the northern part of San Diego Co". (Warner Hot Springs.)] Slender Basketbush. Infrequent, in Chaparral; foothill, 300-1000m: Oak Grove; Warner Springs; Banner; Alpine, Campo; Jamul. Mar.-May.

Schinus L. Pepper-Tree

Branchlets pendent; leaflets to 0.5cm wide . . . . *S. molle*
Branchlets stout; leaflets to 1.5cm wide. . . . . *S. terebinthifolius*

***S. molle** L. Peruvian Pepper-Tree. Native of South America. Infrequent, persisting around old dwellings and escaping from cultivation; coastal, below 300m: Torrey Pines State Reserve; Dehesa; Telegraph Canyon; San Diego. Mar.-June. $2n$=28.

***S. terebinthifolius** Raddi. Brazilian Pepper-Tree. Native of Brazil. Local, escaping from cultivation and naturalized in canyon bottoms; coastal, below 200m: Mission Valley; Chollas Valley; Sweetwater Valley. Mar.-June.

Toxicodendron Tourn.

T. radicans (L.)Kuntze ssp. **diversilobum** (T.& G.)Thorne. [*Rhus d.* T. & G.; *T. comarophyllum* Greene, the type from "Tighe's, near San Diego".] Poison-Oak. Prevalent, in canyons and shaded slopes; cismontane: e.g. San Mateo Creek; Palomar Mountain; Escondido; Valley Center; Oak Grove; Grapevine Canyon; Banner; Cuyamaca Peak; Campo; Alpine; Ramona; Jamul; Balboa Park; Rose Canyon; Kearny Mesa; Florida Canyon; Chollas Valley; Storm Canyon; Laguna Mountains. Mar.-Apr. $2n$=15 pairs.

APIACEAE. Carrot Family
(Umbelliferae)

KEY TO GENERA

Inflorescence capitate, not umbellate . . . . *Eryngium*
Inflorescence a distinct umbel, not capitate
Ovary and fruit bearing prickles, bristles, tubercles or scales
Leaves simple, entire, usually the basal with parallel veining and the upper broad and perfoliate . . . . . *Bupleurum*
Leaves compound
Ovary and fruit armed with spines, hooked bristles, or tubercles
Plants perennial or biennial; flowers perfect and staminate . . . . *Sanicula*
Plants annual; flowers all perfect
Plants glabrous; leaf-divisions elongate, filiform . . . . *Apiastrum*
Plants mostly pubescent; leaf-divisions shorter
Involucre of conspicuous foliaceous bracts; leaves 3-4-pinnatisect; fruit bristly on ribs only . . . . . *Yabea*
Involucre none or of linear bracts; leaves pinnate to 3-pinnatisect; fruit bristly or tubercled throughout . . . . *Torilis*
Ovary and fruit armed with bristles that are never hooked
Fruit several times longer than broad; oil-tubes absent or obscure
Plants annual; the fruit with a beak longer than the body . . . . *Scandix*
Plants perennial; fruit not beaked or if so, the beak shorter than the body. . . . . *Osmorhiza*
Fruit not more than twice as long as broad; oil-tubes evident
Foliage glabrous; fruit armed with unequal subulate bristles. . . . . *Ammoselinum*
Foliage pubescent; fruit armed with barbed bristles . . . . *Daucus*

Ovary and fruit not bearing prickles, bristles, tubercles or scales
Leaves simple; umbels simple or proliferous
Ovary and fruit glabrous; leaves glabrous; plants aquatic . . . . *Hydrocotyle*
Ovary and fruit covered with stellate hairs; leaves stellate-pubescent; plants of dry land . . . . *Bowlesia*
Leaves and umbels variously compound
Ribs of the fruit not prominently winged; fruit terete in cross-section or flattened somewhat laterally
Flowers yellow . . . . *Foeniculum*
Flowers not yellow
Fruit elongate, several times longer than wide
Fruit without ridges, except in the usually well-developed beak . . . . *Anthriscus*
Fruit with ridges . . . . *Osmorhiza*
Fruit roundish to oblong, not more than twice as long as broad
Plants annual, or with celery odor and taste if perennial
Seed globose; mericarps not separating at maturity . . . . *Coriandrum*
Seed ovoid; mericarps separating at maturity . . . . *Apium*
Plants perennial or biennial, not with celery odor and taste
Plants caulescent, mostly rather tall; involucre usually present
Bracts 3-parted to the middle into filiform divisions, closely reflexed . . . . *Ammi*
Bracts entire or toothed, mostly spreading, sometimes none
Oil-tubes none or obscure; leaves decompound into small divisions
Fruit smooth; large herb to 2m . . . . *Conium*
Fruit tuberculate or bristly; herb to 3dm . . . . *Spermolepis*
Oil tubes present; leaves pinnately or ternate-pinnately divided into mostly larger divisions
Leaves all once-pinnate . . . . *Berula*
Leaves pinnately or ternate-pinnately divided or the upper once-pinnate
Stylopodium prominent; ribs not corky; roots tuberous or fusiform . . . . *Perideridia*
Stylopodium none or low; ribs corky
Styles 1/5 - 1/3 as long as the fruit; fruit broadly ovoid to roundish . . . . *Cicuta*
Styles ca. 1/2 as long as the fruit; fruit subcylindric . . . . *Oenanthe*
Plants acaulescent, low; involucre absent . . . . *Tauschia*
Ribs of the fruit (all or some of them) winged; fruit dorsally compressed
Lateral ribs winged; dorsal ribs filiform
Leaves simply pinnate into ovate divisions . . . . *Pastinaca*
Leaves pinnately or ternate-pinnately divided into mostly linear or filiform divisions
Plants annual, with leafy stems; leaf-divisions filiform; plants with anise odor . . . . *Anethum*
Plants perennial, acaulescent or short-stemmed; leaf-divisions often wider; no anise odor . . . *Lomatium*
Lateral, dorsal and intermediate ribs winged or prominent . . . . *Angelica*

Ammi L. Bishop's Weed

***A. majus** L. Larger Bishop's Weed. Native of Eurasia. Infrequent, weed in fields and waste places; coastal, below 300m: San Marcos; San Diego. June. n=11.

Ammoselinum T. & G.

A. giganteum Coult. & Rose. Western Sand-Parsley. Rare, known in the county only from a collection from near Borrego Springs, 150m. Mar.-Apr.

Anethum L. Dill

***A. graveolens** L. Common Dill. Native of Europe. Infrequent, escaping from cultivation; coastal, below 400m: Lake Hodges; Silver Strand; Harbison Canyon. June-Aug. $2n$=22.

Angelica L.

A. tomentosa Wats. California Angelica. Infrequent, in shaded places of Montane Coniferous Forest; montane, above 1350m: Palomar Mountain; Middle Peak. July-Aug. n=11.

Anthriscus Hoffm.

***A. scandicina** Mansfeld. [*A. neglecta* Boiss. & Reuter var. *scandix* (Scop.)Hylander; *A. caucalis* Bieb.]. Bur Chevil. Native of Europe. Rare adventive in McGonigle Canyon, Del Mar. Apr.-June.

Apiastrum Nutt. in T. & G.

A. angustifolium Nutt. in T. & G. [*A. a.* var. *tenellum* Nutt. in T. & G., the type from St. Diego; *A. latifolium* Nutt. in T. & G., the type from St. Diego.] Mock Parsley. Common, in sandy valleys and on slopes; cismontane, below 1100m: e.g. Fallbrook; Red Mountain Grade; Pala; Lake Henshaw; Oak Grove; Borrego Palm Canyon; Banner; Sentenac Canyon; Mount Woodson; Lake Hodges; La Jolla; Murphy Canyon; Mission Hills; San Diego; Silver Strand; Otay Mountain; Tecate Peak. Mar.-May. Type from "St. Diego". n=11.

Apium L.

Plants perennial; leaf-divisions ovate to suborbicular or cuneate . . . . *A. graveolens*
Plants annual; leaf-divisions linear to filiform . . . . *A. leptophyllum*

***A. graveolens** L. Common Celery. Native of Europe. Frequent, in wet places; coastal, foothill, and desert, below 1100m: e.g. Lone Palm Spring; Santa Margarita River ford; Encinitas; Lake Henshaw; Ramona; Sentenac Canyon; Vallecito Stage Station; Jacumba; Clover Flat; Mission Valley; Point Loma; Otay Ranch. May-July. $2n$=22.

***A. leptophyllum** (Pers.)F. Muell. ex Benth. & Muell. Marsh Parsley. Native of Tropics. Infrequent, weed in disturbed and cultivated areas; coastal, below 200m: San Diego; Mission Hills. Apr.-Aug. n=7.

Berula Hoffm. Water-Parsnip

B. erecta (Huds.)Cov. Cut-Leaf Water-Parsnip. Infrequent, in wet places; cismontane, below 1200m: Pala; Palomar Mountain; Lake Henshaw; Ramona; Clover Flat; Alvarado Canyon. July-Oct. 2*n*=18.

Bowlesia R. & P.

B. incana R. & P. American Bowlesia. Frequent, on shaded slopes and burns; cismontane, below 1300m: e.g. San Onofre Mountain; Moosa Canyon; San Luis Rey; Vista; Clevenger Canyon; Ramona; Julian; Mission Gorge; Mahogany Canyon; Sloane Ranch; Jacumba. Mar.-Apr. *n*=16.

Bupleurum L.

***B. lancifolium** Hornem. [*B. subovatum* Link, pro parte.] Lance-leaf Hare's Ear. Native of Europe. Infrequent, weed in cultivated areas; coastal, below 200m: La Jolla. May-June. *n*=14.

Cicuta L. Water-Hemlock

C. douglasii (DC.)Coult. & Rose. Douglas' Water-Hemlock. Infrequent, in wet areas; foothill, montane and desert, above 600m: Palomar Mountain; Lake Henshaw; Cuyamaca Lake; Canebrake Canyon; Potrero. May-July. *n*=11, 22.

Conium L. Poison-Hemlock

***C. maculatum** L. Common Poison-Hemlock. Native of Europe. Frequent, along stream banks and moist areas; coastal, below 120m: Bonsall; San Luis Rey; Olivenhain; San Clemente Canyon; Lake Hodges; Old Town. Apr.-July. 2*n*=22.

Coriandrum L.

***C. sativum** L. European Coriander. Native of Europe. Infrequent, escaping from cultivation: San Diego. May-July. *n*=11.

Daucus L.

Bracts pinnately divided into short linear or lanceolate divisions; rays usually 0.4-4cm long; carpel usually broadest below the middle; central flower of the umbellet white . . . . . *D. pusillus*
Bracts pinnately divided into elongate filiform divisions; rays 3-7cm long; carpel broadest at middle; central flower of the umbellet usually rose or purple . . . . . *D. carota*

***D. carota** L. Queen Anne's Lace, Wild Carrot. Native of Europe. Infrequent, escaping from cultivation in moist areas: Pine Hills; Cuyamaca Lake. Apr.-July. 2*n*=18.

D. pusillus Michx. Rattlesnake Weed. Frequent, on dry slopes and burns; cismontane, below 700m: e.g. San Onofre Bluffs; De Luz; Fallbrook; south of Temecula; Eagles Nest; Mount Woodson; San Pasqual; Barona Valley; Miramar Naval Air Station; Mission Hills; La Jolla; Point Loma; Otay Ranch; Otay Mountain; Dulzura; Alpine; Campo. Also Vallecito Stage Station. Apr.-June.

Eryngium L.

E. aristulatum Jeps. ssp. **parishii** (Coult. & Rose)Beauchamp. [*E. a.* var. *p.* (Coult.& Rose)Math. & Const.; *E. p.* Coult. & Rose] San Diego Coyote-Thistle. Occasional, in vernal pools; coastal, 20-150m: Camp Pendleton; San Marcos; Miramar Naval Air Station; Clairemont Mesa; College Park; East San Diego; Otay Mesa. Mar.-July. Type from Oceanside.

Foeniculum Adans.

***F. vulgare** Mill. Sweet Fennel. Native of Europe. Abundant, in disturbed areas; coastal, below 400m: e.g. San Onofre Bluffs; Oceanside; Fallbrook; Encinitas; La Jolla; Bostonia; Spring Valley; El Cajon; Dulzura. May-Sep. 2*n*=22.

Hydrocotyle L. Marsh Pennywort

Leaves peltate
  Inflorescence a simple umbel . . . . . *H. umbellata*
  Inflorescence an interrupted spike . . . . . *H. verticillata*
Leaves not peltate . . . . . *H. ranunculoides*

H. ranunculoides L. f. Floating Marsh-Pennywort. Occasional, in streams, ponds, and seepage areas; foothill and montane, above 300m: Lake Henshaw; Warner Ranch; Eagles Nest; Lake Hodges; Ramona; Pine Valley; Little Laguna Lake; Glen Cliff Campground; White Star. Also Borrego; Collins Valley. Mar.-Aug. *n*=24.

H. umbellata L. Many-Flower Marsh-Pennywort. Rare, known in the county only at Santa Margarita River ford, 240m. Mar.-July. 2*n*=48.

H. verticillata Thunb. [*H. cuneata* Coult. & Rose] Whorled Marsh-Pennywort. Occasional, in ponds and streams; coastal, below 300m: Pala; Lake Hodges; Lake Murray; Jamul. Apr.-Sep. 2*n*=88 +/− 4.

Lomatium Raf.

Fruit notched at each end, the wings distinct on each side of the body; leaflets cuneate to obovate . . . . . *L. lucidum*
Fruit not or scarcely notched; leaves mostly dissected into numerous narrow segments
  Bractlets broad, rounded to obovate; flowers yellow
    Fruit ovate to obovate, 9-15mm. long; sepals prominent in young fruits . . . . . *L. vaseyi*
    Fruit oblong to ovate, 5-11mm. long; sepals obsolete . . . . . *L. utriculatum*

Bractlets narrow, linear to lanceolate; flowers white, purple or yellow
Plants short-caulescent; petals densely pubescent over entire upper surface; cismontane. . . . . *L. dasycarpum*
Plants acaulescent; petals glabrous or sometives pubescent mostly along the margins; largely desert *L. mohavense*

L. dasycarpum (T. & G.)Coult. & Rose. [*Peucedanum pringlei* Coult. & Rose, the type from San Diego County.] Woolly-Fruit Lomatium. Frequent, on rocky slopes; cismontane: e.g. De Luz; Moro Hill; Red Mountain Grade; Henshaw Dam; Love Valley; Warner Ranch; Banner; San Felipe; Inspiration Point; Julian; Cuyamaca Lake; Stonewall Mine; Laguna Public Camp; Jacumba; Alpine; Canyon City; Lyons Valley; Mission Valley; Emerald Hills; Miramar Naval Air Station; Mount Soledad; San Diego. Apr.-June. n=11.

L. lucidum (Nutt.)Jeps. Shiny Lomatium. Occasional, on rocky slopes; cismontane, below 1400m: e.g. Encinitas; Solana Beach; Torrey Pines State Reserve; Palomar Mountain; Oriflamme Canyon; Harper Ranch; Jacumba; Point Loma; San Miguel Mountain; Otay Mountain. Jan.-May. Type from "woods of St. Diego". n=11.

L. mohavense (Coult. & Rose)Coult. & Rose ssp. **longilobum** Theobald. Infrequent, on rocky slopes; desert, 600-1100m: San Felipe Hill; Blair Valley; Kentwood-in-the-Pines; Jacumba; southeast corner of county. Mar.-Apr. Type from 2.6 miles west of Jacumba School, along U.S. Hwy. 80. n=11.

L. utriculatum (Nutt.)Coult. & Rose. Common Lomatium. Occasional, on grassy slopes; southwest cismontane, below 1100m: e.g. Descanso; Fletcher Hills; Adobe Falls; Paradise Hills; Paradise Valley; San Diego. Feb.-May. n=11.

L. vaseyi (Coult. & Rose)Coult. & Rose. Vasey's Lomatium. Rare, known in the county only from a collection from El Cajon, 140m. Mar.-Apr.

Oenanthe L.

O. sarmentosa Presl in DC. Pacific Oenanthe. Occasional, in marshes and seepage areas; foothill and montane, above 550m: e.g. Palomar Mountain; Lake Henshaw; Hot Springs Mountain; San Felipe; upper Borrego Palm Canyon; Cuyamaca Mountains; Cuyamaca Lake; Green Valley; Dehr Creek; Ballena; Viejas Valley; Laguna Mountains; Clover Flat; Campo. Jan.-Oct. n=22.

Osmorhiza Raf. Sweet-Cicely

Involucels of several bractlets, conspicuous . . . . *O. brachypoda*
Involucels lacking or rudimentary . . . . *O. chilensis*

O. brachypoda Torr. ex Durand. California Sweet-Cicely. Infrequent, in shaded areas, Montane Coniferous Forest; montane, above 900m: Doane Valley; Fry Creek; Eagles Nest; Mesa Grande; Witch Creek; Clevenger Canyon; Descanso Junction; Cuyamaca Lake; Garnet Peak; Mount Laguna. Apr.-June. n=11.

O. chilensis H. & A. [*O. nuda* Torr.] Mountain Sweet-Cicely. Infrequent, in shaded areas, Montane Coniferous Forest; montane, above 1350m: Palomar Mountain; Doane Valley; Eagles Nest; Pine Hills; Cuyamaca Lake; Stonewall Mine; Laguna Camp. May-July. n=11.

Pastinaca L.

***P. sativa** L. Common Parsnip. Native of Europe. Infrequent, escaping from cultivation: Julian. July. 2n=22.

Perideridia Rchb.

Bractlets usually setaceous; fruit ca. as broad as long. . . . . *P. gairdneri*
Bractlets scarious or with scarious margins; fruit longer than broad . . . . *P. parishii*

P. gairdneri (H. & A.)Mathias ssp. **gairdneri**. Gairdner's Yumpah. Rare, known in the county only from collections from Ramona and Sweetwater. June-July. n=17.

P. parishii (Coult. & Rose)Nels. & Macbr. ssp. **latifolia** (Gray)Chuang & Const. Southern Yumpah. Scarce, in wet areas, Montane Coniferous Forest; montane, above 1350m: Cuyamaca Lake; Laguna Meadow. July-Sep. n=19.

Sanicula L. Snakeroot

Basal leaves leaves ternately or palmately divided, rarely entire
Fruits prickly on upper part only, pedicellate or stipitate . . . . *S. arguta*
Fruits prickly to the base . . . . *S. crassicaulis*
Basal leaves pinnatifid or pinnately or ternate-pinnately dissected or decompound
Leaves once or twice pinnatifid with a toothed rachis . . . . *S. bipinnatifida*
Leaves 1- to 3-pinnate or pinnately or ternate-pinnately decompound, without a margined or toothed rachis
Stems from fusiform taproots . . . . *S. graveolens*
Stems from globose or irregular tuberous root . . . . *S. tuberosa*

S. arguta Greene ex Coult. & Rose. Sharp-Tooth Sanicle. Common, on grassy slopes; cismontane, below 1500m: e.g. Fallbrook; Lilac; Buena; Cardiff; Escondido; Julian; Anahuac School; Ramona; Murray Dam; Adobe Falls; Lee Valley; El Cajon; Woodwardia Canyon; Emerald Hills; National City; Mount Soledad; Miramar. Mar.-Apr. Type from "hills near San Diego". n=8.

S. bipinnatifida Dougl. ex Hook. Purple Sanicle. Occasional, on grassy slopes; cismontane, below 1500m: e.g. Palomar Mountain; Mesa Grande; Henshaw Dam; Pine Hills; Julian; Cuyamaca Lake; Azalea Creek; Harper Ranch; El Cajon Mountain; Sloane Ranch; Fletcher Hills; Poway Junction; Lyons Valley; El Cajon; Grossmont; National City. Mar.-May. n=8.

S. crassicaulis Poepp. ex DC. var. **crassicaulis**. Pacific Sanicle. Common, on grassy slopes; foothill and coastal, below 1200m: e.g. San Onofre; De Luz; Fallbrook; La Costa; Palomar Mountain; El Cajon Mountain; Ramona; Glen Lonely; Alpine; Descanso; Tecate Peak; Point Loma. Mar.-May. n=8, 12, 16.

S. graveolens Poepp. ex DC. [*S. nevadensis* Wats.] Sierra Sanicle. Uncommon, in Chaparral and Montane Coniferous Forest; montane, above 1300m: Garnet Peak. Apr. n=8.

S. tuberosa Torr. Turkey-Pea Sanicle. Frequent, on grassy slopes; cismontane, below 1600m: e.g. Fallbrook; Moosa Canyon; Pine Hills; Julian; Cuyamaca Lake; Inspiration Point; El Cajon Mountain; Alpine; Foster; Mission Gorge; Santee; San Miguel Mountain; Lyons Peak; Otay Mountain. Mar.-June. n=8.

Scandix L.

***S. pecten-veneris** L. Weedy Shepherd's Needle. Native of Eurasia. Infrequent, weed in waste places; coastal, below 150m: San Diego; Ocean Beach; Mission Valley; Apr.-June. n=14; 2n=16.

Spermolepis Rafin.

S. echinata (Nutt.)Heller. Uncommon, in sandy areas; desert, below 1000m: south of Vallecitos Stage Station; lower Box Canyon. Apr.

Tauschia Schlecht.

Leaves simply pinnate; fruit with conspicuous ribs . . . . *T. arguta*
Leaves ternate-pinnate or bipinnate . . . . *T. parishii*

T. arguta (T. & G.)Macbr. [*Velaea a.* var. *ternata* Coult. & Rose, the type from Cuyamaca Mountains.] Southern Tauschia. Frequent, in Chaparral; cismontane, below 1400m: e.g. Rainbow; Moosa Canyon; Oak Grove; Eagles Nest; Montezuma View Point; Culp Valley; Banner; Julian; Kentwood-in-the-Pines; Pine Hills; Inaja Memorial; Middle Peak; Vallecito Canyon; Campo; Mountain Springs; Lyons Peak; Sloane Ranch; Ramona; El Cajon Mountain; Encinitas. Apr.-June. Type from "woods of St. Diego". n=11.

T. parishii (Coult. & Rose)Macbr. Parish's Tauschia. Uncommon, on dry slopes; montane, above 1350m: Palomar and Laguna Mountains. May. n=11.

Torilis Adams. Hedge-Parsley

Umbels mostly lateral; leaf-opposed peduncles to 5cm long . . . . *T. nodosa*
Umbels mostly terminal on peduncles usually more than 5cm long . . . . *T. japonica*

***T. japonica** (Houtt.)DC. Japanese Hedge-Parsley. Native of Japan. Local introduction at Dyche Valley, Palomar Mountain.

***T. nodosa** (L.)Gaertn. Knot Hedge-Parsley. Native of Europe. Uncommon, weed in moist places; coastal, below 700m: upper San Onofre Canyon. Apr.-June. 2n=22. n=12.

Yabea Koso-Pol.

Y. microcarpa (H. & A.)Koso-Pol. [*Caucalis m.* H.& A.] California Hedge-Parsley. Occasional, on open, shaded slopes; cismontane: e.g. Moro Hill; Fallbrook; Doane valley; Eagles Nest; Clevenger Canyon; Cuyamaca Lake; Middle Peak; Laguna Camp; Laguna Lakes; Garnet Peak; Jacumba; Otay Ranch; San Diego; Miramar Naval Air Station. Apr.-June. n=6, 11.

APOCYNACEAE. Dogbane Family

KEY TO GENERA

Plant creeping or trailing; flowers blue; seeds naked . . . . *Vinca*
Plant not trailing; flowers not blue; seeds comose . . . . *Apocynum*

Apocynum L.

Corolla 4-6mm long, 2-3 times as long as calyx . . . . *A. pumilum*
Corolla 2-3mm long; leaves ascending
  Bracts of the inflorescence scarious and aristate; leaves evidently petiolate; follicles 9-15cm long. . . *A. cannabinum*
  Bracts of the inflorescence semi-foliaceous or blade-bearing leaves nearly or quite sessile; follicle 4-10cm long . . . . *A. sibiricum*

A. cannabinum L. [*A. suksdorfii* Greene]; Indian Hemp. Infrequent, along streams and in damp areas; cismontane, below 1500m: De Luz; Palomar Mountain; Montezuma Valley; Lake Henshaw; San Ignacio; Sentenac Canyon; Banner Grade; Dehr Creek; Vallecitos Canyon; Jamacha; Japatul Valley; Tijuana River Valley. May-Aug.

A. pumilum (Gray)Greene var. **rhomboideum** (Greene)Beg. & Bel. Pubescent Dogbane. Infrequent, in shaded areas, Montane Coniferous Forest; montane, above 1200m: Palomar Mountain; Pine Hills; Cuyamaca Mountains. June-Aug.

A. sibiricum Jacq. var. **salignum** (Green)Fern. [*A. suksdorfii* Greene] Willow-leaf Indian Hemp. Infrequent, in moist areas; montane, above 1100m: Horse Camp at Cuyamaca Lake. June-Aug.

Vinca L. Periwinkle

***V. major** L. Blue Periwinkle. Native of Europe. Uncommon, persisting around old dwellings and naturalized in shaded areas along streams; cismontane: Fallbrook; Harbison Canyon; Clevenger Canyon; Rose Canyon; Jamul; Cuyamaca North Peak. Mar.-July. 2n=92.

ARALIACEAE. Ginseng Family

KEY TO GENERA

Plant large perennial herb, with large compound leaves . . . . *Aralia*
Plant sprawling or climbing by aerial roots; leaves simple, lobed . . . . *Hedera*

Aralia L.

A. californica Wats. California Spikenard. Local, on shaded slopes on Palomar Mountain; 1200-1400m: Pauma Creek; Azalea Creek. June-Aug. 2*n*=48.

Hedra L. Ivy

***H. helix** L. English Ivy. Native of Eurasia. Uncommon, persisting around old dwellings and at dumps; coastal, below 200m: Encinitas; Paradise Hills. June-Sep. 2*n*=48.

ASCLEPIADACEAE. Milkweed Family

Stems not twining . . . . *Asclepias*
Stems twining
  Crowns wanting; corolla urn-shaped or campanulate; pollinia strictly pendulous. . . . . *Cynanchum*
  Crowns present; corolla rotate or open-campanulate; pollinia pendulous or horizontal
    Flowers solitary; pollinia horizontal . . . . *Matelea*
    Flowers borne in umbels or racemes
      Flowers white . . . . *Araujia*
      Flowers not white. . . . . *Sarcostemma*

Araujia Brot.

***A. sericofera** Brot. White Bladder-Flower. Native of South America. Uncommon, escaping from cultivation; coastal, below 150m: San Diego; National City. July-Sep.

Asclepias L. Milkweed

The purplish hoods without horns on inner surface; herbage densely tomentose . . . . *A. californica*
The hoods with horns on inner surface
  The hoods 2-3 times as long as stamens and stigma; filament-column almost lacking; pedicels deflexed in fruit; stems woody below, 1-1.5m tall . . . . *A. subulata*
  The hoods ca. as long as stamens and stigma; filament-column well developed
    Leaves 1cm or less wide, often early deciduous; pedicels erect in fruit
      Plants herbaceous, rarely 1m high, with persistent green leaves; stems green
        Leaves mostly in whorls corollas lobes white . . . . *A. fascicularis*
        Leaves opposite; corolla lobes orange . . . . *A. curassavica*
      Plants shrubby, 1-2m. high, the leaves soon deciduous; stems white . . . . *A. albicans*
    Leaves mostly 2-7cm wide; pedicels deflexed in fruit
      Leaves usually 3 or more at a node; horns mostly included within the hoods; cismomtane. . . . . . . *A. eriocarpa*
      Leaves opposite; horns exserted; desert . . . . *A. erosa*

A. albicans Wats. Wax Milkweed. Occasional, on rocky slopes; desert, below 950m: e.g. The Narrows; near Vallecito Stage Station; mouth of Canebrake Canyon; Mountain Palm Springs; Dos Cabezas; InKoPah Gorge. Mar.-May.

A. californica Greene. California Milkweed. Occasional on dry slopes; foothill and montane, above 750m: e.g. Nate Harrison Grade; Doane Valley; Eagles Nest; Banner Grade; San Felipe Valley; Mesa Grande; Ballena; Cuyamaca Lake; Viejas Grade; Buckman Spring. Apr.-July. Type from "mountains east of San Diego".

***A. curassavica** L. Bloodflower Milkweed. Native of Neotropics. Uncommon, escaping from cultivation; coastal, below 100m: San Ysidro; National City. Apr.-June. 2*n*=22

A. eriocarpa Benth. Indian Milkweed. Frequent, on dry or grassy slopes; foothill and montane, above 350m: e.g. Chihuahua Valley; Doane Valley; Wyss Ranch; Eagles Nest; Lake Henshaw; Cuyamaca Lake; Pine Hills; Santa Ysabel; Barona Valley; Viejas Valley; Laguna Mountains. June-Aug.

A. erosa Torr. Desert Milkweed. Occasional, on dry slopes and in washes; desert, 200-950m: Oak Grove; Warner Springs; San Felipe Valley; San Felipe Wash; Agua Caliente Spring; Dubber; Jacumba. May-July.

A. fascicularis Dcne. in A. DC. Narrow-Leaf Milkweed. Occasional, in open fields and along roadsides; cismontane, below 1550m: e.g. San Marcos; Kearny Mesa; Mission Valley; Spring Valley; Otay; Campo; Pine Valley; Hipass; Buckman Spring; Laguna Mountains; Paso Picacho; Oriflamme Canyon; San Felipe Valley; Witch Creek. May-Aug.

A. subulata Dcne. in A. DC. Rush Milkweed. Infrequent, in rocky washes and sandy places; desert, below 600m: Arroyo Salado; Palm Spring; Mountain Palm Springs; Rockhouse Canyon; Mountain Springs. Apr.-Dec. 2*n*=22.

Cynanchum L.

C. utahense (Engelm.)Woodson. Deboltia. Rare, on bajada slopes; deserts below 800m: San Felipe; Blair Valley Ranger Station; below Coyote Mountain, southeast of Dolomite Mine. Apr.-June.

Matelea Aubl.

M. parvifolia (Torr.)Woodson. [*Gonolobus californicus* Jeps., the type from Ironwood (Yaqui) Well; *Gonolobus hastulatus* (Gray) Gray] Rare, known in the county from Plum Canyon and Yaqui Well, 500-400m. Mar.-May.

Sarcostemma R. Br.

Corolla purplish; plant glabrous or with subappressed hairs . . . . *S. cynanchoides*
Corolla greenish-yellow; plant with short spreading hairs . . . . *S. hirtellum*

S. cynanchoides Dcne. ssp. **hartwegii** (Vail)R. Holm. [*S. heterophyllum* auct. non Engelm.] Hartweg's Milkvine. Infrequent, on shrubby slopes; cismontane and desert, below 450m: Pauma Valley; San Pasqual; Lakeside; Paradise Valley; Torrey Pines; Harbison Canyon; San Ysidro; Borrego Palm Canyon; south base of Pinyon Mountain; Box Canyon. Apr.-July.

S. hirtellum (Gray)R. Holm. [*S. heterophyllum* var. *h.* Gray] Ramblng Milkvine. Frequent, in washes; desert, below 450m: e.g. Borrego Palm Canyon; The Narrows; Fish Creek; Vallecito Stage Station; Carrizo Stage Station. Apr.-May.

ASTERACEAE. Sunflower Family
(Compositae)

KEY TO MAJOR GROUPS AND GENERA

Anthers tailed or sagittate at their base
  Some or all of the corollas appearing definitely bilabiate . . . . **Group A**
  Corollas not at all or obscurely bilabiate
    Style without ring of hairs or distinct thickened ring below the branches; plants not prickly; receptacle naked . . . . **Group B**
    Style with a ring of hairs or with a thickened ring below the papillate branches; plants usually prickly; receptacle usually densely bristly . . . . **Group C**
Anthers not tailed or sagittate at their base
  Heads with flowers all perfect and corollas strap-shaped, 5-toothed at apex (true ligules) . . . . **Group D**
  Heads with flowers tubular when perfect, and usually regular; marginal flowers often female or neutral and often strap-shaped, 2-3 toothed (ray-flowers or rays)
    Pappus none or vestigial, sometimes a mere crown
      Rays none or vestigial. . . . . **Group E**
      Rays evident . . . . **Group F**
    Pappus evident on some or all of the akenes
      Rays none or vestigial
        Akenes with a pappus of scales or awns or both, these sometimes united into a low crown . . . . **Group G**
        Akenes with a pappus of capillary bristles, rarely with additional outer scales . . . . **Group H**
      Rays evident, present
        Akenes with a pappus of firm awns or chaffy scales
          Rays yellow to orange or brown . . . . **Group I**
          Rays white to pink, red or purplish . . . . **Group J**
        Akenes with a pappus of capillary bristles, rarely with additional outer scales . . . . **Group K**

Group A. (Corollas bilabiate; anthers tailed)
[Mutisieae]

Plants herbaceous; corollas pink or whitish; involucre well imbricated . . . . *Acourtia*
Plants shrubby; corollas yellow; involucre in only 2 distinct rows . . . . *Trixis*

Group B. (Heads discoid, anthers tailed; plants not prickly)
[Inuleae]

Receptacle naked; phyllaries many, not scalelike; often perennial
  Phyllaries dry but not scarious; plants not tomentose
    Large herbs. . . . . *Pluchea*
    Shrubs . . . . *Tessaria*
  Phyllaries scarious; herbage woolly; flowers all fertile, perfect and pistillate in the same head; taprooted *Gnaphalium*
Receptacle chaffy, at least near the margin; phyllaries several or none, when present saclike (open in *Evax*) bearing pistillate (fertile) flowers; small woolly annuals
  Fruit-bearing bracts open, merely subtending the pistillate flowers, persistent . . . . *Evax*
  Fruit-bearing bracts conduplicate or saclike, enclosing the pistillate flower falling with the akene
    Akenes gibbous, the style lateral; true involucre below the fertile bracts of ca. 5 scarious scales . . . . *Micropus*
    Akenes straight, the style apical or nearly so
      Leaves alternate; receptacle convex to almost linear
        Fertile pistillate flowers all epappose, completely enclosed in woolly hyaline-appendaged phyllaries; receptacle chaffy throughout; true involucre none . . . . *Stylocline*
        Fertile pistillate flowers of 2 sorts, the outer epappose and bract-enclosed, the inner pappose but not bract-subtended; true involucre scanty . . . . *Filago*
      Leaves opposite; receptacle subglobose to obpyriform; true involucre and pappus none; inner flowers bractless . . . . *Psilocarphus*

Group C. (Anthers tailed; style with a ring of hairs or thickened ring below the papillate branches; plants usually prickly)
[Cynareae]

Akenes basifixed; marginal flowers not enlarged
  Filaments united below; leaves white-mottled along the veins; pappus several-seriate, not plumose . . . . *Silybum*

Filaments separate; leaves not white-mottled
Pappus of short scales in several series; outer phyllaries leaflike; yellow- or orange-flowered annuals *Carthamus*
Pappus of elongate slender bristles, these sometimes united at base; outer phyllaries not leaflike; flowers not bright yellow or orange
Phyllaries enlarged at base, broad; flowers violet; receptacle fleshy . . . . *Cynara*
Phyllaries narrower; flowers white, pink, or purplish; receptacle not fleshy
Pappus-bristles plumose (or those of marginal flowers sometimes merely barbellate) . . . . *Cirsium*
Pappus-bristles merely barbellate . . . . *Carduus*
Akenes obliquely inserted on the receptacle; marginal flowers often enlarged
Leaves not prickly on the margin; heads not subtended by the upper leaves; flowers in most species not yellow
Outer florets sterile . . . . *Centaurea*
All florets hermaphroditic . . . . *Acroptilon*
Leaves prickly on the margin; heads subtended by the upper leaves; flowers yellow . . . . *Cnicus*

Group D. (Corollas all strap-shaped)
[Cichorieae]

Herbage usually coarsely hairy; flowers blue; akenes short, truncate at apex; pappus paleaceous or none; pollen pale yellow; style branches mostly elongate . . . . *Cichorium*
Not in all respects as above
Low, acaulescent, glabrous; pollen orange; style branches short and blunt
Akenes not beaked; pappus (at least in part) broad at base, tapering into bristlelike awns . . . . *Microseris*
Akenes beaked; pappus of white capillary bristles . . . . *Agoseris*
Not in all respects as above
Annuals to shrubs, mostly glabrous, some appressed-tomentose, none hirsute with spreading hairs; flowers pink, white or sometimes yellow; akenes cylindrical, fusiform, or beaked, never flattened; pappus setose; style-branches mostly short
Plants non-scapose, usually branched
Pappus not plumose
Heads relatively slender . . . . *Prenanthella*
Heads subglobose . . . . *Malacothrix*
Pappus plumose, at least above
Akenes beakless; flowers rose or flesh-colored; annual or usually perennial . . . . *Stephanomeria*
Akenes beaked; flowers white or nearly so; fistulous-stemmed annual . . . . *Rafinesquia*
Plants scapose or subscapose; heads many-flowered; annuals except in a few species of *Malacothrix*.
Pappus of plumose bristles . . . . *Anisocoma*
Pappus of capillary bristles, never plumose, quickly deciduous
Akenes beakless, truncate . . . . *Malacothrix*
Akenes beaked . . . . *Calycoseris*
Not in all respects as above
Herbs with grasslike clasping leaves; heads large, solitary, yellowish or purplish; pappus of somewhat flattened plumose bristles . . . . *Tragopogon*
Not in all respects as above
Herbs with coarse spreading hirsute pubescence; flowers mostly yellow; akenes mostly fusiform or beaked; pappus mostly of coarse plumose setae
Receptacle chaffy-bracted; at least the inner akenes beaked; leaves mostly basal . . . . *Hypochoeris*
Receptacle naked
Akenes terete, truncate, the outer ones enveloped by the hardening phyllaries . . . . *Hedypnois*
Akenes beaked; stems leafy; outer phyllaries broader than the inner . . . . *Picris*
Herbs of various habit; pappus of numerous nonplumose setae
Akenes flattened; leafy-stemmed herbs; pappus of soft and numerous bristles
Involucre campanulate; akenes not beaked . . . . *Sonchus*
Involucre cylindrical; akenes beaked . . . . *Lactuca*
Akenes not obviously flattened; leaves mostly basal or subbasal; pappus various or none
Akenes truncate; pappus fragile, usually sordid or fuscous; caulescent . . . . *Hieracium*
Akenes beaked; pappus mostly bright white; heads solitary on fistulous naked scapes; pappus not early deciduous . . . . *Taraxacum*

Group E. (Rays and pappus none or vestigial)

Heads unisexual; both staminate and pistillate on the same plant; involucre of pistillate heads closed and somewhat burlike; staminate heads in a raceme or spike, with open involucres
Phyllaries of staminate heads united
Fruiting involucre winged with broad scarious scales; leaves or their divisions filiform . . . . *Hymenoclea*
Fruiting involucre not so winged; leaves toothed or pinnatifid . . . . *Ambrosia*
Phyllaries of staminate heads distinct; involucre of pistillate heads becoming a stout spiny bur . . . . *Xanthium*
Heads not unisexual; pistillate involucre not becoming a bur
Phyllaries in a single series or lacking
Leaves opposite; phyllaries several; head several-flowered . . . . *Lasthenia*
Leaves mostly alternate
Leaves or their lobes linear-filiform; akenes long-villous; plant woody . . . . *Iva*
Leaves or their lobes not linear-filiform; akenes not villous; plant not woody
Akenes obovoid or pyriform, not winged; perennial herbs; phyllaries ca. five . . . . *Iva*
Akenes flattened, pectinate-winged; annual herbs; phyllaries 6-7, the 1 or 2 enlarged . . . . *Dicoria*
Phyllaries in 2 or more usually imbricated series
Heads in spikes, recemes or panicles . . . . *Artemisia*
Heads solitary or in capitate clusters or in corymbs
Akenes, especially the marginal, stipitate . . . . *Cotula*
Akenes sessile
Receptacle convex or hemispheric, not chaffy . . . . *Matricaria*
Receptacle flat, receptacle not chaffy . . . . *Soliva*

Group F. (Rays evident; pappus none or vestigial)

Receptacle chaffy with bracts at base of at least some of the flowers or with stiff bristles
  Phyllaries usually imbricated, or at least in more than 1 series
    Leaves opposite
      Heads (including rays) barely 1cm in diameter; rays 1-5, white . . . . *Eclipta*
      Heads (including rays) ca. 3cm in diameter; rays yellow . . . . *Guizotia*
    Leaves alternate (at least in upper 2/3 of plant) or largely basal
      Rays white
        Heads relatively large, solitary or few . . . . *Anthemis*
        Heads very small, many, in dense flattish or rounded panicles . . . . *Achillea*
      Rays yellow
        Inner involucre cup-like, its phyllaries united to or above the middle; akenes terete, beakless . . *Thelesperma*
        Inner involucre with phyllaries free nearly to the base; akenes flat or angled
          Involucre of 2 very unlike series of phyllaries; herbs or somewhat woody and with fleshy stems *Coreopsis*
          Involucre of 2 or more series of similar phyllaries, often rather suffrutescent . . . . *Encelia*
  Phyllaries in a single series
    Receptacle bristly, honeycombed; phyllaries not enclosing the ray-akenes; plant rather tomentose, annual . . . . *Eriophyllum*
    Receptacle not bristly
      Ray-akenes enclosed in the obcompressed phyllaries, of which the infolded lateral margins meet
        Disk-flowers 6; ray-akenes obcompressed . . . . *Lagophylla*
        Disk-flower 1 and the ray-akene obcompressed, or more than 1 and the ray-akenes not obcompressed *Madia*
      Ray-akenes not completely enclosed, the phyllaries not obcompressed
        Rays 3-toothed or 3-lobed, the lobes subparallel; herbage without tack-shaped glands
          Upper leaves and phyllaries not terminated by open pit glands . . . . *Hemizonia*
          Upper leaves and phyllaries terminated by open pit glands, leaf tip truncate . . . . *Holocarpha*
        Rays 3-cleft or -parted into palmately spreading lobes; herbage mostly with tack-shaped glands *Calycadenia*
Receptacle naked, without bristles or chaffy bracts
  Phyllaries graduated in several, unequal series, closely imbricate, at least the inner somewhat scarious . . . . *Chrysanthemum*
  Phyllaries in 1-2 series, or at least not imbricate
    Rays white or pink, low plant in cultivated areas . . . . *Bellis*
    Rays yellow or orange
      Leaves opposite
        Phyllaries in 1 series; plants mostly annual. . . . . *Lasthenia*
        Phyllaries in 2 or more series; plants perennial, succulent . . . . *Jaumea*
      Leaves alternate
        Phyllaries broad, in 2 distinct series (the outer somehwat reflexed, the inner erect); sunflower-like perennial . . . . *Venegasia*
        Phyllaries not as above
          Rays persistent and becoming papery; phyllaries many; woolly perennials of the deserts . . . . *Baileya*
          Rays not as above
            Plants rather succulent, glabrous except for a microscopic glandular pubescence and very sparse flocs of wool in the leaf-axils and below the heads . . . . *Blennosperma*
            Plants not as above
              Herbage not white-woolly; annual herbs; leaves coarsely sinuate-dentate; plants 1-3dm high . . . . *Dimorphotheca*
              Herbage white-woolly at least when young; rays yellow or white . . . . *Eriophyllum*

Group G. (Rays none or vestigial; pappus of paleae or flattened scales or stiff bristles or awns or both)

Receptacle chaffy
  Pappus of plumose awns; plant a low almost leafless shrub . . . . *Bebbia*
  Pappus not of plumose awns
    Akenes with pectinately toothed wings; annuals . . . . *Dicoria*
    Akenes not with pectinately toothed wings
      Awns or teeth of pappus not retrorsely barbed, shrubby . . . . *Encelia*
      Awns or teeth retrorsely barbed, herbaceous . . . . *Bidens*
Receptacle not chaffy
  Leaves, at least all the lower, opposite
    Plant woody; leaves deltoid, small, on long petioles, deserts . . . . *Pleurocoronis*
    Plant herbaceous . . . . *Lasthenia*
  Leaves mostly alternate
    Pappus a mere crown; phyllaries imbricate, somewhat scarious or scarious-margined. . . . . *Matricaria*
    Pappus of well developed paleae or awns or both
      Leaves and involucre punctate with translucent oil-glands . . . . *Dyssodia*
      Leaves and involucre not punctate with translucent oil-glands
        Pappus of bristles or these alternating with minute paleae or awns
          Pappus of 2-8 deciduous awns; plants perennial, usually strongly viscid, mostly cismontane . . . . *Grindelia*
          Pappus of ca. 4 bristles alternating with ca. 4 minute paleae; annual, not glutinous; desert. . . . . *Malperia*
        Pappus of flat or awnlike paleae
          Pappus of 20-30 silvery awns; plant woody; deserts . . . . *Acamptopapppus*
          Pappus of mostly fewer parts
            Phyllaries in a single series
              Pappus-paleae erose or fimbriate; leaves 3-lobed at apex . . . . *Eriophyllum pringlei*
              Pappus-paleae nearly or quite entire
                Flowers whitish or purplish
                  Pappus-paleae linear, with a strong midrib; leaves entire . . . . *Palafoxia*
                  Pappus-paleae nerveless; leaves toothed to pinnatifid . . . . *Chaenactis*

Flowers yellow
Pappus-paleae 4; leaves dissected. . . . . *Chaenactis glabriscula*
Pappus-paleae more numerous
Plants annual; viscid; leaves mostly linear, entire; coastal . . . . . *Amblyopappus*
Plants perennial, not viscid; leaves mostly dissected . . . . . *Eriophyllum*
Phyllaries in 2 or more series
Pappus-paleae 5, dissected into bristles; white-woolly annual; desert. . . . . *Trichoptilium*
Pappus-paleae 10-22, not so dissected
Pappus-paleae obtuse; leaves in a basal rosette . . . . . *Hymenopappus*
Pappus-paleae lanceolate, each ending in an awn; leaves both basal and cauline . . . . . *Hymenothrix*

Group H. (Rays none or vestigial; pappus of capillary bristles)

Heads unisexual; plants dioecious, usually woody at base . . . . . *Baccharis*
Heads not strictly unisexual
Plants well developed shrubs
Phyllaries proper, 4-7, in a single series, of equal length; leaves narrow, entire . . . . . *Tetradymia*
Phyllaries more numerous
Leaves scalelike, at least on flowering branches; akenes glabrous; heads closely placed in larger clusters . . . . . *Lepidospartum*
Leaves not scalelike
Phyllaries arranged in rather distinctly vertical ranks . . . . . *Chrysothamnus*
Phyllaries not so arranged
Flowers whitish
Pappus double, the inner of bristles, the outer of short paleae; petioles longer than the small leaf blades . . . . . *Pleurocoronis*
Pappus simple, of capillary bristles only; petioles mostly shorter than leaf blades . . . . . *Brickellia*
Flowers yellow
Phyllaries merely of 1 series, subequal, subulate; desert shrub with terete resinous-punctate leaves . . . . . *Peucephyllum*
Phyllaries of more than 1 series, unequal; widely distributed . . . . . *Isocoma*
Plants herbaceous
Plants annual
Herbage white-woolly, aromatic; leaves broad, rounded; heads solitary in forks of branches; deserts *Psathyrotes*
Herbage and leaves not as above
Phyllaries in a single series, equal or nearly so; leaves dentate to pinnatifid . . . . . *Senecio vulgaris*
Phyllaries in more than 1 series, unequal
Outer corollas enlarged, more deeply cleft on inner side; herbage often strong-smelling . . . . . *Lessingia*
Outer corollas not enlarged, very slender; herbage not usually strong-smelling . . . . . *Conyza*
Plants biennial to perennial
Principal phyllaries essentially equal and in 1 series, although some basal outer reduced bracts may be present
Leaves opposite; yellow flowered plants of moist mountain sites. . . . . *Arnica*
Leaves alternate or basal; pappus-bristles barbellate, not truly plumose; flowers yellow to orange . . . *Senecio*
Principal phyllaries more or less imbricate in 2-several series
Phyllaries and leaves with conspicuous translucent oil-glands . . . . . *Porophyllum*
Phyllaries and leaves not having translucent oil-glands
Flowers not yellow
Akenes 5-angled or -ribbed; leaves often opposite . . . . . *Ageratina*
Akenes 10-ribbed; leaves alternate . . . . . *Brickellia*
Flowers yellow
Style-appendages very short, 0.5mm long or less; phyllaries scarcely imbricate . . . . . *Erigeron*
Style-appendages mostly 0.7mm or more long; phyllaries plainly imbricate
Plant with squarrose phyllaries
Herbaceous biennials or perennials. . . . . *Machaeranthera*
Shrubby perennials . . . . . *Xylorhiza*
Plant not as above
Akenes somewhat flattened; involucre 9-12mm high; plants herbaceous. . . . . *Chrysopsis*
Akenes terete or angled; involucre mostly 12mm or higher; plants mostly woody at base . . *Isocoma*

Group I. (Rays well developed, yellow to orange to brown, at least when fresh; pappus of well developed paleae or scales or of stiff awns)

Receptacle chaffy or bristly throughout or with a circle of chaffy bracts surrounding the disk-flowers
Involucre of 1-several series of phyllaries, none enfolding the ray-akenes
Phyllaries in 2 distinct dissimilar series; akenes flattened at right angles to the radius of the head
Pappus of 2-6 firm mostly retrorsely barbed awns; leaves entire to pinnately compound; akenes flattened *Bidens*
Pappus of 2 minute teeth; leaves pinnately dissected . . . . . *Coreopsis*
Phyllaries in 2 or more similar series
Rays pistillate . . . . . *Wyethia*
Rays neutral, the akenes not developing
Akenes somewhat thickened
Pappus of 2 awns and several short squamellae . . . . . *Viguiera*
Pappus of 2, rarely many, deciduous paleaceous awns. . . . . *Helianthus*
Akenes very flat
Plants annual; awns 2, strong; akene-body conspicuously white-margined. . . . . *Geraea*
Plants perennial, forming low rounded shrubs; awns 1-2, weak. . . . . *Encelia*
Involucre with 1 series of phyllaries, each partly or completely enclosing an akene
Ray-akenes completely enfolded by the phyllaries
Flowers red-brown; pappus silvery-scarious, the longer paleae longer than the akenes; akenes 10-costate and tuberculate-scabrous . . . . . *Achyrachaena*

Flowers yellow; pappus and akenes not as above . . . *Layia*
Ray-akenes half-enclosed by the phyllaries . . . *Hemizonia*
Receptacle not chaffy or bristly
Involucre and leaves with translucent oil-glands
Akenes pubescent; leaves lacking stiff spreading basal bristles; phyllaries generally 2-seriate . . . *Dyssodia*
Akenes glabrous; leaves with a few stiff spreading bristles at base; phyllaries 1-seriate . . . *Pectis*
Involucre and leaf lacking translucent oil-glands
Involucre of several series of graduated phyllaries
Pappus of 2-8 slender deciduous awns; heads rather large, gummy; coarse herbs . . . *Grindelia*
Heads small; pappus of disk-flowers of 1 series of straight paleae; leaves mostly linear . . . *Gutierrezia*
Involucre of ca. 1-2 equal or subequal but not graduated series of phyllaries
Leaves opposite; mostly low annuals . . . *Lasthenia*
Leaves mostly alternate
Phyllaries in 1 series . . . *Eriophyllum*
Phyllaries in more than 1 series
Phyllaries reflexed; rays short, broad . . . *Helenium*
Phyllaries not reflexed; receptacle flat; akenes linear, flattened or somewhat 3-angled; pappus-paleae mostly 4 . . . *Hulsea*

Group J. (Rays well-developed, white to pink, red or purplish; pappus of paleae or scales or of stiff awns)

Heads 5-7cm across; peduncles 1.5-3dm long; white-woolly perennial . . . *Arctotis*
Heads smaller
Outermost akenes usually enfolded by their subtending phyllary
Ray-akenes half enclosed by the phyllaries; disk-akenes sterile, undeveloped, glabrous, epappose . . . *Hemizonia*
Ray-akenes completely enclosed by the phyllaries . . . *Layia*
Outermost akenes not so enfolded
Leaves opposite; rays 4-5, short; akenes black . . . *Galinsoga*
Leaves alternate or basal
Akenes compressed, 2-edged or 2-nerved . . . *Perityle*
Akenes somewhat thickened, 4-5 angled
Herbage hispid-pilose; upper leaves subtending the heads . . . *Monoptilon*
Herbage woolly; upper leaves not subtending the heads . . . *Eriophyllum*

Group K. (Rays well developed; pappus of soft capillary bristles, mostly without any paleae)

Rays yellow
Phyllaries many, usually more or less imbricated; style-tips not truncate
Plants annual . . . *Chaetopappa*
Plants perennial
Ray-akenes without pappus; tall leafy herbs with broad leaves . . . *Heterotheca*
Ray-akenes with pappus
Pappus in 1 series
Pappus-bristles unequal; involucre mostly 6-18mm high, if shorter, mostly on shrubby plants
Akenes turbinate, 2-3mm long; leaves dentate; herbaceous perennial . . . *Machaeranthera*
Akenes nearly prismatic, subcylindric or fusiform, more than 3mm long
Involucres hemispheric, 10-18mm wide . . . *Ericameria*
Involucres turbinate or subcylindric
Heads large, the 8-15mm high involucres tightly imbricate, mostly 6-8-seriate; herbage without glandular dots . . . *Hazardia*
Heads small, the 3-8mm high involucres loosely imbricate, 2-6-seriate; herbage with distinct resin pits . . . *Ericameria*
Pappus-bristles equal; involucre mostly 4-5cm high, or if larger, on herbaceous plants . . . *Solidago*
Pappus in 2 series, the outer much shorter than the inner and of short linear scales or bristles
Leaves oblong-spatulate to oblanceolate or elliptic, well distributed on the stems . . . *Chrysopsis*
Leaves linear, mostly basal . . . *Erigeron*
Phyllaries fewer, or if more, at least not strongly imbricated, mostly subequal and in 1 series; leaves mostly opposite . . . *Arnica*
Rays white to pink, blue or purple
Pappus of ray-flowers much reduced . . . *Corethrogyne*
Pappus well developed in both ray- and disk-flowers
Heads nodding in bud; leaves narrow; annuals with slender stems . . . *Chaetopappa*
Heads erect in bud
Rays very numerous, filiform, scarcely exceeding the disk-flowers; annual weeds with numerous sublinear leaves . . . *Conyza*
Rays much surpassing the disk
Phyllaries subequal or mostly imbricate, neither leafy nor with chartaceous base and herbaceous green tip; style-appendages 0.5mm long or less, lanceolate or broader . . . *Erigeron*
Phyllaries either subequal and the outer leafy, or usually imbricate, with chartaceous and green tips; style-appendages lanceolate or narrower, usually more than 0.5mm long
Plants annual or biennial or perennial from a distinct taproot; leaves mostly toothed; phyllaries chartaceous or coriaceous toward base, often spreading or recurved at the herbaceous tips
Plants herbaceous . . . *Machaeranthera*
Shrubs with a woody tap root . . . *Xylorhiza*
Plants mostly perennial and if so, rhizomatous or fibrous-rooted; phyllaries mostly herbaceous; style-appendages lanceolate to subulate, acute; leaves mostly 3-20cm long; plants often tallish; mostly montane and cismontane . . . *Aster*

Acamptopappus Gray Goldenhead
(Astereae - Solidagininae)

A. sphaerocephalus (Harv. & Gray)Gray var. **sphaerocephalus**. Desert Goldenhead. Occasional, in sandy areas; desert, below 900m: e.g. Borrego Palm Canyon; Sentenac Canyon; lower San Felipe Valley; Banner; Pinyon Mountain; Box Canyon; La Puerta; Vallecito Valley. Dec.-May. *n*=9.

Achillea L. Yarrow
(Anthemideae - Anthemidinae)

Herbage gray, villous to woolly; margins of phyllaries light brown to straw-colored; Montane Coniferous Forest . . . . . *A. millefolium* var. *lanulosum*
Herbage usually green, moderately villous
  Leaf-segments oblanceolate, all segments tending to be oriented in same plane. Possible escape in lawns, etc. . . . . . *A. millefolium* var. *m.*
  Leaf-segments linear, spreading and ascending; native
    Involucre generally 5-6mm high; Montane Coniferous Forest . . . . . *A. millefolium* var. *pacifica*
    Involucre generally 6-9mm high; mostly below 1000m, cismontane . . . . . *A. millefolium* var. *californica*

A. millefolium L. var. **californica** (Pollard)Jeps. [*A. borealis* ssp. *c.* (Pollard)Keck] California Yarrow. Common, on shaded slopes; cismontane, below 800m: El Cajon; Rolando; Pacific Beach; San Miguel Mountain; Otay Mountain; Monument School. Jan.-June. Type from "sea coast at Santa Ysabel" (a vexing localty!). *n*=27, 54.

A. millefolium var. **lanulosum** (Nutt.)Piper. Woolly Yarrow. Common, on shaded slopes; montane, above 900m: e.g. Palomar, Cuyamaca, and Laguna Mountains; Rattlesnake Valley; Descanso. Also Clover Flats; Otay Mountain. May-Oct. *n*=18.

***A. millefolium** var. **millefolium**. Common Yarrow. Native of Europe. Infrequent, weed in lawns; coastal: San Diego. June-Sep. *n*=27.

A. millefolium var. **pacifica** (Rydb.)G.N. Jones. Pacific Yarrow. Uncommon, in shaded places of Montane Coniferous Forest; montane, above 1500m: Palomar Mountain. May-Sep. *n*=18.

Achyrachaena Schauer
(Heliantheae - Madinae)

A. mollis Schauer. Blow-Wives. Infrequent, on grassy slopes; cismontane, below 700m: e.g. Ysidora; Vista; Ramona; Montezuma Mesa; Paradise Valley; Otay Ranch; Echo Valley. Apr.-May. *n*=8.

Acourtia D. Don
(Mutisieae)

A. microcephala DC. [*Perezia microcephala* (DC.) Gray non Molina] Sacapellote. Common, on dry slopes and in Coastal Sage Scrub and Chaparral; cismontane, below 1300m: e.g. Oak Grove; Mesa Grande; Descanso; Laguna Junction; San Diego; Encinitas; Del Mar; Viejas Valley; Campo; La Posta; Jacumba. Also Palomar Mountain. June-Aug. *n*=27.

Acroptilon Cass.
(Cynareae)

***A. repens** (L.)DC. [*Centaurea r.* L., *C. picris* Pallas ex Willd.] Russian Knapweed. Native of Asia. Uncommon, weed in cultivated areas; below 200m: Rancho Santa Fe; El Cajon; Chula Vista.

Ageratina Spach
(Eupatorieae)

***A. adenophora** (Spreng.)King & Robins. [*Eupatorium a.* Spreng.] Sticky Eupatorium. Native of Mexico. Uncommon, escaping from cultivation; coastal, below 200m: Sweetwater Reservoir. Dec.-May.

Agoseris Raf. Mountain Dandelion
(Cichorieae - Microseridinae)

Plants annual; akene-beak mostly 2-3 times as long as akene-body; involucre 12-18mm high . . . . . *A. heterophylla*
Plants perennial; akene-beak variable; involucre mostly higher.
  Akene-body tapering at apex, gradually beaked; leaf-segments not retrorse; phyllaries mostly in 3 sets *A. grandiflora*
  Akene-body truncate at apex, abruptly beaked; leaf-segments retrorse; phyllaries in 2 quite unlike sets . . *A. retrorsa*

A. grandiflora (Nutt.)Greene. [*A. plebia* Greene] Large-Flower Agoseris. Infrequent, in Montane Coniferous Forest; montane, above 900m: Mesa Grande; Julian; Pine Hills; Cuyamaca Lake; Middle Peak; Descanso; Laguna Mountains; Laguna Lakes. May-July. 2*n*=18.

A. heterophylla (Nutt.)Greene. Woodland Agoseris. Occasional, on grassy slopes and burns; cismontane, above 300m: Palomar Mountain; Pala; Henshaw Dam; Mesa Grande; San Felipe Valley; Harper Ranch; Cuyamaca Lake; Laguna Mountains; Hauser Creek; Jacumba. Apr.-May. 2*n*=18, 36.

A. retrorsa (Benth.)Greene. Spear-Leaf Agoseris. Frequent, in Montane Coniferous Forest; montane, above 900m: Palomar Mountain; Hot Springs Mountain; Pine Hills; Cuyamaca Mountains; Harper Ranch; Laguna Mountains; Descanso. Also Tecate Summit. May.

Amblyopappus H. & A.
(Helenieae - Heleniinae)

A. pusillus H. & A. [*Aromia tenuifolia* Nutt.] Pineapple Weed. Common, in Sage Scrub and Coastal Salt Marsh; coastal, below 100m: e.g. San Onofre Bluffs; Carlsbad; Torrey Pines State Reserve; La Jolla; Mission Valley; Point Loma; Telegraph Canyon; Silver Strand; Imperial Beach. Mar.-May. *n*=8.

Ambrosia L.
(Heliantheae - Ambrosiinae)

Fruiting involucre unarmed or with a single row of short prickles below the single beak
  Plants with leaves once-pinnatifid into coarse lobes, greenish, rough-pubescent . . . . *A. psilostachya*
  Plants with leaves twice-pinnatifid into minute obtuse lobes 1mm wide, very canescent with soft hairs. . . *A. pumila*
Fruiting involucre armed with several rows of spines below the 1-4 beaks
  Plants herbaceous
    Bur 2-4mm long, its spines mostly hooked. . . . . *A. confertiflora*
    Bur 7-10mm long, its spines flattened and channeled, not hooked
      Plants decumbent, littoral perennial herbs . . . . *A. bipinnatifida*
      Plants erect weedy annuals . . . . *A. acanthicarpa*
  Plants shrubby or at least woody at the base
    Bur with straight spines
      Leaves once to thrice pinnately divided into very small obtuse lobes; fruit globular, 4-6mm long, 2-beaked, the spines not villous . . . . *A. dumosa*
    Bur with hooked spines; leaves not pinnatifid or divided . . . . *A. chenopodiifolia*

A. acanthicarpa Hook. [*Franseria a.* (Hook.)Cov.; *F. palmeri* Rydb., the type from San Diego.] Annual Bur-Weed. Common, weed in disturbed areas; cismontane: Harrison Grade; Warner Hot Springs; Oak Grove; Rincon; Escondido; Santa Ysabel; Laguna Junction; Pine Valley; Boulder Creek; Cottonwood Valley; Lakeside; Boulder Peak. July-Oct. *n*=18.

A. bipinnatifida (Nutt.)Greene. [*Franseria b.* Nutt.; *A. chamissonis* (Less.)Greene, in part; *F. c.* ssp. *bipinnatisecta* (Less.)Wiggins & Stockwell] Bather's Delight, Beach Burweed. Common, in dunes and other sandy areas of Coastal Strand; coastal, below 5m: e.g. San Onofre; Del Mar; La Jolla; Mission Beach; Ocean Beach; Coronado; Imperial Beach; Initial Boundary Monument. Jan.-July. *n*=16.

A. chenopodiifolia (Benth.)Payne. [*Franseria c.* Benth.] San Diego Burbush. Common, on mesas and open slopes; southern coastal, below 150m: Otay; Otay Mesa; Greg Rogers Park, Chula Vista; San Ysidro. Mar.-June. 2*n*=36.

A. confertiflora DC. [*Franseria c.* (DC.)Rydb.] Weak-Leaf Burbush. Frequent, weed of disturbed areas; cismontane, below 1200m: e.g. Oak Grove; Banner; San Felipe; Jamacha; Boulevard; Barrett Junction; Sweetwater Valley; Telegraph Canyon; Otay; San Diego. Jan.-Oct. *n*=36, 54.

A. dumosa (Gray)Payne. [*Franseria d.* Gray] White Bursage. Abundant, in sandy areas; desert, below 1000m: e.g. Borrego Palm Canyon; Borrego Springs; Borrego Mud Hills; Fish Creek; Banner; Dos Cabezas; Mountain Springs. Mar.-May. *n*=18, 36, 54, 63, 72.

A. psilostachya DC. var. **californica** (Rydb.)Blake in Tidestr.. Western Ragweed. Common, weed of disturbed areas; cismontane, below 1200m: e.g. Palomar Mountain; De Luz; Valley Center; Oak Grove; Witch Creek; Warner Springs; Santa Ysabel; Pine Hills; Pine Valley; Laguna Mountains; Dulzura; National City; San Diego; Lemon Grove; Encinitas. Aug.-Dec. *n*=36, 54, 72.

A. pumila (Nutt.)Gray. San Diego Ragweed. Occasional, in valleys and disturbed areas; coastal, below 150m: San Luis Rey; Bonsall; Lake Hodges; Bostonia; Santee; El Cajon; Padre Dam; Spring Valley; Mission Valley; National City; Sweetwater Dam. May-Oct. Type from "near St. Diego".

Anisocoma T. & G. Scale-Bud
(Cichorieae - Stephanomeriinae)

A. acaulis T. & G. Scale-Bud. Infrequent, in sandy areas; foothill, montane, and desert; 800-1500m: e.g. Sunshine Summit; Warner Ranch; Hot Springs Mountain; Lake Henshaw; Laguna Mountain; Tecate Summit; Boulevard; Jacumba. Apr.-May. 2*n*=14.

Anthemis L.
(Anthemideae - Anthemidinae)

***A. cotula** L. Dog Mayweed. Native of Europe. Common, weed of roadsides and disturbed areas; cismontane, below 1500m: e.g. San Onofre Bluffs; Palomar Mountain; Cardiff; San Marcos; Cuyamaca Mountains; Cuyamaca Lake; San Clemente Canyon; Alvarado Canyon; Lakeside; Chula Vista; Lee Valley; Otay; Hauser Creek; Descanso. May-Oct. 2*n*=18.

Arctotis L.
(Arctotideae)

***A. stoechadifolia** Berg. var. **grandis** (Thunb.)Less. Large-Flower African-Daisy. Native of South Africa. Infrequent, escaping from cultivation; coastal, below 150m: Pacific Beach; Murray Dam; Balboa Park. July. 2*n*=18.

Arnica L.
(Senecioneae)

A. cordifolia Hook. var. **cordifolia**. Heart-Leaf Arnica. Rare, known in the county only from an 1899 collection from the Cuyamaca Mountains; above 1400m. May-Aug.

Artemisia L. Sagebrush; Wormwood
(Anthemideae - Chrysantheminae)

Flowers all perfect; shrubs (except *A. palmeri*)
  Receptacle naked; plants shrubby, mostly canescent . . . . *A. tridentata*
  Receptacle with chaff; leaves green above . . . . *A. palmeri*
Flowers not all perfect, the marginal pistillate, the central ones fertile or sterile
  Plants definitely woody at base, not rhizomatous . . . . *A. californica*
  Plants not woody at base
    Herbage glabrous, rarely pubescent, never tomentose . . . . *A. dracunculus*
    Herbage usually pubescent, mostly white-tomentose beneath
      Principal leaves narrow, not more than 1cm. wide (exclusive of lobes when present), tomentose on both sides or green above; stems rarely over 1m high
        Leaves 1-2cm long . . . . *A. ludoviciana* ssp. *albula*
        Leaves 3-11cm long . . . . *A. ludoviciana* ssp. *l.*
      Principal leaves 1-5cm wide exclusive of lobes when present, frequently entire; leaves greener above than beneath; stems to 3m high . . . . *A. douglasiana*

A. californica Less. Coastal Sagebrush. Abundant, coastal, below 400m: e.g. San Onofre Bluffs; Santa Margarita River; Fallbrook; De Luz; Pala; Rincon; Escondido; Torrey Pines; Foster; Mission Gorge; La Jolla; Pacific Beach; Point Loma; Chollas Valley; Jamacha; Telegraph Canyon; Otay Mesa; Dulzura. Nov.-Feb. $2n=18$.

A. douglasiana Bess. in Hook. [*A. heterophylla* Nutt.] Douglas Mugwort. Occasional, in disturbed areas; cismontane, below 1100m: e.g. San Onofre Bluffs; Palomar Mountain; Oak Grove; Warner Springs; San Felipe; Valley Center; Oceanside; Ramona; Old Town; Sweetwater Valley; Jamacha; Descanso; Guatay; Buckman Spring. July-Dec. $n=27$.

A. dracunculus L. Tarragon, Dragon Sagewort. Frequent, in Chaparral; foothill and montane, above 200m: e.g. Palomar Mountain; Chihuahua Valley; Oak Grove; Mesa Grande; Santa Ysabel; Warner Springs; Witch Creek; San Felipe; Cuyamaca and Laguna Mountains; Pine Valley; Boulevard; Potrero Grade; Lakeside; Jamacha; Fernbrook; Paradise Valley. July-Dec. $2n=18$.

A. ludoviciana Nutt. ssp. **albula** (Woot.)Keck. Pinyon Wormwood. Infrequent, on dry slopes; montane and desert, below 1100m: Borrego Palm Canyon; Santa Ysabel. May-Nov.

A. ludoviciana ssp. **ludoviciana**. [*A. l.* var. *gnaphalodes* (Nutt.) T. & G.] Western Mugwort. Occasional, in dry, open places; montane and desert, above 600m: e.g. San Felipe; Sentenac Canyon; Santa Ysabel; Witch Creek; Cuyamaca Mountains; Laguna Mountains; Buckman Spring; Mountain Springs. $n=18$.

A. palmeri Gray. San Diego Sagewort. Frequent, in low places; south coastal, below 600m: Mount Woodson; San Vicente Creek; Foster; Del Dios; Poway; Miramar; La Jolla; Pacific Beach; San Diego; Old Town; Bostonia; National City; Oneonta; Otay; Telegraph Canyon; Jamul; Harbison Canyon; Alpine; Whispering Oaks. June-Sep. Type from "Jamuel [Jamul] Valley".

A. tridentata Nutt. ssp. **tridentata**. [*A. t.* var *angustifolia* Gray] Great Basin Sagebrush. Frequent, in valleys; foothill and montane, above 250m: e.g. De Luz; Fallbrook; Pauma; Palomar Mountain; Eagles Nest; Montezuma View Point; Chihuahua Valley; Warner Springs; Descanso; Laguna Junction; Boulevard; McCain Ranch: Hipass; Jacumba; Potrero; Campo; Dulzura; Del Mar Heights. July-Dec. $2n=18, 36$.

Aster L.
(Astereae - Asterinae)

Plants annual . . . . *A. exilis*
Plants perennial
  Plants woody toward the base; stems rushlike, almost leafless, the leaves inconspicuous at anthesis, often spinose in the leaf-axils; heads radiate, white; akenes glabrous . . . . *A. spinosus*
  Plants herbaceous; stems leafy or subscapose, the leaves obvious.
    Pubescence of stem in lines below the leaf-bases, commonly neither uniform under the heads nor confined to the inflorescence . . . . *A. hesperius*
    Pubescence of stem uniform, or if in lines, uniform under the heads or scanty and confined to the inflorescence
      Herbage, including phyllaries, glabrate, or if pubescent, the hairs short or appressed; inflorescence an open often divergent panicle . . . . *A. chilensis*
      Herbage, including phyllaries, cinereous with short spreading hairs; inflorescence a close panicle or raceme . . . . *A. bernardinus*

A. bernardinus Hall. San Bernardino Aster. Rare, in wet meadows, Southern Oak Woodland and Montane Coniferous Forest; montane, above 1200m: Palomar Mountain; Chihuahua Valley; Boulder Creek below Cuyamaca Lake. July-Nov. $2n=18$ pairs.

A. chilensis Nees var. **chilensis**. [*A. menziesii* Lindl. in Hook.] Chile Aster. Occasional, in meadows and other open areas, Chaparral and Montane Coniferous Forest; foothill and montane, above 900m: Palomar Mountain; Chihuahua Valley; Cuyamaca Lake, Cuyamaca Mountains; Pine Hills; Witch Creek; Pine Valley; Boulevard. June-Oct. $n=13,32$; $2n=24$ pairs, 32 pairs.

A. exilis Ell. [*A. subulatus* var. *ligulatus* Shinners] Slim Aster. Occasional, in moist areas; coastal, below 100m: Santa Margarita Ranch; Lakeside; Lemon Grove; Spring Valley. Aug.-Oct. $2n=10$.

A. hesperius Gray. Western Aster. Uncommon, in wet areas; foothill and montane, above 800m: Chimney Creek; Ballena; Mesa Grande; Witch Creek; Julian; Banner; falls of the San Diego River; Laguna Meadow. Aug.-Oct. $n=24, 32$.

A. spinosus Benth. Mexican Devil-Weed. Occasional, in sandy or disturbed places; south coastal, below 100m: Mission Gorge; Sweetwater Valley; Sunnyside; Bonita; San Ysidro; Tijuana River Valley. June-Sep. $n=9$.

Baccharis L.
(Astereae - Baccharidinae)

Herbage puberulent or pubescent; akenes puberulent . . . . *B. brachyphylla*
Herbage not hairy
  Stems herbaceous to base; leaves ovate-lanceolate; phyllaries thin, viscid-ciliate; akenes puberulent . . . *B. douglasii*
  Stems woody below; plants shrubby; phyllaries not viscid-ciliate
    Akenes puberulent . . . . *B. vanessae*
    Akenes glabrous
      Plants persistently leafy, not broomlike (or sometimes somewhat so in *B. emoryi)*
        Leaves linear or linear-lanceolate, entire or evenly toothed, 5-15cm long; akenes 5-nerved . . . . . *B. glutinosa*
        Leaves cuneate to oblong-lanceolate, few-toothed above the middle or entire; akenes 10-nerved
          Leaves, at least the upper, narrowly oblong; inflorescence sparsely leafy . . . . *B. emoryi*
          Leaves all ovate or broadly cuneate; infloresence densely leafy . . . . *B. pilularis*
      Plants with leaves mostly deciduous before anthesis, broomlike with numerous sulcate fastigiate branches; heads scattered; akenes 10-nerved
        Pistillate pappus very short, ca. 3mm long in fruit; larger leaves obovate . . . . *B. sergilodes*
        Pistillate pappus to 1cm or more in fruit; larger leaves linear . . . . *B. sarothroides*

B. brachyphylla Gray. Short-Leaf Baccharis. Uncommon, in upper desert drainages and rocky areas, 500-1000m: Jacumba; Mountain Spring; Scissors Crossing. Aug.-Nov.

B. douglasii DC. Douglas Baccharis. Rare, in drainage channel and streams, interior region above 500m: Oak Grove; Scissors Crossing. July-Oct. *n*=9.

B. emoryi Gray. Emory's Baccharis. Occasional, along streams and in canyons; foothill and coastal, below 900m: e.g. Pala; near Beauty Mountain; Fallbrook; Scripps Ranch; Telegraph Canyon; Potrero. Also Sentenac Canyon. Aug.-Dec.

B. glutinosa (Ruiz & Pavon)Pers. [*B. viminea* DC.] Mule-Fat, Seep-Willow. Common, along water courses; coastal, foothill, and desert, below 1100m: e.g. San Onofre; San Luis Rey; Fallbrook; Pala Creek; Summers Ranch; Volcan Indian Reservation; Borrego Palm Canyon; Banner; Scissors Crossing; Mason Valley; Bow Willow Canyon; Poway; El Cajon; Mission Gorge; Barrett Junction; Otay Mountain; San Diego; Tijuana River Valley. Mar.-July. *n*=9.

B. pilularis DC. ssp. **consanguinea** (DC.)Wolf. Coyote Bush, Chaparral Broom. Infrequent, in Sage Scrub; coastal, below 200m: San Onofre Bluffs; Oceanside; Fallbrook; Valley Center; Torrey Pines State Reserve; Point Loma; Tijuana River Valley. Aug.-Dec. *n*=9.

B. sarothroides Gray. Broom Baccharis. Common, along washes and in disturbed, sandy places; coastal, below 800m: e.g. Fallbrook; Mussey Grade; Alpine; Potrero; Barrett; Del Mar; Point Loma; Pacific Beach; Mission Valley; Bostonia; Murphy Canyon; Paradise Valley; Sweetwater Valley; Marron Valley. Sep.-Dec. Type from "southern borders of California, San Diego Co., near the old Mission Station, the boundary monument, etc."

B. sergiloides Gray. Squaw Waterweed. Infrequent, along washes and moist areas; foothill and desert, 600-1200m: Oak Grove; Borrego Palm Canyon; Banner; Oriflamme Canyon; Potrero Canyon; Laguna Junction; Bankhead Spring; Alpine; Campo. Feb.-Oct. 2*n*=18.

B. vanessae Beauchamp. Endemic to San Diego County. Local in Chaparral; coastal, below 300m: Encinitas; Mira Mesa; Poway; Mount Israel; Del Dios. Sep.-Nov. *n*=9.

Baileya Harv. & Gray
(Helenieae - Riddelliinae)

Ray-flowers 5-7; heads relatively small, loosely cymose; involucre ca. 6mm broad . . . . *B. pauciradiata*
Ray-flowers 20-50; heads larger, mostly solitary on elongate peduncles; involucre 1cm or more broad *B. pleniradiata*

B. pauciradiata Harv. & Gray. Colorado-Desert Marigold. Occasional, in washes; desert, below 200m: Borrego Valley; Borrego Oil Well; Rockhouse Canyon; Vallecito Valley; Palm Spring; Carrizo Creek; Jan.-May. *n*=16.

B. pleniradiata Harv. & Gray. Woolly Marigold. Rare, known in the county only from a collection from Borrego Valley, 160m. June. *n*=16.

Bebbia Greene
(Heliantheae - Galinsoginae)

B. juncea (Benth.)Greene. [*Bebbia juncea* var. *aspera* Greene; *B. a.*(Greene)A. Nels.] Rush Sweetbush. Frequent, on rocky slopes; desert, below 800m and occasionally in cismontane below 300m: e.g. Borrego Palm Canyon; The Narrows; Fish Creek; Pinyon Mountain; Agua Caliente Hot Springs; Canebrake Canyon; Mountain Palm Springs; Bow Willow Canyon; Dos Cabezas; Fallbrook; San Vicente Creek; Cottonwood; Dehesa; Mission Gorge; Otay. Jan.-July. *n*=9.

Bellis L.
(Astereae-Bellidinae)

***B. perennis** L. English Daisy. Native of Europe. Uncommon, weed in lawns: Balboa Park; East San Diego. Apr.-Sep. 2*n*=18.

Bidens L. Bur-Marigold
(Heliantheae - Coreopsidinae)

Akenes flat, cuneate to obovate . . . . . *B. laevis*
Akenes narrow, linear-tetragonal; rays inconspicuous or lacking . . . . . *B. pilosa*

B. laevis (L.)B.S.P. Smooth Beggar's-Tick. Rare, known in the county only in a slough at Carlsbad. Sep.-Nov. *n*=11, 12

***B. pilosa** L. [*B. californica* Nutt., the type from St. Diego and St. Barbara.] Common Beggar's-Tick. Native of Neotropics. Infrequent, weed in disturbed areas; cismontane, below 1100m: Hormiguero Ranch; Lake Murray; San Diego; Pine Valley. Mar.-Aug. *n*=12, 14, 36; 2*n*=72.

Blennosperma Less.
(Helenieae - Heleniinae)

B. nanum (Hook.)Blake var. **nanum**. [*B. californicum* (DC.)T. & G.] Common Blennosperma. Uncommon, in vernally moist places; cismontane, below 1400m: Julian; Cuyamaca Lake; Fletcher Hills; Kearny Mesa. Feb.-Apr. *n*=7.

Brickellia Ell.
(Eupatorieae)

Heads clustered, usually racemose-paniculate, not in corymbs
  Heads mostly 9-10mm high, involucre puberulous . . . . . *B. desertorum*
  Heads mostly 12-14mm high, involucre essentially glabrous . . . . . *B. californica*
Heads solitary or at the ends of corymbosely arranged branchlets
  Outer phyllaries broadly ovate, leaflike; leaves ovate, usually sharply toothed . . . . . *B. arguta*
  Outer phyllaries linear to narrowly oblong, not leaflike . . . . . *B. frutescens*

B. arguta Rob. var. **odontolepis** Rob. Colorado Brickellbush. Scarce, in rocky areas; desert, 600-900m: San Felipe; Sentenac Canyon; Oriflamme Mountains; Vallecito Valley; Mountain Springs; Mortero Palms; Dubber. Apr.-May. Type collected "in desertis Colorado nominatis".

B. californica (T.& G.)Gray var. **californica**. California Brickellbush. Occasional, on rocky slopes; cismontane, below 900m: San Onofre Bluffs; San Luis Rey; Rincon; Oak Grove; Sumners Ranch; Warner Springs; Hot Springs Mountain; San Felipe; San Marcos; Mussey Grade; Foster; Ramona; Alpine; Jamul; Potrero; Campo; Lake Murray. Aug.-Oct. *n*=9.

B. desertorum Cov. Desert Brickellia. Rare, in rocky desert areas 800-400m: Dos Cabezas; Bow Willow Palms. Aug.-Nov. *n*=9.

B. frutescens Gray. Rigid Brickellbush. Occasional, on dry, rocky slopes; desert, below 1200m: e.g. Angelina Spring; San Felipe; The Narrows; Box Canyon; Mason Valley; Mountain Springs; Jacumba. Apr.-June. Type from Mountain Springs.

Calycadenia DC. Rosin Weed
(Heliantheae - Madiinae)

C. tenella (Nutt.)T. & G. [*Hemizonia t.* (Nutt.)Gray, *Osmadenia t.* Nutt.] Southern Calycadenia. Prevalent, on open slopes and mesas; cismontane, below 1200m: e.g. San Onofre Canyon; O'Neill Lake; Fallbrook; Lilac; Harrison Grade; Pine Hills; Ramona; Kearny Mesa; Flynn Spring; Alpine; Clover Flat; Tecate; Otay; Pacific Beach. Mar.-Aug. Type from St. Diego. *n*=9

Calycoseris Gray
(Cichorieae - Stephanomeriinae)

Flowers yellow; akenes smooth, ca. 8mm long including the beak . . . . . *C. parryi*
Flowers white; akenes rugulose, ca. 6mm long including the beak . . . . . *C. wrightii*

C. parryi Gray. Yellow Tack-Stem. Uncommon, in open areas; cismontane and desert: Hot Springs Mountain; San Felipe; Oriflamme Mountains; Vallecito Stage Station; Jacumba; Tecate Summit. Apr.-June.

C. wrightii Gray. [*C. w.* var. *californica* Bdg., type taken "in sand, near San Felipe".] White Tack-Stem. Uncommon, in sandy areas; desert, below 300m: Borrego Palm Canyon; San Felipe; Yaqui Well; near Ocotillo Wells. Mar.-May. 2*n*=14.

Carduus L.

Heads usually 1-5 at the ends of branches; phyllaries with small rough hairs on margin . . . . . *C. pycnocephalus*
Heads usually 5-20 at the ends of branches; phyllaries glabrous or subciliate . . . . . *C. tenuiflorus*

***C. pycnocephalus** L. Italian Thistle. Native of Europe. Rare, naturalized in disturbed cismontane areas: Sweetwater Lake; Marron Valley. May-July. 2*n*=ca. 54.

***C. tenuiflorus** Curt. Native of Europe. Introduced at the campground at Dos Picos County Park.

Carthamus L.
(Cynareae)

***C. baeticus** (Boiss. & Reut.)Nym. Smooth Distaff-Thistle. Native of Mediterranean region. Rare introduction in disturbed areas: San Diego. July-Aug.

Centaurea L. Star-Thistle
(Cynareae)

Phyllaries not spinose at the tips. . . . . *C. cyanus*
Phyllaries ending in definite terminal, but sometimes short, spines
  Terminal spine of phyllary less than 5mm long; flowers purplish to white . . . . *C. diluta*
  Terminal spine of phyllary more than 5mm long
    Flowers purplish or pinkish; stems wingless
      Akenes with definite pappus . . . . *C. iberica*
      Akenes without pappus or with merely a vestige . . . . *C. calcitrapa*
    Flowers yellow; stem-wings with decurrent leaf-bases
      Spines slender, purplish, 1cm long or less; plants branched mostly above the base; corolla glandular *C. melitensis*
      Spines stout, yellow, 1-2cm long; plants branched from the base; corolla without glands . . . . *C. solstitialis*

***C. calcitrapa** L. Purple Star-Thistle. Native of Europe. Infrequent, introduced weed in disturbed areas: Springhill; Viejas Valley. July-Oct. 2*n*=20.

***C. cyanus** L. Bachelor's-Button. Native of Europe. Uncommon, escaping from cultivation: Alpine; Bratton Valley. May-Aug. 2*n*=24.

***C. diluta** Ait. Pale-Flower Centaurea. Native of Europe. Rare, introduced weed: Vista, 100m. May-July. 2*n*=20 pairs.

***C. iberica** Trev. Iberian Star-Thistle. Native of Asia Minor. Infrequent, weed in disturbed fields and waste places, above 300m: Ramona; Rancho Santa Teresa; Ballena; Crouch Ranch. Aug.-Nov. 2*n*=16, 20.

***C. melitensis** L. Tocalote. Native of Europe. Prevalent, weed in disturbed fields and waste places, below 800m: e.g. San Onofre Bluffs; De Luz; Barona Valley; Alpine; Dulzura; Miramar; Point Loma. Apr.-June. 2*n*=22, 24, 36.

***C. solstitialis** L. Barnaby's Thistle. Native of Europe. Uncommon, weed in cultivated areas: Mesa Grande; Wynola. May-Oct. 2*n*=16.

Chaenactis DC.
(Helenieae - Heleniinae)

Pappus-paleae 10 or more in 2 series; perennials or biennials; leaf-segments linear, mostly entire; suffruticose *C. parishii*
Pappus-paleae 4-5 in 1 series or with outer reduced paleae, or obsolete; annuals
  Flowers yellow
    Peduncles very long, subscapose and slender . . . . *C. glabriuscula* var. *lanosa*
    Peduncles relatively short, usually much less than half the length of the stem
      Phyllaries oblong, plane, with thick obtuse tips; marginal corollas mostly with palmate limb
        Phyllaries 8-10mm high, the heads medium large; lowlands . . . . *C. glabriuscula* var. *g.*
        Phyllaries 5-7mm high, the heads small; montane . . . . *C. glabriuscula* var. *curta*
      Phyllaries boat-shaped, narrowly linear; marginal corollas only slightly enlarged, seldom with palmate limb
        Leaf-segments very narrow, rather short and recurved. . . . . *C. glabriuscula* var. *tenuifolia*
        Leaf-segments succulent, short, obtuse . . . . *C. glabriuscula* var. *orcuttiana*
  Flowers white or tinged with purple
    Plants 3-15dm high; pappus none or rudimentary; akenes flattened . . . . *C. artemisiaefolia*
    Plants mostly lower than 3dm; pappus of persistent entire or erose paleae; akenes subterete
      Phyllaries attenuate into a bristle-tip
        Pappus-paleae not more than 1/3 the length of the corolla . . . . *C. carphoclinia* var. *attenuata*
        Pappus- paleae of central flowers almost as long as the corolla
          Leaves distributed along stems, but reduced upward . . . . *C. carphoclinia* var. *c.*
          Leaves mostly gathered into a dense, basal cluster . . . . *C. carphoclinia* var. *peirsonii*
      Phyllaries obtuse or acute
        Leaves entire or once-parted into few linear lobes, green; plants soon glabrate
          Phyllaries 8-10mm long, essentially glabrous including the short straight tip; marginal corollas much enlarged; pappus uniseriate . . . . *C. fremontii*
          Phyllaries 12-16mm long, at least the curving tip densely puberulent; marginal corollas scarcely enlarged; pappus biseriate . . . . *C. xantiana*
        Leaves bipinnatifid with short thick segments, grayish; plants usually persistently tomentose; phyllaries 6-9mm long, glandular-puberulent . . . . *C. stevioides*

C. artemisiaefolia (Harv. & Gray)Gray. Artemisia Pincushion. Frequent, on open slopes and burns; cismontane, 400-1400m: e.g. Rainbow Creek; Palomar Mountain; Henshaw Dam; Hot Springs Mountain; San Felipe Valley; Banner; Ramona; Lake Hodges; Mount Woodson; Sweetwater Valley; Dulzura; Tecate Mountain; Campo; Buckman Spring; White Star; Jacumba. Apr.-July. *n*=8.

C. carphoclinia Gray var. **attenuata** (Gray)Jones. Death Valley Pincushion. Infrequent, in sandy areas; desert, below 400m: Borrego Springs; Yaqui Well; northeast corner of county; Vallecito Stage Station. Mar.-June. *n*=8.

C. carphoclinia var. **carphoclinia**. Pebble Pincushion. Infrequent, in sandy areas; desert, below 400m: Borrego Palm Canyon; San Felipe; Borrego Springs; Kane Springs Road; Palm Spring. Mar.-May. *n*=8.

C. carphoclinia var. **peirsonii** (Jeps.)Munz. Peirson's Pincushion. Uncommon, in sandy areas; desert, below 100m: Benson Dry Lake; Palm Wash; northeast corner of county; Silent Canyon. Mar.-May. Type from "Silent Canyon, in the desert foothills, southeast of the Santa Rosa Mountains".

C. fremontii Gray. Desert Pincushion. Frequent, on slopes and mesas; desert, below 1200m: e.g. Borrego Valley; northeast corner of county; Borrego Palm Canyon; Banner; San Felipe Valley; Scissors Crossing; Sentenac Canyon; Mason Valley; Carrizo Stage Station; Dos Cabezas; Jacumba; southeast corner of county. Mar.-May. 2*n*=10.

C. glabriuscula DC. var. **curta** (Gray)Jeps. Small-Flower Pincushion. Occasional, on open slopes; foothill and montane, above 1000m: Warners Ranch; San Felipe Valley; Hot Springs, Cuyamaca, and Laguna Mountains. Apr.-Aug. *n*=6.

C. glabriuscula var. **glabriuscula**. Yellow Pincushion. Uncommon, on open slopes; cismontane: San Diego; Rainbow; Laguna Lakes. Mar.-May. *n*=6, 2*n*=5 pairs.

C. glabriuscula var. **lanosa** (DC.)Hall. Woolly Pincushion. Uncommon, in dry valleys; foothill, 700-1300m: Lake Henshaw; Montezuma Valley; Moretti Junction; Barrett. Mar.-June. *n*=6

C. glabriuscula var. **orcuttiana** (Greene)Hall. Orcutt's Pincushion. Frequent, in sandy areas; coastal, below 150m: e.g. Oceanside; Carlsbad; Solana Beach; Encinitas; Del Mar; Torrey Pines; Pacific Beach; Point Loma; Ocean Beach; North Island; Initial Boundary Monument. Jan.-June. Type from "along the beaches about San Diego". *n*=6.

C. glabriuscula var. **tenuifolia** (Nutt.)Hall. San Diego Pincushion. Common, on open and disturbed slopes and mesas and burns; cismontane: e.g. San Onofre Canyon; Moro Hill; Bonsall; Rainbow; Pala; Lilac; San Luis Rey; Bernardo; Lake Hodges Grove; La Jolla; Miramar; Ocean Beach; Point Loma; Mission Hills; La Mesa; Dulzura; Witch Creek; Barrett; Ramona; Campo; Laguna Camp; Rattlesnake Valley; Montezuma View Point. Mar.-June. Type from St. Diego. *n*=6.

C. parishii Gray. Parish's Pincushion. Uncommon, in Chaparral, montane; above 1500m: Garnet Peak; Cuyamaca Peak; Stonewall Peak; Garnet Mountain. June-July. *n*=6.

C. stevioides H. & A. var. **stevioides**. [*C. brachypappa* Gray; *C. latifolia* Stockwell, the type from Jacumba Springs; *C. mexicana* Stockwell] Broad-Flower Pincushion. Infrequent, in sandy areas; desert, below 1000m: Borrego Palm Canyon; Borrego Springs; Jacumba. Mar.-Apr. *n*=5.

C. xantiana Gray. Rare, known in the county only at upper Mason Valley, 700m. Apr. *n*=7.

Chaetopappa DC.
(Astereae-Solidagininae)

C. aurea (Nutt.)Keck. [*Pentachaeta a.* Nutt.] Golden Daisy. Frequent, in open areas; cismontane: e.g. San Onofre Canyon; De Luz; Moro Hill; Banner; Rattlesnake Valley; Pine Hulls; Cuyamaca Lake; Cuyamaca Mountains; Laguna Lakes; Morris Ranch; Descanso; Otay Mountain; Kearny Mesa; Torrey Pines; La Jolla; Point Loma. Mar.-June. Type from "dry plains near the sea, in the vicinity of St. Diego". *n*=9.

Chrysanthemum L.
(Anthemideae - Chrysantheminae)

Plants annual
  Heads with a tricolored effect, the disk purple, the rays white or red or purple with yellow or purple ring . . . . . *C. carinatum*
  Heads not tricolored, light yellow . . . . . *C. coronarium*
Plants perennial; rays white; heads borne singly . . . . . *C. leucanthemum*

***C. carinatum** L. Tricolor Chrysanthemum. Native of Morocco. Rare, escaping from cultivation: Silver Strand. Apr.-Juiy. *n*=9.

***C. coronarium** L. Garland Chrysanthemum. Native of Europe. Common weed in disturbed areas, especially along roadways; coastal, below 200m: Oceanside; Santee; San Diego; Ocean Beach; Silver Strand; Glen View. Apr.-Aug. 2*n*=18, 36.

***C. leucanthemum** L. [*Leucanthemum vulgare* Lam.] Common Ox-eye Daisy. Native of Old World. Local, naturalized in Spencer Valley, 1080m. July. 2*n*=18, 36, 54.

Chrysopsis (Nutt.)Ell. Golden-Aster
(Astereae - Solidagininae)

Heads rather many, cymose or paniculate, not subtended by leafy bracts; mostly away from immdiate coast
  Leaves pubescent with spreading hairs . . . . . *C. villosa*. var. *echioides*
  Leaves pubescent with silky appressed hairs . . . . . *C. villosa* var. *fastigiata*
Heads large, solitary or few in a cluster, closely subtended by leafy bracts; herbage obviously glandular; mostly coastal . . . . . *C. villosa* var. *sessiliflora*

C. villosa (Pursh)Nutt. var. **echioides** (Benth.)Gray. [*Heterotheca e.* (Benth.)Shinners] Frequent, in open areas; cismontane, below 1500m: e.g. Doane Valley; Witch Creek; Ramona; Buckman Spring; Potrero; Torrey Pines; San Diego; Initial Boundary Monument. July-Oct. *n*=9.

C. villosa var. **fastigiata** (Greene)Hall. [*Heterotheca f.* (Greene)Harms] Rare, known in the county only at Warner Hot Springs, 930m. Oct. *n*=9.

C. villosa var. **sessiliflora** (Nutt.)Gray. [*Heterotheca s.* (Nutt.)Shinners] Occasional, on slopes and in washes; coastal, below 150m: e.g. San Elijo Lagoon; Torrey Pines; Pacific Beach; Ocean Beach; North Island; Silver Strand; San Diego; Boundary Monument. Also Palomar Mountain. Mar.-July.

Chrysothamnus Nutt. Rabbit-Brush.
(Astereae - Solidagininae)

Leaves resinous-punctate, terete . . . . . *C. paniculatus*
Leaves not resinous-punctate . . . . . *C. nauseosus*

C. nauseosus (Pall.)Britt. ssp. **consimilis** (Greene)Hall & Clem. Uncommon, in open areas; montane and desert, below 1200m: Descanso Junction; The Narrows. Oct.-Nov. *n*=9.

C. paniculatus (Gray)Hall. Punctate Rabbit-Brush. Occasional, in washes; desert, below 600m: The Narrows; Vallecito and Carrizo Stage Stations; Canebrake Canyon; Palm Spring; Carrizo Gorge. Also Dodge Valley and Chihuahua Valley. June-Dec. 2*n*=18.

Cichorium L. Chicory
(Cichorieae - Cichoriinae)

*__C. intybus__ L. Mediterranean Chicory. Native of Europe. Occasional, naturalized in moist, disturbed areas; cismontane, below 1400m: Palomar Mountain; Lake Henshaw; San Marcos; Fletcher Hills; Santee; Viejas Valley. June-Oct. 2*n*=18.

Cirsium L. Thistle
(Cynareae)

Stem spinose-winged by decurrent leaf-bases almost as long as internodes; leaves scabrous-hispid above . . . *C. vulgare*
Stems not spinose-winged by decurrent leaf bases or only shortly so; leaves not scabrous-hispid above
  Middle or outer phyllaries with a conspicuous glutinous dorsal ridge
    Stems 1-4dm high; spines of the phyllaries not more than 5-6mm long . . . . *C. undulatum*
    Stems 5-15dm high: spines of the phyllaries ca. 1cm long; introduced from elsewhere in the state *C. ochrocentrum*
  Middle or outer phyllaries not bearing a conspicuous glutinous dorsal ridge
    Heads leafy-bracteate (the spiny foliaceous subtending bracts usually more than 3), usually clustered
      Plants acaulescent, 2.5-4.5cm high. . . . . *C. tioganum*
      Plants caulescent, 5-20dm high . . . . *C. brevistylum*
    Heads not subtended by leafy bracts, usually not much clustered
      Stems coarse, the peduncles mostly 3-5mm in diameter . . . . *C. occidentale*
      Stems more slender, the peduncles mostly 2-2.5mm in diameter . . . . *C. californicum*

C. brevistylum Cronq. [*C. edule* auct. non Nutt.] Indian Thistle. Infrequent, on dry slopes; coastal, below 200m: Camp Pendleton; San Diego. Apr.-Aug. 2*n*=34.

C. californicum Gray. [*C.c.* var. *bernardinum* (Greene) Petrak] California Thistle. Common, on grassy slopes; cismontane, below 1500m: e.g. Chariot Canyon; La Posta; Herron Ranch; Hormiguero Ranch; Palomar Mountain; Eagles Nest; Cuyamaca Lake; Desert View; Ramona; Oceanside; Lake Hodges Grove; Mount Soledad; Mission Hills; Point Loma; San Diego. Mar.-May. 2*n*=30.

C. occidentale (Nutt.)Jeps. Cobweb Thistle. Frequent, on grassy slopes; coastal, below 800m: San Onofre Canyon; Moro Hill; Hormiguero Ranch; Olivenhain; Lake Hodges; San Diego; Chollas Valley; Sweetwater. Mar.-May. 2*n*=30.

*__C. ochrocentrum__ Gray. Yellow-Spine Thistle. Native of Great Basin and elsewhere. Local, introduced in disturbed areas; cismontane, below 200m: San Marcos; Escondido; Otay Mesa. June-July. 2*n*=30, 31, 32.

C. tioganum (Congdon)Petrak. [*C. drummondii* auct. non T. & G.; *C. acaulescens* K. Schum. in Just.] Bird's-Nest Thistle. Frequent, on slopes and in grassy meadows; montane, above 900m: Palomar Mountain; Doane Valley; Eagles Nest; Montezuma View Point; Volcan Mountain; Witch Creek; Santa Ysabel; Pine Hills; Julian; Harper Ranch; Campo; Cuyamaca Lake; Laguna Mountains. May-Aug. 2*n*=17 pairs.

*__C. undulatum__ (Nutt.)Spreng. [*C. megacephalum* (Gray)Ckll.] Wavy-Leaf Thistle. Native of Arizona and elsewhere. Local, naturalized in waste places: Escondido; Otay Mesa. May-Sep. 2*n*=26.

*__C. vulgare__ (Savi)Ten. Bull Thistle. Native of Europe. Uncommon, weed in disturbed areas; cismontane, below 800m: Vista; Warner Hot Springs; Montezuma Mesa. May-Sep. 2*n*=68.

Cnicus L.
(Cynareae)

*__C. benedictus__ L. Blessed Thistle. Native of Europe. Frequent, weed of fields and other disturbed areas; cismontane, below 1200m: e.g. Rainbow Valley; Vista; Paradise Canyon; Poway; Hidden Glen; Laguna Junction; Olivenhain; Jamacha. May-July. n=11; 2*n*=22.

Conyza L., emend. Less.
(Astereae - Asterinae)

Upper and lower leaves coarsely dentate; ray-flowers without ligules; herbage glandular-viscid . . . . *C. coulteri*
Upper leaves, at least, entire; ray-flowers with ligules
  Involucre copiously hairy, the outer phyllaries herbaceous . . . . *C. bonariensis*
  Involucre subglabrous, the phyllaries with scarious margins . . . . *C. canadensis*

*__C. bonariensis__ (L.)Cronq. [*Erigeron linifolius* Willd.] Flax-Leaf Fleabane. Native of South America. Occasional, weed of waste places; coastal, below 150m: Oceanside; Lake Hodges; San Diego; Ocean Beach. June-Dec. 2*n*=54.

*__C. canadensis__ (L.)Cronq. [*Erigeron c.* L.] Common Horseweed. Native of eastern United States. Common weed in disturbed areas; cismontane, below 1200m: e.g. San Luis Rey; Fallbrook; Valley Center; Palomar Mountain; Oak Grove; Mesa Grande; San Felipe Valley; Ballena; Foster; Bostonia; Alpine; Boulevard; White Star; Canyon City; Potrero Grade; Point Loma; Silver Strand; La Jolla; Encinitas. Aug.-Nov. 2*n*=18.

C. coulteri Gray. Coulter's Fleabane. Uncommon, in alkaline areas; coastal, below 200m: Lake Murray; Silver Strand; Meadowbrook School, Poway; Rose Canyon. May-Oct.

Coreopsis L.
(Heliantheae - Coreopsidinae)

Perennials; stems stout; heads 5-8cm across; coastal
 Heads numerous, cymosely clustered on leafy peduncles; introduced in disturbed areas . . . . *C. gigantea*
 Heads few, on naked peduncles. . . . . *C. maritima*
Annuals; stems slender; heads 2-5cm across
 Rays yellow
  Akenes nonciliate; pappus-paleae none. . . . . *C.californica*
  Akenes of disk ciliate, with 2 pappus-paleae; ray-akenes glabrous, epappose . . . . *C. bigelovii*
 Rays purple-brown . . . . *C. basalis*

***C. basalis** (Otto & Dietr.)Blake var. **wrightii** (Gray)Blake. Wright's Coreopsis. Native of Texas. Local, adventive about Pechstein Reservoir, 120m. May.

C. bigelovii (Gray)Hall. Bigelow's Coreopsis. Rare, known in the county only at Mason Valley, 700m. Mar.

C. californica (Nutt.)H. K. Sharsm. California Coreopsis. False Tidy-Tips. Frequent, on open slopes; cismontane, below 1100m: e.g. De Luz; Fallbrook; San Luis Rey; Rainbow; Moosa Canyon; Encinitas; Cardiff; Del Mar; La Jolla; Kearny Mesa; Point Loma; Tecate Junction; Witch Creek; Julian; Jacumba. Mar.-May. Type from "near St. Diego".

***C. gigantea** (Kell.)Hall. Giant Sea-Dahlia. Native of California Channel Islands and from Point Dume, Los Angeles County, north to Oso Flaco Lake, San Luis Obispo County. Local, established by hydroseeding on road cuts; coastal, below 50m: La Costa; Ponto; Solana Beach. Mar.-May.

C. maritima (Nutt.)Hook. f. San Diego Sea-Dahlia. Frequent, on bluffs; coastal, below 150m: e.g. Oceanside; Cardiff; San Elijo Lagoon; San Marcos Creek; Del Mar; Torrey Pines State Reserve; La Jolla; Pacific Beach; Point Loma; Silver Strand; Imperial Beach; Initial Boundary Monument; Tijuana Hills. Mar.-May. Type from "shelving rocks, near the sea; at St. Diego".

Corethrogyne DC. Sand-Aster
(Astereae - Asterinae)

Involucre glandular at anthesis
 Involucre 10-12mm high, hemispheric, with long-stalked glands; coastal . . . . *C. filaginifolia* var. *incana*
 Involucre 6-7mm high; stems fairly slender, quite erect; montane
  Stems usually under 4dm high. . . . . *C. filaginifolia* var. *glomerata*
  Stems usually over 4dm high
   Tomentum extending up to involucre; involucre turbinate, 5-7mm high. . . . . . *C. filaginifolia* var. *bernardina*
   Tomentum not extending up to involucre, but upper parts glandular . . . . *C. filaginifolia* var. *virgata*
Involucre tomentose at anthesis
 Leaves linear . . . . *C. filaginifolia* var. *linifolia*
 Leaves not linear . . . . *C. filaginifolia* var. *sessilis*

C. filaginifolia (H. & A.)Nutt. var. **bernardina** (Abrams)Hall. Foothill Cudweed-Aster. Infrequent, on open slopes; foothill and montane, 800-1300m: Palomar Mountain; Oak Grove; Warner Hot Springs. Sep.-Oct. n=5.

C. filaginifolia var. **glomerata** Hall. [*C. brevicula* Greene, the type from mountains of San Diego County; *C. f.* var. *brevicula* (Greene)Canby; *C. racemosa* Greene, the type from mountains of San Diego County] Mountain Cudweed-Aster. Occasional, on open slopes; cismontane, 300-1400m: e.g. Palomar Mountain; Oak Grove; San Felipe; Santa Ysabel; Julian; Pine Hills; Ramona; Mussey Grade; Guatay; Cuyamaca Peak; Laguna Mountains; Boulevard; Canyon City; El Cajon; Jamul. July-Oct. n=5.

C. filaginifolia var. **incana** (Nutt.)Canby. [*C. filaginifolia* var. *pacifica* Hall, the type from Pacific Beach; *C. i.* Nutt] San Diego Sand-Aster. Occasional, in sandy and disturbed areas; coastal, below 100m: near Del Mar; Pacific Beach; Ocean Beach; Point Loma; North Island; Boundary Monument No. 258. Jan.-June. Type from "near St. Diego".

C. filaginifolia var. **linifolia** Hall. [*C. l.* (Hall)Ferris] Del Mar Sand-Aster. Endemic to San Diego County. Occasional, in sandy and disturbed places; coastal, below 150m: Carlsbad; Encinitas; Del Mar; Del Mar Heights; Torrey Pines State Reserve; Fort Rosecrans; San Diego Bay. July-Sep. Type from "about one kilometer south of Del Mar".

C. filaginifolia var. **sessilis** (Greene)Canby. [*C. s.* Greene] San Bernardino Cudweed-Aster. Rare, known in the county only from a collection at Sunday School Flats on Palomar Mountain, 1300m. July-Oct.

C. filaginifolia var. **virgata** (Benth)Gray. Virgate Cudweed-Aster. Common, in open areas; coastal and foothill, below 1100m: O'Neill Lake; Fallbrook; Oak Grove; Torrey Pines; Rose Canyon; Point Loma; Otay; Chollas Valley; Silver Strand; Japatul Valley; Descanso; Campo; Boulevard. July-Oct. n=5.

Cotula L.
(Anthemideae - Chrysantheminae)

Annual; leaves finely pinnate; heads 2-5mm broad . . . . *C. australis*
Perennial; leaves entire or pinnatifid; heads 8-10mm broad . . . . *C. coronopifolia*

***C. australis** (Sieber ex Spreng.)Hook. f. Australian Brass-Buttons. Native of Australia. Common, weed of disturbed areas; cismontane, below 200m: e.g. De Luz; Miramar; East San Diego; San Diego; Silver Strand; National City; Dulzura. Jan.-May.

***C. coronopifolia** L. African Brass-Buttons. Native of South Africa. Common, in wet places, below 700m: e.g. O'Neill Lake; San Marcos; Point Loma; Kearny Mesa; Poway Junction; Ramona; Lakeside; Chollas Heights; Sweetwater Lake; Sears Ranch; Banner; Coyote Creek. Mar.-Dec. 2*n*=20.

Cynara L. Artichoke
(Cynareae)

***C. cardunculus** L. [*C. scolymus* L.] Globe Artichoke, Cardoon. Native of Europe. Occasional, escaping from cultivation and naturalized in open fields; coastal, below 200m: Camp Pendleton; Vista; La Costa; Bernardo; Pacific Beach; Rancho Santa Fe. May-July. 2*n*=34.

Dicoria T. & G.
(Heliantheae - Ambrosiinae)

D. canescens T. & G. ssp. **canescens**. Desert Dicoria. Infrequent, in washes and other sandy areas; desert, below 200m: Borrego Valley; The Narrows; Vallecito and Carrizo Stage Stations; Carrizo Creek. Nov.-Jan. *n*=18.

Dimorphotheca Vaill. Cape-Marigold
(Calenduleae)

***D. sinuata** DC. Blue-Eye Cape-Marigold. Native of South Africa. Local, escaping from cultivation and persisting about dumps: Ramona; Escondido; Encinitas; San Diego; Borrego Springs. Nov.-Apr.

Dyssodia Cav.
(Helenieae - Tagetininae)

Involucre 12-16mm high; leaves alternate, inconspicuous . . . . *D. porophylloides*
Involucre 5-7mm high; leaves opposite, mostly longer than the internodes . . . . *D. pentachaeta*

D. pentachaeta (DC.)Robbins var. **belenidium** (DC.)Strother. [*Thymophylla p.* (DC.)Small] Thurber's Dyssodia. Rare, known in the county only from a collection from the southeast slope of Pinyon Mountain, 900m. Jan. *n*=13,16.

D. porophylloides Gray. San Felipe Dyssodia. Frequent, in washes and on mesas; desert, below 1100m: e.g. Borrego Palm Canyon; San Felipe Wash; Fish Creek Mine; Sentenac Canyon; upper Blair Valley; lower Box Canyon; Vallecito Stage Station; Canebrake Canyon; Mountain Palm Springs; Jacumba; Mountain Springs. Nov.-Apr. Type from "sandy hills, near San Felipe". *n*=13.

Eclipta L.
(Heliantheae - Verbesininae)

***E. prostrata** (L.)L. [*E. alba* auct.] False-Daisy. Native of Neotropics. Local, naturalized in moist areas; coastal, below 150m: below Lake Hodges; near Otay Lake; Sweetwater Reservoir; Chula Vista. Jan.-Sep. *n*=11.

Encelia Adans.
(Heliantheae - Verbesininae)

Heads in cymose panicles; peduncles glabrous; leaves white-tomentulose
  Disc corollas yellow . . . . *E. farinosa* var. *f.*
  Disc corollas purple . . . . *E. farinosa* var. *phenicodonta*
Head solitary at tips of pubescent peduncles; leaves not tomentulose
  Disks purple; rays 18-30mm long . . . . *E. californica*
  Disks yellow; interior and deserts
    Rays none; leaves scabrous with scattered pustulate-based hairs . . . . *E. frutescens*
    Rays present, 5-15mm long; leaves finely appressed-pubescent, sometimes with slender scabrous hairs between . . . . *E. virginensis*

E. californica Nutt. California Encelia. Prevalent on bluffs and slopes; coastal, below 150m: e.g. San Onofre Bluffs; Cardiff; Lake Hodges; San Pasqual; Del Mar; Torrey Pines State Reserve; Pacific Beach; Point Loma; Mission Valley; Murphy Canyon; Rolando; Jamul; Emerald Hills; Silver Strand; Boundary Monument No. 258; Otay Ranch. Jan.-July.

E. farinosa Gray ex Torr. var. **farinosa**. Incienso, Desert Encelia. Abundant, on dry slopes; desert, below 1100m: e.g. Lone Palm; Borrego Palm Canyon; northeast corner of county; Warner Springs; Hellhole Canyon; The Narrows; Yaqui Well; Box Canyon; Campbell Ranch; Dos Cabezas; Mountain Springs; Jacumba. Mar.-May. Type from vicinity of Cariso (Carrizo) Creek. *n*=17, 18.

E. farinosa var. **phenicodonta** (Blake)Jtn. Purple-eyed Incienso. Rare, known in the county only from a collection from Hellhole Canyon, 350m. Apr.

E. frutescens (Gray)Gray ex Torr. var. **frutescens**. Rayless Encelia. Occasional, on rocky slopes and mesas; desert, below 200m: northeast corner of county; Borrego Mud Hills; Fish Creek; Ocotillo Wells; The Narrows; Moonlight Canyon; Palm Spring; Carrizo Stage Station. Nov.-Apr. *n*=17, 2*n*=18 pairs.

E. virginensis A Nels. ssp. **actoni** (Elmer)Keck. [*E. frutescens* var. *a.* (Elmer)Blake] Acton's Encelia. Occasional, on slopes and in valleys; desert and foothill, below 1100m: Oak Grove; Warner Hot Springs; Warner Ranch; Eagles Nest; Borrego Palm Canyon; Grapevine Spring; Scissors Crossing; Jacumba; Pinyon Mountain; Mountain Springs. Oct.-July.

Ericameria Nutt. Goldenbush
(Astereae - Solidagininae)

Involucres hemispheric, 10-18mm wide; leaves entire . . . . . *E. linearifolia*
Involucres turbinate or subcylindric
  Ray-flowers present; leaves filiform; heads mostly paniculate
    Outer phyllaries mostly caudate-tipped . . . . . *E. ericoides*
    Outer phyllaries obtuse
      Leaves mostly 8-12mm long . . . . . *E. pachylepis*
      Leaves 1.5-4cm long . . . . . *E. palmeri*
  Ray-flowers reduced or wanting
    Leaves filiform to linear, less than 3mm wide . . . . . *E. brachylepis*
    Leaves oblanceolate to obovate, 3-10mm wide
      Heads 9-12 flowered, discoid; erect shrub 2-5m high; leaves 2-6cm long, mostly tapering to base and apex . . . . . *E. arborescens*
      Heads more than 16-flowered, sometimes radiate; spreading subshrubs mostly less than 1m high; leaves cuneate
        Disk flowers 33 or fewer per head . . . . . *E. cuneata* ssp. *spathulata*
        Disk flowers 36 or more per head . . . . . *E. cuneata* ssp. *macrocephala*

E. arborescens (Gray)Greene ssp. **parishii** (Greene)Urbatsch. [*Haplopappus p.* (Greene) Blake] Parish's Goldenbush. Occasional in Chaparral; foothill and montane, often abundant after a burn or brush clearing; above 800m: e.g. Deer Flat; Bear Valley; Cuyamaca Peak; Harper Ranch; Viejas Valley; Oakzanita; Guatay; Laguna Junction; Wildwood Glen; Otay Mountain. July-Oct. $2n$=18.

E. brachylepis (Gray)Hall. [*Haplopappus propinquus* Blake; *H. b.* (Gray) Hall non Phil.] Boundary Goldenbush. Occasional, on arid slopes; foothill and desert, 200-1000m: Fallbrook; San Pasqual; Dodge Valley; Poway; Lakeside; Angelina Spring; upper Blair Valley; Box Canyon; La Puerta; Carrizo Gorge; Mountain Springs; Table Mountain; Jacumba; Campσ; Potrero. Sep.-Dec. Type from "Larken's Station (Jacumba).

E. cuneata (Gray)McClatchie var. **macrocephala** Urbatsch. Laguna Goldenfleece. Endemic to San Diego County. Occasional, on rock outcrops; montane and transmontane, above 1000m: Desert View, Laguna; Ivy Point; Garnet Peak. Type from Desert View Point, Laguna Mountains. Sep.-Dec.

E. cuneata var. **spathulata** (Gray)Hall. [*Haplopappus c.* var. *s.* (Gray)Hall] Wedgeleaf Goldenbush. Frequent, on rocky slopes and outcrops; desert and montane, above 800m: e.g. Hot Springs Mountain; Montezuma Valley; Jacumba; southeast corner of county. Apr.-Nov. n=9.

E. linearifolia (DC.)Urbatsch & Wussow. [*Stenotopsis l.* (DC.) Rydb.; *Haplopappus l.* DC.; *H. l.* ssp. *interior* (Cov.) Hall] Narrowleaf Goldenbush. Occasional, on slopes; desert and foothill; 300-1200m: Montezuma View Point; San Felipe; Banner; Shaw Canyon; Hidden Glen; Witch Creek; Viejas Grade; Barrett; Hipass; Campbell Ranch; Dos Cabezas; Dubber; Mountain Springs. Mar.-June. n=9.

E. pachylepis (Hall)Urbatsch & Bondy. [*Haplopappus palmeri* ssp. *pachylepis* Hall] Box Springs Goldenbush. Rare, known in the county only at Oak Grove, 840m. Sep. n=9.

E. palmeri (Gray)Hall. [*Haplopappus p.* Gray] Palmer's Goldenbush. Uncommon, in dry valleys and plains, coastal, below 200m: Mission Valley; Mahogany Canyon; Balboa Park; Jamacha; Cottonwood; Dulzura; Telegraph Canyon. Aug.-Dec.

E. ericoides (Lessing)Jeps. ssp. **pinifolia** (Gray)Urbatsch & Bondy. [*Haplopappus p.* Gray; *H. illinitus* Eastw.] Pine Goldenbush. Frequent, in Chaparral, foothill, 500-1600m: e.g. Pala; Palomar Observatory; Eagles Nest; Warners Pass; Santa Ysabel; Anahuac; Oakzanita; Glen Cliff Campground, Pine Valley; Guatay; High Bridge; Boulevard; Mountain Springs. Sep.-Apr.

Erigeron L.
(Astereae - Asterinae)

Internodes very numerous and short; leaves linear or narrowly oblong, uniform from base to near top of plants; phyllaries markedly graduate
  Foliage mostly grayish . . . . . *E. foliosus* var. *covillei*
  Foliage mostly green
    Leaves filiform, mostly glabrous . . . . . *E. foliosus* var. *stenophyllus*
    Leaves narrowly oblong, somewhat strigose . . . . . *E. foliosus* var. *f.*
Internodes not excessively numerous or usually very short; leaves variously shaped, sometimes linear, but then the basal obviously larger than the cauline; phyllaries equal or imbricate
  Cauline leaves ample, sessile and clasping, 8-15cm long; rays 150-400, 5-10mm long; moist places below 1200m . . . . . *E. philadelphicus*
  Cauline leaves mostly not very well developed, commonly linear or narrow, not as above . . . . . *E. divergens*

E. divergens T. & G. [*E. incomptus* Gray] Diffuse Daisy. Uncommon, in sandy places; cismontane: Laguna Lakes; Potrero; Otay Creek; Otay Mesa; Tijuana River Valley. Mar.-Nov. $2n$=ca. 27, 36.

E. foliosus Nutt. var. **covillei** (Greene)Compton. Rare, known in the county only from a collection from San Felipe Valley, 800m. May-Aug.

E. foliosus var. **foliosus**. Leafy Daisy. Occasional, on grassy slopes and in meadows; foothill and montane, above 150m: e.g. Doane Valley; Palomar Mountain; Oak Grove Trail; San Felipe; Cuyamaca Lake; Laguna Public Camp; Hauser Creek; Jacumba; Campo; Descanso; Potrero; Del Dios. June-July.

E. foliosus var. **stenophyllus** (Nutt.)Gray. Occasional, on grassy slopes; cismontane, below 1200m: e.g. San Onofre Canyon; De Luz; Fallbrook; Pala; Ramona; Barona Valley; Poway Grade; Lake Hodges; Olivenhain; Otay Mountain; Jamul Mountain; El Capitan; Dulzura; Viejas Grade; Dubber; Jacumba. Apr.-June. n=9.

***E. philadelphicus** L. Philadelphia Daisy. Native of eastern United States. Local, naturalized in moist places: Balboa Park, 40m. Apr. 2*n*=18.

Eriophyllum Lag.
(Helenieae - Heleniinae)

Perennial (or biennials), somewhat woody at base
  Leaves green above; heads in compact corymbose clusters . . . . *E. confertiflorum* var. *c.*
  Leaves gray-tomentose above; heads in mostly open corymbs . . . . *E. confertiflorum* var. *laxiflorum*
Annuals, 2-20cm high
  Heads mostly sessile in leafy-bracteate terminal clusters; pappus-paleae lacerate-fimbriate
    Rays present; pappus paleae shorter than corolla-tube . . . . *E. multicaule*
    Rays lacking; pappus-paleae equaling the corolla-tube . . . . *E. pringlei*
  Heads pedunculate; pappus-paleae if present not lacerate-fimbriate
    Anthers with acute to obtuse tips; peduncles usually 2-10cm long; ligules yellow; pappus less than 0.4mm long . . . . *E. ambiguum*
    Anthers with linear-subulate tips
      Pappus of 4-5 unequal paleae . . . . *E. lanosum*
      Pappus of 6-10 paleae
        Rays yellow . . . . *E. wallacei* var. *w.*
        Rays white to reddish . . . . *E wallacei* var. *rubellum*

E. ambiguum (Gray)Gray var. **ambiguum**. Desert Woolly-Daisy. Scarce, on rocky slopes and washes; desert, 900-1300m: Hot Springs Mountain; Mountain Springs; Bow Willow Palms; Mortero Palms. May. *n*=7.

E. confertiflorum (DC.)Gray var. **confertiflorum**. Long-Stem Golden-Yarrow. Prevalent, in Sage Scrub, Chaparral and Montane Coniferous Forest; cismontane: e.g. San Onofre Bluffs; De Luz; Rainbow; Moosa Canyon; Palomar Mountain; Eagles Nest; North Peak; Cuyamaca Lake; Laguna Lakes; Descanso; Poway Cutoff; Olivenhain; Miramar; San Pasqual; Lake Hodges; Solana Beach; Point Loma; Emerald Hills; Lee Valley; El Capitan; Otay Mountain. Jan.-July. n=8, 16; 2*n*=24 pairs.

E. confertiflorum var. **laxiflorum** Gray. Gray-Leaf Golden-Yarrow. Uncommon, on rocky slopes; desert, 800-1500m: Dubber; Table Mountain. Jan.-July.

E. lanosum (Gray)Gray. Woolly Frocks. Uncommon, in sandy areas; desert, below 800m: one mile west of Vallecito Stage Station; Volcanic Hills; southwest of Lava Flow Wash. *n*=4.

E. multicaule (DC.)Gray. Many-Stem Woolly Daisy. Infrequent, in sandy places; cismontane and desert, below 900m: Fallbrook; Rincon Indian Reservation; Borrego Valley. Dec.-Apr. *n*=7, 9.

E. pringlei Gray. Pringle's Woolly-Daisy. Infrequent, in sandy places; desert, below 900m: San Felipe; Scissors Crossing; Vallecito Stage Station; Inner Pasture. Apr.-June. *n*=7.

E. wallacei Gray var. **rubellum** (Gray)Gray. Pinkish Woolly-Daisy. Infrequent, in sandy places; desert, below 1400m: Sentenac Canyon; San Felipe Wash; San Felipe Valley; Pinyon Mountain; Whale Peak; Box Canyon; Vallecito Stage Station; Cuyamaca Lake; foot of Garnet Peak; Harper Ranch; Vallecito Canyon. Apr.-May. Type from "a dry valley, near San Felipe". *n*=5.

E. wallacei var. **wallacei**. Wallace's Woolly-Daisy. Frequent in open areas; foothill and desert, 700-1200m: e.g. Chihuahua Valley; Sheep Canyon; Warner Ranch; Montezuma View Point; Laguna Camp; Buckman Spring; Tecate Mountain; Hidden Glen; Campo; Boulevard; Hipass; Jacumba; Dubber; Mountain Springs; Vallecito and Carrizo Stage Stations; Canebrake Canyon. Apr.-May. *n*=5.

Evax Gaertn.
(Inuleae)

Leaf blades about 7-15mm wide; heads in terminal clusters if stems developed . . . . *E. caulescens*
Leaf blades about 3-5mm wide; heads solitary in axils of cauline leaves if stems developed . . . . *E. sparsiflora*

***E. caulescens** (Benth.)Gray. Involucrate Evax. Native of northern California. Local, naturalized on clay areas: Sweetwater Dam; western Otay Mesa. Mar.

E. sparsiflora (Gray)Jeps. [*E. s.* var. *brevifolia* Gray] Erect Evax. Scarce, in vernally moist places; coastal, below 100m: Encinitas; Mission Gorge; San Diego; Paradise Valley; Rice Canyon; Sweetwater Lake. Apr.-May.

Filago L.

Uppermost leaves much longer than the heads
  Leaves linear; receptacle obconic or convex . . . . *F. arizonica*
  Leaves subulate with broadish base; receptacle low, nearly flat . . . . *F. gallica*
Uppermost leaves scarcely or not longer than the heads
  Plants erect, 6-25cm high . . . . *F. californica*
  Plants depressed, spreading, the stems 3-10cm long . . . . *F. depressa*

F. arizonica Gray. [*F. depressa* auct. non Gray] Arizona Filago. Uncommon, in sandy areas; desert, below 1100m: Scissors Crossing; Mason Valley; Vallecito Stage Station; southeast corner of county. Also San Diego. Mar.-May. 2*n*=14 pairs.

F. californica Nutt. California Filago. Common, in alkaline and open places; cismontane and desert, below 1400m: San Onofre Bluffs; Lilac; Rainbow; Pala; Love Valley; Montezuma View Point; Borrego Valley; Lone Palm; Banner; Mount Woodson; Moosa Canyon; Foster; El Cajon Mountain; Barona Valley; Poway Grade; La Jolla; San Diego; San Miguel Mountain; Barrett; Harbison Canyon; southeast corner of county; Box Canyon. Mar.-June.

F. depressa Gray. Dwarf Filago. Uncommon, in sandy areas; desert, below 800m: Borrego Palm Canyon; Yaqui Well; San Felipe; upper Mason Valley. Mar.-May.

***F. gallica** L. [*Logfia g.* Cosson & Germ.] Narrow-Leaf Filago. Native of Europe. Occasional, weed in disturbed areas and burns; foothill and coastal, below 800m: e.g. Twin Oaks Valley; Lakeside; Alpine; Japatul Valley; Dulzura. Apr.-June.

Galinsoga R. & P.
(Heliantheae - Galinsoginae)

***G. parviflora** Cav. Small-Flower Galinsoga. Native of Neotropics. Scarce, weed in cultivated areas. Coastal, below 200m: Balboa Park. Aug. n=8, 16.

Geraea T. & G.
(Heliantheae - Verbesininae)

Heads radiate; phyllaries densely ciliate . . . . *G. canescens*
Heads discoid; phyllaries densely glandular . . . . *G. viscida*

G. canescens T. & G. Desert Sunflower. Common, on sandy flats; desert, below 200m: e.g. Borrego Valley; The Narrows; Benson Dunes; Oil Well; Kane Springs Road; Carrizo Stage Station. Oct.-Nov. $2n$=18 pairs.

G. viscida (Gray)Blake. Sticky Geraea. Occasional, on dry slopes and open areas; southern foothill, 900-1200m: e.g. Cottonwood Grade; Potrero; Campo; Hipass; Buckman Spring; Boulder Oaks; La Posta; Boulevard; McCain Valley; Clover Flat; Jacumba; Dubber. May-June. Type from "near Larken's Station [Jacumba]".

Gnaphalium L. Cudweed; Everlasting
(Inuleae)

Pappus-bristles united at base, falling away in a ring; phyllaries but little imbricate
  Lower leaf-surface loosely villous-lanate; involucre nearly buried in wool, 3-4mm high . . . . *G. pensylvanicum*
  Lower leaf-surface closely white-pannose, the subappressed hairs tightly enmeshed; involucre densely wooly at base only, 4-6mm high . . . . *G. purpureum*
Pappus-bristles not or only partly united at base, deciduous separately or in small groups
  Heads small, the involucre mostly 2-4mm high; glomerules of heads leafy-bracted; plants less than 2dm high, usually much branched . . . . *G. palustre*
  Heads larger, mostly 4-7mm high; glomerules of heads not leafy-bracted; plants mostly 2-10dm high, the stems often simple below the inflorescence
    Leaves green in age, at least on the upper surface
      Mature leaves usually green on both surfaces; plants with rather strong balsamlike odor; annual to biennial
        Inflorescence corymbose; involucre rounded, white; leaves lanceolate to oblong; common, widespread . . . . *G. californicum*
        Inflorescence paniculate; involucre narrower, often pinkish; leaves lance-linear . . . . *G. ramosissimum*
      Mature leaves green above, white beneath; plants not strongly scented, perennial
        Leaves oblong to broadly linear, broadly auriculate at base; stems tending to branch; 6-9dm high . . *G. bicolor*
        Leaves narrowly linear, with short-decurrent bases; stems mostly simple; 3-6dm high . . . . *G. leucocephalum*
    Leaves permanently tomentose
      Leaves not or scarcely decurrent
        Middle phyllaries about twice as long as broad, rounded-obtuse at apex . . . . *G. microcephalum*
        Middle phyllaries about 3 times as long as broad, acutish . . . . *G. canescens*
      Leaves strongly decurrent
        Stems usually simple above; heads in dense terminal clusters; phyllaries all decidedly obtuse *G. straminium*
        Stems usually branched above; heads in rather open corymbose panicles, at least the inner phyllaries abruptly pointed . . . . *G. beneolens*

G. beneolens Davids. Fragrant Everlasting. Common, on open and disturbed slopes; foothill and coastal, below 1400m: e.g. Palomar Mountain; Oak Grove; Julian; Point Loma; El Monte; Lake Murray; Dulzura; Canyon City; Pine Valley; Hipass; White Star. July-Nov.

G. bicolor Bioletti. Bicolor Cudweed. Common, on open slopes and in disturbed areas; coastal, below 800m: e.g. San Onofre Bluffs; San Mateo Creek; Rainbow; Hormiguero Ranch; San Marcos Creek; Vista; Del Mar; Lake Hodges; El Capitan; Sloane Ranch; Dehesa; Lee Valley; Otay; Lake Murray; San Diego; Point Loma. Jan.-May.

G. californicum DC. California Everlasting. Common, on disturbed and open slopes; foothill and coastal, below 1500m: e.g. San Onofre Bluffs; De Luz; Rainbow; Lilac; Del Mar School; Lake Hodges; Olivenhain; Barona Valley; Julian; Tighe Ranch; Mount Soledad; Point Loma; Chollas Heights; Sloane Ranch; Jamul Mountain; Otay Mountain; White Star; Dubber. Feb.-June.

G. canescens DC. [*G. wrightii* Gray] Felty Everlasting. Rare, known in the county form a collection from Campo, 760m. July-Oct.

G. leucocephalum Gray. White-head Cudweed. Uncommon, in sandy areas; cismontane, below 500m: lower Cottonwood Canyon at De Luz Creek.

G. microcephalum Nutt. White Everlasting. Occasional, on slopes; cismontane: De Luz; Palomar Mountain; Oakzanita; Monument School; Dulzura; Laguna Mountains. Also Pinyon Mountain. July-Oct. Type from "St. Diego". $2n$=14 pairs.

G. palustre Nutt. Lowland Cudweed. Frequent, in wet habitats: e.g. Doane Valley; San Diego; National City; Laguna Meadow; Live Oak Springs; Cuyamaca Lake; Bow Willow Canyon. May-Oct.

***G. pensylvanicum** Willd. [*G. peregrinum* Fern.] Wandering Cudweed. Native of subtropical regions but apparently introduced into our area. Uncommon, weed in disturbed areas; coastal, below 150m: San Luis Rey; Lakeside; Home Avenue Canyon; San Diego. Mar.-June.

***G. purpureum** L. Rare, known in the county from a collection at San Diego. March.

G. ramosissimum Nutt. Pink Everlasting. Scarce, on dry slopes; foothill, below 800m: Valley Center; Bear Valley. Mar.-Sep.

G. straminium H.B.K. [*G. chilense* Spreng., *G. c.* var. *confertiflorum* Greene] Cotton-Batting Plant. Common, in moist habitats; cismontane: e.g. Richman Ranch; San Onofre Bluffs; Red Mountain Grade; Lilac; Doane Valley; Bear Valley; Olivenhain; Miramar; Witch Creek; La Jolla; Point Loma; Murphy Canyon; Silver Strand; Campo; Espinosa Trail; Hauser Creek; Pine Valley; Laguna Public Camp; Clover Flat. Apr.-Aug. $2n=14$ bivalents.

Grindelia Willd. Gumplant
(Astereae - Solidagininae)

Involucres mostly 1cm in diameter; ligules 5-8mm long . . . . . *G. hallii*
Involucres 12-25mm in diameter; ligule (when present) mostly 8-15mm long
  Ray flowers present . . . . . *G. robusta* var. *r.*
  Ray flowers lacking . . . . . *G. robusta* var. *bracteosa*

G. hallii Steyerm. San Diego Gumplant. Endemic to San Diego County. Frequent, in open areas, Chaparral and Montane Coniferous Forest; montane, above 800m: e.g. Lake Henshaw; Mesa Grande; Pine Hills; Julian; Cuyamaca Lake; Paso Picacho; Shrine Camp; Crouch Meadow; Corte Madera. July-Nov. Type from open meadows about Cuyamaca Lake. $n=6$.

G. robusta Nutt. var. **bracteosa** (J.T. Howell)Keck. [*G. rubicaulis* var. *b.* (J. T. Howell)Steyerm.] Rayless Gumplant. Rare, known in the county only at Pine Hills, 1280m. May-July.

G. robusta var. **robusta**. [*G. rubicaulis* var. *elata* Steyerm.] Big Gumplant. Occasional, on open slopes and in fields; cismontane, below 1200m: e.g. Julian; Torrey Pines; La Jolla; San Onofre; Oceanside; Point Loma; Mission San Diego; Silver Strand; Otay Mesa; Sunnyside; El Cajon; National City; San Ysidro; Ramona. Apr.-Aug. $n=6$, 12.

Guizotia Cass.
(Heliantheae - Coreopsidinae)

***G. abyssinica** (L. f.)Cass. Ethiopian Nigerseed. Native of Africa. Infrequent, naturalized in disturbed areas and in fallow fields; coastal, below 200m: Escondido; Chula Vista. July-Sep. $n=15$.

Gutierrezia Lag. Matchweed
(Astereae - Solidagininae)

Heads with only 2-3 flowers; involucre very narrow, to 1.5mm wide; akenes aborted . . . . . *G. microcephala*
Heads with more than 4 flowers; involucre turbinate, 1.5mm or more wide; akenes of disk-flowers fertile
  Heads mostly solitary at the ends of branchlets; flowers usually more than 10; involucre 2-5mm wide . . *G. sarothrae*
  Heads clusterd at the ends of branchlets; flowers 5-10; involucre less than 5mm wide . . . . . *G. californica*

G. californica (DC.)T.& G. [*G. sarothrae* auct.non (Pursh)Britt. & Rusby] Broom Matchweed. Prevalent, in Chaparral; cismontane and occasionally desert, below 1500m: e.g. De Luz; Oak Grove; Warner Springs; Banner Grade; Cuyamaca Lake; Descanso Junction; Rancho Santa Fe; Mission Valley; Otay Mountain; Dulzura; Pine Valley; Mountain Springs; Pinyon Peak. Feb.-Nov. $n=4$.

G. microcephala (DC.)Gray. Native to desert areas north of the county. Introduced by CalTrans hydroseeding at Buckman Springs.

G. sarothrae (Pursh)Britt. & Rusby. [*G. bracteata* Abrams] San Joaquin Matchweed. Occasional, in open Sage Scrub and Chaparral areas; foothill, 300-1500m: e.g. Chihuahua Valley; San Felipe; Santa Ysabel; Witch Creek; Mussey Grade; Alvarado Canyon; Casa de Oro; Dulzura; Alpine; Laguna Grade. July-Sep. $n=8$, 12.

Hazardia Greene

Heads large, discoid . . . . . *H. squarrosus*
Heads small, radiate; ray florets conspicuous and much longer than the disk . . . . . *H. orcuttii*

H. orcuttii(Gray)Greene. [*Haplopappus o.* Gray] Orcutt's Goldenbush. Known in the county only in Chaparral at Encinitas.

H. squarrosus (H. & A.)Greene ssp. **grindelioides** (DC.)Clark. [*Haplopappus s.* H.& A. ssp. *g.* (DC.)Keck] Sawtooth Goldenbush. Common, on dry slopes; cismontane: e.g. Fallbrook; Carlsbad; Fernbrook; Poway Grade; San Felipe Valley; Inspiration Point; Desert View; Ballena; Miramar; Cowles Mountain; Point Loma; Spring Valley; Cottonwood; Tecate; Campo; Descanso Junction; Guatay. June-Oct.

Hedypnois Schreb.
(Cichorieae - Leontodontinae)

***H. cretica** (L.)Willd. [*Rhagadiolus hedypnois* All.] Crete Hedypnois. Native of Eurasia. Frequent, weed in disturbed areas; coastal, below 200m: e.g. Pacific Beach; Rancho Santa Fe; San Diego; Ostrich Farm; Lakeside. Apr.-May. $2n=8$.

Helenium L. Sneezeweed
(Helenieae - Heleniinae)

H. puberulum DC. Rosilla. Occasional, in wet places; cismontane: San Mateo Creek estuary; Mission Valley; Julian; Cuyamaca Lake; Greene Valley; Boulder Creek. Feb.-Oct. n=29; 2*n*=58.

Helianthus L. Sunflower
(Heliantheae - Verbesininae)

Plants annual
  Outer phyllaries ovate, abruptly caudate, strongly ciliate . . . . . *H. annuus*
  Outer phyllaries lanceolate, gradually attenuate, not obviously ciliate . . . . . *H. niveus*
Plants perennial
  Plants strongly glaucous and rhizomatous . . . . . *H. ciliaris*
  Plants not strongly glaucous and rhizomatous
    Outer phyllaries shorter than the disk, dry places . . . . . *H. gracilentus*
    Outer phyllaries longer than the disk, wet places . . . . . *H. californicus*

H. annuus L. ssp. **lenticularis** (Dougl.)Ckll. Western Sunflower. Common, ruderal weed; cismontane, below 1200m: e.g. Santa Margarita Ranch; Oak Grove; Santa Ysabel; Point Loma; Descanso; Buckman Spring. Feb.-Oct. 2*n*=34.

H. californicus DC. California Sunflower. Occasional, in wet places; cismontane: e.g. Palomar Mountain; North Peak; Cuyamaca Lake; Jamul; National City; Campo. Aug.-Oct. *n*=51.

***H. ciliaris** DC. Blueweed. Native of Arizona and elsewhere. Local, naturalized on Otay Mesa, 120m. June-Nov. *n*=51.

H. gracilentus Gray. Slender Sunflower. Common, on open slopes; cismontane, above 250m: e.g. San Onofre Canyon; Mussey Grade; Mount Woodson; Palomar Mountain; Ramona; Banner Grade; Cuyamaca Peak; Garnet Peak; Monument Peak; Laguna Junction; Lawson Peak; Tecate Mountain; Otay Mountain. Apr.-Sep. Type from "the mountains 45 miles north-east of San Diego". *n*=17.

H. niveus (Benth.)Bdg. ssp. **canescens** (Gray)Heiser. [*H. petiolaris* var. c. Gray; *H. canus* (Britt.)Woot. & Standl.] Gray Sunflower. Scarce, in sandy areas; desert, below 200m: Borrego Valley; Clark Dry Lake; Yaqui Well. Mar.-June. *n*=17.

Hemizonia DC. Tarweed
(Heliantheae - Madiinae)

Leaves not spine-tipped; receptacular bracts confined to a row surrounding the outer disk-flowers and united into a cup; pappus of quadrate or oblong paleae
  Disk-akenes sterile
    Phyllaries not keeled; anthers yellow . . . . . *H. kelloggii*
    Phyllaries keeled; anthers black
      Ray-flowers 5; disk-flowers 6. . . . . . *H. fasciculata*
      Ray-flowers 8, disk flowers 13-21. . . . . . *H. conjugens*
  Disk-flowers mostly fertile
    Radical leaves pinnatifid or bipinnatifid; herbage grayish-hirsute; ray-flowers 8-13, the ligules as broad as long . . . . . *H. paniculata*
    Radical leaves merely lobed; herbage bright green, soft-pubescent; ray-flowers 13-20, ligules half as broad as long . . . . . *H. floribunda*
Leaves tipped with a rigid spine or apiculation; receptacular bract subtending each disk-flower free, persistent; pappus none or of very narrow paleae
  Pappus none; anthers yellow
    Receptacular bracts pungent. . . . . . *H. pungens*
    Receptacular bracts obtuse or acute, not cuspidate . . . . . *H. laevis*
  Pappus present; anthers black . . . . . *H. australis*

H. australis (Keck)Keck. Southern Spikeweed. Rare, in grassy fields; coastal, below 200m: Escondido; Ramona; San Dieguito Valley. June-Sep. *n*=11.

H. conjugens Keck. Otay Tarweed. Uncommon, on clay slopes and mesas; south coastal, below 100m: San Diego; La Presa; Sweetwater Dam; Telegraph Canyon; Otay. May-June. Type taken "in river-bottom land near Otay".

H. fasciculata (DC.)T. & G. [*H. ramosissima* Benth., the type from San Diego, *H. fasciculata* var. *r.* (Benth.)Gray] Fascicled Tarweed. Common, on slopes and mesas; coastal and foothill, below 900m: e.g. San Onofre Bluffs; Las Flores; La Presa; Pala; Lake Henshaw; Oceanside; Olivenhain; Poway Grade; Mesa; Cedar Creek Falls; Alpine; Kearny Mesa; Jamul; Alvarado Canyon; Telegraph Canyon; Otay Mountain; Potrero; Dehesa; Balboa Park; La Mesa; La Jolla; Point Loma; Silver Strand. Mar.-July. *n*=12.

H. floribunda Gray. Tecate Tarweed. Infrequent, in stream channels in Chaparral; south foothill, 300-700m: Dulzura; Tecate Junction; Potrero; Canyon City; Hipass; McCain Valley; Bankhead Spring; Live Oak Springs; Jacumba. Sep.-Oct. Type from "California, near the southern boundary, on the Fort Yuma Road, 80 miles east of San Diego". *n*=13.

H. kelloggii Greene. Kellogg's Tarweed. Uncommon, in open areas; cismontane, below 1000m: Palomar Mountain; Japatul Valley; La Mesa; Kensington Park. Apr.-July. *n*=9.

H. laevis (Keck)Keck. Smooth Spikeweed. Rare, in grassy fields; coastal, below 150m: San Luis Rey; Santee. Apr.-Sep. *n*=9.

H. paniculata Gray. San Diego Tarweed. Frequent, on slopes and mesas; south coastal, below 200m: e.g. Spring Valley; Paradise Valley; Telegraph Canyon; Otay Ranch; San Ysidro. Also Las Flores and Bonsall. Apr.-July. *n*=12.

***H. pungens** (H. & A.)T.& G. [*Centromadia p.* (H. & A.)Greene] Common Spikeweed. Native of northern California. Occasional, in disturbed areas, below 150m: e.g. Oceanside; Borrego Spring; Santee; National City; Chula Vista. May-Aug. *n*=9.

Heterotheca Cass.
(Astereae - Solidagininae)

H. grandiflora Nutt. Telegraph Weed. Common, in sandy and disturbed areas; coastal and foothill, below 800m: e.g. De Luz; Carlsbad; Harrison Grade; Ramona; Santa Ysabel; Campo; National City; San Diego; Imperial Beach. Jan.-Dec. *n*=9.

Hieracium L. Hawkweed
(Cichorieae - Crepidinae)

Ligules white or ochroleucous; stem (except at base) and involucre glabrous, or involucre with a few long hairs . . . . . *H. albiflorum*
Ligules yellow; stems and involucres pubescent or hirsute . . . . . *H. argutum*

H. albiflorum Hook. White Hawkweed. Occasional, in Montane Coniferous Forest; montane, about 1400m: Palomar Mountain; Hot Springs Mountain; North Peak and Middle Peak, Cuyamaca Mountains. Aug.-Sep. 2*n*=18.

H. argutum Nutt. var. **parishii** (Gray) Jeps. Uncommon, in Chaparral; foothill and montane; 800m to 1200m: Corte Madera; Lawson Peak.

Holocarpha (DC.)Greene
(Helenieae - Heleniiae)

H. virgata (Gray)Keck ssp. **elongata** Keck. Curving Tarweed. Frequent, in open areas; coastal and foothill, below 1100m: e.g. Inaja Memorial; Poway; Santee; Lakeside; Jamacha; Lower Otay Dam; Proctor Valley; La Mesa; El Cajon; San Diego. July-Oct. Type from San Diego.

Hulsea T. & G.

Cauline leaves saliently dentate
- Rays reddish, ciliate; akenes 6-7mm long . . . . . *H. heterochroma*
- Rays yellow, not ciliate; akenes 4-5mm long . . . . . *H. mexicana*

Cauline leaves mostly entire; basal or lower leaves white-villous, the upper somewhat glandular
- Plants annual or biennial; heads mostly 2-3.5cm across; phyllaries hirsute . . . . . *H. californica*
- Plants perennial; heads mostly 1-2cm across; phyllaries subglabrate . . . . . *H. vestita*

H. californica T. & G. ex Gray in Torr. San Diego Hulsea. Endemic to San Diego County. Occasional, on burns or clearings; foothill and montane, above 1000m: e.g. Palomar Mountain; Eagles Nest; San Felipe; Chariot Canyon; Stonewall Peak; Harper Ranch; Desert View; Green Valley Falls; Cherry Flat; Laguna Junction; Buckman Spring; Live Oak Springs; Espinosa Trail; Campo. Apr.-Aug. Type from mountains east of San Diego. 2*n*=19 pairs.

H. heterochroma Gray. Great Hulsea. Rare, in open areas, montane: Highpoint Truck Trail, Palomar Mountain, 1200m.

H. mexicana Rydb. Mexican Hulsea. Rare, in open, sandy areas; transmontane: north of Mountain Springs and on Table Mountain, 750-825m. Apr. 2*n*=19 pairs.

H. vestita Gray ssp. **callicarpha** (Hall)Wilken. Beautiful Hulsea. Infrequent, in open areas of Chaparral; foothill and montane, 1000-1400m: Palomar Mountain; Lost Valley; Hot Springs Mountain; Cuyamaca Mountains. May-July. 2*n*=19 pairs.

Hymenoclea T. & G.
(Heliantheae - Ambrosiinae)

Wings of the fruiting involucre cuneate or obovate, in a single whorl of 7-12, 1-2mm long . . . . . *H. monogyra*
Wings of the fruiting involucre reniform-orbicular, spirally arranged, 3.5-8mm long . . . . . *H. salsola*

H. monogyra T. & G. ex Gray. Desert Fragrance; Leafy Burrobush. Frequent, in sandy washes; south coastal, below 100m: e.g. Mission Valley; Murphy Canyon; Chollas Valley; Jamul; Dulzura; National City; Marron Valley; Tijuana River Valley. Sep.-Jan. *n*=18.

H. salsola T.& G. ex Gray var. **salsola**. White Burrobush. Common, in sandy areas; desert, below 900m: e.g. Borrego Palm Canyon; Borrego Valley; Banner; Scissors Crossing; Agua Caliente Springs; Mountain Palm Springs; Mountain Springs. Mar.-June. *n*=18.

Hymenopappus L'Her.
(Anthemideae - Chrysantheminae)

H. filifolius Hook. var. **lugens** (Greene)Jeps. [*H. l.* Greene] Pine Cutleaf. Infrequent, in Montane Coniferous Forest; montane, above 1300m: Inspiration Point; Harrison Park; Warner Ranch; Cuyamaca Lake; Cuyamaca Mountains. June-Aug. *n*=17, 34.

Hymenothrix Gray
(Anthemideae - Chrysantheminae)

H. wrightii Gray. [*Hymenopappus w.* var. *viscidulus* Jeps., the type from Cuyamaca Mountains.] Wright's Hymenothrix. Infrequent, on dry slopes, Montane Coniferous Forest; montane, above 1400m: Palomar Mountain; Stonewall Mine; Cuyamaca Lake; Cuyamaca Lodge; Shrine Camp; Vallecito Trail; Laguna Camp; Wooded Hill; Pine Valley. June-Sep.

Hypochoeris L. Cat's Ear
(Cichorieae - Leontodontinae)

Annual, usually glabrous; outer akenes truncate, inner beaked . . . . *H. glabra*
Perennial, pubescent; akenes all beaked . . . . *H. radicata*

***H. glabra** L. Smooth Cat's-Ear. Native of Europe. Common, in disturbed areas; cismontane, below 800m: e.g. San Onofre Bluffs; Santa Margarita Ranch; Lilac; Palomar Mountain; Ramona; Barona Valley; Miramar; Lakeside; San Diego; Dehesa; Tecate Mountain. Mar.-June. $2n$=10.

***H. radicata** L. Hairy Cat's-Ear. Native of Europe. Uncommon weed, as at Pacific Beach. May-Nov.

Isocoma Nutt.

Phyllaries with green but thin tips; stems brownish; largely cismontane
  Inflorescences terminal, round and compact . . . . *I. veneta* var. *vernonioides*
  Inflorescence open to loose cymose, not in large glomerules
    Robust shrub 1-2m high . . . . *I. veneta* var. *oxyphylla*
    Shrub of slender stems 3-5dm long . . . . *I. veneta* var. *furfuracea*
Phyllaries with a conspicuous thickened gland or resin-pocket near the tip; stems white and shining; deserts . . . . *I. acradenia*

I. acradenia Greene ssp. **eremophila** (Greene)Beauchamp. [*Haplopappus a.* (Greene)Blake ssp. *e.* (Greene)Hall, *I.e.* Greene] Alkali Goldenbush. Occasional, in alkaline places; desert, below 400m: Mountain Palm Springs; Sentenac Canyon; Carrizo Gorge; Canyon Sin Nombre. Aug.-Nov. $2n$=12. Type from "southwestern part of Colorado Desert".

I. veneta (H.B.K.)Greene var. **furfuracea** (Greene)Beauchamp. [*Haplopappus v.* ssp. *f.* (Greene)Hall; *I. decumbens* Greene, the type from clay depressions on the mesas near San Diego.] Spreading Goldenbush. Infrequent, in sandy areas; coastal, below 300m: Encinitas; San Miguel Mountain; La Presa; Jamul; Chula Vista; base of Otay Mountain. Apr.-Nov.

I. veneta var. **oxyphylla** (Greene)Beauchamp. [*Haplopappus v.* var. *o.* (Greene)Hall] San Diego Goldenbush. Uncommon, in sandy areas; cismontane, below 200m: Solana Beach; Rancho Santa Fe; Poway Junction; Mission Valley; Silver Strand; Dehesa; Jamul Valley; Marron Valley; Chula Vista. June-Oct. Type from Jamul Valley.

I. veneta var. **vernonioides** Jeps. [*Haplopappus venetus* var. *vernonioides* Jeps.; *I. leucanthemifolia* Greene, the type from "Warren, in the Mountains of San Diego County (Warner Ranch)".] Coastal Goldenbush. Common, on dry slopes and sandy areas; coastal and foothill, below 1100m: e.g. San Onofre Bluffs; Fallbrook; Oak Grove; Oceanside; Carlsbad; Ballena; Encinitas; Bernardo; Torrey Pines Beach; Point Loma; Old Town; Mission Valley; Balboa Park; Glenview; Dehesa; Lemon Grove; Spring Valley; Imperial Beach; Dulzura. Also Pine Valley. Apr.-Dec. n=6.

Iva L. Poverty Weed
(Astereae - Solidagininae)

Plants slightly woody or almost wholly herbaceous . . . . *I. axillaris*
Plants frutescent perennial shrubs . . . . *I. hayesiana*

I. axillaris Pursh ssp. **robusitor** (Hook.)Bassett. Western Marsh-Elder. Uncommon, in alkaline or saline areas; coastal, below 50m: San Mateo Creek estuary; Mission Valley. May-Sep. n=18, 27.

I. hayesiana Gray. San Diego Marsh-Elder. Frequent, in moist or alkaline places; coastal, below 200m: e.g. Rancho Santa Fe; Miramar Reservoir; Penasquitos Canyon; Alvarado Canyon; Proctor Valley; La Presa; Otay; Tijuana River Valley; Otay Mesa. Apr.-Sep. Type from "Jamuel [Jamul] Valley".

Jaumea Pers.
(Helenieae - Jaumeninae)

J. carnosa (Less.)Gray. Salty Susan; Fleshy Jaumea. Prevalent, in Coastal Salt Marsh; coastal, below 5m: e.g. San Mateo and Santa Margarita River estuaries; Oceanside; Agua Hedionda; San Dieguito Slough; Coronado; Imperial Beach. May-Oct. n=19.

Lactuca L. Lettuce (Cichorieae - Crepidinae)

Cauline leaves held vertically, spinulose on midrib; involucral bracts patent or deflexed in fruit . . . . *L. serriola*
Cauline leaved not held vertically, smooth on midrib, involucral bracts erect in fruit . . . . *L. sativa*

***L. sativa** L. [*L. scariola* L.] Prickly Lettuce. Native of Europe. Common, weed in disturbed areas; cismontane, below 1100m: e.g. Oceanside; Palomar Mountain; Ramona; La Jolla; East San Diego; Pine Valley; Boulevard; National City. May-Sep. $2n$=18.

***L. serriola** L. var. **integrata** Gren. & Godr. Wild Lettuce. Native of Europe. Occasional, weed in disturbed and sandy areas; cismontane: e.g. Santee; San Diego; Julian; Cuyamaca and Laguna Mountains. May-Sep.

Lagophylla Nutt. Hareleaf
(Heliantheae - Madiinae)

L. ramosissima Nutt. Common Hareleaf. Prevalent, in open spaces; cismontane: e.g. Warner Hot Springs; Mesa Grande; Santa Ysabel; Paso Picacho; Cuyamaca Mountains; San Diego. July-Oct. *n*=7.

Lasthenia Cass. Goldfields
(Helenieae - Heleniinae)

Phyllaries united into a cup over 2/3 their length . . . . *L. glabrata*
Phyllaries free from each other
  Phyllaries ca. 4; anther tips subulate or deltoid with wartlike glands . . . . *L. microglossa*
  Phyllaries ca. 6-14; anther tips not as above
    Leaves all essentialy entire; corollas turning a deep red in dilute aqueous alkali . . . . *L. californica*
    Leaves, especially the middle ones, usually pinnately lobed or cleft; corollas remaining yellow in dilute aqueous alkali . . . . *L. coronaria*

L. californica DC. ex Lindley. [*L. chrysostoma* (Fisch. & Mey.)Greene; *Baeria c.* Fisch. & Mey.; *B. c.* var. *gracilis* (DC.)Hall; *B. c.* var. *gracilis* forma *crassa* Hall, the type from Ocean Beach; *B. clevelandii* Gray, the type from near San Diego; *Burrielia parviflora* Nutt., the type from St. Diego".] Common Goldfields. Frequent, in open areas; cismontne and occasionally on desert edge: e.g. Santa Margarita Ranch; Moro Hill; Fallbrook; Pala; Chihuahua Valley; Warner Springs; Culp Valley; Montezuma View Point; San Felipe; Banner; Pine Hills; Cuyamaca Lake; Rattlesnake Valley; Laguna Camp; Boulevard; Mountain Springs; Jacumba; Sloane Ranch; Chollas Valley; Sweetwater Dam; Silver Strand; Ocean Beach; Oceanside. Feb.-May. *n*=8, 16.

L. coronaria (Nutt.)Ornduff. [*Baeria aristata* (Nutt.)Cov., the type from "near St. Diego"; *Ptilomeris anthemoides* Nutt., the type from "near St. Diego"; *P. mutica* Nutt., the type from "near St. Diego."] Southern Goldfields. Frequent, on open slopes; coastal, below 300m: e.g. Fallbrook; Lilac; Bernardo; Torrey Pines; Kearny Mesa; Del Mar; Pacific Beach; Mission Hills; Point Loma; University Heights; Lake Hodges; Whispering Oaks; Proctor Valley. Also Oak Grove. Feb.-May. *n*=5.

L. glabrata Lindl. ssp. **coulteri** (Gray)Ornduff. Coulter's Saltmarsh-Daisy. Infrequent, in Coastal Salt Marsh and alkaline places; coastal, below 10m: Santa Margarita River estuary; Sorrento; Mission Bay; San Diego River at Old Town; San Diego; Chula Vista; Tijuana River estuary. Feb.-May. *n*=7.

L. microglossa (DC.) Greene. Small-rayed Goldfield. Rare, known in the county only along the shore of Corte Madera Lake.

Layia Hook. & Arn.
(Heliantheae - Madiinae)

Anthers black; ligule yellow with white tip or yellow throughout; pappus mostly scabrid or, if plumose, also with inner wool . . . . *L. platyglossa*
Anthers yellow; ligule all white or all yellow; pappus plumose, lacking inner woolly hairs . . . . *L. glandulosa*

Layia carnosa (Nutt.)T. & G. Beach Layia. Although Nuttall gave the type locality as "St. Diego . . . growing in the sands of the sea-coast", the nearest known point in the range of the plant is north of Point Conception.

L. glandulosa (Hook.)H. & A. ssp. **glandulosa**. White Tidy-Tips. Occasional, on arid slopes; desert and montane, 300-1500m: Warner Ranch; Hot Springs Mountain; Culp Valley; San Felipe Valley; Scissors Crossing; Pinyon Mountain; Vallecito Stage Station; Whale Peak; Cuyamaca Peak; Buckman Spring; Boulevard; Jacumba. Mar.-June. *n*=8.

L. platyglossa (Fisch. & Mey.)Gray ssp. **campestris** Keck. Common Tidy-Tips. Common, on grassy slopes; cismontane: e.g. San Onofre; O'Neill Lake; Moro Hill; Fallbrook; Pala; Palomar Mountain State Park; Cuyamaca Lake; Laguna Lake; Rattlesnake Valley; Cardiff; Kearny Mesa; Del Mar; Foster; Poway Grade; Point Loma. Feb.-June. *n*=7.

Lepidospartum Gray
(Senecioneae)

L. squamatum (Gray)Gray. Scale-Broom. Local, in washes and gravelly places; cismontane, below 900m: Fallbrook; Oak Grove; Encinitas; Dodge Valley; Batiquitos Lagoon. Aug.-Dec. *n*=30, ca. 45.

Lessingia Cham.
(Astereae - Asterinae)

Leaves tomentose or glabrate, glandular . . . . *L. glandulifera.* var. *g.*
Leaves persistently white-tomentose, glandular along margin only . . . . *L. glandulifera* var. *tomentosa*

L. glandulifera Gray var. **glandulifera**. [*L. germanorum* Cham. var. *glandulifera* (Gray)J. T. Howell] Valley Lessingia. Local, in valleys and on open slopes; foothill and montane, 700-1400m: Doane Valley; Oak Grove; Pine Valley; Cottonwood Valley; La Posta; Boulevard; McCain Valley at Lark Canyon; Banner; northeast of Hot Springs Mountain. July-Oct.

L. glandulifera var. **tomentosa** (Greene)Ferris. Ranchita Lessingia. Endemic to San Diego County. Rare, known only from Ranchita and Warner Ranch, 800-1500m. Oct. Type from "southwest part of the Colorado Desert, San Diego County".

Machaeranthera Nees
(Astereae - Asterinae)

Plants herbaceous; heads usually several or many; biennials or short-lived perennials; flowers bluish
  Phyllaries narrowly linear, the long narrow loose green tips longer than the indurated base . . . . *M. tephrodes*
  Phyllaries not narrowly linear, the erect or squarrose tips shorter than the indurated base . . . . *M. lagunensis*
Plants shrubby or woody at base; heads large, usually solitary at branch-tips; flowers yellow . . . . *M. juncea*

M. juncea (Greene)Hartman. [*Haplopappus j.* Greene] Rush Chaparral-star. Occasional, on dry hillsides; foothill, 300-800m; e.g. Alpine; Bear Valley; Twin Peaks; Cowles Mountain; Potrero; Jamul; Lindero. June-Oct. Type fron San Diego County.

M. lagunensis Keck. Laguna Mountains Aster. Infrequent, in Montane Coniferous Forest; montane, above 1500m: Laguna Mountains; below Desert View. Aug. Type "from 2 miles south of the main recreation area in the Laguna Mountains, . . . at 5200 feet elevation . . . on dry slopes under pines".

M. tephrodes (Gray)Greene. [*Aster canescens* auct. non Pursh; *M. hiemalis* A. Nels., the type from Devil's Canyon, near Jacumba.] Ash-Color Aster. Occasional, on dry slopes; montane and desert, 800-1600m: Julian; Chariot Canyon; Middle Peak; Boulder Oaks; Boulevard; Jacumba. Apr.-May; Sep.-Dec. n=4, 5.

Madia Mol. Tarweed
(Heliantheae - Madiinae)

Disk-flowers 1 (or 2), glabrous, phyllary embracing the akene and sulcate to tip; tiny herbs
  Ray-akenes laterally compressed, with sharp ventral angle, glabrous; involucre strongly stipitate-glandular, the glands yellow . . . . *M. exigua*
  Ray-akenes obocompressed, not angled, pubescent; involucre minutely stipitate-glandular, the glands black . . . . *M. minima*
Disk-flowers 8-16, pubescent; phyllary with empty flat tip
  Disk-akenes sterile; receptacle pubescent; ray-flowers conspicuous
    Basal leaves forming a large rosette, the lower cauline leaves closely imbricated . . . . *M. elegans* ssp. *densifoia*
    Basal leaves few or in a small rosette, the lower cauline leaves often crowded, the upper well spaced . . . . *M. elegans* ssp. *e.*
  Disk-akenes fertile; receptacle glabrous; ray-flowers inconspicuous
    Glandular above the middle, slender and flexuous; heads small, to 9mm high . . . . *M. gracilis*
    Glandular well down toward the base; stem stout; heads larger, the involucre 8-12mm high . . . . *M. sativa*

M. elegans D. Don ssp. **densifolia** (Greene)Keck. Autumn Madia. Occasional, in grassy areas; montane, above 1300m: Lone Fir Point; Middle Peak; Corte Madera. Apr.-Nov. n=8.

M. elegans ssp. **elegans**. Elegant Madia. Infrequent, in grassy areas; montane, above 1400m: Palomar Mountain; Cuyamaca Lake; Middle Peak. June-Aug. n=8.

M. exigua (Sm.)Gray. Pygmy Madia. Occasional, on grassy slopes; foothill, 200-800m: e.g. De Luz; Barona Indian Reservation; Big Boulder Ranch; Glen Lonely; Cottonwood Grade. Apr.-June. n=16.

M. gracilis (Sm.)Keck ssp. **gracilis**. Slender Madia. Occasional, on grassy slopes; montane, above 1100m: e.g. Palomar Mountain; Mesa Grande; Warner's Ranch; Witch Creek; Julian; Cuyamaca Lake. May-June. n=16, 24.

M. minima. (Gray)Keck. [*Hemizonella m.* Gray] Small Madia. Uncommon, on open slopes; foothill and montane, above 400m: Silverwood Wildlife Refuge; Cuyamaca Lake; Cuyamaca Mountains; Laguna Lakes; Laguna Camp. May-June. 2n=32.

***M. sativa** Mol. Coast Madia. Native of Chile. Rare, known in the county only from a collection from the Cuyamaca Mountains, above 1400m. May-Oct. n=16.

Malacothrix DC.
(Cichorieae - Stephanomeriinae)

Outer phyllaries orbicular, 4-5mm wide; with broad scarious margins; persistent pappus-bristles 1-4 . . . . . . *M.coulteri*
Outer phyllaries lanceolate to linear, narrower; with narrow pale margins
  Plants annual
    Leaf-segments linear-filiform, elongate; persistent pappus-bristles 2 or more
      Stems simple, without leaves and ending in solitary heads; involucre 12-15mm high . . . . . *M. californica* var. *c.*
      Stems few-branched, each branch ending in a head; stems with a few leaves; involucres 5-12mm high . . . . *M. californica* var. *glabrata*
    Leaf-segments lanceolate to oblong, short, usually toothed; persistent pappus-bristles 1 or 0
      Akenes brown or straw-colored; cauline leaves often toothed; plants usually un-branched below *M. clevelandii*
      Akenes more than 1.7mm long, subcylindrical, gray-brown to straw-colored, with 15 equally prominent ribs; involucre 7-10mm high . . . . *M. stebbinsii*
  Plants perennial, suffrutescent; leaves linear to lanceolate or oblanceolate
    Herbage densely white-tomentose when young; inner phyllaries obtuse . . . . *M. incana*
    Herbage essentially glabrous; inner phyllaries attenuate . . . . *M. saxatilis*

M. californica DC. var. **californica**. Desert Dandelion. Occasional, on sandy areas; cismontane but infrequent in the desert, below 1200m: e.g. Pala; Lake Henshaw; Hot Springs Mountain; Warner Ranch; Boulevard; Jacumba; Campo; Morena Valley; Tijuana River Valley. Mar.-May. 2n=14.

M. californica var. **glabrata** Gray in D.C. Eaton. [*M. g.* (Gray)Gray] Glabrous Malacothrix. Common, in sandy areas; desert, below 1300m: e.g. northeast corner of county; Montezuma View Point; Borrego Valley; Borrego Palm Canyon; San Felipe Wash; San Felipe Valley; Yaqui Well; Vallecito Stage Station; Jacumba. Also Lakeside. Mar.-May. 2n=14.

M. clevelandii Gray. Cleveland's Malacothrix. Occasional, on open slopes and burns; cismontane, below 1400m: e.g. Hot Springs Mountain; Henshaw Grade; Montezuma View Point; San Felipe Valley; Rattlesnake Valley; Vallecito Canyon; Dulzura; Tecate Mountain; Alpine; Campo; Boulevard. Apr.-July. Type from near San Diego. n=7; 2*n*=14.

M. coulteri Harvey and Gray var. **cognata** Jeps. Snake-heads. Rare, known in the county only from San Diego and San Ysidro; coastal, below 100m. Mar.-May.

M. incana (Nutt)T.& G. [*Malacomeris i.* Nutt.] Dune Malacothrix. The type is cited from an island in the bay at San Diego (North Island?) but the plant has not been collected in the county since 1836.

M. saxatilis (Nutt.)T.& G. var. **tenuifolia** (Nutt.)Gray. Cliff Malacothrix. Uncommon on bare slopes, coastal, below 200m: south of Oceanside; north bank of San Luis Rey River.

M. stebbinsii Davis & Raven. Stebbins' Malacothrix. Scarce, on slopes and in canyons; desert, below 800m: Borrego Palm Canyon; San Felipe Hill. Apr. 2*n*=28.

Malperia Wats.
(Eupatorieae)

M. tenuis Wats. Brown Turbans. Uncommon, on alkaline, clay slopes; desert, below 100m: Fish Creek; Split Mountain. Mar. 2*n*=20.

Matricaria L.
(Anthemideae - Chrysantheminae)

M. matricarioides (Less.)Porter. Common Pineapple-Weed. Common, weed in disturbed and moist habitats; below 1100m: e.g. San Onofre; Banner; Point Loma; San Diego; National City. Mar.-May. 2*n*=18.

Micropus L.
(Inuleae)

M. californicus Fisch. & Mey. Slender Cottonweed. Occasional, in open places; cismontane and desert: e.g. Palomar Mountain; Henshaw Dam; Warner Pass; Ramona; Scissors Crossing; Box Canyon; Cuyamaca Lake; Cuyamaca Mountains; Descanso; Viejas Valley; El Cajon Mountain; San Diego; Lyons Valley; Otay Mountain. Mar.-May.

Microseris D. Don
(Cichorieae - Microseridinae)

Heads always erect; pappus-paleae deciduous; awns filiform, white . . . . *M. lindleyi*
Heads nodding or semierect, at least in the bud; pappus-paleae persistent; awns stouter, brown or yellow
  Paleae narrowly lanceolate, bifid at apex, one or more of outer phyllaries lanceolate after anthesis. . *M. heterocarpa*
  Paleae not bifid at apex; outer phyllaries ovate or deltoid after anthesis
    Paleae villous or scabrous, 2-5 in number, 5-13mm long
      Paleae 3-4mm shorter than akene. . . . . *M. douglasii* ssp. *d.*
      Paleae 0.5-2mm shorter than akene . . . . *M. douglasii* ssp. *platycarpha*
    Paleae relatively smooth, 5 in number, 3.5-7mm long . . . . *M. elegans*

M. douglasii (DC.)Sch.-Bip. ssp. **douglasii**. [*M. parishii* Greene, the type from near San Luis Rey.] Douglas' Microseris. Uncommon, in grassy places; coastal, below 300m: Fallbrook; San Luis Rey; San Diego. Apr. 2*n*=18.

M. douglasii ssp. **platycarpha** (Gray)Chamb. [*M. p.* (Gray)Sch.-Bip.; *M. breviseta* Greene, the type from San Diego.] Small-Flower Microseris. Frequent, on grassy slopes; coastal, below 400m: e.g. San Luis Rey; Rancho Santa Fe; Ramona; Kearny Mesa; Point Loma; Mission Valley; Mission Hills; Sweetwater Valley; National City. Mar.-May. Type from San Luis Rey. 2*n*=18.

M. elegans Greene ex Gray. Elegant Microseris. Infrequent, on grassy slopes and mesas; south coastal, below 200m: Bernardo; Lakeside; San Diego; National City. Apr.-May. 2*n*=18.

M. heterocarpa (Nutt.)Chamb. [*M. lindleyi* (DC.)Gray var. *clevelandii* (Greene)Hall; *Calais parryi* Gray, the type from near San Diego.] Derived Microseris. Frequent, on grassy slopes; coastal, below 200m: e.g. Ysidora; De Luz; Rancho Santa Fe; Bernardo; Torrey Pines; Kearny Mesa; Lakeside; San Diego; Pacific Beach; Home Avenue Canyon; El Cajon; San Miguel Mountain. Also Santa Ysabel. Mar.-May. The type from "St. Diego". *n*=18.

M. lindleyi (DC.)Gray. [*M. linearifolia* (Nutt.)Sch.-Bip.; *Uropappus l.* Nutt.] Silver Puffs. Common, in sandy areas; cismontane and desert, below 1400m: e.g. De Luz; Lilac; Ysidora; Pala; Palomar Mountain; Bear Valley; Warner Ranch; Hot Springs Mountain; Borrego Palm Canyon; Vallecito Stage Station; San Felipe Hill; Harper Ranch; Bernardo; Pacific Beach; Torrey Pines State Reserve; Point Loma; El Cajon; Barrett Dam; Dulzura; Pine Valley. Feb.-May. 2*n*=18.

Monoptilon T. & G. Desert Star
(Astereae - Asterinae)

M. bellioides (Gray)Hall. Mohave Desert Star. Frequent, in sandy areas; desert, below 300m: e.g. Coyote Creek; Borrego Palm Canyon; Borrego Valley; northeast corner of county; Ocotillo Wells; Box Canyon; Carrizo Stage Station; Dos Cabezas; Pioneer Mail Campground; Laguna Mountain. Mar.-May. 2*n*=8.

Palafoxia Lag.
(Helenieae - Heleniinae)

P. arida Turner & Morris var. **arida**. [*P. linearis* auct. non (Dav.)Lag.] Desert Spanish-Needles. Common, in sandy areas; desert, below 400m: e.g. Collins Valley; Borrego Valley; Borrego Palm Canyon; Fish Creek; Yaqui Well; Palm Spring; Carrizo Stage Station. Nov.-Apr. *n*=10, 12.

Pectis L.
(Helenieae - Tagetininae)

P. papposa Harv. & Gray ex Gray var. **papposa**. Chinch Weed. Frequent, on flats; desert, below 800m: e.g. Fish Creek; La Puerta; Mason Valley; Vallecito Valley; Mountain Palm Spring; Carrizo Stage Station. June; Sep.-Dec. *n*=12.

Perityle Benth.
(Helenieae - Peritylinae)

P. emoryi Torr. in Emory. Emory's Rock-Daisy. Occasional, in canyons; coastal and desert, below 300m: e.g. Borrego Palm Canyon; Borrego Valley; northeast corner of county; Tamarisk Grove; Vallecito Stage Station; Palm Spring; Carrizo Stage Station; San Onofre Bluffs; Point Loma; Sweetwater Dam. Oct.-July. Type from "Cordilleras of California". *n*= ca. 36, 50-57.

Peucephyllum Gray
(Senecioneae)

P. schottii Gray in Torr. Pigmy-Cedar. Occasional, on rocky slopes and cliffs; desert, below 400m: northeast corner of county; Coyote Mountain; Calcite Canyon; Fish Creek; Mortero Palms; Volcanic Hills; southwest of Lava Flow Wash. Dec.-May. *n*=20.

Picris L. Ox-Tongue
(Cichorieae - Leontodontinae)

***P. echioides** L. Bristly Ox-Tongue. Native of Europe. Occasional, weed in disturbed areas; cismontane, below 300m: e.g. Santa Margarita River Estuary; Twin Oaks; Del Dios; National City; Otay Mesa. Dec.-June. 2*n*=10.

Pleurocoronis King & Robinson

P. pluriseta (Gray)King & Robinson. . [*Hofmeisteria p.* Gray] Desert Arrow-Leaf. Frequent, in rocky canyons; desert, below 700m: e.g. Borrego Palm Canyon; Rockhouse Canyon; Fish Creek; The Narrows; San Felipe; Mountain Palm Springs; Carrizo Stage Station. Nov.-Apr. 2*n*=18.

Pluchea Cass.
(Inuleae)

P. odorata (L.)Cass. var. **odorata**. [*P. purpurascens* (Sw.)DC.; *P. camphorata* auct. non DC.] Salt-Marsh Fleabane. Occasional, in moist and alkaline habitats; cismontane and desert, below 1300m: e.g. San Mateo and Santa Margarita River estuaries; Oceanside; Fallbrook; Julian; Agua Caliente Hot Springs; Canebrake Canyon; San Dieguito River estuary; San Diego; Silver Strand. July-Jan. *n*=10.

Porophyllum (Vaill.)Adans.
(Senecioneae)

P. gracile Benth. [*P. caesium* Greene, the type from Cajon Hills; *P. vaseyi* Greene, the type from Mountain Springs.] Odora. Common, on dry slopes; cismontane and desert, below 1200m: e.g. San Onofre Canyon; De Luz; Fallbrook; San Marcos; Moosa Canyon; Rincon Grade; Angelina Spring; Sentenac Canyon; The Narrows; San Felipe Wash; Earthquake Valley; Box Canyon; Vallecito Valley; Mountain Palm Springs; Mountain Springs; Jacumba; Viejas Valley; Tecate; Dulzura; Dehesa; Mission Gorge. Oct.-June. *n*=24.

Prenanthella Rydb.
(Cichorieae - Stephanomeriinae)

P. exigua (Gray)Rydb. [*Lygodesmia e.* Gray] Egbertia. Rare, known in the county only at Mountain Springs, 730m. May.

Psathyrotes Gray
(Senecioneae)

P. ramosissima (Torr.)Gray. Velvet Rosettes. Occasional, in sandy areas; desert, below 300m: Borrego Springs; San Felipe Wash; The Narrows; Carrizo Stage Station; Bow Willow Canyon; Imperial Highway Monument; Ocotillo Wells; June Wash; Lava Flow Wash; East Dolomite Mine Road. Dec.-Apr. *n*=17.

Psilocarphus Nutt.
(Inuleae)

Well developed receptacular bracts ca. 3mm long at maturity; leaves oblong to linear . . . . *P. brevissimus*
Well developed receptacular bracts ca. 2mm long; leaves spatulate or oblanceolate . . . . *P. tenellus*

P. brevissimus Nutt. [*P. globiferus* Nutt.] Woolly Marbles. Frequent, on shores of drying ponds: e.g. San Marcos; Ramona; San Diego; San Ysidro; Upper Otay Lake; Little Laguna Lake. Apr.-June.

P. tenellus Nutt. var. **tenellus**. Slender Woolly-Heads. Occasional, on shores of drying ponds; coastal, below 200m: Rancho Santa Fe; San Diego; Kearny Mesa; Otay Lake. Mar.-May.

Rafinesquia Nutt.
(Cichorieae - Stephanomeriinae)

Rays white, 5-8mm long; akene-beak slender, ca. as long as the body; pappus dull or brownish, the bristles plumose to the tip with straight hairs . . . . . . *R. californica*
Rays white but veined outside with rose-purple, 10-15mm long; akene-beak shorter than the body; pappus bright white, the bristles not plumose near the tips, with soft, subarachnoid hairs . . . . . . *R. neomexicana*

R. californica Nutt. [*Nemoseris c.* (Nutt.)Greene] California Chicory. Frequent, on open slopes and burns; cismontane, below 1200m: e.g. De Luz; Lilac; Agua Tibia Mountains; Montezuma View Point; Santa Teresa Valley; Santa Ysabel; La Jolla; San Diego; Mission Hills; Otay Ranch; Campo. Apr.-May. Type from "near the seacoast, in the vicinity of St. Diego". $2n$=16.

R. neomexicana Gray. [*Nemoseris n.* (Gray)Greene] Desert Chicory. Common, in sandy areas; desert, below 700m: e.g. Borrego Palm Canyon; Borrego Valley; Borrego Springs; San Felipe Wash; northeast corner of county; Box Canyon; Vallecito and Carrizo Stage Stations; southeast corner of county. Mar.-Apr. $2n$=16.

Senecio L. Groundsel
(Senecioneae)

Plants annual
  Rays well developed, conspicuous, native. . . . . . *S. californicus*
  Rays lacking or very inconspicuous
    Bracteoles at base of involucre well developed, black-tipped . . . . . . *S. vulgaris*
    Bracteoles at base of involucre none or poorly developed, not black-tipped
      Phyllaries ca. 8, near coast . . . . . . *S. aphanactis*
      Phyllaries ca. 13, deserts . . . . . . *S. mohavensis*
Plants perennial
  Stems climbing; leaves ivy-like, introduced . . . . . . *S. mikanioides*
  Stems not climbing; leaves not ivy-like
    Leaves pinnatifid into narrowly linear segments or sometimes entire and linear; plants somewhat woody at base
      Herbage white-tomentose when young . . . . . . *S. douglasii* var. *d.*
      Herbage glabrous, yellow-green . . . . . . *S. douglasii* var. *monoensis*
    Leaves not pinnatified, mostly basal, herbaceous perennial. . . . . . *S. ganderi*

S. aphanactis Greene. California Groundsel. Occasional, on open and grassy slopes, below 800m: e.g. Kearny Mesa; Pacific Beach; Silver Strand; San Diego; San Miguel Mountain; Jacumba. Mar.-Apr. n=20.

S. californicus DC. California Butterweed. Common, on open slopes; cismontane and desert, below 1200m: e.g. San Onofre Canyon; Fallbrook; Encinitas; lower Doane Valley; Oak Grove; Montezuma View Point; Del Mar; Kearny Mesa; Ocean Beach; Silver Strand; San Diego; Imperial Beach; Jacumba; Mountain Springs; Box Canyon; San Felipe Valley; San Felipe Wash; Borrego Palm Canyon. Apr.-Oct. n=20.

S. douglasii DC. var. **douglasii**. Sand-Wash Butterweed. Occasional, along washes and canyons; coastal and desert, below 900m: e.g. San Mateo Creek; Sweetwater Valley; Tijuana River Valley; Jacumba; Canebrake Canyon; Vallecito Canyon Wash; San Felipe Wash. June-Oct. n=20.

S. douglasii var. **monoensis** (Greene)Jeps. [*S. m.* Greene] Mono Butterweed. Frequent, along washes and canyons; foothill and desert, below 900m: e.g. Oak Grove; Cedar Creek Falls; Vallecito Stage Station; Bow Willow Canyon; Jacumba; Live Oak Springs; Dulzura. Mar.-May. n=20.

S. ganderi Barkley & Beauchamp. [*S. ionophyllus* auct. non Greene] San Diego Butterweed; Post-Toasties. Endemic to San Diego County. Local, in Chaparral; foothill, 400-1200m: El Cajon Mountain; McGinty Mountain; Lawson Peak; Tecate Mountain. Mar.-May. Type from Lawson Peak.

***S. mikanioides** Otto. German Ivy. Native of South Africa. Rare, adventive in moist canyons: Oceanside; Fallbrook; San Elijo Canyon, west of Escondido. Dec.-Mar.

S. mohavensis Gray. Mojave Groundsel. Infrequent, in washes; desert, below 800m: Borrego Palm Canyon; northeast corner of county; San Felipe; Agua Caliente Hot Springs; Vallecito Stage Station; Agua Caliente Hot Springs; Mountain Palm Springs; Mountain Springs. Jan.-Apr. n=20. Some botanical workers believe this species is an Old World taxon, perhaps *S. flavus* (Decne)Sch.-Bip. in Webb & Berth.

***S. vulgaris** L. Common Groundsel. Native of Europe. Occasional, weed in disturbed areas; coastal and foothill, below 400m: e.g. De Luz; Rainbow; Bernardo; Tecate Peak. Jan.-May. n=20.

Silybum Adans.
(Cynareae)

***S. marianum** (L.)Gaertn. Milk Thistle. Native of Europe. Occasional, weed on grassy slopes; coastal, below 300m: e.g. San Onofre; Bonsall; Jamacha; Monument School. May-July. $2n$=34.

Solidago L. Goldenrod
(Astereae - Solidagininae)

Plants with well developed creeping herbaceous rhizomes; stems rather equably leafy, the lowest leaves not prominently different from the upper and at length deciduous
  Leaves glandular-punctate, lance-linear; inflorescence ample, copiously bracteate, the heads in terminal cymose clusters; ray-flowers 15-25 . . . . . . *S. occidentalis*

Leaves not punctate, broader; inflorescence more compact, not interrupted, the heads not glomerate in cymes; ray-flowers 8-13 . . . . *S. californica*
Plants with shorter rather woody rhizomes or a branched caudex; stems with lower and basal petiolate leaves much larger than the upper reduced sessile ones; inflorescence without recurving branches . . . . . . *S. confinis*

S. californica Nutt. California Goldenrod. Common, in Chaparral, Southern Oak Woodland, and Montane Coniferous Forest; foothill and montane, above 400m: e.g. De Luz; Palomar Mountain; lower Doane Valley; Oak Grove; near Beauty Mountain; Mesa Grande; Hot Springs Mountain; Borrego Palm Canyon; San Felipe; Pine Hills; Mussey Grade; Boulder Creek; Laguna Mountains. May-Nov. n=9.

S. confinis Gray. Marsh Goldenrod. Occasional, in wet places of Southern Oak Woodland and Montane Coniferous Forest; foothill and montane, above 400m: e.g. Henshaw Dam; Palomar Mountain; Santa Ysabel; Boulevard. Aug.-Nov. Type from "southern borders of California". n=9.

S. occidentallis (Nutt.)T. & G. Western Goldenrod. Common, in wet places; cismontane: e.g. Oceanside; San Luis Rey River; Palomar Mountain; Mesa Grande; Cuyamaca Lake; Boulevard. Also Sentenac Canyon. Aug.-Nov. n=9.

Soliva R. & P.
(Anthemideae - Chrysantheminae)

Akenes with wings reduced to a hardened marginal callus, or if toothed, the teeth minute and little or not projecting above the akene-body . . . . *S. daucifolia*
Akenes with broad membranous entire wings, each wing projecting above the body of the akene as a tooth 1-1.5mm long . . . . *S. sessilis*

***S. daucifolia** Nutt. Coast Soliva. Native of northern California. Uncommon, weed in lawns; coastal, below 200m: Escondido; Point Loma. Apr.-July.

S. sessilis R. & P. Common Soliva. Scarce, in moist disturbed places; coastal, below 200m: San Diego. Apr.-July. n=59 or 60.

Sonchus L. Sow-Thistle
(Cichorieae - Crepidinae)

Leaves serrate or with broad lobes; plants stout; akenes broad, flat, thin-edged
  Basal auricles of the leaves rounded; akenes 3-nerved on each side, otherwise smooth . . . . . . *S. asper*
  Basal auricles of the leaves acute; akenes striate, transversely wrinkled . . . . . . *S. oleraceus*
Leaves pinnately parted into narrow lobes; plant slender; akenes narrow, thickish . . . . . . *S. tenerrimus*

***S. asper** (L.)Hill. Spiny-Leaf Sow-Thistle. Native of Europe. Common, weed in disturbed places; cismontane: e.g. De Luz; Palomar Mountain; Hot Springs Mountain; Laguna Mountains; Laguna Lakes; Point Loma; Miramar. Mar.-Aug. 2n=18.

***S. oleraceus** L. Common Sow-Thistle. Native of Europe. Common, weed in disturbed places; foothill and coastal, below 500m: e.g. Ramona; Pacific Beach; Point Loma; Spring Valley; San Diego; Silver Strand. Feb.-July. 2n=32.

***S. tenerrimus** L. [*S. tenuifolius* Nutt., the type from "round St. Diego, in shady ravines, among rocks".] Slender Sow-Thistle. Native of Europe. Scarce, weed in disturbed places; coastal, below 200m: San Diego. Apr.-June. 2n=14.

Stephanomeria Nutt.
(Cichorieae - Stephanomeriinae)

Plants annual
  Akenes groved longitudinally
    Heads in small panicles along branches; peduncles 1-4cm long
      Peduncles and involucres glabrous or sparsely glandular . . . . . . *S. exigua* ssp. *e.*
      Peduncles and involucres conspicuously glandular . . . . . . *S. exigua* ssp. *deanei*
    Heads solitary or clustered on short divaricate peduncles less than 1cm long . . . . . . *S. diegensis*
  Akenes not grooved
    Calyculate bracts reflexed; 8-9 florest per head . . . . . . *S. virgata* ssp. *v.*
    Calyculate bracts appressed; 5-6 florets per head . . . . . . *S virgata* ssp. *pleurocarpa*
Plants perennial . . . . . . *S pauciflora*

S. diegensis Gottlieb. San Diego Wreath-Plant. Frequent, in disturbed or open areas; coastal, below 300m: e.g. De Luz; Carlsbad; Del Mar; Torrey Pines State Reserve; Bernardo; Fernbrook; Padre Dam; Bostonia; Lakeside; Flynn Spring; Witch Creek; Coronado; Silver Strand; San Ysidro; Otay Mountain; Tecate Mountain; Potrero Grade. Aug.-Nov. Type from near the entrance to Torrey Pines State Park. n=8.

S. exigua Nutt. ssp. **deanei** (Macbr.)Gottlieb. Deane's Wreath-Plant. Frequent, in sandy fields and on open slopes; cismontane, below 1500m: e.g. San Luis Rey; Rincon; Palomar Mountain; Warner Hot Springs; San Felipe; Borrego Palm Canyon; Julian; Shrine Camp; Pine Hills; Witch Creek; Ballena; Lakeside; Harbison Canyon; White Star; Boulder Oaks; Tierra del Sol. June-Oct. Type from Sweetwater Valley, Las Paderas Ranch. n=8.

S. exigua ssp. **exigua**. Small Wreath-Plant. Infrequent, in sandy areas; desert, below 300m: The Narrows. Apr.-May. n=8.

S. pauciflora (Torr.)A. Nels. Few-Flower Wreath-Plant, Desert-Straw. Frequent, in washes and open places; desert, below 900m: e.g. northeast corner of county; Borrego Palm Canyon; Borrego Valley; Fish Creek; The Narrows; Vallecito Stage Station; Mountain Palm Springs; Jacumba. May-Sep. 2n=16.

S. virgata Benth. ssp. **pleurocarpa** (Greene)Gottlieb. Tall Wreath-Plant. Occasional, in open areas; cismontane above 500m: e.g. Palomar Mountain; Palomar Mountain State Park; Inaja Indian Reservation; Pine Hills; Middle Peak; Japatul Valley; Laguna Public Camp; Boulder Oaks. July-Nov. *n*=8.

S. virgata ssp. **virgata**. Virgate Wreath-Plant. Occasional, in open, often disturbed areas; cismontane, below 1200m: e.g. San Onofre; Fallbrook; Escondido; Poway; San Diego; Chula Vista. July-Oct. *n*=80.

Stylocline Nutt.
(Inuleae)

Pistillate flowers enclosed in the central portion of wholly hyaline bracts which have winglike margins, cismontane . . . . *S. gnaphalioides*
Pistillate flowers enfolded by the basal part of the bracts which do not have wing-like margins, deserts *S. micropoides*

S. gnaphalioides Nutt. Everlasting Nest-Straw. Common, on open slopes; cismontane, and at the desert edge, below 1300m: e.g. Love Valley; Borrego Palm Canyon; Mesa Grande; Banner; Scissors Crossing; Ramona; Bernardo; Kearny Mesa; Torrey Pines; San Diego; Dulzura; southeast corner of county. Feb.-June. 2*n*=14 pairs.

S. micropoides Gray. Desert Nest-Straw. Infrequent, in washes and on slopes; desert, below 300m: Borrego Valley; San Felipe; Vallecito and Carrizo Stage Stations. Mar.-Apr. 2*n*=14 pairs.

Taraxacum Zinn. Dandelion
(Cichorieae - Crepidinae)

All or some of the phyllaries corniculate-appendaged; akenes somewhat reddish at maturity; leaves tending to be deeply cut their whole length . . . . *T. laevigatum*
All or nearly all of the phyllaries unappendaged; akenes pale-gray-brown to olive-brown when mature; leaves tending to have a petiolar base . . . . *T. officinale*

***T. laevigatum** (Willd.)DC. Red-Seed Dandelion. Native of Old World. Local, adventive in moist places; cismontane: Balboa Park; Cedar Creek; Shrine Camp; Laguna Lakes. Feb.-May. 2*n*=24.

***T. officinale** Weber in Wiggers. [*T. vulgare* (Lam.)Schrank.] Common Dandelion. Native of Europe. Common, weed in moist areas: e.g. Oceanside; Doane Valley; Eagles Nest; Harper Ranch; Laguna Camp; San Diego. Jan.-Dec. 2*n*=16, 24, 48.

Tessaria Ruiz & Pavon
(Inuleae)

T. sericea (Nutt.)Shinners. [*Pluchea s.* (Nutt.)Cov., *Polypappus s.* Nutt.] Desert Arrow-weed. Common, along water courses; cismontane and desert, below 1050m: e.g. Bonsall; Carlsbad; Borrego Palm Canyon; Sentenac Canyon; Canebrake Canyon; Del Mar; San Diego; Tijuana River Valley; Barrett Dam. Nov.-Aug. *n*=10.

Tetradymia DC.
(Senecioneae)

T. comosa Gray. Cotton-Thorn. Occasional, in open areas, Chaparral and Southern Oak Woodland; foothill, 300-900m: Palomar Mountain; Oak Grove; Buckman Spring; Campo; Potrero; Dulzura; Rice Canyon. Feb.-Oct. Type from "southeast borders of California", also "Campo, 60 miles east of San Diego". 2*n*=60.

Thelesperma Less.
(Heliantheae - Coreopsidinae)

***T. megapotamicum** (Spreng.)Ktze. American Hipple-seed. Native of central North America and southward. Local, naturalized on Clemson Ranch, San Marcos. July.

Tragopogon L. Goat's-Beard.
(Cichorieae - Stephanomeriinae)

***T. porrifolius** L. Purple Salsify. Native of Europe. Occasional, weed in disturbed and waste places; coastal, below 1300m: e.g. Escondido; Vista; National City; Julian. Apr.-June. 2*n*=12.

Trichoptilium Gray
(Helenieae - Heleniinae)

T. incisum (Gray)Gray. Desert Yellow-Head. Frequent, in sandy areas and desert pavement; desert, below 800m: e.g. northeast corner of county; Ocotillo Wells; Fish Creek; Borrego Palm Canyon; Borrego Valley; Yaqui Well; Vallecito Stage Station; Mountain Springs. Apr.-June. *n*=13.

Trixis P. Br.
(Mutisieae)

T. californica Kell. var. **californica**. California Trixis. Occasional, on rocky slopes; desert, below 800m: Salvador Canyon; Coyote Canyon; San Felipe; Borrego Valley; Yaqui Pass; Fish Creek; Sentenac Canyon; San Felipe Wash; Vallecito and Carrizo Stage Stations; Mountain Palm Springs; Mountain Springs. Nov.-June. n=13.

Venegasia DC.
(Helenieae - Jaumeninae)

V. carpesioides DC. Jesuit Flower. Infrequent, on shaded canyon slopes; coastal and foothill, below 900m: San Onofre Bluffs; Horno Canyon; Temecula Canyon; San Marcos Mountains; Cole Canyon; Del Mar School; Otay Mountain. Mar.-May. *n*=19.

Viguiera H.B.K.
(Heliantheae - Verbesininae)

Leaves ovate; involucre canescent . . . . . *V. parishii*
Leaves lanceolate; involucre green . . . . . *V. laciniata*

V. laciniata Gray. San Diego Sunflower. Common, on open slopes; south coastal and foothill, below 400m: e.g. Mission Gorge; Mission Valley; Flynn Spring; Potrero; Tecate; Telegraph Canyon; Barrett; Dehesa; Spring Valley; Mission Hills; Otay Lake; La Mesa; Dulzura. Also San Onofre; Bonsall. Jan.-July. Type from "Rancho Gamacha [Jamacha]".

V. parishii Greene. [*V. deltoidea* Gray var. *parishii* (Greene)Vasey & Rose. Parish's Golden-Eye. Frequent, on dry slopes; desert, below 1100m: e.g. Coyote Creek; Culp Valley; Box Canyon; lower Vallecito Canyon; Vallecito Stage Station; Mountain Springs; Jacumba. Jan.-June. The type is cited from San Luis Rey, which seems curious in view of the presently known range. 2*n*=18 pairs.

Wyethia Nutt.
(Heliantheae - Verbesininae)

W. ovata T.& G. ex Torr. in Emory. [*W. coriacea* Gray, the type from Mesa Grande.] Southern Mule's-Ears. Common, in open areas; foothill and montane, above 600m: e.g. Boucher Hill; Palomar Mountain; Mesa Grande; Pine Hills; Ballena; Santa Ysabel; Wynola; Inspiration Point; Cuyamaca Lake; Descanso; Alpine; Laguna Mountains; Laguna Camp. May-Sep. Type from the "western side of the Cordilleras of California".

Xanthium L. Cocklebur
(Heliantheae - Ambrosiinae)

Leaves lanceolate, tapering at both ends, short-petiolate, with conspicuous 3-forked spines in their axils *X. spinosum*
Leaves broadly ovate or cordate, long-petiolate, without spines . . . . . *X. strumarium*

***X. spinosum** L. Spiny Clotbur. Native of South America. Common, weed of moist places: Montezuma Valley; San Pasqual; Murphy Canyon; Encinitas; Spring Valley; July-Oct. *n*=18.

***X. strumarium** L. var. **canadense** (Mill.)T.& G. [*X. pennsylvanicum* Wallr.] Eastern Cocklebur. Native of eastern United States. Occasional, weed of waste places: e.g. Oak Grove; Mesa Grande; Encinitas; San Diego; Campo. July-Oct. *n*=18.

Xylorhiza Nutt.

X. orcuttii (Vasey & Rose)Greene. [*Machaeranthera o.* (Vasey & Rose)Cronq. & Keck, *Aster o.* Vasey & Rose] Orcutt's Aster. Infrequent, in sandy areas of gypsum soils; desert, below 400m: Borrego Valley; Borrego Springs; Benson Dry Lake; Oil Well; Split Mountain; Banner Grade; Carrizo Stage Station; Palm Spring; Canyon Sin Nombre. Mar.-Apr. Type from "Cariso [Carrizo] Creek Wash". *n*=6.

AVICENNIACEAE.

Avicennia L.

***A. marina** (Forsr.)Vierh. var. **resinifera** (Forst. f.)Bakh. Manawa. Native of New Zealand. Introduced into Coastal Salt Marsh at Kendall Frost-Mission Bay Reserve.

BASELLACEAE.

Boussingaultia H.B.K.

***B. gracilis** Miers var. **pseudo-baselloides** Bailey. Madeira Vine. Native of Neotropics. Local, escaping from cultivation; coastal, below 150m: Balboa Park; San Diego. July-Sep.

BATACEAE.

Batis L. Saltwort

B. maritima L. American Saltwort. Common, in Coastal Salt Marsh; coastal, below 5m: e.g. San Elijo Lagoon; Sorrento estuary; Mission Bay; San Diego Bay; Tijuana River estuary. July-Oct.

BERBERIDACEAE. Barberry Family

Mahonia Nutt. Barberry

Racemes densely many-flowered, the floral bracts deltoid-ovate, obtuse or acute; leaflets 2-3.5cm wide
  Leaflets glossy-green above and beneath, the lower surface not microscopically papillate, the leaflets crowded on the rachis, often overlapping
    Leaflets 2-3.5cm wide . . . . . *M. pinnata*
    Leaflets 1-2cm wide. . . . . *M.* sp. nov.
  Leaflets dull above and beneath, the lower surface microscopically papillate
    Leaflets cordate at base , suborbicular, with 8-20 teeth on each edge . . . . . *M. amplectens*
    Leaflets not cordate at base, oblong-ovate, mostly 3-10 teeth on each edge . . . . . *M. dictyota*

Racemes loosely 3-9 flowered, the floral bracts mostly lanceolate, subacuminate; leaflets mostly 1-2cm wide
Leaflets stiff-rigid, somewhat folded along the midribs, the 2 sides elevated and strongly crisped with some of the stout maringal spines almost vertical to the leaf surface; berries yellowish-red. *M. higginsiae*
Leaflets less rigid, almost plane, the bristle-like spines almost in the plane of the leaf surfaces; berries red *M. nevinii*

M. amplectens Eastw. [*Berberis a.* (Eastw.)Wheeler] Dull-Leaf Mahonia. Scarce, in Chaparral and Montane Coniferous Forest; montane, above 1050m: Dyche Valley; Banner; Inspiration Point; Laguna Mountains; Guatay. Apr.-May. 2*n*=14 pairs.

M. dictyota (Jeps.)Fedde. [*Berberis d.* Jeps.; *B. californica* Jeps.] California Mahonia. Infrequent, in Chaparral and Montane Coniferous Forest; montane, above 1000m: Palomar Mountain; Santa Ysabel; Pine Hills; Cuyamaca Dam; Inspiration Point; Green Valley Falls; Guatay; Descanso. Apr.-May.

M. higginsiae (Munz)Adrendt. [*Berberis h.* Munz; *B. fremontii* auct. non Torr.; *B. haematocarpa* auct. non Woot.] San Diego Mahonia. Infrequent, in open areas and among rocks; foothill and desert, 750-1100m: Boulevard; Dubber; Walker Ranch. May-July. Type from a "dry rocky point at Boulevard".

M. nevinii (Gray)Fedde. [*Berberis n.* Gray] Nevin's Mahonia. Rare, in Chaparral; cismontane and transmontane,300-700m: San Felipe Wash; Dripping Springs Trail, Agua Tibia Wilderness Area. Also persisting from cultivation, Spring Valley, Torrey Pines State Reserve. Mar.-Apr.

M. pinnata (Lag.)Fedde. ssp. **pinnata**. [*Berberis p.* Lag.] Shinyleaf Mahonia. Occasional, on exposed slopes; coastal and foothill, below 1200m: e.g. Wohlford Dam; Julian; Santa Maria Mountain; Pine Valley; Rancho Santa Fe; Poway; Mission Valley; Chollas Valley; Alpine; Tecate Peak; Japatul Valley; Campo. Mar.-May.

M. sp. nov. Poway Mahonia. A rhizomatous plant which resprouts after fire and grows under chaparral in coastal canyons and on Chaparral-covered slopes; coastal, below 400m: Olivenhain, Poway; Black Mountain; Penasquitos Canyon.

BETULACEAE. Birch Family

Alnus Mill. Alder

A. rhombifolia Nutt. White Alder. Common, along streams; cismontane, below 1550m: e.g. Doane Valley; San Mateo Creek; Hot Springs Mountain; Volcan Mountain; Borrego Palm Canyon; San Felipe Valley; Banner Grade; Oriflamme Canyon; Cuyamaca Mountains; San Luis Rey River; Mission Valley; Wildwood Glen; Descanso; Laguna Mountains. Jan.-Apr.

BIGNONIACEAE. Bignonia Family

KEY TO GENERA

Leaves linear-lanceolate . . . . *Chilopsis*
Leaves boad-ovoid. . . . . *Catalpa*

Chilopsis D. Don

C. linearis (Cav.)Sweet. [*C. l.* var. *arcuata* Fosb.] Desert Willow. Common, along washes; desert, below 920m: e.g. Oak Grove; San Felipe; Borrego Valley; Borrego Palm Canyon; The Narrows; Chariot Canyon; Mason Valley; Vallecito Valley; Dubber; Manzanita Indian Reservation. May-July. 2*n*=40.

Catalpa Scop.

***C. bignonioides** Walt. Native of southeast United States. Local, persisting from cultivation and naturalized along roads and water courses: Olivenhain; San Luis Rey. Apr.-June.

BORAGINACEAE. Borage Family

KEY TO GENERA

Style deeply cleft, the branches each with a capitate stigma . . . . *Tiquilia*
Style entire, terminated by a single simple or obscurely lobed stigma
  Style borne on the summit of the fruit on the pericarp, falling away with the nutlets; stigma discoid, usually surmounted with a short sterile appendage . . . . *Heliotropium*
  Style borne between the lobes of the fruit (nutlets) and attached directly and independently to the gynobase or receptacle; stigma capitate, lacking a sterile apical appendage
    Mature calyx very irregular, burlike, two of the lobes united and enclosing the fruit, becoming horned with 7-9 long glochidiate appendages; ovules two . . . . *Harpagonella*
    Mature calyx not conspicuously irregular, not armed with barbed hornlike appendages; ovules usually 4
      Nutlets widely spreading in fruit, flat, armed with hooked prickles . . . . *Pectocarya*
      Nutlets erect, often not armed with prickles
        Corolla rotate; anthers connivent into a cone . . . . *Borago*
        Corolla tubular, funnelform or salverform; anthers not forming a cone
          Corolla bright blue, locally adventive plants, 4-9dm high . . . . *Echium*
          Corolla white or yellow or orange, natives
            Flowers white
              Nutlets attached laterally along a usually partly opened and generally basally forked groove or slit in the pericarp. *Cryptantha*
              Nutlets lacking a distinct ventral groove or slit in the pericarp, this replaced by an elevated keel and conspicuous attachment-scar . . . . *Plagiobothrys*
            Flowers yellow or orange
              Plants perennial . . . . *Cryptantha*
              Plants annual . . . . *Amsinckia*

Amsinckia Lehm. Fiddleneck

Calyx 3-lobed by fusion of 4 of the 5 members in pairs . . . . *A. tessellata*
Calyx 5-lobed, or the axial pair of lobes fused; nutlets muricate and variously rugose
Axial pair of calyx-lobes usually united; leaves mostly erose-dentate; nutlets dark, 1.5-2mm long . . . . . *A. spectabilis*
Axial pair of calyx-lobes not united
Corolla 2-3 times as long as calyx, orange . . . . *A. eastwoodiae*
Corolla 1-2 times as long as calyx, orange to orange-yellow
Corolla 5-10mm broad, 1-2 times as long as calyx . . . . *A. intermedia*
Corolla 2-3mm broad, 1-1.5 times the calyx . . . . *A. menziesii*

A. eastwoodiae Macbr. [*A. intermedia* var. *e.* (Macbr.)Jeps.; *A. jonesii* Suksd., the type from Fallbrook.] Large-Flower Fiddleneck. Uncommon, in grassy fields; cismontane, below 1100m: Fallbrook; Paradise Hills; La Jolla; San Diego; Jacumba. Mar.-June. *n*=12.

A. intermedia F. & M. [*A. albicarpa* Suksd., the type from the Colorado Desert; *A. angustata* Suksd., the type from Cottonwood Valley; *A. caduca* Suksd., the type from San Diego; *A. curvata* Suksd., the type from "Warrens Campo"; *A. decumbens* Suksd., the type from "grassy places in park, San Diego"; *A. diversifolia* Suksd., the type from La Jolla; *A. laxa* Suksd., the type from Lakeside; *A. palmeri* Suksd., the type from "southern part of San Diego County"; *A. spencerae* Suksd., the type from the Colorado Desert.] Rancher's Fiddleneck. Common, in open areas; cismontane and desert, below 1600m: e.g. San Onofre; De Luz; Fallbrook; Palomar Mountain; Montezuma View Point; Borrego Valley; Mason Valley; Cuyamaca Lake; Campo; Pine Valley; Jacumba. Mar.-June. *n*=15, 17, 19.

A. menziesii (Lehm.)Nels. & Macbr. [*A. retrorsa* Suksd.] Rigid Fiddleneck. Occasional, on slopes and in fields; cismontane, below 1400m: e.g. Mesa Grande; Cuyamaca Lake; Alpine; Lyons Valley; El Cajon Valley; Jamul Valley; Jamacha; Imperial Beach; Campo. Mar.-June. *n*=8, 13, 17.

A. spectabilis F. & M. var. **spectabilis**. [*A. littoralis* Greene ex Suksd., the type from the "peninsula of San Diego [Point Loma]". Seaside Fiddleneck. Occasional, in sandy areas; coastal, below 50m: e.g. Agua Hedionda; Pacific Beach; Ocean Beach; Coronado; Imperial Beach. Feb.-June. *n*=5.

A. tessellata Gray. [*A. deltoidea* Suksd., the type from between Jacumba and Mountain Springs.] Checker Fiddleneck. Occasional, in sandy areas; desert, below 1300m: Borrego Palm Canyon; San Felipe; Yaqui Pass; Sentenac Canyon; Mountain Springs; Table Mountain. Mar.-June. *n*=12.

Borago L. Borage

***B. officinalis** L. Common Borage. Native of Europe. Rare, escaping from cultivation: San Diego; National City. May.

Cryptantha Lehm.

Calyx circumscissile at maturity; low diffuse plants; inflorescence compact, each flower in axil of leafy bract . . . . *C. circumscissa*
Calyx not circumscissile
Gynobase subulate, protruding beyond the nutlets, bearing a sessile stigma on its top; root and base of plant with a purple dye; each flower in the axil of a leafy bract
Corolla 0.5-1.2mm broad . . . . *C. micrantha* ssp. *m.*
Corolla 1.5-3.5mm broad . . . . *C. micrantha* ssp. *lepida*
Gynobase shorter than the nutlets; style developed; root and herbage usually with very little or no purple dye; flowers all or in part bractless except in *C. maritima*
Nutlets roughened or in *C. maritima* at least one of them so
Margin of nutlets decidedly winged or knifelike
Pedicels usually evident, slender, 1-4mm long; lateral angles of nutlets distinctly winged
Nutlets homomorphic, broadly winged . . . . *C. holoptera*
Nutlets heteromorphic, narrowly winged . . . . *C. racemosa*
Pedicels obscure or none, less than 1mm long
Nutlets 1 or 2; odd nutlet axial . . . . *C. utahensis*
Nutlets 4
Nutlets homomorphous, obscurely rugose, the back high-convex . . . . *C. costata*
Nutlets heteromorphous, verrucose or muricate; lateral margins of the nutlets distinctly winged; corolla 1mm broad
Lateral margins on all nutlets . . . . *C. pterocarya* var. *cycloptera*
Lateral margins not on all nutlets
Axial nutlet unmargined, the others broad-winged . . . . *C. pterocarya.* var. *p.*
One nutlet unmargined, the others wth narrow, knife-like margins. . . . . *C. pterocarys* var. *purpusii*
Margins of the nutlets rounded or obtuse
Nutlets decidedly heteromorphous, 1-4, the large nutlet axial and sometimes less roughened
Nutlets 4, triangular-ovate, 0.7-0.9mm long; mature calyx subglobose, minute, the lobes scarcely longer than the nutlets . . . . *C. micromeres*
Nutlets 1-2, oblong-lanceolate, 1-2mm long; mature calyx oblong, the lobes surpassing the nutlets . . . . *C. maritima*
Nutlets homomorphic
Nutlets lanceolate
Nutlets only 1, rarely 2, in a normal fruit; style not more than half as long as nutlet, corolla less than 1mm broad, deserts . . . . *C. decipiens*
Nutlets 4 in normal fruit; style often more than half as long as the nutlets
Corolla conspicuous, 2-8mm broad; plant hirsute, mostly cismontane . . . . *C. intermedia*
Corolla inconspicuous, 1-2mm broad; desert plants
Fruiting calyx 6-12mm long
Plant bristly with spreading hairs. . . . . *C. barbigera*

Plant strigose with appressed hairs . . . . *C. nevadensis*
Fruiting calyx 3-4mm long . . . . *C. angustifolia*
Nutlets ovate to triangular-ovate
Dorsal side of nutlets obtuse and with at least a faint median ridge; style surpassing the nutlets
Corolla 2-6mm broad . . . . *C. muricata* var. *m.*
Corolla 1-2.5mm broad . . . . *C. muricata* var. *jonesii*
Dorsal side of nutlets flat or low-convex, without median ridge; style not surpassing nutlets *C. simulans*
Nutlets smooth and shining, not roughened
Calyx armed with pale encrusted spreading hooked or curved hairs . . . . *C. flaccida*
Calyx armed with straight tawny unencrusted hairs
Nutlets with excentric goove, one side of nutlet on lower surface appearing as if somewhat deformed, montane . . . . *C. affinis*
Nutlets with a centrally-placed groove
Dorsal side of nutlets low-convex or flat; nutlets 1-2, homomorphic . . . . *C. maritima*
Dorsal side of nutlets rounded-convex, the lateral angles rounded or obtuse
Style less than half as long as nutlets; fruiting calyx 1.5-2mm long . . . . *C. microstachys*
Style almost as long as nutlets; calyx longer
Fruiting calyx 2-4mm long, moderately white-bristly
Corolla 2-5mm broad . . . . *C. clevelandii* var. *florosa*
Corolla 1-2mm broad . . . . *C. clevelandii* var. *c.*
Fruiting calyx 6-10mm long, conspicuously tawny-bristly . . . . *C. ganderi*

C. affinis (Gray)Greene. Common Cryptantha. Rare, in Montane Coniferous Forest; montane, above 1300m: Cuyamaca Lake; Middle Peak; Cuyamaca Peak. June-Aug.

C. angustifolia (Torr.)Greene. Narrowleaf Cryptantha. Frequent, in sandy areas; desert, below 400m: e.g. Borrego Valley; Seventeen Palms; Benson Dry Lake; Vallecito Stage Station; Mountain Palm Springs; Carrizo Stage Station. Dec.-Apr.

C. barbigera (Gray)Greene. Bearded Cryptantha. Frequent, in sandy areas; desert, below 700m: e.g. Hellhole Canyon; Borrego Palm Canyon; Borrego Valley; San Felipe Wash; Scissors Crossing; Choia Ranch; Box Canyon; La Puerta; Dubber; Mountain Springs. Also near Beauty Mountain. Dec.-May.

C. circumscissa (H.& A.)Jtn. Western Cryptantha. Scarce, in sandy areas; desert, 600-1000m: Jacumba; Mountain Springs. Apr.-Aug. n=12.

C. clevelandii Greene var. **clevelandii**. Cleveland's Cryptantha. Frequent, in grassy areas; coastal, below 200m: e.g. Pala; Monserrate; Mission Gorge; Chollas Valley; Point Loma; Mission Hills; Otay Ranch. Apr.-June. Type from "shaded places along streamlets in the hills back of San Diego".

C. clevelandii var. **florosa** Jtn. Cleveland's Large Cryptantha. Uncommon, in sandy places; coastal, below 50m: San Onofre; San Elijo Lagoon; Linda Vista; Murphy Canyon. Apr.-June. Type from "roadside, Linda Vista".

C. costata Bdg. Ashen Cryptantha. Uncommon, in sandy areas; desert, below 50m: Borrego Springs; Oil Well. Mar.-Apr. Type from "Borrego Springs, Colorado Desert". Feb.-May.

C. decipiens (Jones)Heller. Gravel Cryptantha. Infrequent, in sandy areas; desert, below 1100m: San Felipe Ranch; Borrego Palm Canyon; Borrego Valley; Choia Ranch; San Felipe Hill; Mountain Springs; southeast corner of county; north Mortero Wash; Canyons; southwest of Lava Flow Wash; The Potrero; Plum Canyon. Mar.-May.

C. flaccida (Dougl.)Greene. Flaccid Cryptantha. Rare, dry ridges, known in the county only at Cuyamaca Lake, 1400m. Apr.-June.

C. ganderi Jtn. Gander's Cryptantha. Scarce, in sandy areas; desert, 130-170m: Borrego Valley. Feb.-May. Type from "near school at Borrego Valley, Larrea-Franseria association, 500 feet altitude".

C. holoptera (Gray)Macbr. Roughstem Cryptantha. Scarce, in gravelly places; desert, below 200m: Split Mountain. Mar.-Apr.

C. intermedia (Gray)Greene. Nievitas Cryptantha. Common on open slopes and burns; cismontane, below 1600m: e.g. Fallbrook; Bonsall; Palomar Mountain; Oak Grove; Warner Springs; San Felipe Valley; Julian; Descanso; San Pasqual; Jacumba; Campo; Kearny Mesa. Mar.-July. n=12.

C. maritima (Greene)Greene. [*C. m.* var. *pilosa* Jtn.] Whitehair Cryptantha. Frequent, in sandy areas; desert, below 400m: e.g. Borrego Palm Canyon; Borrego Valley; Borrego Mud Hills; Yaqui Well; Mountain Palm Springs; Carrizo Stage Station. Jan.-Apr.

C. micrantha (Torr.)Jtn. ssp. **lepida** (Gray)Mathew & Raven. Hairy Purpleroot Cryptantha. Frequent, in sandy places; foothill and montane, above 1000m: e.g. Morgan Hill; Hot Springs Mountain; Chihuahua Valley; Montezuma View Point; San Felipe Valley; Banner; Rattlesnake Valley; Laguna Lakes; Campo; Jacumba. Apr.-July. Type from San Diego County.

C. micrantha ssp. **micrantha**. Purpleroot Cryptantha. Frequent, in sandy areas; desert, below 1900m: e.g. Coyote Canyon; Borrego Palm Canyon; Clark Dry Lake; San Felipe Ranch; San Felipe Wash; Vallecito and Carrizo Stage Stations; Pinyon Point on Mount Laguna. Mar.-June. n=12.

C. micromeres (Gray)Greene. Minute-Flower Cryptantha. Occasional, on open slopes and burns; foothill and coastal, below 700m: e.g. Red Mountain Grade; Lilac; Valley Center; Poway Grade; Kearny Mesa; Point Loma; Home Avenue Canyon; Harbison Canyon; Alpine. Mar.-May. Type from Soledad.

C. microstachys (Greene ex Gray)Greene. Tejon Cryptantha. Scarce, in dry openings; foothill and coastal, below 800m: north of Mira Mesa; San Vicente Valley; Mission Hills; Barona Valley; Potrero; west of Campo. Mar.-June.

C. muricata (H.& A.)Nels. & Macbr. var. **jonesii** (Gray)Jtn. Jones' Cryptantha. Occasional, on slopes; coastal and foothill, below 1300m: Pala; Doane Valley; Rincon; Mesa Grande; Hot Springs Mountain; Montezuma View Point; Laguna Mountain; El Capitan; Barona Valley; Mount Woodson; Point Loma; Jamacha; Barrett; Jacumba. Apr.-Aug.

C. muricata var. **muricata**. Prickly Cryptantha. Infrequent, in open places; cismontane: Harrison Grade; near Beauty Mountain mine; Rattlesnake Valley; Cuyamaca Peak; Pine Valley; Del Dios; Los Terranitos; Otay Mountain. May-Sep.

C. nevadensis Nels. & Kenn. var. **nevadensis**. Nevada Cryptantha. Occasional, in sandy areas; desert, below 700m: Borrego Valley; Yaqui Pass; San Felipe; Vallecito and Carrizo Stage Stations; Mountain Palm Springs; Mountain Springs; Coyote Mountain; Lava Flow Wash; Agua Caliente Springs; Canyon Sin Nombre. Dec.-Apr.

C. pterocarya (Torr.)Greene var. **cycloptera** (Greene)Macbr. Single-winged Cryptantha. Rare, in sandy areas; desert, below 700m: Banner Grade; Mountain Springs; northeast corner of county. Mar.-Apr.

C. pterocarya var. **pterocarya**. Wing-nut Cryptantha. Scarce, in sandy areas; desert, 500m-1200m: Warner Ranch; Coyote Creek; Scissors Crossing; Jacumba. Mar.-Apr.

C. petrocarya var. **purpusii** Jeps. Purpus' Wing-nut Cryptantha. Rare, in sandy areas; desert, 600-1300m: Montezuma View Point; Mountain Springs; southeast corner of county. Apr.

C. racemosa (Wats.)Greene. Woody Cryptantha. Uncommon, in sandy areas; desert, below 600m: Borrego Palm Canyon; Yaqui Pass; San Felipe Creek; Fish Creek; Mountain Springs; Smugglers Canyon; Alta Bisnaga Wash. Nov.-Apr.

C. simulans Greene. Pine Cryptantha. Uncommon, in Montane Coniferous Forest; montane, above 1200m: Doane Valley; Sunday School Flats; Palomar Mountain; Laguna Mountain; Hot Springs Mountain; Middle Peak. May-June.

C. utahensis (Gray)Greene. Scented Cryptantha. Rare, in sandy areas; desert, 300-900m: San Felipe Hill; Oriflamme Mountains; InKoPah County Park; Potrero. Mar.-Apr.

Echium L. Viper's Bugloss

***E. plantagineum** L. Native of southern Europe. Local, naturalized at Fallbrook and De Luz. May. $2n=16$.

Harpagonella Gray

H. palmeri Gray. [*Pectocarya p.* (Gray)Veno, nom. nud.] Palmer's Grappling-hook. Occasional, on open clay slopes and burns; coastal and desert, below 1000m: e.g. Guajome Mesa; Rancho Santa Fe; Olivenhain; Poway Grade; Kearny Mesa; Emerald Hills; Mission Gorge; Otay; Dehesa; Rice Canyon; Table Mountain; Box Canyon. Mar.-Apr.

Heliotropium L. Heliotrope

H. curvassavicum L. ssp. **oculatum** (Heller)Thorne. Salt Heliotrope. Common, weed in disturbed and alkaline areas; below 1400m: Carlsbad; Encinitas; Doane Valley; Warner Hot Springs; San Felipe Valley; Scissors Crossing; Carrizo Stage Station; Jacumba; Otay. Mar.-Oct.

Pectocarya DC. ex Meissn.

Nutlet-margins entire, lacking fringelike teeth and uncinate hairs; nutlet-body cuneate-obovate, plant erect or nearly so . . . . . *P. setosa*
Nutlet-margins lacerate or undulate or uncinate-bristly; nutlet-body linear or oblong
  Nutlets with margins entire or undulate along sides, armed only at distal end where densely uncinate-bristly
    Nutlets all wing-margined . . . . . *P. penicillata*
    Nutlets heteromorphic, 1 of each divergent pair winged, the other wingless or merely margined
      Nutlets of each fruit in two planes, one pair curved inward, the other pair recurved, all oblong when viewed from above . . . . . *P. heterocarpa*
      Nutlets of each fruit all spreading in one plane, ovate-elliptic when viewed from above . . . . . *P. peninsularis*
  Nutlets with margins pectinately lacerate or toothed most of their length, also usually uncinate-bristly about the distal end.
    Nutlet-margins narrow and inconspicuous, the teeth nearly or quite distinct
      Nutlets straight or somewhat falcate; marginal appendages stout . . . . . *P. linearis*
      Nutlets conspicuously recurved; marginal appendages very elongate . . . . . *P. recurvata*
    Nutlet-margins broad and conspicuous, the teeth obviously confluent . . . . . *P. platycarpa*

P. heterocarpa (Jtn.)Jtn. Chuckwalla Pectocarya. Uncommon, in sandy areas; desert, below 800m: Coyote Creek; Borrego Valley; Benson Dry Lake; Mason Valley; Mountain Palm Springs; Carrizo Stage Station. Mar.-Apr.

P. linearis (R.& P.)DC. ssp. **ferocula** (Jtn.)Thorne. Slender Pectocarya. Occasional, in open areas; cismontne and desert, below 900m: e.g. San Mateo Creek; Fallbrook; Bonsall; Cuca Indian Reservation; Escondido; San Felipe Valley; The Narrows; Dubber; Lakeside; Point Loma; Jamacha; Harbison Canyon; Tecate Junction. Mar.-May. $2n=24$ pairs.

P. penicillata (H.& A.)A. DC. Winged Pectocarya. Occasional, in dry sandy places; cismontane and desert, below 1000m: e.g. Fallbrook; Pala; Rincon; Lake Henshaw; Montezuma View Point; San Felipe Valley; Banner; Mason Valley; Buckman Spring; Anahuac; Campo; Lyons Valley; Balboa Park. Mar.-May. $2n=12$ pairs.

P. peninsularis Jtn. Peninsular Pectocarya. Rare, washes, roadsides and open areas; desert, below 300m: northwest of Narrows; Borrego Valley.

P. platycarpa (Munz & Johnstn.)Munz & Johnstn. Broad-Fruited Pectocarya. Occasional, in sandy areas; desert, below 800m: e.g. Borrego Palm Canyon; Sentenac Canyon; Vallecito and Carrizo Stage Stations; Mountain Palm Springs; Mountain Springs. Mar.-Apr.

P. recurvata Jtn. Recurved Pectocarya. Uncommon, in sandy areas; desert, below 900m: Borrego Palm Canyon; Scissors Crossing; Box Canyon; Carrizo Stage Station; southeast corner of county. Mar.-Apr. n=12.

P. setosa Gray. [*P. s.* var *aptera* Jtn., the type from "dry canyon floor near Campo".] Bristly Pectocarya. Occasional, in sandy places; desert and foothill, below 1100m: Montezuma View Point; San Felipe Valley; San Felipe Wash; Banner; Box Canyon; Buckman Spring; Campo; Jacumba; Mountain Springs; McCain Valley Recreation Area. Apr.-May.

Plagiobothrys F. & M.

Leaves all alternate
  Caruncle of nutlet elongate, apparently extending along crest of ventral keel; nutlets trigonous; corolla 1-2.5mm broad; nutlets tessellate . . . . *P. jonesii*
  Caruncle round or nearly so, at or below end of ventral keel
    Lowest leaves not in a rosette; caruncle weakly developed, borne at tip of a short ventral stipe; nutlets lacking a broad transverse ventral groove, montane and cismontane . . . . *P. californicus*
    Lowest leaves mostly in a rosette; caruncle well developed, sessile on nutlet; nutlet with a shallow transverse ventral groove
      Calyx circumscissile in fruit; less than 4mm long; lobes usually connivent over fruit; mature nutlets usually only 1 or 2
        Inflorescence a long simple bracted raceme; nutlets highly arched in lateral outline, 1-2.5mm long; corolla 2-3mm broad . . . . *P. arizonicus*
        Inflorescence forked, bracted at base only if at all; nutlets low and flattened in lateral outline, 2-3mm long; corolla 3-9mm broad . . . . *P. nothofulvus*
      Calyx not circumscissile, or if so, the strongly accrescent calyx over 4mm long; calyx-lobes erect or spreading; nutlets usually 4
        Transverse dorsal crests of nutlets very narrow and sharp, with median keel enclosing polygonal granulate areoles; stems rather coarse . . . . *P. canescens*
        Transverse dorsal crests of nutlets very low and broad, separated only by low lineate ridges
          Nutlets cruciform due to the abrupt constrictions at apex and base, glossy; plant rarely dye-stained *P. tenellus*
Leaves opposite, at least below
  Nutlets attached to gynobase by a somewhat well developed stipelike ventral projection
    Nutlets 1-1.5mm long . . . . *P. californicus* var. *gracilis*
    Nutlets 1.5-2mm long
      Stems erect compact, 2-8cm long . . . . *P. californicus* var. *ursinus*
      Stems spreading or prostrate, 1-4dm long
        Herbage with spreading-hirsute pubescence, corolla 4-7mm broad . . . . *P. californicus* var. *c.*
        Herbage rougher, short-hispid, corolla 2-3mm broad . . . . *P. californicus* var. *fulvescens*
  Nutlets attached directly to gynobase, without a stipelike projection
    Scar large, deeply excavated, 1/4 to 1/2 the length of the nutlet; nutlets with glochidiate appendages . . . . *P. acanthocarpus*
    Scar small, not much if at all excavated, 1/5 the length of the nutlet or less
      Nutlet attachment basal, frequently substipitate; plants somewhat succulent; calyx-lobes strongly ribbed
        Plants prostrate; calyx-lobes connivent and together directed off toward one side of the flower *P. leptocladus*
        Plants erect; clayx-lobes spreading equally . . . . *P. stipitatus*
      Nutlet attachment lateral to obliquely basal; calyx-ribs only rarely thickened
        Scar linear or nearly so, usually borne on a knifelike attachment . . . . *P. undulatus*
        Scar broad, the attachment not at all knifelike, or if so the nutlets rather asymmetric
          Scar near or at base of nutlet, oblique to the ventral keel . . . . *P. bracteatus*
          Scar distinctly lateral, parallel with the ventral keel . . . . *P. hispidulus*

P. acanthocarpus (Piper)Jtn. Adobe Popcornflower. Infrequent, on mesas and in vernal pools; coastal, below 500m: San Marcos; Soledad Mountain; Kearny Mesa; Poway Grade; Grossmont; Ramona; Point Loma; Sweetwater and Otay Reservoirs; Otay Ranch; Otay Mesa. Mar.-May.

P. arizonicus (Gray)Greene ex Gray. Arizona Popcornflower. Occasional, in dry places; foothill and montane, above 300m: e.g. Fallbrook; Warner Ranch; Lake Henshaw; Santa Ysabel; Montezuma View Point; Puerta de la Cruz; San Felipe Valley; Banner; Cuyamaca; Green Valley; Rattlesnake Valley; Garnet Peak; Mason Valley. Mar.-May.

P. bracteatus (Howell)Jtn. Bract Popcornflower. Infrequent, in vernally moist places; cismontane: Kearny Mesa; National City; Pine Hills; Mount Laguna. Apr.-June.

P. californicus (Gray)Greene var. **californicus**. [*P. allocaryoides* Brand, the type from San Diego.] California Popcornflower. Common, in grassy places; coastal and foothill, below 500m: e.g. Fallbrook; Las Flores; Moosa Canyon; La Jolla; Kearny Mesa; Mission Valley; Barona Valley; Potrero; San Miguel Mountain; Silver Strand. Mar.-May.

P. californicus var. **fulvescens** Jtn. Rough Popcornflower. Frequent, in dry places; cismontane, below 1600m: e.g. Lilac; Morgan Hill; Lake Henshaw; San Pasqual; Rattlesnake Valley; Cuyamaca Lake; Buckman Spring; Jacumba; Lyons Valley; Barrett Junction; Otay Mountain; Point Loma. Mar.-May.

P. californicus var. **gracilis** Jtn. San Diego Popcornflower. Infrequent, on open slopes; coastal, below 600m: Rancho Santa Fe; Soledad; Point Loma; Home Avenue Canyon; Lyons Valley; Dehesa; Otay Mesa. Mar.-May. Type from San Diego.

P. californicus var. **ursinus** (Gray)Jtn. Bear Valley Popcornflower. Rare, known in the county only from a collection at Tecate Mountain, 1000m. Apr.

P. canescens Benth. Valley Popcornflower. Scarce, in grassy places and burns; foothill and coastal, below 1000m: Soledad; Foster; National City; Dulzura; San Felipe Ranch. Mar.-May.

P. hispidulus (Greene)Jtn. Harsh Popcornflower. Infrequent, along the shoreline of ponds and lakes; montane, above 1200m: Julian; Cuyamaca Lake; Laguna Lake. June-Sep.

P. leptocladus (Greene)Jtn. Wire-Stem Popcornflower. Rare, known in the county only from a collection from Point Loma. Mar.-May.

P. jonesii Gray. Jones' Popcornflower. Uncommon, sandy areas; desert, 500m-1000m: southwest of Lava Flow Wash.

P. nothofulvus (Gray)Gray. Rusty Popcornflower. Frequent, in grassy places; cismontane, below 1200m: e.g. Palomar Mountain; Escondido; Ramona; Julian; Lake Henshaw; Warner Hot Springs; Grossmont; Earthquake Valley; Santa Ysabel; San Miguel Mountain; Point Loma. Mar.-May. *n*=12.

P. stipitatus (Greene)Jtn. var. **micranthus** (Piper)Jtn. Small-Flower Popcornflower. Rare, known in the county only from a collection from Campo, 800m. Mar.-May.

P. tenellus (Nutt.)Gray. Slender Popcornflower. Occasional, in grassy places; foothill and montane, above 600m: e.g. Palomar Mountain; Julian; Cuyamaca Lake; Laguna Lakes; Lyons Valley; Campo. Mar.-may.

P. undulatus (Piper)Jtn. Coast Popcornflower. Rare, known in the county only from a collection from a vernal pool near Ramona, 450m. Mar.-June.

Tiquilia Pers.

Fruit globose until completely mature, not lobed, bearing the style on its rounded summit, at last breaking into quarters to form the 4 nutlets; leaves obscurely veined; flowers borne singly in leaf axils or at forks of stems . . . *T. canescens*
Fruit early divided into distinct nutlets and bearing the style between the apices; leaves with evidently impressed veins; flowers in clusters at forks
  Leaves with 2-3 pairs of shallowly impressed veins; calyx glabrous or short-pubescent within . . . . *T. palmeri*
  Leaves with about 6 pairs of deeply impressed veins; calyx long-villous within . . . . *T. plicata*

T. canescens (DC.)A. Richardson var. **canescens**. [*Coldenia c.* DC.] Shrubby Coldenia. Rare, known in the county only at Yaqui Pass, 600m. Mar.-May. *n*=9.

T. palmeri (Gray)A. Richardson. [*Coldenia p.* Gray] Palmer's Coldenia. Frequent, in sandy areas; desert, below 600m: Coyote Creek; San Felipe; Yaqui Well; The Narrows; Ocotillo Wells; Carrizo Stage Station. Apr.-June. *n*=8, 9.

T. plicata (Torrey)A. Richardson. [*Coldenia p.* (Torr.)Cov.] Plicate Coldenia. Frequent, in sandy areas; desert, below 400m: e.g. Coyote Creek; Benson Dry Lake; San Felipe; Agua Caliente; Canebrake Canyon; Palm Spring; Carrizo Stage Station. Apr.-July. *n*=8.

BRASSICACEAE. Mustard Family (Cruciferae)

KEY TO GENERA

Fruit indehiscent or tardily dehiscent, moderately to strongly compressed, 1 or 6-12 seeded
  Ovary 2-celled, 6-12 ovuled; pod hispidulous, the hairs straight, the pod ultimately dehiscent, seeds hispidulous . . . . *Heterodraba*
  Ovary 1-celled, 1 or 2 ovuled, mature pod 1-seeded; seeds glabrous
    Silicles not winged, with hooked hairs; pubescence of stems of branched hairs . . . . *Athysanus*
    Silicles wing-margined all around, suborbicular, not with hooked hairs; stems glabrous or with simple hairs . . . . *Thysanocarpus*
Fruit dehiscent by valves or breaking transversely into joints, normally containing 2 or more seeds
  Siliques at length breaking transversely into seed-bearing indehiscent joints
    Fruits transversely 2-jointed, 2-seeded; petals 5-10mm long
      Flowers purplish or whitish; plants of sea-beaches . . . . *Cakile*
      Flowers yellow with darker veins . . . . *Rapistrum*
    Fruit breaking irregularly into joints, several-seeded; petals 15-20mm long . . . . *Raphanus*
  Siliques or fruits not breaking transversely into seed-bearing joints, but dehiscent by valves
    Fruit a silique, elongate, several times longer than wide or thick
      Siliques long-stipitate, the stipe 1-2cm long; anthers twisted; sepals spreading or recurved at anthesis *Stanleya*
      Siliques sessile or with much shorter stipes; anthers usually not twisted; sepals usually erect
        Siliques compressed contrary to the narrow partition; racemes leafy . . . . *Tropidocarpum*
        Siliques not compressed contrary to the partition; racemes leafless
          Stigma-lobes situated over the valves; anthers sagittate at base; petals usually linear
            Calyx urn-shaped or flask-shaped at anthesis; siliques sessile or subsessile
              Siliques conspicuously beaked, pendent, sessile, strongly compressed; valves of mature fruit remaining attached at tip; seeds flat, winged; annual . . . . *Streptanthella*
              Siliques not or inconspicuously beaked, erect to deflexed, the valves separating completely; annual to perennial
                Siliques flattened; seeds flat, usually winged . . . . *Streptanthus*
                Siliques terete or 4-sided; seeds wingless or nearly so . . . . *Caulanthus*
            Calyx open at anthesis, not urn-shaped; siliques sometimes short-stipitate, mostly terete or 4-sided when mature . . . . *Caulanthus*

Stigma-lobes situated over the placentae; anthers usually not sagittate at the base; petals frequently wider than linear
Basal leaves forming basal rosettes, from which arise the flowering stems
Flowers yellow; stems angled . . . . . *Barbarea*
Flowers white; stems terete
Leaves simple; undivided to sinuate or lyrate . . . . . *Arabis*
Leaves deeply pinnatifid or pinnate
Leaves pinnate with definite leaflets; seeds wingless . . . . . *Cardamine*
Leaves pinnatifid; seeds winged . . . . . *Sibara*
Basal leaves not in well-formed rosettes
Fruits with a long indehiscent conical to flattened beak
Seeds globose, in 1 row in each locule . . . . . *Brassica*
Seeds ellipsoid to ovoid, in 2 rows in each locule . . . . . *Eruca*
Fruits not with indehiscent beak, but dehiscent to tip
Pubescence simple or wanting
Siliques strongly flattened
Valves of silique nerved; leaves simple . . . . . *Arabis*
Valves of silique not nerved; leaves mostly pinnate . . . . . *Cardamine*
Siliques terete or 4-angled
Plants with creeping base, perennial; seeds biseriate
Flowers yellow . . . . . *Rorippa*
Flowers white . . . . . *Nasturtium*
Plants without creeping bases, mostly annual or biennial
Pods somewhat 4-angled; seeds uniseriate; biennial or perennial of wet places . . . . *Barbarea*
Pods terete; annuals
Siliques 2-10cm long (if shorter, closely appressed); seeds uniseriate . . . . . *Sisymbrium*
Siliques 0.4-1.2cm long, divaricate; seeds biseriate . . . . . *Rorippa*
Pubescence of forked or stellate hairs
Leaves pinnatifid to pinnate . . . . . *Descurainia*
Leaves entire or nearly so
Flowers yellow to orange . . . . . *Erysimum*
Flowers white to rose or purple
Petals 3-15mm long; stigmas not horned on back; epidermal cells of septum of silique not reticulate . . . . . *Arabis*
Petals ca. 20mm long; stigmas horned on back; epidermal cells of silique-septum reticulate; naturalized along seacoast . . . . . *Matthiola*
Fruit a silicle, 1-2 (-3) times longer than wide
Silicles evidently flattened
Silicles flattened parallel to the broad partition
Pubescence of appressed 2-branched hairs attached at middle; silicles orbicular with flat margins *Lobularia*
Pubescence stellate or forked, but not as above; silicles not margined
Petals entire to emarginate . . . . . *Draba*
Petals bifid . . . . . *Erophila*
Silicles flattened contrary to the narrow partition
Silicles twinlike, the 2 locules somewhat rounded with deep notch between both at summit and base
Silicles 6-18mm wide; pubescence stellate . . . . . *Dithyrea*
Silicles ca. 2mm wide; pubescence not stellate . . . . . *Coronopus*
Silicles not twinlike
Silicles inverted-triangular, broad at summit; annuals; flowers ca. 2mm long . . . . . *Capsella*
Silicles orbicular to oval or cuneate
Petals 18-20mm long; suffrutescent perennial . . . . . *Lyrocarpa*
Petals 1-5mm long
Cells of the silicle 1-seeded . . . . . *Lepidium*
Cells of the silicles 2-many seeded
Silicles notched to subtruncate at apex, 4-18mm long; petals 3-5mm long . . . . . *Thlaspi*
Silicles rounded at apex, 2-4mm long; petals 1mm long . . . . . *Hutchinsia*
Silicles turgid or inflated, not flattened
Plants in scapose tufts, the leaves mostly basal . . . . . *Draba*
Plants with leafy stems
Petals white, 2-3mm long; long-lived perennials . . . . . *Cardaria*
Petals yellow; annual to perennial
Valves of the silicles 1-nerved; cauline leaves sagittate-clasping, not lobed . . . . . *Camelina*
Valves of the silicles not nerved; cauline leaves not sagittate-clasping; leaves pinnatifid; plants of wet places . . . . . *Rorippa*

Arabis L. Rock-Cress

Basal leaves obovate to broadly oblanceolate, obtuse and rounded at apex, often forming a flat rosette at base of stems . . . . . *A. glabra*
Basal leaves linear to linear-oblanceolate, acute to rarely obtuse (if broader, then minutely pubescent or with reflexed siliques or both)
Plant hoary with a very minute pubescence
Pedicles arched downward . . . . . *A. pulchra* var. *gracilis*
Pedicels geniculately reflexed . . . . . *A. pulchra* var. *p.*
Plant greenish, not hoary
Mature fruiting pedicels erect to ascending; siliques erect or ascending, sometimes arcuate
Pubescence on stems with fine dendritic hairs . . . . . *A. sparsiflora* var. *californica*
Pubescence on lower portion of stems hirsute with large simple or branched hairs . . *A. sparsiflora* var. *arcuata*
Mature fruiting pedicels diverging at right angles to strictly reflexed; siliques straight to arcuate, diverging at right angles to stem or even deflexed
Seeds biseriate; cauline leaves linear; petals 8-20mm long . . . . . *A. pulchra*

Seeds uniseriate; cauline leaves wider, petals less than 12mm long
  Mature fruiting pedicels descending to strictly deflexed, straight, not widely spreading; siliques mostly straight, pendulous to appressed . . . . *A. holboellii*
  Mature fruiting pedicels spreading at right angles to stem, straight or arched downward; siliques straight and spreading at right angles or arcuate
    Pedicels slender, glabrous, 10-20mm long; outer basal leaves broadly oblanceolate, obtuse; petals 6-9mm long . . . . *A. perennans*
    Pedicels stout, pubescent, 5-12mm long; outer basal leaves narrowly oblanceolate, acute; petals 8-12mm long . . . . *A. sparsiflora*

***A. glabra** (L.)Bernh. Tower Mustard. Native of eastern North America and elsewhere. Occasional, in canyons and on shaded slopes; foothill and montane, above 300m: e.g. Harbison Canyon; Lyons Valley; El Cajon Mountain; Mesa Grande; Palomar and Cuyamaca Mountains. Mar.-July.

A. holboellii Hornem. var. **pinetorum** (Tidestr.)Roll. Rare, known in the county only from Laguna Public Campground, 1850m. May-July. $2n=21$.

A. perennans Wats. Nevada Rock-Cress. Occasional, on rocky slopes; montane and desert, above 300m: e.g. Morena Dam; Montezuma View Point; Borrego Palm Canyon; San Felipe Valley; Box Canyon; Desert View; Monument Peak; southeast corner of county. Mar.-May.

A. pulchra Jones var. **gracilis** Jones. [*A. p.* var. *glabrescens* Wiggins, the type from gravelly slope between Julian and Banner; *A. p.* var. *viridis* Jeps., the type from the summit of Mountain Springs grade.] Infrequent, on rocky slopes; foothill and desert, 600-1200m: Descanso Junction; Table Mountain; Jacumba; Banner Grade; San Felipe Valley. Apr.-May.

A. pulchra var. **pulchra**. Beautiful Rockcress. Infrequent, on rocky slopes; foothill and desert, 500-1200m: Indian Canyon; Banner; Mountain Springs; Dubber; Jacumba; Boulevard. Mar.-May.

A. sparsiflora Nutt. in T. & G. var. **arcuata** (Nutt.)Roll. Local, on dry slopes; foothill, 750-900m: Witch Creek; Santa Ysabel; Campo. Apr.-May.

A. sparsiflora var. **californica** Roll. [*A. maxima* auct. non Greene] Occasional, on dry slopes; foothill and montane, above 300m: e.g. Palomar Mountain State Park; Harrison Grade; Castro Canyon; Oak Grove; Hot Springs Mountain; Guymon Lodge; Banner; Cuyamaca Lake; Pine Valley; Hipass; Campo; Lyons Valley; southeast corner of county. Mar.-May. Type from near Campo. $2n=22$.

Athysanus Greene

A. pusillus (Hook.)Greene. Dwarf Athysanus. Common, in grassy and open areas; cismontane, below 1400m: e.g. Montezuma View Point; Lake Henshaw; Witch Creek; Moosa Canyon; Fallbrook; Laguna and Cuyamaca Mountains; Jacumba; Pine Valley; Tecate Junction; Lyons Valley; San Diego. Feb.-May.

Barbarea Scop. Winter-Cress

B. orthoceras Ledeb. Erect-Pod Winter-Cress. Common, in wet places; foothill and montane, above 700m: e.g. Palomar Mountain; Wynola; Witch Creek; Inspiration Point; Pine Hills; Cuyamaca Lake; Laguna Mountains; Japatul Valley; Hidden Glen; Corte Madera Lake. Apr.-Sep. $n=8$.

Brassica L. Mustard

Upper stem leaves with clasping base; leaves thin, the basal with scattered hairs; petals 6-8mm long; siliques 3-6cm long . . . . *B. rapa*
Upper stem-leaves not clasping
  Beak of silique terete or conic, often seedless
    Pedicels mostly shorter than sepals; siliques appressed against the stem
      Plant annual; petals ca. 8mm long . . . . *B. nigra*
      Plant perennial; petals ca. 5mm long . . . . *B. geniculata*
    Pedicels longer than sepals; siliques divergent . . . . *B. tournefortii*
  Beak of silique flat or conspicuously angled; beak usually 1-seeded . . . . *B. kaber*

***B. geniculata** (Desf.)J. Ball. Shortpod Mustard. Native of Europe. Prevalent, weed in disturbed and overgrazed areas; coastal and foothill, below 1200m: e.g. San Mateo Creek; De Luz; San Luis Rey; Pala; Oak Grove; Santa Ysabel; Pine Hills; Pine Valley; White Star; Campo; The Willows; Barrett Junction; National City; Escondido; Bernardo; San Marcos. Apr.-Nov.

***B. kaber** (DC.)L.C. Wheeler. Field Charlock. Native of Europe. Infrequent, weed in disturbed areas; coastal, below 200m: e.g. Silver Strand; Lake Murray; Chula Vista; Spring Valley; Otay Ranch. Jan.-Apr. $n=9$.

***B. nigra** (L.)Koch in Rohling. Black Mustard. Native of Europe. Common, weed in disturbed areas; coastal, below 200m: e.g. San Onofre; San Luis Rey; Lakeside; Glen Abbey. Mar.-June. $2n=16$.

***B. rapa** L. ssp. **sylvestris** (L.)Janchen. [*B. campestris* L.] Field Mustard. Native of Europe. Common, weed in disturbed and overgrazed areas; coastal and foothill, below 800m: e.g. San Onofre; Rincon; Buena; Witch Creek; Vista Grande; San Marcos; Escondido; San Pasqual; Viejas Valley; Poway Grade; Solana Beach; Mission Valley; Bonita; Otay; Dulzura. Jan.-May. $n=10$.

***B. tournefortii** Gouan. Wild Turnip. Native of North Africa. Rare adventive, usually desert waste areas. Camp Minnewawa; Borrego Springs. Jan.-June. $n=10$.

Cakile Mill. Sea-Rocket

Leaves sinuate-dentate; petals ca. 6mm long; pods without hornlike processes at apex of lower joint . . . . . *C. edentula*
Leaves pinnatifid; petals 9-10mm long; pods having 2 triangular protuberances at apex of lower joint. . . . *C. maritima*

***C. edentula** (Bigel.)Hook. [*C. e.* ssp. *californica* (Heller)Hult.] Native of eastern United States. Formerly prevalent, on sandy beaches and salt marshes, but now rare; coastal, below 5m: San Onofre; Pacific Beach; Ocean Beach; Imperial Beach; Monument Beach. Feb.-Sep. *n*=9.

***C. maritima** Scop. ssp. **maritima**. Native of Europe. Occasional, on beaches and salt marshes; coastal, below 5m: e.g. San Onofre; Encinitas; Solana Beach; Monument Beach. Feb.-May. *n*=9.

Camelina Crantz False-Flax

***C. microcarpa** Andrz. Bigseed False-Flax. Native of Europe. Infrequent, in disturbed areas: Chula Vista, 25m. June. 2*n*=40.

Capsella Medic. Shepherd's-Purse

***C. bursa-pastoris** (L.)Medic. Shepherd's-Purse. Native of Europe. Common, weed in distrubed and cultivated areas; cismontane: e.g. San Mateo Creek; Moosa Canyon; Rainbow; Vista; San Marcos; Doane Valley; Garnet Peak; Borrego Palm Canyon; Banner; Witch Creek; San Pasqual; Lakeside; San Diego; Dulzura. Feb.-May. *n*=8; 2*n*=32.

Cardamine L.

Basal leaves forming basal rosettes, from which arise the flowering stems
  Plants with slender rootstocks; petals 4-6mm long . . . . *C. gambelii*
  Plants from a slender taproot; petals about 2mm long . . . . *C. oligosperma*
Basal leaves from a thickened fleshy horizontal rhizome; petals mostly 9-14mm long . . . . *C. californica*

C. californica (Nutt.)Greene. [*Dentaria integrifolia* var. *c.* (Nutt.)Jeps.] Milkmaids. Frequent, on shaded slopes; cismontane, below 1400m: e.g. Moosa Canyon; Agua Tibia Mountains; Harrison Park; San Marcos; Del Dios; Miramar; Del Mar; La Jolla; Point Loma; Mount Soledad; San Miguel Mountain; Otay Mountain. Feb.-May. *n*=8, 16; 2*n*=32.

C. gambelii Wats. Gambel's Bittercress. Uncommon, in wet places; montane, above 1000m: Julian. Apr.-June.

C. oligosperma Nutt. Few-seeded Bitter-cress Rare, known from a collection at Del Mar, 20m. Mar.-July.

Cardaria Desv.

***C. draba** (L.)Desv. [*Lepidium d.* L.] Hoary Cress. Native of Europe. Occasional, in disturbed areas; coastal, below 300m: Mission Valley; Chula Vista. Apr.-May. *n*=32.

Caulanthus Wats.

Calyx open at anthesis . . . . *C. lasiophyllum*
Calyx urn-shaped or flask-shaped at anthesis
  Cauline leaves sessile and auriculate at the base
    Plants glabrous or inconspicuously pubescent . . . . *C. cooperi*
    Plants evidently hirsute or pilose
      Stigma distinctly 2-lobed; calyx yellowish; stigma shallowly lobed . . . . *C. simulans*
      Stigma very small, entire, calyx purple . . . . *C. stenocarpus*
  Cauline leaves sessile or petioled; but not auriculate . . . . *C. hallii*

C. cooperi (Wats.)Pays. Cooper's Caulanthus. Occasional, on dry, rocky slopes; desert, below 1000m: e.g. Yaqui Well; Box Canyon; Vallecito Stage Station; Mountain Springs. Mar.-May. 2*n*=14 pairs.

C. hallii Pays. Hall's Caulanthus. Occasional, in sandy areas; north desert, below 1300m: Chihuahua Valley; Borrego Palm Canyon; Rockhouse Canyon. Mar.-May.

C. lasiophyllum (H.& A.)Pays. [*Thelypodium l.* (H. & A.)Greene; *Thelypodium l.* var. *utahense* (Rydb.)Jeps.] California Mustard. Occasional, on open slopes; coastal, foothill and desert, below 1200m: e.g. Lilac; Foster; Poway Grade; San Miguel Mountain; Adobe Falls; Emerald Hills; Pacific Beach; Potrero; Montezuma View Point; Coyote Canyon; Borrego Valley; Rockhouse Canyon; Tamarisk Grove; Wagon Wash; Blair Valley; Mason Valley; Vallecito Valley.. Mar.-June. *n*=14.

C. simulans Pays. Payson's Caulanthus. Infrequent, in rocky places; foothill and desert, below 1500m: Lilac; Monserrate; Pala; Twin Oaks Valley; Montezuma View Point; Grapevine Canyon; San Felipe Valley; San Felipe Wash; Rattlesnake Valley; Bankhead Springs. Also Stonewall and Cuyamaca Peaks. Mar.-May.

C. stenocarpus Pays. Slender-Pod Caulanthus. Infrequent, on open slopes and burns; foothill, 300-900m: Bernardo; Dehesa; Harbison Canyon; Kimball Grade; Tecate Mountain. Mar.-May. Type from dry hillsides near Bernardo.

Coronopus Trev. Wart-Cress

***C. didymus** (L.)Sm. Lesser Wart-Cress. Native of Europe. Infrequent, weed in disturbed areas: Balboa Park; San Diego. Mar.-July. 2*n*=32.

Descurainia Webb & Berth. Tansy-Mustard

Upper leaves 2-3-pinnate, mostly into linear segments; siliques 10-30 (typically ca. 20)mm long; silique-septum with 2-3 longitudinal nerves . . . . . *D. sophia*
Upper leaves pinnate, the leaflets often deeply incised; siliques mostly less than 15mm long; silique-septum 1-nerved
  Seeds in 1 row in each locule of silique; style conspicuous; siliques 9-15mm long; stems glandular-pubescent, especially above
    Plants glandular-pubescent . . . . . *D. richardsonii* ssp. *viscosa*
    Plants subglabrous to moderately pubescent . . . . . *D. richardsonii* ssp. *incisa*
  Seeds in 2 rows in each locule; style nearly or quite obsolete; siliques clavate to oblong-elliptic, mostly less than 12mm long
    Segments of upper leaves narrowly oblong to linear; racemes mostly glandular-pubescent
      Petals 2-2.5mm long, bright yellow, plants strict . . . . . *D. pinnata* ssp. *menziesii*
      Petlas 1-3mm long, whitish to yellow; plants branches below . . . . . *D. pinnata* ssp. *halictorum*
    Segments of upper leaves ovate to oblanceolate; racemes glabrous
      Petals scarcely 2mm long, yellow . . . . . *D. pinnata* ssp. *glabra*
      Petals over 3mm long, cream-colored . . . . . *D. pinnata* ssp. *brachycarpa*

***D. pinnata** (Walt.)Britt. ssp. **brachycarpa** (Richard.)Detl. [*D. b.* (Richard.)O. E. Schulz.] Glandular Tansy-Mustard. Native of central and eastern United States. Rare, adventive at Carrizo Stage Station, 150m. Apr.

D. pinnata ssp. **glabra** (Woot.& Standl.)Detl. Uncommon, in sandy areas; north desert, below 1200m: Borrego Springs; Borrego Palm Canyon; The Narrows; Banner Grade. Mar.-May. *n*=14.

D. pinnata ssp. **halictorum** (Ckll.)Detl. Occasional, on open slopes; foothill and desert, below 1600m: e.g. Chihuahua Valley; Borrego Palm Canyon; Montezuma View Point; northeast corner of county; Banner; Laguna Mountains; Boulevard; Live Oak Springs; Jacumba; southeast corner of county. Mar.-July. *n*=7, 14, 21.

D. pinnata ssp. **menziesii** (DC.)Detl. [*D. p.* auct. non Britt.] Occasional, on open slopes; cismontane, below 1500m: e.g. San Felipe Creek; Moosa Canyon; Palomar Mountain; Eagles Nest; Little Laguna Lake; Laguna Junction; Jacumba; Witch Creek; Canyon City; Emerald Hills; Mission Hills; Point Loma. Feb.-June.

D. richardsonii (Sweet)O.E. Schulz ssp. **incisa** (Engelm.)Detl. [*D. i.* (Engelm.)Britt.] Scarce, in Montane Coniferous Forest; montane, above 1200m: Doane Valley. Aug. *n*=21.

D. richardsonii ssp. **viscosa** (Rydb.)Detl. [*D. v.* Rydb.] Scarce, known in the county only at Pine Valley, 1100m. June. *n*=7.

***D. sophia** (L.)Webb. Fine-Leaf Tansy-Mustard. Native of Europe. Uncommon, weed in dry waste places: Chula Vista; Pine Valley; Jacumba. May-Sep. 2*n*=28.

Dithyrea Harv. Spectacle-Pod

D. californica Harv. Californica Spectacle-Pod. Occasional, in sandy areas; desert, below 1000m: e.g. Lone Palm; Borrego Valley; Borrego Palm Canyon; Benson Dunes; Dry Lake School; San Felipe; Yaqui Well; Box Canyon; Vallecito and Carrizo Stage Stations. Mar.-Apr. *n*=10.

Draba L.

Pedicels mostly pubescent; leaves usually dentate
  Pods stellate . . . . . *D. cuneifolia* var. *sonorae*
  Pods nearly or quite glabrous . . . . . *D. cuneifolia* var. *integrifolia*
Pedicels always glabrous; leaves usually entire . . . . . *D. reptans*

D. cuneifolia Nutt. ex T. & G. var. **integrifolia** Wats. [*D. i.* var. *brevifolia* Wats. ex Munz] Desert Whitlow. Occasional, on moist slopes and clay areas; coastal, foothill, and desert, below 1200m: e.g. Fallbrook; Pala; Witch Creek; Warner Hot Spring; Borrego Palm Canyon; Montezuma View Point; Lone Palm; Sentenac Canyon; Banner; Mountain Palm Springs; Vallecito Stage Station; Mason Valley; southeast corner of county; Harbison Canyon; Mahogany Canyon; Rice Canyon; Point Loma. Feb.-Apr.

D. cuneifolia var. **sonorae** (Greene)Parish. Sonora Whitlow. Uncommon in shaded places: La Jolla; Chollas Valley; Borrego Valley. Feb.-Apr.

D. reptans (Lam.)Fern. California Whitlow. Rare, known in the county only from a collection from Vallecito Stage Station, 550m. Apr. *n*=15.

Erophila DC.

***E. verna** (L.)Chevall. [*Draba v.* L.] Vernal Whitlow. Native of Europe. Local, naturalized in meadows; montane, above 1000m: Doane Valley; east grade of Palomar Mountain; Cutca Valley. Mar. *n*=7, 15, 32.

Eruca Mill. Rocket

***E. vesicaria** (L.)Cav. ssp. **sativa** (Mill.)Thell. [*E. s.* Mill.] Garden Rocket. Native of Europe. Uncommon, weed in waste places: Bankhead Springs; Balboa Park. May-Nov. 2*n*=22.

Erysimum L. Wallflower

Plants not suffrutescent; caudex not noticeably elongate or long-branched above ground, erect, or at least not sprawling or widely spreading, mostly without or with only very short sterile branches; lowest leaves mostly not marcescent
  Seeds distally winged; siliques tetragonal, fairly protrusively keeled on flatter surfaces . . . . . *E. capitatum*
  Seeds distally winged and along one side; siliques strongly compressed . . . . . *E. ammophilum*

Plants suffrutescent; caudex elongate, long-branched above ground, sprawling or widely spreading, with elongate sterile branches; lower leaves mostly marcescent . . . . . *E. suffrutescens*

E. ammophilum Heller. Infrequent, in sandy areas; coastal, below 50m: Solana Beach; Del Mar; Torrey Pines; Point Loma; San Diego Bay. Feb.-May. *n*=18.

E. capitatum (Dougl.)Greene. [*E. asperum* auct. non DC.] Douglas' Wallflower. Frequent, in open areas, Chaparral and Montane Coniferous Forest; foothill, montane, and desert, above 600m: e.g. Palomar, Hot Springs, Laguna, and Cuyamaca Mountains; Chihuahua Valley; Culp Valley; Indian Canyon; Vallecito Stage Station; Pine Valley; Corte Madera; Canyon City; Encinitas. Apr.-July. 2*n*=36.

E. suffrutescens (Abrams)G. Rossb. Bushy Wallflower. Uncommon, in sandy areas; coastal, below 50m: Torrey Pines; Ocean Beach.

Heterodraba Greene

H. unilateralis (Jones)Greene. Rare, known in the county only from a collection from near San Ysidro. Mar.-Apr.

Hutchinsia R. Br.

H. procumbens (L.)Desv. [*Capsella p.* (L.)Fries; *Lepidum p.* L.; *Hymenolobus p.* (L.)Nutt. in T.& G.] Prostrate Hutchinsia. Infrequent, weed in alkaline areas; coastal and foothill, below 800m: Escondido; Oak Grove; La Costa; Torrey Pines Mesa; San Diego; Imperial Beach. Feb.-June. 2*n*=12.

Lepidium L. Peppergrass

Leaves perfoliate or with clasping bases . . . . . *L. campestre*
Leaves not perfoliate or with clasping bases
  Styles none, or less than 0.3mm long, usually shorter than notch of fruit
    Silicles notched at apex, but not or slightly winged
      Pedicels terete or only slightly flattened
        Sepals persisting until fruits are almost mature; pedicels somewhat narrowly wing-margined; cauline leaves pinnatifid to lobed . . . . . *L. oblongum*
        Sepals deciduous along with petals and stamens or soon after; pedicels not wing-margined; cauline leaves mostly entire
          Petals wanting or shorter than sepals; cotyledons incumbent in the seed (with the back of 1 against the radicle).
            Branches bearing many short corymbiform racemes in the leaf-axils as well as the longer terminal ones . . . . . *L. ramosissimum*
            Branches bearing simple, long, naked racemes. . . . . *L. densiflorum*
          Petals equalling or exceeding sepals; cotyledons incumbent or accumbent (the edges against the radicle)
            Plants 1.5-6dm tall; cauline leaves incised to simple . . . . . *L. virginicum* var. *pubescens*
            Plants 1-2dm tall; cauline leaves parted or with narrow lobes . . . . . *L. virginicum* var.*robinsonii*
      Pedicels strongly flattened
        Plant freely branched from base, pubescent; stamens 2 or 4; petals sometimes lacking
          Plants hirsute-hispid, pedicles pubescent on lower side . . . . . *L. lasiocarpum* var. *l.*
          Plants less pubescent, pedicles glabrous on lower side. . . . . *L. lasiocarpum* var. *georginum*
        Plant usually simple below, glabrous; stamens 6; petals evident . . . . . *L. nitidum*
    Silicles plainly winged at apex with 2 prominent divergent lobes or teeth
      Petals not more than 1.2mm long
        Silicles ca. 2-3mm long. . . . . *L. ruderale*
        Silicles over 3mm long
          Silicles mostly over 4mm long, with acuminate divergent winged apices over 1mm long . . . . . *L. dictyotum* var. *acutidens*
          Silicles about 3.5mm long, with winged apices less than 1mm long . . . . . *L. dictyotum* var. *d.*
      Petals 2-4mm long; silicles 5.5-7mm long. . . . . *L. latipes*
  Styles well developed, at least 0.3mm long and exceeding the notch of the fruits
    Silicles roundish, 3-4.5mm long. . . . . *L. flavum* var. *felipense*
    Silicles oblong, 1-1.3mm long. . . . . *L. flavum*. var. *f.*

***L. campestre** (L.)R. Br. Cow-Cress, Field Peppergrass. Native of Europe. Local, naturalized at Wynola, 1100m. June. *n*=8.

L. densiflorum Shrad. var. **pubicarpum** (A. Nels.)Thell. Uncommon, in dry areas; below 1000m: Escondido; San Diego; Potrero; Puertacitos. May-Aug. 2*n*=16 pairs.

L. dictyotum Gray var. **acutidens** Gray. [*L. acutidens* var. *microcarpum* Thell., the type from San Diego.] Sharp-Tooth Peppergrss. Rare, known in the county from a collection from San Diego. Apr.

L. dictyotum var. **dictyotum**. Alkali Peppergrass. Rare, in alkaline areas; coastal, below 150m: Sweetwater Valley. Apr.

L. flavum Torr. var. **felipense** C. L. Hitchc. Borrego Peppergrass. Local, in sandy areas; north desert, below 900m: San Felipe; Borrego Palm Canyon; Little Blair Valley. Mar.-June. Type from San Felipe.

L. flavum var. **flavum**. Yellow Peppergrass. Rare, in washes; desert, below 900m: San Felipe; Yaqui Well. Mar.-May.

L. lasiocarpum Nutt. ex T.& G. var. **georginum** (Rydb.)C. L. Hitchc. Hairy-Pod Peppergrass. Uncommon, on sandy flats; desert, below 300m: Borrego Springs. Apr.-May.

L. lasiocarpum var. **lasiocarpum**. Sand Peppergrass. Occasional, on grassy slopes and open areas; coastal, foothill, and desert, below 1000m: e.g. Pauma Valley; San Luis Rey; Fallbrook; Cardiff; San Diego; Chollas Valley; Imperial Beach; northeast corner of county; Lone Palm; Borrego Palm Canyon; Benson

Dunes; Vallecito and Carrizo Stage Stations; Mountain Palm Springs; Dubber; southeast corner of county. Mar.-May.

L. latipes Hook. Dwarf Peppergrass. Rare, in ephemeral ponds; coastal, below 150m: San Diego. Mar.-May.

L. nitidum Nutt. var. **nitidum**. Shining Peppergrass. Common, in open places; cismontane, below 1400m: e.g. Fallbrook; San Mateo Creek; Witch Creek; Volcan Indian Reservation; Springhill; Julian; Cuyamaca Lake; Harper Ranch; Lake Hodges; Dulzura; Sweetwater Valley; Rolando; Point Loma. Mar.-June.

***L. oblongum** Small. Wayside Peppergrass. Native of South America. Uncommon, weed in disturbed areas; south coastal, below 200m: San Diego; Chollas Valley; Silver Strand. Mar.-May.

***L. ramosissimum** A. Nels. Branched Peppergrass. Native of Rocky Mountain region. Local, adventive in Chollas Valley, 70m. Feb.-Apr.

***L. ruderale** L. Roadside Peppergrass. Native of Europe. Local, adventive on Point Loma, 50m. Mar.

L. virginicum L. var. **pubescens** (Greene)Thell. Common, in disturbed areas; cismontane: e.g. Fallbrook; Pala; Bonsall; Palomar, Hot Springs, Cuyamaca, and Laguna Mountains; Valley Center; Clover Flats; Potrero; Barrett; Olivenhain; Jamacha; Dehesa; San Diego. Also Agua Caliente Hot Springs. Mar.-Sep.

L. virginicum var. **robinsonii** (Thell.)C.L. Hitchc. [*L. r.* Thell.] Robinson's Pepper-Grass. Occasional, on grassy slopes and waste places; coastal and foothill, below 700m: e.g. Fallbrook; Rainbow; San Luis Rey; Pala; Rancho Santa Fe; Bernardo; Harbison Canyon; Potrero Grade; Barrett; Whispering Oaks; La Jolla Farms; Emerald Hills; San Diego. Jan.-Apr. Type from San Diego.

Lobularia Desv.

***L. maritima** (L.)Desv. [*Alyssum m.* (L.)Lam.] Sweet Alyssum. Native of Europe. Occasional, escaping from cultivation; coastal, below 150m: e.g. Solana Beach; Torrey Pines Beach; National City. Also Coyote Creek. Jan.-Dec. n=12.

Lyrocarpa Hook. & Harv.

L. coulteri Hook. & Harv. var. **palmeri** (Wats.)Roll. Palmer's Lyre-Pod. Infrequent, among rocks and in canyons; desert, below 700m: Hellhole Canyon; Borrego Springs; Borrego Palm Canyon; San Felipe Wash; The Narrows; Yaqui Well; Agua Caliente Hot Springs; Vallecito Stage Station; Canebrake Canyon. Feb.-May. 2n=20.

Matthiola R. Br. Stock

***M. incana** (L.)R. Br. Common Stock. Native of Europe. Occasional, naturalized in sandy areas; coastal, below 150m: e.g. Oceanside; Carlsbad; Ponto; Cardiff; Solana Beach; Torrey Pines. Nov.-May. n=7.

Nasturtium R. Br.

***N. officinale** R. Br. [*Rorippa nasturtium-aquaticum* (L.)Schinz & Thell.] White Water-Cress. Native of Europe. Occasional, in wet places: Palomar, Hot Springs, Cuyamaca, Laguna, and Corte Madera Mountains; Lake Henshaw; O'Neill Lake; Ramona; Descanso; Lake Murray; White Star; Sweetwater Lake; Vallecito Stage Station; Borrego Palm Canyon. Mar.-Nov. 2n=32.

Raphanus L. Radish

Fruit markedly constricted between the seeds and breaking readily into 1-seeded joints; beak slender *R. raphanistrum*
Fruit not markedly constricted between the seeds and not breaking up into joints; beak long-conical . . . . . . *R. sativus*

***R. raphanistrum** L. Jointed Charlock. Native of Europe. Infrequent, weed in waste places; coastal, below 200m: San Onofre; Las Flores; San Luis Rey; Pacific Beach; National City; Jamacha Junction; Monument Beach. Jan.-Apr. n=9.

***R. sativus** L. Wild Radish. Native of Europe. Occasional, weed in waste places; coastal and foothill, below 1000m: e.g. San Onofre; Herron Ranch; San Luis Rey; Ponto; Witch Creek; San Pasqual; El Cajon; Jamul; Rancho Santa Fe; San Marcos; Mission Valley; National City; Silver Strand. Jan.-June. n=9.

Rapistrum Crantz

***R. rugosum** (L.)All. Native of Europe. Local, adventive in fallow fields; coastal, below 200m: Whalen Ranch; Oceanside; Vista; Buena. May-Aug. 2n=16.

Rorippa Scop.

Plants annual
  Petals 1.2-3.5mm long; oldest siliques on lower portion of the terminal raceme . . . . *R. palustris*
  Petals usually less than 1.2mm long; oldest siliques on lower axillary racemes or siliques nearly equal in development at corresponding points on terminal and lateral racemes
    Pod septum margin minutely hirsute and the valves readily dehiscent, or if margin glabrous, the valves not readily dehiscent; petals spreading between the sepals in living plants . . . . *R. curvisiliqua*
    Pod septum margin glabrous; valves readily dehiscent; petals erect in living plants
      Plants mostly shorter than 3dm; cauline leaves usually fewer than 10; silique valves minutely papillose . . . . *R. tenerrima*
      Plants mostly taller than 3dm; cauline leaves usually more than 10; silique-valves glabrous . . . . . . *R. intermedia*
Plants perennial, with creeping rootstocks; petals longer than sepals . . . . *R. sinuata*

R. curvisiliqua (Hook.)Bessey ex Britt. var. **orientalis** Stuckey. Prostrate Yellow-Cress. Scarce, in shallow water; montane, above 1000m: Doane Valley; Laguna Public Camp. May-Sep.

R. intermedia (O. Ktze.)Stuckey. Tall Yellow-Cress. Rare, known in the county only from Cuyamaca Lake, 1400m.

R. palustris (L.)Besser ssp. **occidentalis** (Wats. in Gray)Abrams. Pacific Yellow-Cress. Uncommon, along shorelines of lakes and permanent ponds: Rice Canyon; Sweetwater Lake. May.

R. sinuata (Nutt. in T.& G.)A.S. Hitchc. Spreading Yellow-Cress. Scarce, along shorelines of lakes and permanent ponds: Doane Valley; Laguna Mountains. May-Sep.

R. tenerrima Greene. Lyrate-leaf Yellow-Cress. Scarce, in shallow water; montane, above 1000m: Palomar and Cuyamaca Mountains. May-Aug.

Sibara Greene

S. virginica (L.)Roll. [*Arabis v.* (L.)Trel.] Virginia Rockcress. Scarce, about ephemeral ponds; coastal, below 150m: San Diego; Point Loma; Otay Mesa; San Ysidro. Mar.-May. $2n$=16.

Sisymbrium L.

Siliques closely appressed, 1-1.5cm long; leaves pinnatifid . . . . . *S. officinale*
Siliques not appressed, 4-10cm long
  Fruiting pedicels ca. as thick as the siliques; petals 6-9mm long
    Upper leaves pinnatifid with linear divisions; siliques ca. 1mm wide . . . . . *S. altissimum*
    Upper leaves entire to hastate; siliques 1-2mm wide . . . . . *S. orientale*
Fruiting pedicels more slender than the siliques; petals 3-4mm long . . . . . *S. irio*

***S. altissimum** L. Tumble Mustard. Native of Europe. Common, weed in disturbed areas and burns; foothill and montane, above 100m: e.g. Chihuahua Valley; Mesa Grande; Palomar, Hot Springs, Cuyamaca, and Laguna Mountains; Lilac; Foster; White Star; Mount Woodson; Jamacha. Apr.-July. $2n$=14.

***S. irio** L. London Rocket. Native of Europe. Frequent, in waste places; coastal, below 150m: e.g. San Onofre; Moosa Canyon; San Dieguito Valley; Kearny Mesa; Pacific Beach; Point Loma; San Ysidro. Also Borrego Palm Canyon. Feb.-May. n=7, 14, 21, 28.

***S. officinale** (L.)Scop. Hedge Mustard. Native of Europe. Common, in shaded areas; cismontane, below 1500m: O'Neill Lake; Rainbow; Doane Valley; Eagles Nest; Lake Henshaw; Witch Creek; Pine Hills; El Capitan; Anahuac School; Otay Ranch; Lee Valley. Apr.-June. n=7.

***S. orientale** L. Hare's-Ear Cabbage. Native of Europe. Infrequent, weed along roads and in disturbed areas; coastal and foothill, below 950m: Warner Hot Springs; Oceanside; Moosa Canyon; Point Loma; Chollas Heights. Mar.-Apr. n=7.

Stanleya Nutt. Prince's Plume

S. pinnata (Pursh)Britt. var. **pinnata**. Desert Prince's Plume. Rare, in gypsophilous, sandy areas; north desert, below 1000m: Rockhouse Canyon. Apr. n=12.

Streptanthella Rydb.

S. longirostris (Wats.)Rydb. [*S. l.* var. *derelicta* J. T. Howell.] Long-Beak Twist-Flower. Occasional, in sandy areas; desert, below 600m: e.g. Lone Palm; Borrego Springs; Borrego Palm Canyon; Benson Dunes; San Felipe Wash; Yaqui Pass; Carrizo Stage Station. Mar.-Apr. n=14.

Streptanthus Nutt.

Plants essentially glabrous
  Petals white, 7-9mm long . . . . . *S. bernardinus*
  Petals purple, sometimes with white tips or edges, 10-14mm long . . . . . *S. campestris*
Plants somewhat setose-hispid especially toward the base
  Plants perennial . . . . . *S. bernardinus*
  Plants annual . . . . . *S. heterophyllus*

S. bernardinus (Greene)Parish. [*Agianthus jacobaeus* Greene, the type from the Cuyamaca Mountains.] San Bernardino Jewelflower. Uncommon, on slopes, Montane Coniferous Forest; montane, above 1200m: Julian; North and Middle Peaks; Laguna Public Camp; Laguna Lakes. Also Dulzura. June-Aug.

S. campestris Wats. Southern Jewelflower. Uncommon, on Chaparral slopes; montane and foothill, above 900m: Oriflamme Canyon; Julian; Laguna Mountains; Pine Valley; Campo. May-July. Type from Campo.

S. heterophyllus Nutt. ex T. & G. [*Caulanthus h.* (Nutt.)Pays.] San Diego Jewelflower. Frequent, on open slopes and burns; foothill and coastal, below 1350m: e.g. Fallbrook; Bonsall; Witch Creek; San Felipe; Culp Valley; Julian; Ramona; Barona Indian Reservation; Barrett; Tecate Junction; Lyons Valley; Foster; Poway; San Miguel Mountain; Chollas Heights; Kearny Mesa; Mount Soledad. Feb.-Apr. Type from "bushy hills, near St Diego".

Thlaspi L. Penny-Cress

***T. arvense** L. Field Penny-Cress. Native of Europe. Scarce, weed in waste places: Balboa Park. June. $2n$=14.

Thysanocarpus Hook. Lacepod; Fringepod

Cauline leaves auriculate, the basal leaves rosulate, usually hirsute
  Wing of silicle rayed; plants mostly pubescent below, cismontane
    Silicles 3-4mm wide, the wing usually not perforate . . . . *T. curvipes* var. *c.*
    Silicles 5-6mm wide, the wing often perforate . . . . *T. curvipes* var. *elegans*
  Wing of silicle mostly plane, without rays; plants glabrous and glaucous . . . . *T. curvipes* var. *eradiatus*
Cauline leaves not auriculate, the basal not rosulate, usually glabrous
  Pedicels recurved in fruit; silicles mostly 3-5mm wide
    Wings lacking well-defined rays . . . . *T. lacinatus* var. *l.*
    Wings with well-defined rays . . . . *T. lacinatus* var. *crenatus*
  Pedicels spreading, almost straight; silicles 2.5mm wide . . . . *T. laciniatus* var. *rigidus*

T. curvipes Hook. var. **curvipes**. [*T. c.* var. *pulchellus* Greene] Hairy Lacepod. Occasional, on grassy slopes; cismontane: e.g. Fallbrook; Moosa Canyon; Escondido; Pine Hills; Cuyamaca and Laguna Mountains; Boulevard; Harbison Canyon; Sloane Ranch; Otay Mountain; Miramar. Mar.-May. n=7; $2n$=28.

T. curvipes var. **elegans** (F. & M.)Rob. in Gray. Elegant Lacepod. Infrequent, on grassy slopes; foothill and montane, above 150m: Palomar Mountain; Pine Hills; Escondido; Mussey Grade; Foster. Mar.-May. $2n$=28.

T. curvipes var. **eradiatus** Jeps. Glaucous Lacepod. Infrequent, in washes and canyons; desert, below 1000m: Borrego Valley; Sentenac Canyon; Box Canyon; Blair Valley; Vallecito Canyon; Vallecito Stage Station; Jacumba. Apr.-May.

T. laciniatus Nutt. ex T. & G. var. **crenatus** (Nutt.)Brew. Notch Fringepod. Occasional, on grassy slopes; foothill, 200-1400m: e.g. Chihuahua Valley; El Monte Oaks; Whispering Oaks; San Miguel Mountain; Potrero Grade. Mar.-May.

T. laciniatus var. **laciniatus**. Southern Fringepod. Occasional, on grassy slopes; coastal and foothill, below 950m: e.g. Moosa Canyon; Montezuma View Point; Mission Gorge; Cottonwood. Mar.-May.

T. laciniatus var. **rigidus** Munz. Laguna Fringepod. Local, in Montane Coniferous Forest; montane, above 1300m: Laguna Mountains. May. Type from Laguna Camp.

Tropidocarpum Hook.

T. gracile Hook. Slender Dobie-Pod. Infrequent, on grassy slopes; cismontane, below 1200m: e.g. Pala; Mesa Grande; Warner Hot Springs; Montezuma View Point; Sentenac Canyon; Lakeside; Tecate Junction; San Miguel Mountain; National City; Point Loma. Mar.-Apr. n=8.

BURSERACEAE. Torchwood Family

Bursera Jacq. Elephant-Tree

B. microphylla Gray. Small-Leaf Elephant Tree. Infrequent, on rocky slopes; desert, below 500m: Fish Creek; Canebrake Canyon; Bow Willow Canyon; Indian Gorge; Mountain Palm Springs. May-June.

CACTACEAE. Cactus Family

KEY TO GENERA

Stems jointed, flattened or cylindrical; areoles with numerous minute barbed bristles (glochids) as well as stiffer spines; leaves small, subulate, shed early . . . . *Opuntia*
Stems not jointed, leafless; areoles without glochids
  Flower on an area of stem developed 1-several years earlier, not on the new growth of the current season, therefore below the apex of the stem or branch; fruit spiny
    Stems ribbed, the spines borne in bundles on definite ribs; flower-bud borne within the spiniferous areole or just above it
      Stems 0.5-1.5m long, branching near the base; flowers yellow . . . . *Bergerocactus*
      Stems 1-3dm long, cespitose; flowers red to purplish . . . . *Echinocereus*
    Stems with separate tubercles that bear the spines; flower-bud not associated with the spiniferous areole . . . . *Mammillaria*
  Flower borne on the new growth of the current season, hence near the apex of the stem or branch; fruit scaly but not spiny
    Areoles of ovary not woolly; fruit fleshy for some time after maturing; stems mostly simple . . . . *Ferocactus*
    Areoles of ovary woolly; fruit drying soon after maturity; stems in clumps . . . . *Echinocactus*

Bergerocactus Britt. & Rose

B. emoryi (Engelm.)Britt. & Rose. [*Cereus e.* Engelm.] Golden Snake-Cactus. Uncommon, on dry slopes; coastal, below 150m: Torrey Pines State Reserve; La Jolla Farms; Point Loma; Telegraph Canyon; Otay Mesa; San Ysidro; Tijuana Hills; Boundary Monument No. 258. May-June. Type from "dry hills near the sea shore, about the boundary line". n=22.

Echinocactus Link & Otto

E. polycephalus Engelm. & Bigel. var. **polycephalus**. Mohave Moundcactus. Rare, known in the county only near Palo Verde Canyon, 350m. Mar.-May.

Echinocereus Engelm.

Flowers scarlet; central spine(s) not flattened . . . . *E. triglochidiatus*
Flowers lavender to purple; principal central spine flattened
  Lower spine flattened, curving and twisted . . . . *E. engelmannii* var. *munzii*
  Lower central spine deflexed, stout and flattened . . . . *E. engelamannii* var. *e.*

E. engelmannii (Parry)Ruempler var. **engelmannii**. [*Cereus e.* Parry.] Engelmann's Hedgehogs. Frequent, on rocky slopes, alluvial fans and washes; desert, 600-1300m: e.g. northeast corner of county; Coyote Canyon; Box Canyon; Scissors Crossing; Vallecito Stage Station; Canebrake Canyon; Mountain Springs; Jacumba. Apr.-May. Type from "mountains about San Felipe on the eastern declivity of the cordilleras". *n*=22.

E. engelmannii var. **munzii** (Parish)Pierce & Fosberg. [*Cereus m.* Parish.] Munz's Hedgehog. Scarce, on rocky outcrops; montane, above 1200m: east of Cuyamaca Lake; Chariot Canyon. May-June.

E. triglochidatus Engelm. var. **mojavensis** (Engelm. & Bigel.)L. Benson. Mound Cactus. Cited by Benson from an area apparently near Kane Springs Wash, but this may be an Imperial County collection.

Ferocactus Britt. & Rose Barrel Cactus

Plants cylindrical, 6-25dm high; flowers yellow, deserts . . . . . *F. cylindraceus*
Plants subglobose, 1-3dm high; flowers greenish-yellow, coastal . . . . . *F. viridescens*

F. cylindraceus (Engelm.)Orcutt. [*F. acanthodes* (Lemaire)Britt. & Rose var. *acanthodes*; *Echinocactus a.* Lemaire; *F. a.* var. *lecontei* (Engelm.)Lindsay; *E. viridescens* var. *cylindraceus* Engelm.], California Barrel. Common, on rocky slopes and alluvial fans; desert, below 1200m: e.g. Coyote Canyon; San Felipe Valley, upper Box Canyon; Mountain Springs Grade; Jacumba. Apr.-May. The type from near San Felipe.

F. viridescens (Nutt.)Britt. & Rose. [*Echinocactus v.* Nutt.; *E. limitus* Engelm. in Coulter, the type from the boundary line south of San Diego.] Coast Barrel Cactus. Occasional, on dry slopes; coastal, below 1500m: e.g. north bank of San Luis Rey River at Oceanside; Del Mar; Poway; Kearny Mesa; Rose Canyon; National City; Point Loma; Casa de Oro; Sunnyside; Otay Mesa; Initial Boundary Monument. May-June. Type from "arid hills & c. near St. Diego". *n*=11.

Mammillaria Haw.

Hooked central spine 1; base of seed not corky; radial spines 12-28 per areole; stems of older plants usually branching . . . . . *M. dioica*
Hooked central spine mostly 2-4; base of seed corky, the corky part half as large as the seed body or larger; stems unbranched . . . . . *M. tetrancistra*

M. dioica K. Bdg. [*M. d.* var. *incerta* (Parish)Munz, the type from Vallecito.] Fish-hook Cactus. Occasional, on dry slopes and mesas; coastal, below 150m, and desert, below 1200m: e.g. San Dieguito River estuary; Mount Soledad; Cowles Mountain; San Diego; Paradise Valley; Point Loma; Sunnyside; Boundary Monument No. 258; Coyote Canyon; Mason Valley; Split Mountain; Sentenac Canyon; Vallecito Stage Station; Mountain Springs. Feb.-Apr. *n*=22.

M. tetrancistra Engelm. [*M. pellosperma* Engelm.] Yaqui Mammillaria. Infrequent, on rocky slopes; desert, below 1200m: Scissors Crossing; Box Canyon; San Felipe Hill; Mountain Springs; Canebrake Canyon; Dubber; Table Mountain. Apr.-May. Type "from San Diego to the junction of the Gila with the Colorado".

Opuntia Mill. Prickly-Pear; Cholla

Joints of stem cylindrical or clavate, tuberculate; glochids usually small
 Stem nearly covered with low flat platelike tubercles; spines solitary or wanting; branches slender, 6-10mm thick, the old ones with solid woody axis . . . . . *O. ramosissima*
 Stem either smooth or with simple tubercles; spines more than 1; ultimate branches stouter, the woody axis with a reticulate cylinder
  Ultimate joints disarticulating readily; fruit fleshy when first mature
   Fruits proliferous, persistent; flowers red to red-purple, coastal . . . . . *O. prolifera*
   Fruits not proliferous or persistent after first season; flowers yellowish green, deserts
    Spine sheaths straw-colored when young . . . . . *O. bigelovii* var. *b.*
    Spine sheaths reddish-brown when young . . . . . *O. bigelovii* var. *hoffmannii*
  Ultimate joints firmly attached; fruit dry at maturity
   Terminal joints 7-9mm in diameter; central spine mostly 1, radial spines 6-8; spines of fruit weakly barbed . . . . . *O. wigginsii*
   Terminal joints mostly at least 15mm in diameter; central and radial spines indistinguishable; spines of fruit not barbed
    Spines of fruit few, 1-2 in the areoles of the 2 uppermost nearly horizontal series, mostly cismontane
     Stems prostrate to suberect, coastal . . . . . *O. parryi* var. *serpentina*
     Stems ascending . . . . . *O. parryi* var. *p.*
    Spines of fruit many, several per areole in the upper nearly horizontal series, desert
     Tubercles 1-2 times as long as broad; longer terminal joints 10-15cm long
      Joints 1.5-2cm thick . . . . . *O. echinocarpa* var. *e.*
      Joints 3-4cm thick . . . . . *O. echinocarpa* var. *wolfii*
     Tubercles 3-several times as long as broad; larger joints 15-40cm long . . . . . *O. acanthocarpa*
Joints of stem flattened after the seedling stage, not tuberculate
 Fruit dry on maturity; plants small or low, the areoles 0.5-1.5cm apart
  Joints branching above . . . . . *O. basilaris* var. *ramosa*
  Joints branching from base . . . . . *O. basilaris* var. *b.*
 Fruit fleshy and juicy; plants fairly large, the areoles mostly 15-50mm apart
  Plant treelike, 3-7m tall, cultivated species escaped about old homesites or in deep valley soil . . . . *O. ficus-indica*
  Plant not treelike, rarely 2m tall
   Spines subcircular to elliptic in cross section, usually 1-3 per areole, not all yellow
    Joints green, flowers yellow . . . . . *O. littoralis* var. *l.*
    Joints glaucous, flowers pinkish . . . . . *O. littoralis* var. *vaseyi*

Spines flattened at least basally, usually 3 or more per areole
Spines not all deflexed, spreading in all directions
Spines not all yellow, with at least reddish or brownish bases
Fruit purplish, obovoid to elongate; spines 1-6 (-11) per areole, straight or nearly so
Mature spines not white or pale gray, spreading
Spines brown or brownish-red or sometimes with lighter tips, the longest mostly 3.5-6cm long, deserts . . . . *O. phaeacantha* var. *major*
Spines not necessarily all same color, gray, red, brown, yellow or mixtures of these; longest 1.2-2cm long, cismontane . . . . Hybrid population *O. occidentalis*
Mature spines white or pale gray or younger sometimes brownish, 1 in each areole deflexed, mostly desert regions . . . . *O. phaeacantha* var. *discata*
Fruit red, subglobose; spines mostly 8-16 per areole, the lower ones curving and twisting gently, coastal . . . . *O. oricola*
Spines all yellow, becoming gray or dark in age . . . . Hybrid population *O. demissa*
Spines all deflexed (1-7 per areole), clearly yellow, becoming black in age; fruit grayish, subglobose to ellipsoid, tinged purple; plant 1-2m tall . . . . *O. chlorotica*

O. acanthocarpa Engelm. & Bigel. ssp. **ganderi** Wolf. Gander's Cholla. Occasional, on rocky slopes and alluvial fans; desert, below 800m: San Felipe Valley; Vallecito Wash. Apr.-June. Type from 3 miles below the old Vallecito Stage Station. *n*=11.

O. basilaris Engelm. & Bigel. var. **basilaris**. Common Beavertail Cactus. Frequent, on dry, rocky slopes; desert, below 1200m: e.g. Oak Grove; Coyote Canyon; Lost Valley; Clark Dry Lake; Borrego Palm Canyon; Sentenac Canyon; Banner; Dos Cabezas; Jacumba. Mar.-June. 2*n*=22.

O. basilaris var. **ramosa** Parish. Long-Joint Beavertail. Scarce, on exposed slopes; montane, above 1500m: Volcan Mountain. May-June.

O. bigelovii Engelm. var. **bigelovii**. Teddy Bear Cholla. Prevalent, on rocky slopes and alluvial fans; desert, below 1000m: e.g. Borrego, Mason, and Vallecito Valleys; Canebrake Canyon; Jacumba. Apr.-May.

O. bigelovii var. **hoffmannii** Fosb. [*O. fosbergii* C.B. Wolf.] Hoffmann's Cholla. Endemic to San Diego County. Occasional, on alluvial fans and washes; desert, 150-500m: Box Canyon; Vallecito Wash; Canebrake Canyon. Mar.-Apr. Type from an alluvial fan at the mouth of Cane Brake Canyon.

O. chlorotica Engelm. & Bigel. Pancake Pear. Infrequent, on rocky slopes; foothill and desert, below 1100m: Sentenac Canyon; Buckman Spring; Hipass; Kitchen Creek; Table Mountain. May-June.

O. demissa Griffiths. Occasional, on dry slopes and valleys; cismontane, below 400m: San Pasqual; Rancho Santa Fe. May-June. Type from east of San Diego.

O. echinocarpa Engelm. & Bigel. var. **echinocarpa**. Silver Cholla. Frequent, on rocky slopes and alluvial fans; desert, below 1300m: Clark Dry Lake; San Felipe and Borrego Valleys; Mountain Springs. Apr.-May. *n*=11.

O. echinocarpa var. **wolfii** L. Benson. [*O. e.* var. *parkeri* auct. non Coulter] Wolf's Cholla. Occasional, on rocky slopes and alluvial fans; desert, 300-1200m: San Felipe Valley; below Sentenec Sanyon; Vallecito Canyon; Jacumba. Apr.-May.

*****O. ficus-indica** (L.)Mill. Indian Fig. Native of Neotropics. Occasional, escaping or persisting from cultivation: Encinitas; Mission Valley; San Ysidro; May-June. *n*=44.

O. littoralis (Engelm.)Ckll. var. **littoralis**. Coastal Prickly-Pear. Common, on dry slopes; coastal, below 150m: e.g. San Onofre; Del Mar; Point Loma; Boundary Monument No. 258. May-June.

O. littoralis var. **vaseyi** (Coult.)L. Benson & Walkington. [*O. v.* (Coult)Britt. & Rose.] Mesa Prickly-Pear. Uncommon, on dry slopes; foothill, 800-1400m: Eagles Nest. May-June. *n*=33.

O. occidentalis Engelm. & Bigel. Western Prickly-Pear. Frequent, on dry slopes and valleys; cismontane, below 1000m: Fallbrook; Pala; Lakeside; Point Loma; San Ysidro. May-June. *n*=33.

O. oricola Philbrick. Infrequent, on slopes and low hills; coastal, below 150m: e.g. San Diego; Chula Vista; Imperial Beach; San Ysidro. Apr.-June. *n*=33.

O. parryi Engelm. var. **parryi**. [*O. echinocarpa* var. *parkeri* (Engelm.)Coulter, the type from Campo.] Valley Cholla. Frequent, in open areas; foothill, 150-1000m: Oak Grove; Hot Springs Mountain; Pala; San Felipe Valley; Viejas Grade; Jacumba; Campo. May-June. Type from the "eastern slope of the California mountains, near San Felipe".

O. parryi var. **serpentina** (Engelm.)L. Benson. [*Cereus? californicus* (Nutt.)T. & G., the type from "arid hills and denuded tracts near St. Diego".] Snake Cholla. Infrequent, on dry slopes; coastal, below 150m: Balboa Park; Point Loma; Chollas Valley; Chula Vista; Jamacha; Telegraph Canyon; Boundary Monument No. 258. Apr.-May. Type from "dry hillsides, San Diego".

O. phaeacantha Engelm. var. **discata** (Griffiths)L. Benson & Walkington. Infrequent, on dry, rocky slopes; foothill, 600-1300m: San Felipe Valley; desert slopes of Laguna Mountains. Apr.-June. *n*=33.

O. phaeacantha var. **major** Engelm. Occasional, on dry slopes and valley floors; foothill, 600-1200m: Warner Hot Springs; San Felipe Valley; Julian; Oriflamme Canyon; Buckman Spring; Dehesa. Apr.-June. *n*=33.

O. prolifera Engelm. Coastal Cholla. Abundant, on dry slopes; coastal, below 200m: e.g. San Onofre; Oceanside; Carlsbad; Del Mar; La Jolla; Point Loma; Bonita; Sunnyside; Telegraph Canyon; Boundary Monument No. 258. Apr.-June. Type from San Diego, on arid hills and in dry creek beds.

O. ramosissima Engelm. Diamond Cholla. Occasional, on alluvial fans and washes; desert, below 700m: Borrego Valley; The Narrows; Agua Caliente Hot Springs; Sweeney Pass; Dos Cabezas; Jojoba Wash; Lava Flow Wash. Mar.-June. *n*=11.

O. wigginsii L. Benson. Wiggins' Cholla. Rare, known in the county only at Canebrake Canyon, 150m.

CALLITRICHACEAE. Water-Starwort Family

Callitriche L.

Fruits pedicelled . . . . . *C. longipedunculata*
Fruits sessile or subsessile
  Carpels broadly winged all around the margin . . . . . *C. longipedunculata*
  Carpels wingless or winged at summit, the wing narrower or lacking on the sides
    Margins of carpels narrowly winged at summit . . . . . *C. verna*
    Margins of carpels wingless or nearly so, reticulations usually not in vertical rows . . . . . *C. heterophylla*

C. heterophylla Pursh ssp. **bolanderi** (Hegelm.)Calder & Taylor. Bolander's Water-Starwort. Scarce, in vernally moist places: Julian; Otay Mountain. Apr. 2*n*=20.

C. longipedunculata Morong. [*C. marginata* Torr. var. *l.* (Morong.)Jeps., *C. marginata* Torr.] Long-Stalk Water-Starwort. Frequent, in vernal pools; coastal and foothill, below 500m: e.g. Rainbow Valley; Poway; Ramona; Kearny Mesa; Linda Vista Terrace; East San Diego; Chollas Heights; Otay Mesa. Feb.-May. Type from "mesas, San Diego".

C. verna L. emend. Kutz. [*C. palustris* auct. non L.] Vernal Water-Starwort. Uncommon, along lake shores; montane, above 1000m: Julian; Cuyamaca Lake. May-Aug. *n*=10.

CAMPANULACEAE. Bellflower Family

KEY TO GENERA

Corolla regular; anthers and filaments distinct
  Leaves broad to roundish; capsule dehiscent on the side
    Capsule opening by uplifting of small lids; ovary somewhat elongate . . . . . *Triodanis*
    Capsule opening by irregular fissures; ovary short and broad . . . . . *Heterocodon*
  Leaves narrow, oblong; capsule dehiscent at apex . . . . . *Githopsis*
Corolla irregular
  Anthers all alike, distinct; flowers minute, inconspicuous . . . . . *Nemacladus*
  Anthers united into a tube, 2 shorter than the others; flowers mostly showier
    Flowers sessile in the axils of foliaceous bracts, the ovary linear and simulating a pedicel . . . . . *Downingia*
    Flowers pedicelled; plants perennial; corolla 20-40mm long . . . . . *Lobelia*

Downingia Torr.

Lower lip of corolla without any purple areas; seeds appearing twisted . . . . . *D. cuspidata*
Lower lip of corolla with 1 purple spot; seeds not appearing twisted . . . . . *D. concolor*

D. concolor Greene ssp. **brevior** (McVaugh)Beauchamp. Cuyamaca Downingia. Endemic to San Diego County. Scarce, at the type locality: eastern margins of Cuyamaca Lake, 1400m. May-June.

D. cuspidata (Greene)Greene. [*D. pulchella* var. *arcana* Jeps., the type from La Mesa.] Toothed Downingia. Common, in vernal pools; coastal and foothill, below 350m: San Marcos; Ramona; Poway; Kearny Mesa; East San Diego; Fletcher Hills; Otay Mesa. Mar.-June. n=11.

Githopsis Nutt.

Stems elongate-filiform, the branches few, widely spreading, almost twining; flowers minute, 3-6mm long; ovary 3-5mm long . . . . . *G. filicaulis*
Stems stout to slender but not filiform, few- to many-branched; flowers over 4mm long; ovary 4-12mm long
  Corolla 3-5mm long; capsule linear, slightly enlarged upward, not ribbed . . . . . *G. diffusa*
  Corolla 4-7mm long; capsule inflated above the middle, ribbed
    Corolla blue . . . . . *G. specularioides* ssp. *s.*
    Corolla white . . . . . *G. specularioides* ssp. *candida*

G. diffusa Gray ssp. **diffusa**. Rare, known in the county only from a collection from the north base of El Cajon Mountain, 420m. Apr.-June.

G. filicaulis Ewan. Slender Blue-Cup. Rare, on shaded slopes, coastal, below 1000m: Silverwood; Padre Dam. Type from Mission (Murphy) Canyon. May.

G. specularioides Nutt. ssp. **candida** Ewan. White Blue-Cup. Uncommon, on open slopes and burns; cismontane, below 1200m: Cuyamaca Lake; Julian; Harbison Canyon; Barona Valley; Morena Dam; Santa Ysabel; Dulzura. Apr.-June. Type from a gravelly burn in Chaparral, 6 miles north of Santa Ysabel.

G. specularioides ssp. **specularioides**. Common Blue-Cup. Uncommon, in open areas and on burns; foothill and montane, above 600m: Palomar Mountain; Love Valley; Morettis; Anahuac School; Cuyamaca Lake; Tecate Mountain; Otay Mountain. Apr.-May.

Heterocodon Nutt.

H. rariflorum Nutt. Heterocodon. Occasional, on burns or grassy slopes; cismontane; Palomar Mountain; Spencer Valley; Del Mar; El Cajon Mountain; Cuyamaca Mountain; Anahuac School; Border Field; Hidden Glen; Japatul Valley; Woodwardia Canyon. May.-Aug.

Lobelia L.

Corolla bright red, 2.5-4cm long, tube split to the base on the upper side . . . . *L. cardinalis*
Corolla blue, less than 2cm long, tube not cleft, at least at the upper end . . . . *L. dunnii*

L. cardinalis L. ssp. **graminea** (Lam.)McVaugh. [*L. splendens* Willd.] Western Cardinalflower. Occasional, in wet areas; foothill and montane, above 750m: e.g. Palomar Mountain; Lake Henshaw; Sheep Canyon; Oriflamme Canyon; Cuyamaca Lake; Ballena; Cuyamaca Mountains; Vallecito Canyon; Oasis Spring; north slope Sequan Peak. July-Sep.

L. dunnii Greene var. **serrata** (Gray)McVaugh. [*Palmerella debilis* var. *s.* Gray] Rothrock Lobelia. Infrequent, in shaded, moist canyons; foothill and montane, 600-1500m: Castro Canyon; Pauma Creek; Cottonwood Creek at Cutca Trail; below Lake Henshaw; Boulder Creek. July-Oct.

Nemacladus Nutt.

Corolla-tube mostly 2-5mm long, much exceeding the calyx; capsule superior, free from the calyx its entire length and exceeding it
  Corolla 5-8mm long . . . . *N. longiflorus* var. *l.*
  Corolla 3-3.5mm long . . . . *N. longiflorus* var. *breviflorus*
Corolla-tube 1.5mm long or less, scarcely or not exceeding the calyx; capsule about half inferior, the free part not or scarcely exceeding the calyx
  Corolla-lobes united at base, so that the tube is almost or about as long as the free lobes
    Stems straight, not at all zigzag; seeds subglobose, with about 6 pits in each row . . . . *N. ramosissimus*
    Stems zigzag; seeds longer than thick, with 8-12 pits in a longitudinal row
      Basal leaves usually pinnatifid; bract at base of pedicel sublinear, not or scarcely enfolding the base of the pedicel; mature capsule 3-4mm long . . . . *N. pinnatifidus*
      Basal leaves entire or toothed; bract at base of pedicel ovate to lanceolate, enfolding and concealing the base of the pedicel; capsule 1.5-2.5mm long . . . . *N. sigmoideus*
  Corolla-lobes scarcely united at base, the tube very short
    Anthers 0.4-0.8mm long; filaments 2-3mm long
      Pedicels slender . . . . *N. rubescens* var. *r.*
      Pedicels capillary . . . . *N. rubescens* var. *tenuis*
    Anthers 0.1-0.3mm long; filaments 1.3-2mm long
      Pedicels stiffly spreading-ascending, not or scarcely curved at tip . . . . *N. glanduliferus* var. *orientalis*
      Pedicels spreading, usually curved near the tip . . . . *N. glanduliferus* var. *g.*

N. glanduliferus Jeps. var. **glanduliferus**. Glandular Threadplant. Occasional, in sandy areas; desert, below 900m: e.g. Borrego Palm Canyon; Sentenac Canyon; San Felipe; Box Canyon; Mountain Palm Springs; Mountain Springs; Jacumba. Mar.-May. Type from Wagon Wash, near Sentenac Canyon.

N. glanduliferus var. **orientalis** McVaugh. Rare, in sandy washes; desert, below 1000m: south side of Tierra Blanca Mountains; 6 miles east of Jacumba. Mar.-May.

N. longiflorus Gray var. **breviflorus** McVaugh. Scarce, in sandy areas; desert, 300-1200m: Banner; Vallecito Canyon; Jacumba. Apr.-June.

N. longiflorus var. **longiflorus**. Long-Flower Threadplant. Infrequent, in dry, sandy areas; cismontane, below 1100m: Pala; Palomar Mountain; Oak Grove; Hot Springs Mountain; San Felipe Wash; Banner; Santa Ysabel; Volcan Mountain; Witch Creek; Encinitas; Torrey Pines State Reserve; Potrero; Descanso; Buckman Spring; Corte Madera Ranch; Campo. Apr.-June.

N. pinnatifidus Greene. [*N. ramosissimus* var. *p.* (Greene)Gray] Comb-Leaf Threadplant. Infrequent, on burns and disturbed slopes; foothill, 450-1200m: Palomar Mountain; San Felipe Creek; Viejas Valley; Descanso; Echo Del; Cuyamaca Peak; Potrero; Buckman Spring; Cottonwood; McGinty Mountain; Otay Mountain. May-June.

N. ramosissimus Nutt. Nuttall's Threadplant. Occasional, on burns and disturbed areas; cismontane, below 1200m: e.g. Fallbrook; Lilac; Pauma Valley; Palomar Mountain; Eagles Nest; San Felipe Creek; Mount Woodson; Barona Valley; Santa Ysabel; Jamul; Potrero; Campo; Buckman Spring; Jacumba. Also at hills south of Vallecito Stage Station. Apr.-May. Type from sandy soils near "St. Diego".

N. rubescens Greene var. **rubescens**. [*N. rigidus* var. *r.* (Greene)Munz] Desert Threadplant. Infrequent, in sandy areas; desert, below 600m: Borrego Palm Canyon; Hellhole Canyon; Borrego Valley; Kane Springs Road; Yaqui Well; Vallecito Stage Station. Apr.-May.

N. rubescens var. **tenuis** McVaugh. Capillary Threadplant. Scarce, in sandy areas; desert, below 300m: Borrego Valley; near Carrizo Stage Station. Jan.-Apr.

N. sigmoideus Robbins. Small-Flower Threadplant. Scarce, in sandy places; desert, below 900m: Banner; San Felipe Wash; east of Jacumba. Apr.-June.

Triodanis Raf.

T. perfoliata (L.)Nieuwland var. **biflora** (R.& P.)Bradley. [*Specularia b.* (R. & P.)F. & M., *T. biflora* (R.& P.)Greene] Small Venus-Looking-Glass. Occasional, on burns and disturbed slopes; coastal and foothill, below 1400m: e.g. De Luz; Fallbrook; Lilac; Palomar Mountain; Warner Hot Springs; Barona Valley; Mussey Grade; Kearny Mesa; Lee Valley; Tecate Mountain; Dehesa; Woodwardia Canyon; Old Padre Dam. Apr.-June. $2n=28$.

CANNABACEAE. Hops Family

Cannabis L. Hemp

***C. sativa** L. Native of Old World. Uncommon, disturbed areas, usually collected only once in each location; coastal, below 200m: Del Mar; Poway; Tijuana River Valley. Apr.-July. $2n=20$.

CAPPARACEAE. Caper Family

KEY TO GENERA

Fruit didymous, 2-celled, each value enclosing a seed and falling away with it . . . . *Wislizenia*
Fruit not didymous, 1-celled, several- to many-seeded
  Capsule 12-45mm long, linear to oblong; petals 6-16mm long, generally clawed. . . . . *Cleome*
  Capsule 3-6mm long, often somewhat wider; petals 1.5-7mm long, not clawed . . . . *Cleomella*

Cleome L.

Woody shrub . . . . *C. isomeris*
Herbaceous perennial, flowers with long-exserted stamens . . . . *C. serrulata*

C. isomeris Greene. [*Isomeris arborea* Nutt.; *I. a.* var. *angustata* Parish; *I. a.* var. *globosa* Cov.] Bladderpod. Occasional, on dry slopes; coastal, foothill, and desert, below 700m: e.g. San Onofre; Pala; Solana Beach; La Jolla; San Pasqual; Rancho Santa Fe; Emerald Hills; Boundary Monument No. 258; Chula Vista; Marron Valley; Potrero; Carrizo Gorge; Banner; Sentenac Canyon; Yaqui Well; The Narrows. Jan.-Dec. Type from "St. Diego". *n*=17; 2*n*=20 pairs.

***C. serrulata** Pursh. Mountain Beeplant. Native of Rocky Mountain region. Local, adventive at Mahogany Canyon, Warner Ranch and Descanso. May-Aug. *n*=17.

Cleomella DC. Stinkweed

C. obtusifolia Torr.& Frem. [*C. taurocranos* Nels., *C. o.* var. *pubescens* Nels., the type from the Colorado Desert, San Diego County.] Bushy Stinkweed. Infrequent, in sandy and alkaline areas; desert, below 200m: Borrego Spring; San Felipe Wash; Seventeen Palms; Arroyo Salado; Ocotillo Wells. Mar.-Apr.

Wislizenia Engelm.

W. refracta Engelm. ssp. **palmeri** (Gray)Keller. [*W. divaricata* Greene, the type from "Bonego [-Borrego] Springs".] California Jackass-Clover. Rare, known in the county only from alkaline soils in Borrego Sink, 150m. May. *n*=20

CAPRIFOLIACEAE. Honeysuckle Family

Berry red or black, few seeded; corolla tubular. . . . . *Lonicera*
Berry white, 1-2-seeded; corolla open-campanulate or tubular-funnelform . . . . *Symphoricarpos*

Lonicera L. Honeysuckle

Flowers in peduncled axillary pairs . . . . *L. japonica*
Flowers sessile, in whorls; plants twining or trailing
  Leaves all distinct, the uppermost pair rarely slightly connate; inflorescence glandular-pubescent
    Leaves whitish and pubescent beneath . . . . *L. subspicata* var. *johnstonii*
    Leaves yellowish and subglabrous beneath . . . . *L. subspicata* var. *denudata*
  Leaves mostly with the uppermost connate
    Leaves mostly with stipules; corolla pinkish to purplish; inflorescence glandular-pubescent. . . . . *L. hispidula*
    Leaves without stipules; corolla yellow; inflorescence glabrous . . . . *L. interrupta*

L. hispidula (Lindl.)Dougl. ex T.& G.var. **vacillans** Gray. California Honeysuckle. Rare, known in the county only at Hot Springs Mountain, 1500m. Apr.-July.

L. interrupta Benth. Chaparral Honeysuckle. Scarce, on dry slopes of Palomar Mountain; montane, above 900m: Harrison Grade; Doane Valley. May-July.

***L. japonica** Thunb. Japanese Honeysuckle. Native of Asia. Infrequently escaping from cultivation as in Florida Canyon, Balboa Park, 30m. Mar.-July.

L. subspicata H. & A. var. **denudata** Rehd. San Diego Honeysuckle. Occasional, in Chaparral; coastal and foothill, below 1500m: e.g. De Luz; Rainbow; Pala; Olivenhain; Inspiration Point; Lake Hodges; Poway Grade; Mission Valley; Rolando; Otay Mountain. Apr.-June. Type from San Diego.

L. subspicata var. **johnstonii** Keck. Southern Honeysuckle. Frequent, in Chaparral and Montane Coniferous Forest; foothill and montane, above 400m: e.g. Palomar Mountain; Hot Springs Mountain; San Felipe Valley; Julian; Cuyamaca Peak; Oasis Spring; Barona Valley; Campo; Corte Madera; La Posta; Dulzura. Apr.-June.

Symphoricarpos Duhamel Snowberry

Corolla short-campanulate, the lobes about as long as tube; twigs closely pubescent; leaves round-oval, 1-4cm long
 . . . . *S. mollis*
Corolla elongate-campanulate, to salverform, the lobes shorter than the tube; leaves lanceolate to elliptical, 0.5-1.5cm
  long . . . . *S. parishii*

S. mollis Nutt. in T. & G. Spreading Snowberry. Common, in Chaparral and Montane Coniferous Forest; foothill and montane, above 600m: e.g. De Luz; Rainbow; Palomar Mountain; Hot Springs Mountain; Volcan Mountains; Cuyamaca Mountains; Guatay Mountain; Corte Madera; Campo; Boulevard. Apr.-June.

S. parishii Rydb. Parish's Snowberry. Uncommon, in Montane Coniferous Forest and Chaparral; montane, above 1200m: Palomar Mountain; Middle Cuyamaca Peak; head of Vallecitos Canyon. June-Aug.

CARYOPHYLLACEAE. Pink Family

KEY TO GENERA

Fruit a 1-seeded indehiscent utricle; petals 0; flowers very small, greenish or whitish
  Stipules present, scarious
    Sepals united below into a short tube . . . . . *Achyronychia*
    Sepals quite or almost distinct
      Annual; leaves elliptic-oblanceolate; stipules minute . . . . . *Herniaria*
      Perennial; leaves subulate; stipules conspicuous . . . . . *Cardionema*
  Stipules lacking; annual with subulate leaves; flowers clustered, greenish . . . . . *Scleranthus*
Fruit a several- to many-seeded capsule; petals usually present; flowers small to large
  Sepals distinct or nearly so; petals when present without claws and borne on a basal disc or at base of sessile ovary
    Stipules wanting
      Capsule ovoid or ellipsoid, dehiscent by as many or twice as many valves or teeth as there are carpels
        Styles opposite sepals and usually 3
          Petals deeply notched or bifid, rarely wanting; valves of capsule bifid or 2-parted . . . . . *Stellaria*
          Petals entire or nearly so; valves of capsule entire or bifid or 2-parted . . . . . *Arenaria*
        Styles alternate with sepals, 4-5 . . . . . *Sagina*
      Capsule cylindrical, often bent near summit, dehiscent by twice as many teeth as there are carpels . . *Cerastium*
    Stipules present, scarious
      Styles 3-5, distinct; petals usually present
        Styles and capsule-valves 5; leaves appearing whorled . . . . . *Spergula*
        Styles and capsule-valves 3; leaves opposite . . . . . *Spergularia*
      Style 1, 3-cleft or toothed; petals minute or 0
        Leaves flat, oblong or obovate; stipules scarious . . . . . *Polycarpon*
        Leaves subulate; stipules setaceous . . . . . *Loeflingia*
  Sepals united into a tubular or cuplike calyx; petals clawed and borne on the carpophore (stipe of ovary)
    Styles 3, rarely 4; capsule dehiscent by 6, rarely 3, 4, or 8 apical teeth . . . . . *Silene*
    Styles 2; capsule mostly 4-valved
      Calyx ovoid, 5-ribbed, wing-angled; petals not appendaged . . . . . *Vaccaria*
      Calyx tubular, 20-nerved, not wing-angled; petals appendaged at base of blades . . . . . *Saponaria*

Achyronychia T. & G.

A. cooperi T. & G. Onyx Flower. Occasional, in sandy flats and washes; desert, below 750m: e.g. Borrego Palm Canyon; Borrego Valley; Lone Palm; Benson Dunes; Mason Valley; Agua Caliente Hot Springs; Split Mountain; Vallecito and Carrizo Stage Stations. Apr.-June.

Arenaria L. Sandwort

Plants annual
  Petals shorter than sepals . . . . . *A. pusilla*
  Petals longer than sepals, or at least as long . . . . . *A. douglasii*
Plants perennial . . . . . *A. macrophylla*

A. douglasii Fenzl ex T. & G. Douglas' Sandwort. Occasional, on open slopes and burns; cismontane: e.g. De Luz; Eagles Nest; Banner Grade; Julian; Cuyamaca Lake; Cuyamaca Peak; Lyons Valley; Jacumba; Boulevard; San Diego; San Ysidro. Apr.-June.

A. macrophylla Hook. Large-leaf Sandwort. Uncommon, on shaded slopes in Montane Coniferous Forest in Cuyamaca Mountains; montane, above 1400m: Stonewall Mine; Middle and Cuyamaca Peaks. Apr.-June.

A. pusilla Wats. var. **diffusa** Maguire. Scarce, in moist areas; montane, above 1400m: Doane Valley; Cuyamaca Lake; Laguna Camp; Shrine Camp; Laguna Lakes. Apr.-July.

Cardionema DC.

C. ramosissima (Weinm.)Nels. & Macbr. Tread Lightly, Beach Sand-Mat. Occasional, in sandy open areas; coastal, below 150m: e.g. San Onofre; San Luis Rey; Pala; Poway Grade; Kearny Mesa; Encinitas; Pacific Beach; Torrey Pines State Reserve; Silver Strand; Monument Beach; Tijuana Hills. Mar.-Aug.

Cerastium L. Mouse-Ear Chickweed

Annual, without basal persistent sterile offshoots; petals lacking or about as long as sepals
  Petals present . . . . . *C. glomeratum* var. *g.*
  Petals lacking . . . . . *C. glomeratum* var. *apetalum*
Short-lived perennial, with prostrate or creeping basal branches or offshoots . . . . . *C. fontanum*

*__C. fontanum__ Baumg. ssp. **triviale** (Link)Jalas. [*C. vulgatum* L., nom. dubium.] Native of Eurasia. Infrequent, in wet places and as a lawn weed: Palomar Mountain; San Diego. June.

*__C. glomeratum__ Thuill. var. **apetalum** (Dumort)Rouy & Foucd. [*C. viscosum* var. *a.* (Dumort)Fenzl.] Native of Europe. Occasional, in disturbed areas; coastal and foothill, below 900m: Herron Ranch; Anahuac School; Clevenger Canyon; Lake Murray; Mount Soledad; Balboa Park. Feb.-May.

*__C. glomeratum__ var. **glomeratum**. [*C. viscosum* auct. non L.] Native of Europe. Occasional, in disturbed areas; cismontane, below 1400m: e.g. Doane Valley; Eagles Nest; Cuyamaca Lake; Julian; Witch Creek; Bernardo; Descanso; Harbison Canyon; Dulzura; Otay Mountain; San Diego. Feb.-May. n=36.

Dianthus L.

*__D. barbatus__ L. Sweet William. Native of Europe. Local, garden escape at Summit Grove, Palomar Mountain, 1500m. Aug. 2n=30.

Herniaria L.

***H. cinerea** DC. Gray Herniaria. Native of Europe. Scarce, naturalized in disturbed areas: San Diego. Apr.-Sep.

Loeflingia L.

L. squarrosa Nutt. in T.& G. var. **squarrosa** California Loeflingia. Uncommon, on burns and open areas; coastal, foothill, and desert, below 800m: Pala; Lilac; Pauma; Borrego Valley; San Felipe Valley; Mason Valley; Vallecito Stage Station; Cottonwood; Campo; Kearny Mesa; National City. Mar.-May. Type from "sandy plains, St. Diego".

Polycarpon L.

Sepals about 1mm long; leaves spatulate, opposite . . . . . *P. depressum*
Sepals almost 2mm long; leaves obovate to oblong, often appearing to be in 4's . . . . . *P. tetraphyllum*

P. depressum Nutt. in T. & G. California Polycarp. Occasional, on burns and open areas; foothill and coastal, below 500m: Fallbrook; Rainbow; Pala; Lilac; Encinitas; Rincon; Harbison Canyon; Dulzura; Barrett; Whispering Oaks; La Mesa; Mission Valley; Point Loma. Apr.-June. Type from "bare sand-hills, near St. Diego".

***P. tetraphyllum** (L.)L. Four-Leaf Polycarp. Scarce, weed along roads and in waste places: San Diego; Balboa Park. May-July. $2n=54$.

Sagina L. Pearlwort

Plants annual, without sterile basal rosettes; pedicels straight; stems filiform
  Leaf bases ciliolate; flowers apetalous, usually 4-parted . . . . . *S. apetala*
  Leaf bases not ciliate; flowers with petals, 5-parted . . . . . *S. decumbens*
Plants perennial, with sterile basal rosettes; pedicels often curved at tip; stems somewhat fleshy . . . . . *S. saginoides*

***S. apetala** Ard. var. **barbata** Fenzl. Sticky Pearlwort. Native of Eurasia. Infrequent, weed in disturbed places; cismontane: San Diego. Apr.-June.

S. decumbens (Ell.)T. & G. ssp. **occidentalis** (Wats.)Crow. Western Pearlwort. Uncommon, on slopes; montane and foothill, 300-1500m: Doane Valley; Anahuac School; Hidden Glen; Otay Mountain. May.

S. saginoides (L.)Karst. [*S. s.* var. *hesperia* Fern.] Scarce, in moist areas; montane, above 1300m: French Valley, Palomar Mountain. June.

Saponaria L. Bouncing Bet

***S. officinalis** L. Bouncing Bet. Native of Europe. Local, naturalized in moist places; cismontane: Wynola; Cedar Creek; San Pasqual Valley. June-Sep. $2n=28$.

Scleranthus L. Knawel

***S. annuus** L. German Knotgrass. Native of Europe. Local, adventive in dry places of San Luis Rey River drainage; Lake Henshaw Resort; east of Pala; Agua Tibia Ranch. Mar.-June. $2n=22,48$.

Silene L. Catchfly; Campion

Annuals, mostly introd. weedy plants
  Calyx with 20-30 well developed ribs . . . . . *S. multinervia*
  Calyx usually with 10 ribs, if more, the ribs indistinct
    Plant subglabrous or minutely puberulent, the upper internodes usually with sticky bands
      Plants green . . . . . *S. antirrhina*
      Plants glaucous . . . . . *S. armeria*
    Plant densely pubescent . . . . . *S. gallica*
Perennials
  Corolla bright red; leaves largely lanceolate . . . . . *S. laciniata*
  Corolla white to pink or purplish or yellow
    Petals yellowish, scarcely exserted, 25-35mm long; calyx 15-30mm long . . . . . *S. parishii*
    Petals shorter, not yellow
      Flowers nodding at anthesis; stamens and style long-exserted . . . . . *S. lemmonii*
      Flowers erect or nearly so; stamens included or little exserted . . . . . *S. verecunda*

S. antirrhina L. Snapdragon Catchfly. Occasional, on burns and open areas; cismontane, below 1500m: e.g. De Luz; Rainbow; Doane Valley; Hot Springs Mountain; Cardiff; Boulder Creek; Dulzura; San Diego; Point Loma. Also Borrego Palm Canyon. Apr.-Aug.

***S. armeria** L. Native of Europe. Rare, known in county only on Cuyamaca Peak, 1500m. May.

***S. gallica** L. Common Catchfly. Native of Europe. Occasional, weed in disturbed areas; cismontane, below 1400m: e.g. San Onofre; Rainbow; Palomar Mountain; Moosa Canyon; San Marcos; El Capitan; Ramona; Hipass; Alpine; Otay; Mount Soledad; Kearny Mesa. Feb.-June. $2n=24$.

S. laciniata Cav. ssp. **major** Hitchc. & Maguire. Southern Pink. Occasional, in canyons and on slopes; cismontane, below 1600m: e.g. San Onofre; De Luz; Palomar Mountain; Pala; Banner; Pine Hills; Witch Creek; Bear Valley; Lake Hodges; Ramona; Barona Valley; Espinosa Trail; Jamul; Lee Valley; Chollas Valley; Point Loma. May-July. $n=48$.

S. lemmonii Wats. [*S. palmeri* Wats., the type from the Cuyamaca Mountans.] Lemmon's Pink. Infrequent, in open areas, Montane Coniferous Forest; montane, above 1300m: Pearson Grove; Fry Creek; Cuyamaca Lake; Middle Peak; North Peak. June. $2n=48$.

S. multinervia Wats. Many-Nerve Catchfly. Infrequent, in burns and other disturbed places; foothill, 100-800m: Del Dios; Ramona; Alpine; El Capitan; Kimball Grade; Jamul; Otay Mountain; Tecate Peak. Apr.-May. Type from "Jamuel (Jamul)".

S. parishii Wats. var. **viscida** Hitchc. & Maguire. Parish's Campion. Rare, on forest floor; montane: summit of Hot Springs Mountain, 1500m. May-July.

S. verecunda Wats. ssp. **platyota** (Wats.)Hitchc. & Maguire. Cuyamaca Catchfly. Infrequent, on dry slopes, Montane Coniferous Forest; montane, above 1200m: Harrison Grade; Palomar, Hot Springs, Cuyamaca, and Laguna Mountains; Pine Hills; Cuyamaca Lake. May-Aug. Type from the Cuyamaca Mountains.

Spergula L. Spurry

***S. arvensis** L. Corn Spurry. Native of Europe. Occasional, weed in disturbed areas; foothill and coastal, below 700m: e.g. Encinitas; Escondido; Ramona; San Marcos; San Pasqual; Fletcher Hills; Point Loma. Jan.-Dec. $2n$=18.

Spergularia J. & C. Presl Sand-Spurry

Leaves densely fascicled; stipules lance-acuminate; plants perennial (sometimes annual in *S. rubra*)
  Sepals mostly 5-10mm long; seeds 0.7-0.9mm long; styles 0.6-3mm long
    Petals mostly white . . . . *S. macrotheca* var. *leucantha*
    Petals mostly pink . . . . *S. macrotheca* var. *m.*
  Sepals 3-5mm long; seeds 0.4-0.6mm long; styles 0.2-0.8mm long
    Stem below the inflorescence mostly glabrous; capsule equaling calyx; seeds reticulate, not winged . . . . . *S. rubra*
    Stems below the inflorescence mostly glandular-pubescent; capsule longer than calyx; seeds papillate, usually winged . . . . *S. villosa*
Leaves not densely fascicled; stipules mostly deltoid, not long-acuminate; plants annual
  Stamens 6-10 . . . . *S. bocconii*
  Stamens 2-5
    Sepals 1.6-5mm long; cymes not much compound, leafy; seeds about 0.6-1.4mm long . . . . *S. marina*
    Sepals 0.8-1.6mm long; cymes much compounded, bracteate; seeds about 0.4mm long . . . . *S. platensis*

***S. bocconii** (Scheele)Foucaud. Buccone's Sand-Spurry. Native of Europe. Uncommon, in sandy areas; coastal, below 150m: Torrey Pines State Reserve; San Diego. Apr.-Sep. $2n$=36.

S. macrotheca (Hornem.)Heynh. var. **leucantha** (Greene)Rob. Large-Flower Sand-Spurry. Rare, known in the county only from an old collection from Santa Maria Valley, 420m. Mar.-Oct.

S. macrotheca var. **macrotheca**. Sticky Sand-Spurry. Occasional, in sandy areas; coastal, below 200m: e.g. San Onofre; Oceanside; La Jolla; Point Loma; Foster; Bostonia; San Marcos Creek. Jan.-Dec.

***S. marina** (L.)Griseb. Salt Marsh Sand-Spurry. Native of Eurasia. Occasional, in alkaline places; coastal, below 100m: e.g. O'Neill Lake; Oceanside; Carlsbad; Silver Strand; San Ysidro; Imperial Beach; Spring Valley; Proctor Valley. Also Carrizo Creek. Mar.-Sep. $2n$=36.

***S. platensis** (Camb.)Fenzl. La Plata Sand-Spurry. Native of South America. Uncommon, in alkaline places; coastal and desert, below 200m: Kearny Mesa; Otay Mesa; Carrizo Creek. May.

***S. rubra** (L.)J.& C. Presl. Ruby Sand-Spurry. Native of Europe. Scarce, in disturbed areas; coastal, below 150m: Boulevard; San Diego; Balboa Park; Mission Valley. Apr.-Sep. $2n$=36

***S. villosa** (Pers.)Camb. [*S. clevelandii* (Greene)Rob., the type from "sandy uplands about San Diego."] Villous Sand-Spurry. Native of South America. Occasional, in alkaline and disturbed areas; coastal, below 150m: San Onofre Bluffs; Poway Grade; La Jolla; Silver Strand; Grossmont; Sloane Ranch; Otay Mountain; San Diego; San Ysidro; Point Loma. Apr.-July.

Stellaria L. Chickweed; Starwort

Internodes with longitudinal line of hair; leaves ovate . . . . *S. media*
Internodes lacking line of hair; upper leaves lance-linear . . . . *S. nitens*

***S. media** (L.)Vill. Common Chickweed. Native of Eurasia. Frequent, in shaded areas; coastal and foothill, below 1000m: e.g. Fallbrook; Moosa Canyon; San Luis Rey; Doane Valley; Eagles Nest; Witch Creek; Poway; Barrett; Dulzura; El Capitan; Pacific Beach; Mission Valley. Feb.-Sep. $2n$=28, 40, 42, 44.

S. nitens Nutt. Shiny Chickweed. Occasional, on grassy slopes; foothill and montane, above 900m: Witch Creek; Julian; Shrine Camp. Apr.-May.

Vaccaria Medic. Cow-Herb

***V. pyramidata** Medic. [*Saponaria v.* L.] Native of Europe. Infrequent, weed in fields; cismontane: Pacific Beach; Witch Creek; Dulzura. May-Aug. $2n$=24.

CELASTRACEAE. Staff-Tree Family

Euonymus L. Burning Bush

E. occidentalis Nutt. ex Torr. ssp. **parishii** (Trel.)Beauchamp. [*E. p.* Trel.] San Diego Burning Bush. Occasional, along streams in Montane Coniferous Forest; montane, above 1200m: Doane Valley; Pauma Creek; Cuyamaca Lake; North Peak; Milk Ranch. Apr.-June.

CERATOPHYLLACEAE. Hornwort Family

Ceratophyllum L. Hornwort

C. demersum L. Occasional, in ponds, lakes, and slow-flowing streams; montane: Temecula Canyon; Alvarado Canyon; Mission Valley; Padre Dam. Aug. $2n$=24.

CHENOPODIACEAE. Goosefoot Family

KEY TO GENERA

Sepals strongly imbricate, scarcely united, strongly chartaceous; leaves opposite, united at base . . . . *Nitrophila*
Sepals slightly or not imbricate, herbaceous when young; leaves mostly alternate
  Leaves foliaceous, flattened, but sometimes linear, not particularly fleshy or scaly
    Flowers perfect, sometimes also pistillate, all with calyx and not enclosed in a pair of bracts
      Calyx transversely winged in fruit . . . . *Cycloloma*
      Calyx not transversely winged in fruit
        Calyx of 1 segment; stamens 1-3 . . . . *Monolepis*
        Calyx with 3-5 segments; stamens 1-5
          Stamen 1; flowers axillary, solitary or clustered . . . . *Aphanisma*
          Stamens 4-5; flowers in clusters
            Calyx indurate at base in age, with ovary partly sunk. . . . . *Beta*
            Calyx not indurate, with ovary superior
              Calyx-lobes not with hooked spines; plants glabrous, mealy, or glandular-pubescent . . . *Chenopodium*
              Calyx-lobes with stout hooked spines; plants pilose . . . . *Bassia*
    Flowers imperfect, the pistillate enclosed in 2 accrescent bractlets . . . . *Atriplex*
  Leaves not or scarcely flattened but fleshy and sublinear, or scaly and spiny
    Leaves scalelike; stems and branches fleshy
      Branches opposite; flower-cluster opposite . . . . *Salicornia*
      Branches alternate; flower-cluster alternate . . . . *Allenrolfea*
    Leaves not scalelike, but fleshy and sublinear, or spiny-tipped
      Leaves not tipped with spine or bristlelike hair
        Fruiting calyx transversely winged . . . . *Kochia*
        Fruiting calyx not transversely winged . . . . *Suaeda*
      Leaves tipped with spine or bristly hair . . . . *Salsola*

Allenrolfea Kuntze

A. occidentalis (Wats.)Kuntze. Iodine Bush. Occasional, alkaline areas; desert, below 600m: e.g. Borrego Springs; Vallecito and Carrizo Stage Stations; Dubber. June-Aug.

Aphanisma Nutt. ex Moq. in D.C.

A. blitoides Nutt. ex Moq. in DC. Aphanisma. Frequent, alkaline areas; coastal, below 25m: e.g. San Onofre; San Dieguito Creek; La Jolla; Silver Strand; Imperial Beach. Also Carrizo Creek. Apr.-May. Type from "around San Diego."

Atriplex L. Saltbush

Plants herbaceous, slightly if at all woody at the base, mostly monoecious
  Leaves green or greenish on both surfaces, sparsely mealy and sometimes grayish when young
    Fruiting bracts united only near the base; staminate flowers mixed with pistillate flowers or in very short spikes; leaves mostly hastate . . . . *A. patula*
    Fruiting bracts united to above the middle; staminate glomerules in elongate terminal spikes or panicles; leaves not hastate
      Leaves 1-2cm long . . . . *A. serenana* var. *davidsonii*
      Leaves 2-4cm long . . . . *A. serenana* var. *s.*
  Leaves gray or whitish with a fine scurf, at least on the lower surface
    Bracts thickened, fleshy, ovate, strongly nerved, 3.5-5mm long
      Fruiting bracts turbinate or globoid, spongy, dry, fibrous, 6-12mm long . . . . *A. lindleyi*
      Fruiting bracts ovate, strongly nerved, reddish-fleshy 3.5-5mm long . . . . *A. semibaccata*
    Bracts not thickened or fleshy
      Fruiting bracts broadest below their middle
        Staminate glomerules in long naked terminal spikes; perennial, the plant forming tangled mats; leaves mostly opposite . . . . *A. watsonii*
        Staminate glomerules in upper leaf-axils or in spikes not more than 1cm long (except *A. californica*)
          Leaves coarsely toothed; bracts hard and indurated. . . . . *A. rosea*
          Leaves entire; bracts not becoming hard . . . . *A. californica*
      Fruiting bracts broadest at or above their middle
        Staminate and pistillate flowers mostly mixed in the clusters
          Bracts irregularly toothed at summit. . . . . *A. argentea*
          Bracts evenly toothed all around. . . . . *A. elegans*
        Staminate and pistillate flowers mostly in separate clusters
          Bracts not compressed, elliptic-globose, 5-7mm long; seed 2.5-3mm long . . . . *A. leucophylla*
          Bracts usually somewhat compressed, 1-4mm long; seed 1.5 or less mm long
            Plants annual . . . . *A. pacifica*
            Plants perennial; stems 3-10dm long, slightly woody at base; leaves entire . . . . *A. coulteri*
Plants shrubby, definitely woody, mostly dioecious
  Fruiting bracts with conspicuous extra wings or crests arising at middle of face . . . . *A. canescens*
  Fruiting bracts not with extra lateral wings
    Plants not spiny
      Leaf-margin entire, flat or crisped but not toothed
        Leaf-blades 3-18mm long, oblong or spatulate . . . . *A. polycarpa*
        Leaf-blades 15-50mm long, oblong to ovate-deltoid
          Fruiting bracts 3-4mm long . . . . *A. lentiformis* ssp. *l.*
          Fruiting bracts 4-7mm long . . . . *A. lentiformis* ssp. *breweri*
      Leaf-margin dentate . . . . *A. hymenelytra*
    Plants spiny, the spines consisting of sharp-pointed twigs from which the e leaves and bracts have fallen . . . . *A. lentiformis* et sspp.

A. argentea Nutt. ssp. **expansa** (Wats.)Hall & Clem. [*A. e.* Wats.] Mojave Silver-Scale, Fog Weed. Occasional, alkaline and disturbed places; coastal, below 100m: San Dieguito Creek; Lemon Grove; San Diego Bay; Silver Strand. Also Campo. July-Nov.

A. californica Moq. in D.C. California Saltbush. Frequent, in alkaline and disturbed areas; coastal, below 100m: Encinitas; La Jolla; Ocean Beach; Chula Vista; Silver Strand. Apr.-Nov.

A. canescens (Pursh.)Nutt. ssp. **canescens**. [*Obione tetraptera* Benth., the type "from the coast of California, probably San Diego."] Shad Scale, Hoary Saltbush. Common, in alkaline areas; coastal and desert, below 1200m: e.g. Encinitas; La Jolla; Logan Heights; Coronado; Otay Mesa; Tijuana Hills; Oak Grove; Warner Spring; Borrego Valley; Mountain Springs; Jacumba. June-Aug.

A. coulteri (Moq.)D. Dietr. Coulter's Saltbush. Occasional, in low alkaline places; coastal, below 50m: Santa Margarita Ranch; Balboa Park; Silver Strand; Otay Mesa. Mar.-July.

A. elegans (Moq.)D. Dietr. ssp. **fasciculata** (Wats.)Hall & Clem. Salton Saltbush. Infrequent, in alkaline areas; desert, below 150m: Borrego Springs; Benson Dry Lake; Carrizo Stage Station. Mar.-July.

A. hymenelytra (Torr.)Wats. Desert Holly. Local, in alkaline places; desert, below 800m: Seventeen Palms; Clark Dry Lake; Benson Dry Lake; The Narrows; Mountain Palm Springs; Carrizo Stage Station; Coyote Mountain; Mountain Springs. Jan.-Apr.

A. lentiformis (Torr.)Wats. ssp. **breweri** (Wats.)Hall & Clem. Brewer's Saltbush. Scarce, in alkaline areas; occasional in hydroseed landscaping; coastal, below 100m: Balboa Park. July-Oct.

A. lentiformis ssp. **lentiformis**. Quail Saltbush. Rare, in alkaline areas; desert, below 100m: Carrizo Stage Station. Aug.-Oct.

A. leucophylla (Moq. in DC.)D. Dietr. Seascale, Beach Saltbush. Occasional, on sandy beaches; coastal, below 25m: e.g. Oceanside; Encinitas; Mission Beach; Silver Strand; Imperial Beach. Apr.-Oct.

*__A. lindleyi__ Moq. [*A. halimoides* Lindl. non Raf.] Lindley's Saltbush. Native of Australia. Local, adventive in alkaline places; coastal, below 150m: Lower Otay Lake; Telegraph Creek estuary. May-July.

A. pacifica A. Nels. Pacific Saltbush. Infrequent, on bluffs; coastal, below 100m: San Diego. Mar.-Oct. Type from San Diego.

*__A. patula__ L. ssp. **hastata** (L.)Hall & Clem. [*A. h.* L.] Halberd-Leaf Saltbush. Native of Eurasia. Occasional, in disturbed areas; coastal, below 100m: Santa Margarita Ranch; Encinitas; Carlsbad; Lake Hodges; Bonita. June-Nov. $2n=18$.

A. polycarpa (Torr.)Wats. Many-Fruit Saltbush. Occasional, in alkaline areas; desert, below 1000m: Borrego Valley; Ocotillo Wells; Scissors Crossing; Mountain Palm Springs; Boulevard. July-Oct.

*__A. rosea__ L. Redscale, Red Saltbush. Native of Eurasia. Infrequent, in disturbed areas; cismontane, below 1300m: Oak Grove; Guatay; Pine Valley; Campo; San Diego. July-Oct. $2n=18$.

*__A. semibaccata__ R. Br. Australian Saltbush. Native of Australia. Prevalent, in disturbed and alkaline areas; foothill and coastal, below 900m: e.g. San Mateo Creek; San Marcos; Ramona; Poway Grade; Flynn Springs; La Jolla; Potrero; Imperial Beach. Apr.-Dec.

A. serenana A. Nels. var. **davidsonii** (Standl.)Munz. Davidson's Saltbush. Rare, known in the county only from a single collection from San Diego. May-Oct.

A. serenana var. **serenana**. Bract Saltbush. Infrequent, in alkaline areas; coastal, below 50m: Santa Margarita Ranch; San Diego Bay; Jamacha. May-Oct.

A. watsonii A. Nels. in Abrams. Watson's Saltbush. Common, on sandy beaches; coastal, below 20m: Carlsbad; Point Loma; Silver Strand; Chula Vista; Monument Beach. Mar.-Oct. Type from near San Diego.

Bassia All.

*__B. hyssopifolia__ (Pall.)Kuntze. [*Echinopsilon h.* (Pall.)Moq.] Five-Hook Bassia. Native of Eurasia. Common, in disturbed and alkaline areas; coastal, below 50m: Santa Margarita Ranch; Santee; San Elijo Lagoon; San Diego Bay; Bonita. July-Oct.

Beta L. Beet

*__B. vulgaris__ L. ssp. **maritima** (L.)Arcangeli. Sea Beet. Native of Europe. Infrequent, adventive in moist areas; coastal, below 100m: Cardiff; Pacific Beach; Lakeside; Sweetwater Lake; Chula Vista; Otay. July-Oct. $2n=18$.

Chenopodium L. Goosefoot

Plants somewhat glandular-pubescent or resinous-glandular, especially about the calyx, not mealy or farinose
- Calyx 3-5-toothed, becoming saccate and reticulate; leaves pinnatifid . . . . *C. multifidum*
- Calyx 5 (rarely 4)-parted, not changing in fruit or merely becoming fleshy; leaves various
  - Glomerules small, headlike, all axillary; seed vertical . . . . *C. pumilio*
  - Glomerules in short spikes, the upper in panicles; seed mostly horizontal. . . . . *C. ambrosioides*

Plants mostly mealy or farinose, not glandular-pubescent, but sometimes with nonglandular hairs
- Seeds horizontal or almost entirely so
  - Leaves shining on upper surface, somewhat rhombic; only younger parts mealy; plant branced from base *C. murale*
  - Leaves dull on upper surface, various in shape
    - Pericarp closely adherent to seed and removable with difficulty; at least lower leaves conspicuously sinuate-dentate
      - Pericarp and seed surface essentially smooth
        - Seeds chiefly 1.2-1.5mm broad; sepals largely covering fruit . . . . *C. album*
        - Seeds 0.9-1.2mm broad . . . . *C. missouriense*

Pericarp and seed-surface foveolate-reticulate; seed 1-1.3mm broad
Calyx open in fruit; plant ill-scented . . . . *C. hians*
Calyx closed in fruit
Plants of waste places . . . . *C. berlandieri* var. *sinuatum*
Plants of alkaline places . . . . *C. berlandieri* var. *zschackei*
Pericarp free from or easily removed from the seed, or if adherent, the lower leaves linear or entire or 3-lobed
Main leaf-blades definitely longer than broad . . . . *C. incognitum*
Main leaf-blades scarcely if at all longer than wide . . . . *C. fremontii*
Seeds vertical for the most part
Leaves densely white-farinose beneath, at least when young . . . . *C. macrospermum*
Leaves sparsely or not at all farinose beneath
Plants annual, glabrous; stems up to 6dm long . . . . *C. chenopodioides*
Plants perennial, mealy on upper parts; stems 3-8dm long or more . . . . *C. californicum*

*__C. album__ L. Lamb's Quarters, White Goosefoot. Native of Europe. Occasional, in moist, disturbed areas; cismontane, below 1000m: e.g. De Luz; Pala; San Luis Rey; San Pasqual; Santa Ysabel; Banner; Santee; Jamacha; Tijuana River Valley. June-Oct. 2*n*=26.

*__C. ambrosioides__ L. var. **ambrosioides**. Mexican-Tea. Native of Neotropics. Common, in waste places; coastal, below 100m: e.g. Camp Pendleton; Encinitas; Lake Hodges; Point Loma; Lake Murray; Jamacha. Also Ballena. June-Dec. 2*n*=16, 32, 48.

C. berlandieri Moq. var. **sinuatum** (J. Murr.)H.A. Wahl. Pitseed Goosefoot. Infrequent, in alkaline areas; foothill and montane, above 800m: Valley Center; Shrine Camp; Clover Flat; Dulzura. Mar.-Oct.

C. berlandieri var. **zschackei** (J. Murr.)J. Murr. Infrequent, in alkaline areas; foothill and montane, above 800m: Hatfield Creek; Cuyamaca Lake. Mar.-Oct.

C. californicum (Wats.)Wats. California Goosefoot. Common, on open slopes; cismontane, below 1500m: e.g. San Onofre; Fallbrook; Pala; Montezuma View Point; Witch Creek; Lilac; Escondido; Point Loma; Jamul; Campo; Dubber. Mar.-June.

*__C. chenopodioides__ (L.)Aellen. Prostrate Goosefoot. Native of Europe. Rare, known in the county only along Lake Hodges, 50m. Dec.

C. fremontii Wats. Fremont's Goosefoot. Infrequent, in dry places; foothill and montane, above 900m: Oak Grove; Chihuahua Valley. June-Oct.

*__C. hians__ Standl. Native of Rocky Mountain region. Rare, known in the county only at Pine Valley, 1050m. June-Oct.

C. incognitum H.A. Wahl. Occasional, in Montane Coniferous Forest; montane, above 1300m: Palomar Mountain; Cuyamaca Peak; Shrine Camp. July-Aug.

C. macrospermum Hook. f. var. **farinosum** (Wats.)J.T. Howell. [*C. macrospermum* var. *halophilum* (Phil.)Standl.] Coast Goosefoot. Rare, Coastal Salt Marshes and estuaries: San Elijo Lagoon; Torrey Pines Beach. Also El Nido. Mar.-Apr.

*__C. missouriense__ Aellen. Native of eastern United States. Local, adventive at Ballena and Boulevard. June.

*__C. multifidum__ L. [*Roubieva m.* (L.)Moq.] Cut-Leaf Goosefoot. Native of South America. Prevalent, weed in waste areas; coastal, below 200m: e.g. Carlsbad; La Jolla; Point Loma; East San Diego. June-Nov.

*__C. murale__ L. Nettle-Leaf Goosefoot. Native of Europe. Prevalent, weed in waste places: e.g. Oceanside; Borrego Springs; Oak Grove; Vista; Foster; Jamacha; Otay Mesa. Jan.-Dec. 2*n*=18.

*__C. pumilio__ R. Br. [*C. carinatum* auct. non R. Br.] Tasmania Goosefoot. Native of Australia. Occasional, in waste places: Fallbrook; Mount Woodson; Sweetwater Reservoir; Balboa Park; National City. June-Sep.

Cycloloma Moq. Pigweed

C. atriplicifolium (Spreng.)Coult. Winged Pigweed. Occasional, disturbed and cultivated areas; foothill and coastal, below 400m: San Pasqual; Foster; Mission Valley; Jamacha. May-Sep.

Kochia Roth

*__K. scoparia__ (L.)Schrad. Belvedere, Summer-Cypress. Native of Eurasia. Local, escaping in disturbed areas; coastal, below 20m: San Diego; La Jolla; National City. Aug.-Oct.

Monolepis Schrad.

M. nuttalliana (Schult.)Greene. Nuttall's Monolepis. Occasional, in saline and alkaline areas; desert and south coastal, below 1000m: Borrego Valley; Yaqui Pass; Carrizo Stage Station; Otay Mesa; Balboa Park. Also Morettis. Apr.-Sep.

Nitrophila Wats.

N. occidentalis (Moq.)Wats. American Nitrophila. Scarce, in alkaline areas; foothill, below 1000m: Campo. May-Oct.

Salicornia L. Glasswort

Plants annual
Joints of spike thicker than long, the scales acuminate. . . . . *S. bigelovii*
Joints of spike longer than thick, the scales barely acute. . . . . *S. europaea*
Plants perennial from creeping rootstocks
Seeds glabrous; spikes slender, the upper part sterile . . . . *S. subterminalis*
Seeds pubescent; spikes stoutish, fertile to the tip . . . . *S. virginica*

S. bigelovii Torr. Dwarf Glasswort. Abundant, in Coastal Salt Marsh and tidal flats; coastal, below 5m: e.g. Mission Bay; San Diego Bay; Tijuana River estuary. July-Nov.

S. europaea L. [*S. depressa* Standl., the type from San Diego]. Slender Glasswort. Frequent, in Coastal Salt Marsh and other alkaline places; coastal, below 20m: Oceanside; Mission Bay; San Diego Bay. July-Nov.

S. subterminalis Parish. [*Arthrocnemum s.* (Parish)Standl.] Parish's Glasswort. Common, in Coastal Salt Marsh; coastal, below 20m: e.g. Agua Hedionda; Mission Bay; San Diego Bay; Imperial Beach. Apr.-Sep.

S. virginica L. [*S. ambigua* Michx.] Woody Glasswort. Frequent, in Coastal Salt Marsh and other alkaline areas; coastal, below 200m: e.g. Rancho Santa Fe; Silver Strand; Lemon Grove; Bonsall. Aug.-Nov.

Salsola L. Russian Thistle, Tumbleweed

***S. australis** R. Br. [*S. iberica* Sennen & Pau.; *S. kali* var. *tenuifolia* Tausch.; *S. pestifera* A. Nels.] Russian Thistle. Native of Asia. Abundant, waste places, below 1800m: e.g. Oak Grove; Carlsbad; Santa Ysabel; Valley Center; Del Mar; Mason Valley; Jamul; Santee. July-Oct.

Suaeda Forsk. Sea-Blite

Plants annual; calyx-lobes unequal, corniculately appendaged . . . . *S. calceoliformis*
Plants perennial, somewhat woody at the base; calyx-lobes equal, not appendaged
  Perianth-lobes unequal; stigmas glabrous; flowers (1)-3-5 per leaf axil . . . . *S. esteroa*
  Perianth- lobes equal; stigmas pubescent; 1-9 flowers per leaf axil
    Flowers, including fruiting calyx, 1-2 (-2.5)mm broad; 1-3 flowers per axil. . . . . *S. moquinii*
    Flowers, including fruiting clayx, 1.5-3 (-4)mm broad; (1)-3-9 flowers per axil. . . . . *S. californica*

S. calceoliformis (Hook.)Moq. [*S. depressa* (Pursh)Wats. var. *erecta* Wats.; *S. minutiflora* Wats.] Pursh's Sea-Blite. Scarce, known in the county only at Santa Margarita Ranch, 30m. July-Oct.

S. californica Wats. var. **pubescens** Jeps. [*S. californica* var. *taxifolia* (Standl.)Munz] Occasional, in Coastal Salt Marsh; coastal, below 20m: Cardiff; Del Mar; San Diego Bay. July-Oct. Type from Del Mar.

S. esteroa Ferrin. [*S. californica* Wats. var. *californica* Calif. auct.] California Sea-Blite. Common, in Coastal Salt Marsh; coastal, below 20m: e.g. San Marcos Creek estuary; Del Mar; Mission Bay; San Diego Bay; Imperial Beach. July-Oct.

S. moquinii (Torr.)Greene. [*S.torreyana* Wats.; *S. fruticosa* Wats. non Forsk.; *S. torreyana* Wats. var. *ramosissima* (Standl.)Munz] Torrey's Sea-Blite. Frequent, in Coastal Salt Marsh and other alkaline places; coastal and desert, below 900m: e.g. San Diego Bay; Chula Vista; Otay Valley; Imperial Beach; Borrego Valley; Carrizo Stage Station; Jacumba. May-Oct. $2n=36$.

CISTACEAE. Rock-Rose Family

KEY TO GENERA

Valves of capsule 5 or 10; leaves ovate to oblong; shrub escaped from cultivation . . . . *Cistus*
Valves or capsule 3; leaves linear to lanceolate . . . . *Helianthemum*

Cistus L. Rock-Rose

***C. villosus** L. Purple Rock-Rose. Native of Europe. Local, established along freeway roadcuts; cismontane, below 1500m: Alpine, The Willows. Mar.-June.

Helianthemum Mill. Rush-Rose

Panicle extremely open; petals 8-12mm long. . . . . *H. scoparium* var. *aldersonii*
Panicle usually narrow; petals ca. 4-6mm long . . . . *H. scoparium* var. *vulgare*

H. scoparium Nutt. var. **aldersonii** (Greene)Munz. Frequent, in Chaparral and Sage Scrub; coastal and foothill, below 1000m: e.g. San Onofre Canyon; Olivenhain; De Luz; Solana Beach; Kearny Mesa; Oak Grove; Point Loma; El Cajon Mountain; Lee Valley; Potrero; Tecate Mountain. Feb.-Aug. Type from "mountans of the southern borders of San Diego Co., Calif., among rocks in hard and sterile granitic soils."

H. scoparium var. **vulgare** Jeps. Occasional, in Chaparral and Sage Scrub; coastal and foothill, below 500m: e.g. San Onofre Canyon; Lake Hodges; Torrey Pines State Reserve; Bear Valley; Ramona; Mission Bay; Point Loma; Kearny Mesa; Balboa Park. Mar.-Sep.

CONVOLVULACEAE. Morning-Glory Family

KEY TO GENERA

Leafless twining parasites, with scalelike leaves . . . . *Cuscuta*
Leafy green plants
  Styles 2, distinct or at least partly so; flowers less than 1cm long
    Ovary deeply 2-lobed; creeping matted herb . . . . *Dichondra*
    Ovary entire; erect herb . . . . *Cressa*
  Style entire or cleft at apex only; flowers more than 1cm long
    Stigma capitate . . . . *Ipomoea*
    Stigma filiform or oblong-cylindric
      Stigma oblong, somewhat cylindrical, the stigmatic area and style distinct; ovary with an incomplete septum . . . . *Calystegia*
      Stigma linear, the stigmatic area and the style somewhat continuous; ovary with a complete septum *Convolvulus*

Calystegia R. Br. Morning-Glory

Calyx enclosed or closely subtended by a pair of large sepallike bracts
  Corolla purple to rose . . . . . *C. soldanella*
  Corolla white to cream, sometimes pinkish in age
    Leaves and stems cinereous with dense tomentulose puberulence . . . . . *C. macrostegia* ssp. *arida*
    Leaves and stems glabrous or nearly so
      Middle lobe of leaves narrowly linear; corolla 2-2.5cm long . . . . . *C. macrostegia* ssp. *tenuifolia*
      Middle lobe of leaves narrowly- to deltoid-lanceolate; corolla 3-3.5cm long. . . . . *C. macrostegia* ssp. *longiloba*
Calyx subtended by more remote bracts that are not much like sepals
  Bracts hastately lobed, like the upper leaves . . . . . *C. occidentalis*
  Bracts entire . . . . . *C. longipes*

C. longipes (Wats.)Brummitt. Piute Morning-Glory. Uncommon, dry slopes; foothill and montane, above 700m: Laguna Mountains; Campo; Buckman Springs; Jacumba. May-July.

C. macrostegia ssp. **arida** (Greene)Brummitt. [*C. a.* Greene] Finger-leaf Morning-Glory. Occasional, dry slopes, cismontane, 200-1200m: e.g. Pala; Escondido; Poway. Feb.-May

C. macrostegia (Greene)Brummitt ssp. **longiloba** (Abrams)Brummitt. Infrequent, dry slopes; foothill and coastal below 600m: Fallbrook; Pala; Mount Woodson; San Diego. Mar.-May. The type from San Diego.

C. macrostegia ssp. **tenuifolia** (Abrams)Brummitt. Narrow-leaf Morning-Glory. Frequent, on dry slopes; foothill and coastal, below 600m: Harrison Grade; Bernardo; San Diego; Otay Mesa; Lee Valley; Harbison Canyon. Feb.-May. Type from hill near Bernardo.

C. occidentalis (Gray)Brumitt ssp. **fulcrata** (Gray)Brummitt. [*Convolvulus f.* (Gray)Greene] Sonora Morning-Glory. Occasional, on dry slopes, Montane Coniferous Forest and Chaparral; montane, above 1200m: Palomar Mountain; Eagles Nest; Mesa Grande; Garnet Peak; Oasis Spring; Laguna Public Camp; Cuyamaca Mountains. May-Aug.

C. soldanella (L.)R. Br. [*Convolvulus s.* L.] Beach Morning-Glory. Occasional, sandy beaches; coastal, below 20m: e.g. San Onofre; Oceanside; Carlsbad; Del Mar; Mission Beach; Silver Strand; Imperial Beach. Apr.-Aug.

Convolvulus L. Bindweed

Plants annual; corolla about 6mm long, deeply cleft . . . . . *C. simulans*
Plants perennial; corolla 20-60mm long, not cleft . . . . . *C. arvensis*

***C. arvensis** L. Native of Eurasia. Common Bindweed. Common, disturbed and cultivated areas; foothill and coastal, below 700m: e.g. Lake Hodges; Alpine; Mission Valley; Point Loma; National City; Otay Mesa. May-Oct. $2n=50$.

C. simulans L.M. Perry. Clay Bindweed. Uncommon, slopes and mesas; coastal, below 200m: Vista; La Corta; Rancho Santa Fe; Spring Valley; Sweetwater Valley; Lemon Grove; Otay Valley. Mar.-May.

Cressa L. Alkali Weed

C. truxillensis H.B.K. var. **vallicola** (Heller)Munz. Prevalent, alkaline areas; coastal, below 200m: e.g. Oceanside; Carlsbad; Kearny Mesa; La Jolla; Lindo Lake; National City; Imperial Beach. May-Oct. $2n=28$.

Cuscuta L. Dodder

Capsule globose, somewhat depressed, sometimes thickened about the style-bases
  Ovary without a thickened stylopodium at apex
    Corolla with scalelike appendages attached to the tube below the stamens
      Flowers as broad as or broader than long; corolla subglobose . . . . . *C. pentagona*
      Flowers longer than broad; corolla campanulate to funnelform . . . . . *C. suaveolens*
    Corolla lacking scalelike appendages at base of stamens
      Anthers oval; corolla-lobes spreading, starlike in fruit; flowers sessile . . . . . *C. occidentalis*
      Anthers linear-oblong; corolla not star-shaped; flowers pedicelled
        Flowers densely papillate . . . . . *C. californica* var. *papillosa*
        Flowers lacking papillate vestment . . . . . *C. californica* var. *c.*
  Ovary and capsule with a thickened stylopodium surrounding base of styles . . . . . *C. indecora*
Capsule ovoid, conic or beaked, commonly longer than thick
  Calyx-lobes acute to acuminate, the edges entire
    Calyx shorter than the corolla-tube; flowers about 5mm long; calyx-lobes overlapping. . . . . *C. ceanothi*
    Calyx as long as or longer than corolla-tube; flowers mostly smaller; calyx-lobes not markedly overlapping
      Flowers 3-4.5mm long . . . . . *C. salina* var. *major*
      Flowers 2-3mm long . . . . . *C. salina* var. *s.*
  Calyx-lobes round, obtuse, denticulate. . . . . *C. denticulata*

C. californica H. & A. var. **californica**. Witch's Hair. Abundant, open areas; cismontane, below 1500m: e.g. Palomar Mountain; El Cajon Mountain; Pacific Beach; Mission Hills; Jamul Valley; Mount Laguna; Cottonwood Creek; Dehesa. Also Sentenac Canyon; Jacumba. May-Aug. Known hosts: *Abronia, Eriogonum, Artemisia, Convolvulus, Adenostoma, Ambrosia, Baccharis, Brassica.*

C. californica var. **papillosa** Yuncker. Uncommon, open areas; desert, below 1000m: Borrego Valley; Borrego Palm Canyon. Apr.-July. Known host: *Ambrosia.*

C. ceanothi Behr. [*C. subinclusa* Dur. & Hilg.] Canyon Dodder. Frequent, on open slopes; foothill, montane, and desert, above 400m: e.g. Palomar Mountain; Oak Grove; Eagles Nest; Mount Woodson; Pine Hills; Descanso; Oriflamme Canyon; Canebrake Canyon; Harbison Canyon; Glen Cliff Campground; Tecate Mountain. June-Oct. Known hosts: *Rhus, Malosma, Solidago, Pluchea, Baccharis, Chilopsis, Ceanothus, Prunus, Salix, Quercus*.

C. denticulata Engelm. Rare, known in the county from a single collection along San Felipe Creek at Hwy. 78. May-Oct. Known hosts: *Larrea, Hymenoclea, Ericameria*. $2n$=15 pairs.

C. indecora Choisy. Rare, known in the county only from a collection on Volcan Mountain, 1600m. Oct. n=15.

C. occidentalis Millsp. Western Dodder. Infrequent, cismontane, below 1000m. DeLuz; Poway; Rancho Bernardo; Penasquitos Canyon; Jamul. Jan.-Aug. Known hosts: *Grindelia, Solanum*.

***C. pentagona** Engelm. Field Dodder. Native of eastern United States. Rare, known in the county only from a collection near Lake Henshaw, 825m. Mar.-June. Known host: *Diplacus*. $2n$=ca.56.

C. salina Engelm. var. **major** Yuncker. Salicornia Dodder. Frequent, Coastal Salt Marsh; coastal, below 5m: e.g. San Diego Bay; Coronado. May-Sep. Known hosts: *Frankenia, Salicornia, Suaeda, Jaumea, Triglochin*.

C. salina var. **salina**. [*C. s.* var. *squamigera* (Engelm.)Yuncker] Salty Dodder. Common, Coastal Salt Marsh and other alkaline areas; coastal and desert, below 300m: Mission Bay; Silver Strand; Carrizo Stage Station. May-Sep. Known hosts: *Frankenia, Salicornia, Suaeda*.

***C. suaveolens** Ser. [*C. racemosa* var. *chiliana* Engelm.] Fringe Dodder. Native of South America. Uncommon, disturbed areas; foothill and montane, above 900m: San Felipe; Mount Laguna. Aug.-Oct. Known host: *Ambrosia*.

Dichondra Forst. & Forst. f.

D. occidentalis House. Western Ponyfoot. Uncommon, in understory of Chaparral and other shaded places; foothill and coastal, below 300m: e.g. Agua Hedionda; La Costa; La Jolla; Point Loma; Torrey Pines State Reserve; Poway; La Mesa; Otay Mountain; Tijuana Hills. Mar.-May. Type from San Diego.

Ipomoea L. Morning-Glory

***I. purpurea** (L.)Roth. Native of Neotropics. Uncommon, escaping or persisting from cultivation: National City; San Diego. June-Nov. n=15.

CORNACEAE. Dogwood Family

Cornus L. Dogwood

Flowers in a head or umbel subtended by conspicuous persistent bracts, usually appearing before the leaves *C. nuttallii*
Flowers in a bractless cyme, appearing with or after the leaves
  Veins 3-4 on each side of midrib; leaf-blades 3-5cm long . . . . . *C. glabrata*
  Veins 4-7 on each side of midrib; leaf-blades 5-10cm long . . . . . *C. occidentalis*

C. glabrata Benth. Brown Dogwood. Local, moist places; above 1200m on Hot Springs Mountain. May-June.

C. nuttallii Audubon ex T.& G. Pacific Dogwood. Infrequent, on moist, shaded slopes and along streams; montane, above 1050m: Palomar, Volcan, and Cuyamaca Mountains. Apr.-July.

C. occidentallis (T.& G.)Cov. Creek Dogwood. Occasional, along streams in the Montane Coniferous Forest; montane, above 950m: Palomar; Hot Springs, Cuyamaca, and Laguna Mountains; Pine Valley; near Descanso. Also along the Santa Margarita River, north of Fallbrook. May-July.

CRASSULACEAE. Stonecrop Family

KEY TO GENERA

Plants annual; stamens mostly 3-5 . . . . . *Crassula*
Plants perennial; stamens twice as many as petals
  Flowering stems lateral, arising from axils of leaves of basal rosette . . . . . *Dudleya*
  Flowering stems terminal, not arising in axils of basal rosette . . . . . *Sedum*

Crassula L.

(N.B. The following treatment is based upon a recent revision for which annotated specimens of local collections were not available)

Carpels 1-2 seeded
  Flowers 3-4-merous
    Sepals barely exceeding the carpels; flowers 0.6-1mm long . . . . . *C. connata* var. *erectoides*
    Sepals 1.5 times the length of the carpels; flowers 1-1.5mm long . . . . . *C. connata* var. *eremica*
  Flowers 4-5-merous . . . . . *C. connata* var. *c.*
Carpels 3 or more seeded
  Seed surface rugulose, dull
    Fruiting pedicels less than 1mm long . . . . . *C. aquatica*
    Fruiting pedicels over 1mm long . . . . . *C. saginoides*
  Seed surface smooth, shiny . . . . . *C. solieri*

C. aquatica (L.)Schoenl. in Engl. & Prantl. [*Tillaea a.* L.] Occasional, in vernal pools and other moist places; coastal and foothill, below 600m: San Marcos; Ramona; Kearny Mesa; Otay Mesa. Feb.-Apr.

C. connata (Ruiz & Pav.)Berger in Engl. & Prantl var. **connata**. [*C. erecta* (H. & A.)Berger; *Tillaea e.* H.& A.] Abundant, on open, flat areas; cismontane, below 1000m: e.g. San Onofre; Rainbow; Barona; Valley; San Diego; Bonita; Dulzura; Kearny and Otay Mesas. Feb.-May.

C. connata var. **erectoides** Bywater & Wilkens. Uncommon, cismontane; below 400m: Kearny Mesa.

C. connata var. **eremica** (Jeps.)Bywater & Wickens. [*T. erecta* var. *eremica* Jeps.] Known in the county only from the desert type locality, Vallecito, 400m.

C. saginoides (Maxim.)Bywater & Wilkens. Uncommon, cismontane; below 800m: Kearny Mesa.

C. solieri (Gay)Meigen in Engl. Occasional, in open, flat areas; cismontane, below 1000m: Oceanside; Vista; Poway; La Jolla; San Diego. Feb.-May.

Dudleya Britt. & Rose

Primary stem a somewhat elongating epigaeous caudex; rosette leaves mostly persistent
  Corolla tubular, the petals erect, with only the tips slightly spreading
    Petals united for 1/3 or more of their length
      Flowers spreading to pendant at anthesis, erect in fruit by a sharp bending of the pedicel; corolla red, the lobes about as long as the tube; plant densely white-pulverulent
        Caudex 4-9cm thick; rosette-leaves 8-25cm long; pedicels 5-30mm long . . . . *D. pulverulenta*
        Caudex 1-4cm thick; rosette-leaves 5-15cm long; pedicels 5-15mm long. . . . . *D. arizonica*
      Flowers erect or somewhat spreading, the pedicels not sharply bent. . . . . *D. abramsii*
    Petals united for less than 1/3 of their length
      Pedicels 0.5-8mm long
        Petals thin, often erose, 8-13mm long, 2-3mm wide, pale yellow and with red lines, connate for 1.5-4.5mm; rosette-leaves 1.5-6 (-11)cm long, 0.5-2cm wide . . . . *D. abramsii*
        Petals thick, entire
          Petals 10-16mm long, orange or yellow-tipped to red; inflorescence with flowers loosely arranged on long cincinni, 5-12cm long. . . . . *D. lanceolata*
          Petals 8-12mm long; inflorescence dense, crowded, umbel-like; flowers bright yellow . . . . *D. alainae*
      Pedicels 1-2cm long
        Rosette-leaves 5-30cm long, 1-4cm wide; flower stems 1.5-7.5dm tall; petals orange or yellow-tipped to red, connate for 1-2mm. . . . . *D.lanceolata*
        Rosette-leaves 3-15cm long, 0.5-2.5cm wide; flower stems 0.5-4dm tall; petals yellow-green, connate for 1-4mm . . . . *D. saxosa*
  Corolla with segments widely spreading from near the middle
    Leaves somewhat flattened, viscid. . . . . *D. viscida*
    Leaves terete or nearly so, except at very base
      Branches of inflorescence usually 2 or 3, simple; caudex mostly less than 1cm thick; mature carpels ascending, slightly gibbous . . . . *D. attenuata*
      Branches of inflorescence several, once or twice bifurcate; caudex mostly more than 1cm thick; mature carpels abruptly divergent, strongly gibbous. . . . . *D. edulis*
Primary stems hyogaeous, usually cormlike; rosette-leaves vernal
  Flowers white, with a sweet odor
    Rosette-leaves 0.7-1.5cm long, subglobular. . . . . *D. brevifolia*
    Rosette-leaves 2-7cm long, linear-oblanceolate . . . . *D. blochmanae*
  Flowers yellow, odorless
    Rosette-leaves oblanceolate, strongly narrowed below, 1-7cm long; petals connate for 0.5-1mm. . . . . *D. variegata*
    Rosette-leaves linear-lanceolate, not narrowed at base, 4-15cm long; petals connate for 1-2mm . . . . *D. multicaulis*

D. abramsii Rose. [*Echeveria a.* (Rose)Berger] Abram's Dudleya. Infrequent, in crevices of granitic rocks; foothill and montane, above 700m: Cuyamaca Mountains; Monument Peak; Jacumba; Descanso Junction; Campo. Apr.-June. Type from "wet crevices of rocks west of Jacumba". *n*=17.

D. alainae Reiser. Reiser's Live-forever. Endemic to San Diego County. Rare, on Woodland and Chaparral slopes; transmontane, 800-1200m: Banner; Chariot Canyon. Mar.-May. The type from Banner.

D. arizonica Rose. [*Echeveria a.* (Rose)Kearney & Peebles; *E. lagunensis* Munz, the type from "dry, stony slopes, Campbell Ranch, Vallecito Valley".] Desert Chalk-Lettuce. Infrequent, on dry rocky slopes; desert, 200-300m: Borrego Palm Canyon; Vallecito Stage Station; Shaw Canyon; Dubber; Jacumba. May-July. *n*=17.

D. attenuata (Wats.)Britt & Rose ssp. **orcuttii** (Rose)Moran. [*Echeveria a.* auct. non Berger; *Stylophyllum parishii* Britt. = *Echeveria palensis* Berger, the type supposedly from Pala.] Orcutt's Stylophyllum. Uncommon, sea bluffs about Boundary Monument No. 258, 10m. Records for Pala and Pala Canyon are doubtful. May-July. Type from the "Initial Mexican boundary monument". *n*=17.

D. blochmanae (Eastw.)Moran. [*Sedum b.* Eastw.] Blochman's Hasseanthus. Rare, atop coastal and sea bluffs; coastal, below 100m: Stuart Mesa, Camp Pendleton; Las Flores; La Costa; La Jolla; Pacific Beach. May-June. *n*=17.

D. brevifolia Moran. [*D. blochmaniae* ssp. *b.* (Moran)Moran] Del Mar Hasseanthus. Endemic to San Diego County. Local, on bare sandstone terraces; coastal, below 120m: Del Mar; Del Mar Heights; North City West; McGonigle Canyon; Mesa south of Sorrento Valley; Torrey Pines State Reserve. Apr.-June. Type from open area in chaparral near edge of mesa, Torrey Pines Park. *n*=17.

D. edulis (Nutt.)Moran. [*Echeveria e.* (Nutt.)Berger] Ladies-Fingers. Frequent, on dry slopes and rocky ledges; cismontane, below 1300m: e.g. Pala; Oceanside; Pine Hills; Viejas; southwest slope of Cuyamaca Peak; Otay Mountain; Imperial Beach. May-June. Type from "edges of rocks and ravines, St. Diego". *n*=17.

D. lanceolata (Nutt.)Britt. & Rose. [*Echeveria l.* Nutt.] Coastal Dudleya. Frequent, shaded slopes and open areas; coastal and foothill, below 500m: e.g. Camp Pendleton; San Marcos Villas; Del Mar; Point Loma. May-July. Type from "St. Diego". *n*=34.

D. multicaulis (Rose)Moran. Many-stemmed Hasseanthus. Local, known in the county only on the seaward slope of San Onofre Mountain, 60m. May-June. *n*=17.

D. pulverulenta (Nutt.)Britt. & Rose. [*Echeveria p.* Nutt.] Chalk-Lettuce. Frequent, on dry slopes; cismontane, below 1500m: e.g. Warner Hot Springs; Palomar Mountain; Camp Pendleton; Laguna Mountains; Cuyamaca Peak; Point Loma; Otay Mountain; National City; Bonita; Boundary Monument No. 258. May-July. Type from "St. Diego". *n*=17.

D. saxosa (Jones)Britt. & Rose ssp. **aloides** (Rose)Moran. [*Echeveria lanceolata* var. *a.* (Rose)Munz; *D. delicata* Rose, the type from Spencer Valley] Desert Savior. Uncommon, on rocky slopes; desert, 250-1500m: Banner; San Felipe; Yaqui Wells. Also Coleman Creek, Julian. Apr.-June. Type from San Felipe. *n*=17.

D. variegata (Wats.)Moran. [*Sedum v.* Wats.; *S. oblongorhizum* auct. non Berger] San Diego Hasseanthus. Local, on dry hillsides and mesas; foothill and coastal, below 300m: La Jolla Canyon; Santa Maria Mountain; Kearny Mesa; Dictionary Hill; San Miguel Mountain; El Cajon; La Mesa; Fletcher Hills; Paradise Hills; Rice Canyon; Dehesa; Otay Mountain; Imperial Beach; Otay Border Crossing. May-June. Type from San Diego. *n*=17.

D. viscida (Wats.)Moran. [*Echeveria v.* (Wats.)Berger] San Juan Stylophyllum. Uncommon, on bluffs and rocky cliffs; north coastal, below 100m: Oceanside; bluffs at mouth of Santa Margarita River; Escondido Creek at Olivenhain; San Marcos Creek, east of La Costa. May-June. *n*=17.

Sedum L.

S. spathulifolium Hook. ssp. **yosemitense** (Britt.)Clausen. [*S. s.* ssp. *anomalum* (Britt.)Clausen & Uhl] Yosemite Stonecrop. Rare, known in the county only near Eagles Crag, Agua Tibia Wilderness area. May-July. *n*=15.

CROSSOSOMATACEAE.

Crossosoma Nutt.

C. bigelovii Wats. Uncommon, in canyons and on shaded cllffs; desert, below 1000m: Indian Gorge; Borrego Palm Canyon. Feb.-Apr. *n*=12.

CUCURBITACEAE. Gourd Family

KEY TO GENERA

Fruit fleshy, with hard or firm rind, indehiscent, often dry when mature; flowers yellow . . . . *Cucurbita*
Fruit not fleshy, spinose, dry; flowers white to cream . . . . *Marah*

Cucurbita L. Gourd

Leaf-blades triangular-ovate, 1-2.5dm long; fruit 3-celled, on a slender peduncle without thickened ridges . . . . *C. foetidissima*
Leaf-blades palmately 5-lobed or -cleft, less than 1dm long; fruit seemingly 5-celled, with thickened ridges near the summit of the peduncle
  Leaves lobed to near the base, the lobes very narrow . . . . *C. digitata*
  Leaves lobed about half way, lobes somewhat deltoid . . . . *C. palmata*

***C. digitata** Gray. Finger-leaf Gourd. Native of eastern Sonoran Desert. Local, Buckman Spring, 1000m. Apr.-Sep. 2*n*=40.

C. foetidissima H.B.K. Calabazilla. Occasional, open areas; foothill and coastal, below 1000m: e.g. Fallbrook; Harrison Grade; Oak Grove; Lake Henshaw; Pine Hills; Vista; Witch Creek; The Willows; Viejas Valley; Jamacha Junction; National City. July-Oct. 2*n*=40.

C. palmata Wats. [*C. californica* Torr. ex Wats.] Coyote Gourd. Occasional, dry sandy places; coastal and desert, below 800m: northeast corner of county; Borrego Palm Canyon; Yaqui Well; Fish Creek; The Narrows; Sentenac Canyon; Sweeney Pass; Mountain Springs; Lakeside; Jamacha; Ramona; Potrero; Chollas Valley; Jamul. Apr.-Sep. Type from Cajon Valley. 2*n*=40.

Marah Kell. Wild-Cucumber

M. macrocarpus (Greene)Greene. [*Echinocystis m.* Greene; *Micrampelis m.* (Greene) Greene] Cucamonga Manroot, Chilicothe. Common, on open slopes; cismontane, below 1400m: e.g. San Onofre; Doane Valley; Vista; Hot Springs Mountain; Banner; Culp Valley; Anahuac School; El Cajon; Bankhead Springs; Morena Dam; Potrero; Campo; Dulzura; San Ysidro; San Diego; Pacific Beach. Jan.-Apr. *n*=16, 32.

DATISCACEAE.

Datisca L.

D. glomerata (Presl.)Baill. Durango Root. Occasional, along water courses; cismontane and desert: e.g. De Luz; San Onofre Canyon; Mesa Grande; Witch Creek; Borrego Palm Canyon; San Felipe; Grapevine Canyon; Eagles Nest; Little Laguna Lake; Pine Valley; Barrett; El Cajon Mountain; Alpine; Hauser Creek; Otay Mountain. May-July.

DIPSACACEAE. Teasel Family

KEY TO GENERA

Scales of the elongate receptacle pointed; plant thistlelike . . . . *Dipsacus*
Scales of the receptacle not prickly, but herbaceous,, inconspicuous and concealed among the flowers; plant not thistlelike . . . . *Scabiosa*

Dipsacus L. Teasel

***D. sativus** (L.)Honckeny. [*D. fullonum* L.] Common Teasel. Native of Europe. Occasional, naturalized in low, moist places; coastal, below 200m: Murphy Canyon; Alvarado Canyon; El Cajon; Sweetwater Reservoir; Paradise Valley; San Clemente Canyon. May-July. $2n$=18.

Scabiosa L.

***S. atropurpurea** L. Mourning-Bride. Native of Europe. Rare, escaping from cultivation near Spencer Valley, 1100m. Apr.-Sep.

ELATINACEAE. Waterwort Family

Elatine L. Waterwort

Capsules with 4 carpels; seeds very strongly curved, differently rounded at the two ends . . . . *E. californica*
Capsules with 2-3 carpels; seeds straight to slightly curved, almost alike at both ends . . . . *E. brachysperma*

E. brachysperma Gray. Yerba Fango. Occasional, vernal pools; foothill and coastal, below 500m: Ramona; San Marcos; Kearny Mesa; Fletcher Hills; Linda Vista. Mar.-May.

E. californica Gray. Yerba Estera. Infrequent, along lake margins and vernal pools: Cuyamaca Lake; Little Laguna Lake; Oakzanita; Kearny Mesa. Mar.-May.

ERICACEAE. Heath Family

KEY TO GENERA

Plants somewhat herbaceous, although sometimes with suffrutescent base
  Plants with slender subterranean rootstocks, mostly with green leaves; pollen in tetrads
    Flowers in racemes; valves of capsule with cobwebby margins . . . . *Pyrola*
    Flowers in corymbs; valves without threads on margin . . . . *Chimaphila*
  Plants fleshy-stemmed, without green leaves; pollen simple . . . . *Sarcodes*
Shrubs or trees
  Fruit a dry capsule . . . . *Rhododendron*
  Fruit a berry or drupe
    Ovary superior; fruit reddish
      Berry many-seeded, granular on surface; leaves 7-15cm long, tree . . . . *Arbutus*
      Berry drupe-like; leaves less than 6cm long; shrubs
        Fruit granular on surface or warty . . . . *Comarostaphylis*
        Fruit smooth, not granular or warty
          Leaves revolute, not vertical
            Leaves mostly alternate . . . . *Xylococcus*
            Leaves opposite . . . . *Ornithostaphylos*
          Leaves plane, vertical; filaments dilated at base; fruit usually with separable nutlets, rarely with a solid stone . . . . *Arctostaphylos*
    Ovary inferior; fruit a blue berry . . . . *Vaccinium*

Arbutus L.

A. menziesii Pursh. Pacific Madrone. Scarce, on Chaparral slopes; foothill, 450-1250m: Castro Canyon; Roderick Mountain. Mar.-May. $2n$=26.

Arctostaphylos Adans. Manzanita

Branches prostrate or procumbent, usually less than 6cm high and rooting in contact with the ground . . . . . *A. patula*
Branches erect or ascending, taller, not usually rooting
  Bracts of inflorescence shorter than pedicels, deltoid to almost subulate, even the lower bracts not or scarcely foliaceous
    Pedicels glabrous; basal burl lacking and the plants not stump-sprouting
      Fruit depressed-globose, the nutlets separable . . . . *A. pungens*
      Fruit somewhat ovoid, the nutlets forming a solid stone . . . . *A. patula*
    Pedicels glandular-pubescent
      Basal burl lacking; shrubs not stump-sprouting
        Bracts of inflorescence 5-6mm long, lanceolate, deciduous; pedicels 10-15mm long; sweet-scented shrub from cismontane interior . . . . *A. pringlei*
        Bracts of inflorescence 1-3mm long, deltoid, persistent
          Branchlets glabrous and glaucous . . . . *A. glauca* var. *g.*
          Branchlets pubescent to glandular-puberulent . . . . *A. glauca* var. *puberula*
      Basal burl present; crown-sprouting shrubs
        Branchlets glandular-hairy . . . . *A. glandulosa* ssp. *g.*
        Branchlets not glandular-hairy
          Leaves dark green, plants somewhat prostrate . . . . *A. glandulosa* ssp. *crassifolia*
          Leaves light or yellow-green to pale . . . . *A. glandulosa* ssp. *adamsii*
  Bracts of inflorescence (at least the lower) foliaceous and equaling or exceeding pedicels
    Basal burl present; crown-sprouting shrubs
      Leaves oblanceolate . . . . *A. glandulosa* et sspp.
      Leaves round-ovate . . . . *A. peninsularis*
    Basal burl absent; shrubs not crown-sprouting
      Branchlets not glandular
        Branchlets glabrous to pubescent but not bristly
          Fruit very viscid, globose, 12-15mm in diameter, with a solid stone; corolla 8-9mm long *A. glauca* et vars.
          Fruit with well developed pulp, often depressed-globose, 6-10mm in diameter, the nutlets generally separable; corolla 6-8mm long . . . . *A. glandulosa* et sspp.
        Branchlets bristly with long hairs as well as variously pubescent . . . . *A. glandulosa* et sspp.
      Branchlets with some hairs gland-tipped; stomates on both surfaces of leaves . . . . *A. otayensis*

A. glandulosa Eastw. ssp. **adamsii** (Munz)Munz. Laguna Manzanita. Infrequent, in Chaparral and Montane Coniferous Forest; foothill and montane, above 250m: Jamul Valley; Laguna Mountains. Jan.-Apr. Type from dry granitic slopes and benches, north end of Laguna Mountains on highway to Julian, at 5000 feet.

A. glandulosa ssp. **crassifolia** (Jeps.)Wells. Del Mar Manzanita. Infrequent, in Chaparral; coastal, below 200m: Encinitas; Rancho Santa Fe; Del Mar; Torrey Pines State Reserve. Dec.-Apr. Type from Del Mar.

A. glandulosa ssp. **glandulosa**. [*A. tomentosa* (Pursh)Lindl.; *A. g.* var. *australis* Adams, the type from the west slopes of Otay Mountain.] Eastwood Manzanita. Frequent, in Chaparral and Montane Coniferous Forest; foothill and montane, above 450m: e.g. Palomar Peak; Doane Valley; Warner Hot Springs; Eagles Nest; Julian; Pine Hills; Cuyamaca Peak; Mount Laguna; Laguna Junction; Hipass; Guatay; Mount Woodson; Alpine; El Cajon Mountain; Lyons Peak; Tecate; Otay and San Miguel Mountains. Dec.-Apr. n=26.

A. glauca Lindl. var. **glauca**. [*A. g.* var. *eremicola* Jeps.] Bigberry Manzanita. Frequent, Chaparral; foothill and montane, above 150m: e.g. De Luz; Fallbrook; Palomar Mountain; Oak Grove; Eagles Nest; Montezuma View Point; Oriflamme Canyon; Julian; Boulder Creek; Mussey Grade; Hidden Glen; Japatul Valley; Barrett Dam; Campo; Jacumba; Dubber. Dec.-Mar. n=13.

A. glauca var. **puberula** J.T. Howell. Occasional, in Chaparral; foothill, 800-1500m: The Willows; Alpine; Viejas Grade; Wood Valley; Descanso. Dec.-Mar.

A. otayensis Wies. & Schreib. Otay Manzanita. Infrequent, in Chaparral; foothill, 1100m: Otay, San Miguel, Jamul, and Guatay Mountains. Jan.-Mar. Type from Otay Mountain. n=13.

A. patula Greene ssp. **platyphylla** (Gray)Wells. [*A. parryana* var. *pinetorum* (Roll.)Wies. & Schreib.] Pine Manzanita. Infrequent, Chaparral; foothill and montane, above 600m: Palomar Mountain; Julian; La Posta Creek; Cottonwood Valley; Potrero. May-June.

A. peninsularis Wells. Peninsular Manzanita. Infrequent, in Chaparral; foothill, 200-900m: Santa Margarita Mountains; De Luz; Mount Olympus; Pala Creek; Agua Tibia Mountains. Dec.-Apr.

A. pringlei Parry ssp. **drupacea** (Parry)Wells. Cuyamaca Manzanita. Occasional, Chaparral and Montane Coniferous Forest; montane, above 1300m: Palomar Peak; Hot Springs Mountain; Garrnet Peak; Cuyamaca and Garnet Peak, Laguna Mountains. Feb.-Apr. Type from mountains east of San Diego.

A. pungens H.B.K. Mexican Manzanita. Common, Chaparral and Southern Oak Woodland; foothill and montane, above 600m: e.g. Julian; Cuyamaca Mountains; Buckman Spring; Live Oak Springs; Jacumba; Boulevard; Campo; Guatay; Boulder Creek; Santa Ysabel; Japatul Valley; Viejas Valley. Also Rancho Santa Fe. Feb.-May. $2n$=26.

Chimaphila Pursh   Pipsissewa

C. menziesii (R. Br. ex D. Don)Spreng. Scarce, in shaded areas of Montane Coniferous Forest; montane, above 1350m: Cedar Grove; Doane Valley; Palomar Mountain. June-July.

Comarostaphylis Zucc.

C. diversifolia (Parry)Greene ssp. **diversifolia**. Summer-Holly. Local, in Chaparral; foothill and coastal, below 700m: San Marcos Mountains; Mount Whitney; Rancho Santa Fe; Encinitas; Mount Soledad; Penasquitos Canyon; Del Mar Heights; Iron Mountain; canyons of Mission Valley; Jamul Valley; Otay Mountain. May-June. Type from Jamul Valley.

Ornithostaphylos Small

O. oppositifolia (Parry)Small. Palo Blanca. Rare, known in the county only at Boundary Monument No. 256 in the Tijuana Hills, 125m. Jan.-Apr.

Pyrola L.   Wintergreen

P. picta Sm. [*P. aphylla* Sm.] Frequent, shaded areas, Montane Coniferous Forest; montane, above 1350m: Doane Valley; Hot Springs Mountain; North and Middle Cuyamaca Peaks; Boucher Hill; Cedar Grove; Chimney Creek. May-Aug.

Rhododendron L.

R. occidentale (T. & G.)Gray. Western Azalea. Frequent, shaded slopes and along streams, Montane Coniferous Forest and Southern Oak Woodland; foothill and montane, above 500m: Santa Margarita Mountains; Palomar Mountain; Eagles Nest; Cuyamaca Mountains; El Cajon Mountain. May-July.

Sarcodes Torr.

S. sanguinea Torr. Rare, reported from shaded slopes; Montane Coniferous Forest; montane, above 1350m: Cuyamaca Peak.

Vaccinium L.   Huckleberry

V. ovatum Pursh. California Huckleberry. Scarce, in Chaparral; foothill, 600-800m: Bottle Peak near Lake Wohlford; Big Boulder Creek; El Cajon Mountain. Mar.-May. $2n$=24.

Xylococcus Nutt.

X. bicolor Nutt. [*Arctostaphylos clevelandii* Gray, the type from Potrero.] Mission Manzanita. Common, Chaparral; coastal and foothill, below 800m: e.g. De Luz; Rainbow Valley; Moosa Canyon; Encinitas;

Mussey Grade; Cowles Mountain; Lee Valley; Jamul Valley; Otay Mountain; Potrero; Lyons Peak; Point Loma. Dec.-Feb. The type was said to be from "Monterey", but must be from "St. Diego". $2n=26$.

EUPHORBIACEAE. Spurge Family

KEY TO GENERA

Flowers with a calyx, not in the base of an involucre (cyathium)
  Plants definitely woody
    Leaves palmately lobed, 1-5dm long . . . . . *Ricinus*
    Leaves not palmately lobed, less than 0.5dm long
      Flowers in umbellate axillary clusters; leaves usually entire. . . . . *Tetracoccus*
      Flowers in racemes or spikes; leaves dentate
        Leaves 6-10mm long, gray at least beneath, 1-veined; deserts . . . . . *Bernardia*
        Leaves 10-30mm long, often 3-5-veined from base; cismontane . . . . . *Acalypha*
  Plants herbaceous, at most slightly woody at base
    Plants armed with stiff stinging hairs . . . . . *Eremocarpus*
    Plants not armed with stinging hairs
      Herbage densely stellate-scurfy . . . . . *Croton*
      Herbage not stellate-pubescent
        Petals 5, straw-colored, most evident in staminate flowers; styles 2-cleft . . . . . *Argythamnia*
        Petals lacking; styles entire . . . . . *Stillingia*
Flowers lacking a calyx, included in a cup-shaped and calyxlike involucre (cyathium) which surrounds several staminate flowers and 1 pistillate flower with a 3-lobed pistil
  Leaves opposite, with inequilateral bases, stipules present . . . . . *Chamaesyce*
  Leaves alternate, with symmetrical bases, stipules lacking. . . . . *Euphorbia*

Acalypha L.

A. californica Benth. California Copperleaf. Local, Chaparral and rocky slopes; coastal, foothill, and desert, below 1300m: Ramona; Mussey Grade; Poway; Crest; Harbison Canyon; Jamul; Monument School; Canebrake Canyon; Indian Gorge; Mountain Palm Springs; Bow Willow; Rockhouse Canyon; Mountain Springs. Jan.-June. $2n=10$ pairs.

Argythamnia P. Br.

Pistillate sepals white-margined; plants herbaceous. . . . . *A. serrata*
Pistillate sepals not white-margined; leaves linear-lanceolate, entire; plants woody at base . . . . . *A. lanceolata*

A. lanceolata (Benth.)Muell.-Arg. [*Ditaxis l.* (Benth.)Pax & Hoffm.] Desert Silverbush. Occasional, on rocky slopes and in canyons; desert, below 700m: Borrego Valley; Yaqui Pass; Sentenac Canyon; upper Blair Valley; Vallecito Stage Station; Canebrake Canyon; Mountain Palm Springs; Bow Willow. Dec.-May.

A. serrata (Torr.)Muell.-Arg. [*Ditaxis s.* (Torr.)Heller] Yuma Silverbush. Occasional, slopes and sandy flats; desert, below 500m: Borrego Springs; Benson and Clark Dry Lakes; Fish Creek; The Narrows. Apr.-Nov. $2n=$ca. 26.

Bernardia Houst. ex P. Br.

B. incana Mort. Western Bernardia. Occasional, slopes and valley floors; desert, below 1600m: Borrego Palm Canyon; Montezuma and San Felipe Valleys; Grapevine Canyon; Blair Valley; Box Canyon; Pinyon Mountains; Whale Peak; Agua Caliente Hot Springs. Oct.-Nov.

Chamaesyce S.F. Gray

Ovary and capsule hairy
  Perennial plants
    Involucres or cyathia urceolate; staminate flowers up to 12; appendages entire or crenate. . . . . *C. arizonica*
    Involucres or cyathia not urceolate; staminate flowers 15-60
      Seeds scarcely angled, narrow-ovoid, encircled by 4-5 rounded ridges . . . . . *C. pediculifera*
      Seeds quadrangular, smooth to slightly wrinkled
        Herbage with short straight spreading hairs . . . . . *C. polycarpa* var. *hirtella*
        Herbage with appressed long and weak or matted hairs . . . . . *C. melanadenia*
  Annual plants
    Involucres urceolate; appendages deeply 3-4 parted. . . . . *C. setiloba*
    Involucres obconic to campanulate
      Glands without appendanges or with mere rudiments; seeds smooth; leaves not more than 8mm long . . . . . *C. micromera*
      Glands appendaged; seeds granular or ridged
        Capsules 1.5-1.9mm long; seeds 1-1.4mm long, not transversely ridged; appendages narrower than glands . . . . . *C. serpyllifolia* var. *hirtula*
        Capsules 1-1.4mm long; seeds 0.9-1mm long, transversely ridged . . . . . *C. maculata*
Ovary and capsules glabrous
  Stipules united into a broad white membranous scale . . . . . *C. albomarginata*
  Stipules not so united
    Plants perennial
      Glands discoid, without appendages; capsules 1.7mm long; leaves 2-4mm long . . . . . *C. parishii*
      Glands appendaged; leaves mostly 3-10mm long. . . . . *C. polycarpa* var. *p.*
    Plants annual
      Margins of leaves entire or nearly so. . . . . *C. micromera*
      Margins of leaves toothed, at least toward apex
        Stems mostly erect; leaves oblong to oblong-lanceolate, 8-35mm long seeds with finely rippled surface . . . . . *C. nutans*
        Stems prostrate or nearly so . . . . . *C. serphyllifolia* var. *s.*

C. albomarginata (T. & G.)Small. [*Euphorbia a.* T. & G.] Rattlesnake Spurge. Occasional, in open areas; desert and foothill, 300-800m: Oak Grove; Warner Springs; San Felipe Valley; Mason Valley; Box Canyon; La Puerta Valley; Rattlesnake Valley; Mountain Springs; Jacumba; Cottonwood Valley. Apr.-Nov. n=18, 2n=24 pairs.

C. arizonica (Engelm.)Arthur. [*Euphorbia a.* Engelm.] Arizona Spurge. Rare, known in the county only in Borrego Palm Canyon, 50m. Mar.-Apr. 2n=13 pairs.

***C. maculata** (L.)Small. [*C. supina* auct. non (Raf.)Moldenke; *Euphorbia s.* Raf.] Spotted Spurge. Native of eastern United States. Occasional, weed in disturbed areas; below 1250m: Oceanside; San Luis Rey; Mission Gorge; Balboa Park; Pine Valley; Sentenac Canyon. May-Oct.

C. melanadenia (Torr.)Millsp. [*Euphorbia m.* Torr.; *E. cinerascens* var. *appendiculata* Engelm., the type from San Felipe.] Squaw Spurge. Frequent, on rocky slopes; desert, below 1000m: e.g. Montezuma and Culp Valleys; North Pinyon Mountain; Oriflamme Mountains; upper Blair Valley; Sentenac Canyon; Mason Valley; Mountain Springs; Dubber; Jacumba. Dec.-May. n=8.

C. micromera (Boiss.)Woot. & Standl. [*Euphorbia m.* Boiss. ex Englem.] Sonora Sand Mat. Scarce, in sandy places; foothill, 700-1100m: Oak Grove; Banner; Rockhouse Canyon; Alta Bisnaga Wash; Agua Caliente Springs; Mountain Palm Springs; North Egg Mountain; Mountain Springs; InKoPah. Sep.-Dec.; Apr.-June. 2n=24.

***C. nutans** (Lag.)Small. [*C. maculata* auct. non (L.)Small; *Euphorbia m.* L.] Large Spurge. Native of eastern United States. Infrequent, weed in disturbed places; coastal, below 300m: Bonsall; San Diego. Also Mount Woodson. Apr.-Oct. 2n=28.

C. parishii (Greene)Millsp. in Parish. [*Euphorbia p.* Greene; *E. patellifera* J. T. Howell, the type from near Palm Wash.] Parish's Spurge. Rare, known in the county only near Palm Wash on the eastern flank of the Santa Rosa Mountains, 15m. Apr.-Oct.

C. pediculifera (Engelm.)Rose & Standl. [*Euphorbia p.* Engelm.] Louse Spurge. Scarce, in sandy places; desert, below 500m: Butler Canyon; Vallecito Stage Station. Jan.-Apr. 2n=12 pairs + 1 univalent; 13 pairs + 1 univalent.

C. polycarpa (Benth.)Millsp. in Parish var. **hirtella** (Boiss.)Millsp. [*Euphorbia p.* Benth. var. *hirtella* Boiss.] Desert Sand Mat. Frequent, on sandy plains and washes; desert, below 700m: e.g. Borrego Palm Canyon; Fish Creek; The Narrows; Benson Dry Lake; Canebrake Canyon; Mountain Palm Springs; Bow Willow; Carrizo Stage Station; Dos Cabezas; Carrizo Gorge. Jan.-Dec. 2n=13 pairs.

C. polycarpa var. **polycarpa**. [*Euphorbia p.* Benth.] Small-seed Sandmat. Frequent, in dry areas; cismontane and desert, below 1000m: e.g. De Luz; Pala; Lake Hodges; Rolando; Silver Strand; Boundary Monument; Lakeside; Ramona; Borrego Valley; Vallecito Valley; Jacumba. Jan.-Dec. n=13.

C. serpyllifolia (Pers.)Small var. **hirtula** (Engelm. ex Wats.)Koutnik. [*Euphorbia h.* Engelm. ex Wats.] Pine Spurge. Frequent, in Montane Coniferous Forest; montane, above 900m: upper Doane Valley; Dehr Creek; Julian; Cuyamaca Peak; Laguna Mountains; Pine Valley. Also Jamacha; Mountain Springs; Mount Woodson. June-Sep. Type from "Talley's Ranch in the Cuyamaca Mountains".

C. serphyllifolia var. **serphyllifolia**. [*Euphorbia s.* Pers.] Thyme-leaf Spurge. Occasional, in open areas, Chaparral; foothill, 300-1400m: near Lake Henshaw; Japatul Valley; Glen Cliff Campground; Clover Flats; Jacumba. Aug.-Oct. 2n=22.

C. setiloba (Engelm. ex Torr.)Millsp. in Parish. [*Euphorbia s.* Engelm.] Yuma Spurge. Common, sandy places, after abundant summer rains; desert, below 1000m: Borrego Valley; Rockhouse Canyon; Yaqui Well; Agua Caliente Hot Spring; Mountain Palm Springs; Mountain Springs; Little Blair Valley; Vallecitos; North Egg Mountain; InKoPah. Jan.-Dec. 2n=8 pairs.

Croton L.

Leaf blades mostly 2-4.5cm long; petioles 1-3.5cm long
  Leaves mostly 8-20mm wide; petioles 1-3.5cm long . . . . *C. californicus* var. *c.*
  Leaves mostly 5-10mm wide; petioles 0.5-1cm long . . . . *C. californicus* var. *tenuis*
Leaf blades mostly 0.5-2cm long; perioles 0.3-0.7cm long . . . . *C. californicus* var. *mohavensis*

C. californicus Muell.-Arg. var. **californicus**. Frequent, in sandy areas; cismontane and desert, below 1550m: e.g. Nate Harrison Grade; Oak Grove; San Mateo Creek; Encinitas; San Luis Rey; Mission Valley; Ocean Beach; Imperial Beach; Warner Hot Springs; San Felipe Valley; Vallecito and Carrizo Stage Stations. Mar.-Oct. 2n=14.

C. californicus var. **mohavensis** Ferg. Frequent, in sandy areas; desert, below 1000m: e.g. Coyote Creek; San Felipe Valley; The Narrows; Carrizo Stage Station. Mar.-June.

C. californicus var. **tenuis** (Wats.)Ferg. Occasional, in sandy areas; cismontane and desert, below 1000m: Borrego Valley; Potrero; Tijuana River Valley. Mar.-June. Type from Potrero.

Eremocarpus Benth.

E. setigerus (Hook.)Benth. Doveweed. Common, in dry open or disturbed areas; cismontane, below 1550m: e.g. De Luz; Oak Grove; Palomar Mountain; O'Neill Lake; Valley Center; Lake Hodges; Warner Hot Springs; Cuyamaca Lake; Barona Valley; La Jolla; San Ysidro; Potrero. May-Oct. 2n=20.

Euphorbia L.

Glands of the cyathium without petaloid appendages; leaves essentially symmetrical
Leaves linear; cyathia in terminal headlike clusters, their glands cup-shaped, concealed by the inflexed segments of the involucre
Capsules longer than wide, only slightly if at all lobed . . . . *E. eriantha*
Capsules wider than long, distincly 3-lobed . . . . *E. heterophylla*
Leaves (at least in inflorescence) ovate to rounded; cyathia solitary or in cymes, their glands not cup-shaped
Glands elliptic or transversely oval; leaves serrulate
Floral leaves broad at base; capsules with fleshy ellipsoid-lenticular tubercles . . . . *E. spathulata*
Floral leaves narrowed at base; capsules smooth; seeds subglobose-ovoid . . . . *E. helioscopia*
Glands crescent-shaped or 2-horned; leaves largely entire
Stem-leaves opposite, strongly decussate; capsule 8-12mm high . . . . *E. lathyris*
Stem-leaves alternate; capsule 2-4mm high
Plants annual
Capsule with thin keels; stem leaves petioled; seeds with deep pits . . . . *E. peplus*
Capsule smooth; stem leaves subsessile; seeds vermiculate-ridged to almost smooth . . . . *E. crenulata*
Plants perennial
Woody perennial in montane areas . . . . *E. palmeri*
Leafy stem-succulent in disturbed areas . . . . *E. myrsinites*
Glands of the cyathium with petaloid appendages; leaves alternate or opposite, their bases quite symmetrical; stipules glandlike or none . . . . *E. misera*

E. crenulata Engelm. Chinese Caps. Scarce, in dry places; foothill, 300-1000m: Mesa Grande; Agua Tibia Mountains. Mar.-Aug.

E. eriantha Benth. Desert Poinsettia. Infrequent, on rocky slopes and mesas; desert, below 700m: Borrego Sink; Yaqui Pass; The Narrows; Vallecito Stage Station; Agua Caliente Springs; northeast corner of county. Dec.-Apr. $2n$=14 pairs.

***E. helioscopia** L. Wartweed. Native of Europe. Rare, adventive; coastal, below 200m: north base of Otay Mountain. Apr. $2n$=42.

***E. heterophylla** L. Summer Poinsettia. Native of Neotropics. Uncommon, escape form cultivation; coastal, below 150m: San Diego; San Ysidro. Jan.-Dec. $2n$=14, 28.

***E. lathyris** L. Caper Spurge. Native of Europe. Occasional, waif in disturbed areas and along roadsides: San Felipe Valley; San Onofre. Jan.-Dec. $2n$=20

E. misera Benth. Cliff Spurge. Occasional, on bluffs; coastal, below 150m: Carlsbad; Point Loma; San Diego; Sweetwater Valley; Otay Mesa; San Ysidro; Tijuana Hills. Jan.-Aug. Type from San Diego.

***E. myrsinites** L. Gopher Spurge. Native of Europe. Uncommon, escape in coastal area: Carlsbad.

E. palmeri Engelm. Wood Spurge. Infrequent, in Montane Coniferous Forest; montane, above 1200m: Palomar Mountain; Hot Springs Mountain; Pine Hills; Spencer Valley; Cuyamaca Mountains; Laguna Mountains; Glen Cliff Campground. Apr.-June. Type from "Talley's Ranch in the Cuyamaca Mountains".

***E. peplus** L. Petty Spurge. Native of Europe. Common, in moist, disturbed areas; coastal and foothill, below 400m: e.g. Poway; Point Loma; Flynn Spring; Mahogany Canyon; Sweetwater. Feb.-Aug. $2n$=16.

E. spathulata Lam. [*E. dictyosperma* F. & M.; *Tithymalus s.* (Lam.)W.A. Weber] Reticulate-seed Spurge. Occasional, in dry, open or disturbed areas; coastal and foohill, below 700m: Rainbow Valley; De Luz; Barona Valley; Bernardo; Kearny Mesa; La Jolla; Harbison Canyon; Woodwardia Canyon; San Ysidro. Mar.-June. $2n$=12

Ricinus L.

***R. communis** L. Castor-Bean. Native of the Old World. Common, naturalized in river valleys and disturbed places; coastal, below 300m: e.g. San Onofre; Fallbrook; San Pasqual; Solana Beach; Mission Valley; Logan Heights; Tijuana River Valley. Jan.-Dec. $2n$=20.

Stillingia Garden ex L.

Leaves ovate, sharply serrate, 3-nerved; mostly annual . . . . *S. spinulosa*
Leaves linear, entire or nearly so, 1-nerved; perennial . . . . *S. linearifolia*

S. linearifolia Wats. Linear-leaf Stillingia. Occasional, in dry areas; desert, foothill, and coastal, below 1000m: e.g. Coyote Creek; Borrego Palm Canyon; Sentenac Canyon; San Felipe Valley; upper Blair Valley; Mason Valley; Vallecito Stage Station; Oak Grove; Cottonwood Creek; Barrett; Dehesa; Telegraph Canyon; Del Mar. Mar.-May. Type from "near Boundary Monument, San Diego". $2n$=22.

S. spinulosa Torr. in Emory. Annual Stillingia. Infrequent, in sandy areas; desert, below 1000m: Borrego Springs; The Narrows; near Carrizo Stage Station; southeast of Jacumba; Ocotillo Wells. Mar.-May.

Tetracoccus Engelm.

T. dioicus Parry. San Diego Button Bush. Local, in Chaparral; coastal and foothill, below 1000m: De Luz; Red Mountain Grade; Rainbow Valley; Vista; Agua Tibia Mountains; Barona Valley; Lee Valley; Dehesa; Tecate Junction; Jacumba. Apr.-May.

FABACEAE Pea Family
(Leguminosae)
KEY TO GENERA

Corolla regular or nearly so, valvate in bud; leaves bipinnate, the leaflets mostly small; flowers small, in many-flowered heads or spikes; petals inconspicuous; stamens conspicuous
  Stamens united below into a tube; flowers rose to reddish-purple . . . . . *Calliandra*
  Stamens distinct or nearly so; flowers yellow or whitish
    Petiole with a conspicuous gland . . . . . *Albizia*
    Petiole lacking such a gland
      Stamens more than 10; anthers not gland-tipped . . . . . *Acacia*
      Stamens 10; anthers gland-tipped . . . . . *Prosopis*
Corolla mostly irregular, imbricate in the bud, the petals unlike in size or shape or both; flowers often larger and petals conspicuous; stamens mostly included
  Upper petal internal in bud, enveloped by the lateral ones, therefore the corolla not papilionaceous
    Leaves simple; flowers rose-purple . . . . . *Cercis*
    Leaves compound; flowers yellowish
      Leaves once-pinnate; low subshrubs to herbs . . . . . *Senna*
      Leaves twice-pinnate
        Plants low subshrubs or herbs without spines or thorns
          Herb; leaves with several pairs of pinnae; flowers orange . . . . . *Hoffmannseggia*
          Shrub; leaves with 1 pair of lateral pinnae and the terminal pinna; flowers yellow . . . . . *Caesalpinia*
        Plants large shrubs or trees with spines or thorns
          Rachis of the pinnae flattened, more than 8cm long; leaflets alternate; flowers in racemes up to 20cm long . . . . . *Parkinsonia*
          Rachis of the pinnae terete, not more than 4cm long; leaflets opposite; flowers in short racemes or corymbs . . . . . *Cercidium*
  Upper petal external in bud, the corolla papilionaceous
    Stamens distinct or nearly so
      Flowers yellow; plant an herb . . . . . *Thermopsis*
      Flowers purplish; plant a low shrub . . . . . *Pickeringia*
    Stamens all united or 9 or 5 of them united
      Leaves simple; shrubs and mostly spiny
        Corolla yellow, 15-25mm long
          Calyx deeply 2-lipped; branches spinescent; leaves acicular . . . . . *Ulex*
          Calyx cleft above, hence 1-lipped; branches not spinescent; leaves oblong . . . . . *Spartium*
        Corolla blue to red-purple, 8-10mm long . . . . . *Psorothamnus*
      Leaves 3-many-foliolate
        Leaves palmate or pinnately 1-3-foliolate
          Anthers strongly differentiated, some small and versatile, others larger and basifixed; stamens monadelphous and 10 in number; pods dehiscent
            Leaflets 4-many; herbs or shrubs . . . . . *Lupinus*
            Leaflets 3 or fewer
              Flowers solitary or racemose, terminal . . . . . *Cytisus*
              Flowers solitary or umbellate, axillary . . . . . *Lotus*
          Anthers not much differentiated, or if they are, the plants with prickly pods; stamens 9 and 1, or 5 and monadelphous; pods largely indehiscent
            Plants gland-dotted; leaflets 3-5 . . . . . *Psoralea*
            Plants not gland-dotted; leaflets mostly 3
              Leaves palmately 3-foliate; flowers in ovoid to oblong heads; corolla persistent after flowering . . . . . *Trifolium*
              Leaves pinnately 3-foliate; flowers in spikes or racemes; corolla deciduous
                Pods curved or spirally coiled; style subulate . . . . . *Medicago*
                Pods ovoid, straight; style filiform . . . . . *Melilotus*
        Leaves pinnate, the leaflets more than 3, or sometimes partly replaced by tendrils
          Rachis of leaf prolonged into a tendril or at least a short seta
            Wings of corolla essentially free from keel; abaxial surface of style pubescent . . . . . *Lathyrus*
            Wings of corolla adherent to keel half or more its length; abaxial surface of style pubescent or hairs encircling style or style glabrous . . . . . *Vicia*
          Rachis of leaf not prolonged into a tendril
            Plants gland-dotted; pods indehiscent, 1-2-seeded
              Petal 1; leaflets numerous . . . . . *Amorpha*
              Petals 5
                Pod prickly; flowers yellow-white; a perennial herb . . . . . *Glycyrrhiza*
                Pod not prickly; flowers purplish to rose or blue
                  Flowers in a dense spike; calyx-lobes longer than the tube . . . . . *Dalea*
                  Flowers in loose spikelike racemes; calyx-lobes not longer than the tube . . . . . *Marina*
            Plants not gland-dotted; pods mostly dehiscent
              Leaves ternately-compound . . . . . *Dolichos*
              Leaves pinnately-compound
                Shrubs of spine-tipped branched . . . . . *Caragana*
                Unarmed herbs and subshrubs
                  Flowers in umbels or solitary; leaflets mostly 3-7; flowers mostly yellow or whitish . . . . . *Lotus*
                  Flowers in racemes
                    Plants arboreous with odd-pinnate leaves and usually spiny
                      Flowers whitish, 15-20mm long; pods linear . . . . . *Robinia*
                      Flowers rose-purple, 10mm long; pods thick, torulose . . . . . *Olneya*
                    Plants herbaceous
                      Leaves even-pinnate; annual; leaflets linear-oblong; corolla yellowish, ca. 15mm long; pods 10-20cm long . . . . . *Sesbania*
                      Leaves odd-pinnate . . . . . *Astragalus*

Acacia Mill.
(Mimosoideae)

Leaves (phyllodia) simple
  Phyllodes mostly 2-5-nerved
    Flowers in heads, cream-color; phyllodes somewhat sickle-shaped . . . . *A. melanoxylon*
    Flowers in spikes, yellow; phyllodes curved alike on both edges . . . . *A. longifolia*
  Phyllodes with veination of 1-2 veins or veination inevident or uncertain
    Phyllodia verticillate, linear to subulate, rigid and spinulose-tipped . . . . *A. verticillata*
    Phyllodia alternate
      Phyllodia 1-6cm long; plants armed with spines . . . . *A. armata*
      Phyllodia 8-20cm long; plants unarmed
        Phyllodia distinctly falcate or asymmetrical, 0.8-2.5cm wide; peduncles shorter than the flower heads . . . . *A. pycnantha*
        Phyllodia straight or slightly falcate, 0.3-5cm wide; peduncles longer than flower heads . . . . . . *A. retinodes*

Leaves bipinnate
  Flower heads in axillary racemes or congested at stem tips in termal compound racemes; plants unarmed
    Pinnae 2-4 pairs . . . . *A. baileyana*
    Pinnae 7-20 pairs. . . . . *A.dealbata*
  Flower heads or spikes, some or all, solitary or fascicled from leaf axils or spurs; plants mostly armed
    Legume flat; desert shrub . . . . *A. greggii*
    Legume terete; coastal shrub . . . . *A. minuta*

***A. armata** R. Br. Kangaroo Thorn. Native of Australia. Uncommon, naturalized in waste places; coastal, below 150m: Balboa Park; Telegraph Canyon. Nov.-Feb.

***A. baileyana** F. von Muell. Native of Australia. Occasional, persisting around abandoned dwellings; coastal, below 150m: San Diego; National City. Jan.-Mar. $2n=26$.

***A dealbata** Link. [*A. decurrens* var. *d.* (L.)F. von Muell.] Silver Wattle. Native of Australia. Occasional, escaping from cultivation; coastal, below 150m: Oceanside; San Diego. Jan.-Mar. $2n=26$.

A. greggii Gray var. **arizonica** Isely. Catclaw. Common, in canyons and washes; desert, below 1100m: e.g. Borrego Palm Canyon; Grapevine Canyon; Warner Hot Springs; San Felipe Valley; Sentenac Canyon; Yaqui Well; La Puerta Valley; mouth of Shaw Canyon; Vallecito Canyon; Mountain Springs; Dubber; Jacumba; McCain Ranch. Apr.-June.

***A. longifolia** (Andrews)Willd. [*A. latifolia* Hort.] Golden Wattle. Native of Australia. Uncommon, escaping from cultivation; coastal, below 150m: Encinitas; San Diego. Jan.-Mar. $2n=26$.

***A. melanoxylon** R. Br. Blackwood Acacia. Native of Australia. Infrequent, escaping from cultivation; coastal, below 150m: Point Loma; Balboa Park. Nov.-Mar. $2n=26$.

A. minuta (Jones)Beauchamp. [*A. smallii* Isely; *A. farnesiana* auct. non (L.)Willd.] Local, in Coastal Sage Scrub; coastal, below 50m: San Dieguito Valley; Chollas Valley; Otay. Jan.-May.

***A. pycnantha** Benth. Native of Australia. Uncommon, persisting from cultivation; coastal, below 200m: Proctor Valley. Dec.-Mar.

***A. retinodes** Schlecht. Native of Australia. Infrequent, escaping from highway plantings; coastal, below 150m: Carlsbad; San Diego. Jan.-Dec. $2n=26$.

***A. verticillata** (L'Her.)Willd. Native of Australia. Infrequent, escaping from cultivation; coastal, below 150m: San Diego. Sep.-Apr. $2n=26$.

Albizia Durazz.
(Mimosoideae)

***A. lophantha** (Willd.)Benth. Native of Australia. Local, escaping from cultivation; coastal, below 150m: Balboa Park; Point Loma. Feb.-Apr. $2n=26$.

Amorpha L. False Indigo
(Faboideae)

A. fruticosa L. var. **occidentalis** (Abrams)Kearn. & Peeb. Occasional, along streams and alkaline areas; cismontane, below 1400m: San Mateo Creek; De Luz; Pala; Lake Hodges; San Pasqual; Ramona; Sorrento Valley; Penasquitos Canyon; Mission Gorge; National City; Jamacha; Barrett; Jacumba; Cedar Creek Falls; Santa Ysabel; Cuyamaca Lake; Sloane Canyon. Also San Felipe Valley; Indian Canyon; Borrego Palm Canyon; Oriflamme Canyon. May-July. Type from the San Diego River, near the old San Diego Mission.

Astragalus L. Milkvetch; Locoweed
(Faboideae)

Pod strongly inflated
  Pod 1-celled
    Pod with true stipe or with an elongate gynophore, though these may not be longer than the calyx *A. trichopodus*
    Pod sessile
      Plants annual
        Pods thick-walled, hard, 2-3cm long; corolla 21-28mm long . . . . *A. crotalariae*
        Pods thin-walled, membranous, or if firm, less than 2cm long; flowers smaller
          Flowers purple to pink-purple, drying violet; pods 1cm or more thick . . . . *A. insularis*
          Flowers whitish
            Herbage yellow-hirsutulous; pod stiffly papery, hirsutulous . . . . *A. sabulonum*
            Herbage hairs appressed; pod membranous . . . . *A. aridus*

Plants perennial
  Pods hard, thick-walled, 2-3cm long; flowers 2cm long . . . . *A. crotalariae*
  Pods thin-walled, membranaceous, or if firm, less than 2cm long; flowers smaller
    Flowers purplish-blue or with at least a purplish tinge . . . . *A. palmeri*
    Flowers whitish or ochroleucous
      Pod less than 2cm long, chartaceous and firm, turgid, ovoid, narrowed into acute tip; plant erect . . . . *A. oocarpus*
      Pod usually over 2cm long, membranous, abruptly narrowed at tip; stems decumbent or ascending
        Petals strongly incurved and short-clawed; blades of the keel longer than the claws; flowers and fruits in loose racemes
          Stems procumbent. . . . . *A. douglasii* var. *parishii*
          Stems stout, erect, 4-10dm tall . . . . *A. douglasii* var. *perstrictus*
        Petals not strongly incurved; blades of the keel not longer than their claws
          Peduncles 5.5-14cm long; racemes dense, 3.5-9cm long in fruit . . . . *A. pomonensis*
          Peduncles 12-20cm long; racemes loose, 9-16cm long in fruit. . . . . *A. deanei*
Pod 2-celled
  Terminal leaflets decurrent; herbage canescent or cinereous throughout . . . . *A. magdalenae*
  Terminal leaflets of desert plants not decurrent; herbage variable . . . . *A. lentiginosus*
Pods not strongly inflated
  Plants annuals, perhaps occasionally biennials
    Pods 3-4mm long, transversely wrinkled or ribbed
      Pods deflexed, exserted from the calyx, conspicuously gray-hairy. . . . . *A. gambelianus*
      Pods erect, little exserted from the calyx, usually minutely hairy
        Herbage silvery-gray . . . . *A. didymocarpus* var. *dispermus*
        Herbage glabrous
          Pods strigulous . . . . *A. didymocarpus* var. *d.*
          Pods glabrous. . . . . *A. didymocarpus* var. *obispensis*
    Pods 12-18mm long, not transversely wrinkled or ribbed
      Keel not beaked
        Leaflets hoary; corolla purple, 8-10mm long. . . . . *A. lentiginosus*
        Leaflets greenish-canescent; corolla 4-8mm long, whitish to purplish
          Calyx black-hairy; leaflets retuse to truncate, glabrate above; pods almost completely 2-celled; coastal . . . . *A. tener*
          Calyx white-hairy; leaflets retuse to acute, strigose on both surfaces; pods incompletely 2-celled; deserts
            Calyx 2.5-3mm long . . . . *A. nuttallianus* var. *cedrosensis*
            Calyx 3.2-4.5mm long . . . . *A. nuttallianus* var. *imperfectus*
      Keel produced into a porrect beak; leaflets canescent, retuse; calyx black-hairy; deserts . . . . *A. acutirostris*
  Plants perennial
    Pods not long-hairy; flowers cream-colored . . . . *A. trichopodus*
    Pods long-hairy; flowers crimson . . . . *A. coccineus*

A. acutirostris Wats. Keelback Locoweed. Scarce, in sandy and gravelly areas; desert, 600-1100m: Indian Canyon; San Felipe; Mountain Springs; Jacumba. Mar.-May.

A. aridus Gray. Parch Locoweed. Uncommon, in dunes and open sandy plains; desert, below 400m: Borrego Springs; Silent Canyon; Vallecito and Carrizo Creeks. Apr.

A. coccineus (Perry)Bdg. Crimson Locoweed. Local, in rocky areas; southern desert and foothill, 600-1200m: Bankhead Springs; Jacumba; Dubber; Mountain Springs; Boulder Park. Mar.-May. Type from "western borders of the Colorado Desert". $2n=22$.

A. crotalariae (Benth.)Gray. Salton Sea Locoweed. Occasional, in sandy areas; desert, below 250m: Fonts Point; Borrego Valley; Borrego Springs; San Felipe Creek; Kane Spring Road; Split Mountain; Palm Spring; Carrizo Creek. Jan.-Apr. $2n=24$.

A. deanei (Rydb.)Barneby. Deane's Locoweed. Endemic to San Diego County. Local, in open areas; southern foothill, 250-350m: Dehesa; Whispering Oaks; Rice Canyon; Cottonwood; Barrett; Tecate Road. Mar.-May. Type from Sweetwater Valley.

A. didymocarpus H. & A. var. **didymocarpus**. White Dwarf Locoweed. Frequent, on open slopes; coastal, below 200m: Torrey Pines; Pacific Beach; Silver Strand; Paradise Valley; Chula Vista; Cottonwood; Whispering Oaks. Feb.-May. $2n=12$ pairs.

A. didymocarpus var. **dispermus** (Gray)Jeps. [*A. dispermus* Gray] Desert Dwarf Locoweed. Occasional, on sandy flats; desert, below 1200m: Montezuma View Point; Indian Canyon; Borrego Springs; Borrego Palm Canyon; Blair Valley; Vallecito and Carrizo Stage Stations; Jacumba; Mountain Springs; southeastern corner of county. Apr.-May. $n=13$.

A. didymocarpus var. **obispensis** (Rydb.)Jeps. Large-flower Dwarf Locoweed. Uncommon, on grassy slopes; foothill, 900-1200m: Boulevard; southeastern corner of county. Apr. $2n=13$ pairs.

A. douglasii (T.& G.)Gray var. **parishii** (Gray)Jones. [*Phaca pseudocarpa* Rydb., the type from Cuiamaca [Cuyamaca] Mountains.] Parish's Locoweed. Occasional, on open slopes and fields; foothill and montane, above 300m: Palomar Mountain; Palomar Observatory; Oak Grove; Hot Springs Mountain; Descanso; Laguna Public Camp; Descanso Junction; Hipass; Pine Valley; Boulevard; La Posta Creek; Dulzura. May-Aug.

A. douglasii var. **perstrictus** (Rydb.)Munz & McBurn. Campo Locoweed. Infrequent, in Chaparral; southern foothill, 600-1200m: Campo; Lark Canyon; Boulevard, McCain Valley; Eckener Pass; Jacumba; McCain Valley. Apr.-June. The type from "a hill valley between Campo and Jocumba [Jacumba]".

A. gambelianus Sheld. [*A. nigrescens* Nutt. non Pall.] Gambel's Locoweed. Infrequent, on slopes; coastal and foothill, below 900m: Henshaw Dam; Poway Grade; San Diego; Sloane Ranch; Otay Lake. Mar.-June. $2n=22$.

A. insularis Kell. var. **harwoodii** Munz & McBurn. Harwood's Locoweed. Infrequent, in sandy places; desert, below 400m: Vallecito and Carrizo Stage Stations; Palm Spring. Jan.-May.

A. lentiginosus Dougl. ex Hook. var. **borreganus** Jones. [*A. agninus* Jeps., the type from Borrego Springs.] Borrego Locoweed. Uncommon, on dunes and sandy flats; desert, below 500m: Coyote Creek; Fonts Point Wash; Palo Verde Wash; Borrego Valley; Borrego Springs; Santa Caterina Spring. Feb.-May. Type from "Borregos Springs".

A. magdalenae Greene var. **peirsonii** (Munz & McBurn.)Barneby. Algodones Locoweed. Rare, known in the county only on shifting dunes near Borrego Springs, 150m. Dec.-Apr.

A. nuttallianus DC. var. **cedrosensis** Jones. Scammon's Milkvetch. Uncommon, on rocky flats; desert, below 550m: northeastern corner of county; Borrego Valley; Agua Caliente Hot Springs; Carrizo Stage Station; Volcanic Hills; south of Lava Flow Wash. Dec.-Apr.

A. nuttallianus var. **imperfectus** (Rydb.)Barneby. Rare, known in the county only 3.5km southwest of Carrizo Stage Station, 150m. Apr.

A. oocarpus Gray. San Diego Locoweed. Endemic to San Diego County. Frequent, in open areas of Chaparral and Southern Oak Woodland; foothill and montane, 900-1500m: e.g. Wyss Ranch; Palomar Mountain; Lake Henshaw; Mesa Grande; Santa Ysabel; Witch Creek; Ramona; Ballena; Julian; San Diego River Falls; Pine Hills; Echo Valley; Green Valley; Descanso. May-Aug. Type from mountains east of San Diego.

A. palmeri Gray. [*A. vaseyi* Wats., the type from Mountain Springs; *A. vaseyi* var. *metanus* (Jones)Munz & McBurn.] Palmer's Locoweed. Frequent, on rocky slopes; desert, below 1300m: e.g. Indian Canyon; Lone Palm; Borrego Flats; Borrego Palm Canyon; Montezuma View Point; Grapevine Springs; Yaqui Well; Banner; Sentenac Canyon; San Felipe; Box Canyon; Pinyon Mountain; Agua Caliente Hot Springs; Vallecito Canyon; Mountain Springs; Dubber. Dec.-June.

A. pomonensis Jones. Pomona Locoweed. Occasional, in sandy fields; coastal, below 100m: e.g. Fallbrook; Rainbow; San Diego Bay; Silver Strand. Mar.-May. Type from "the tableland above Fallbrook".

A. sabulonum Gray. Ground Locoweed. Uncommon, in sandy areas; desert, below 700m: Borrego Springs; Split Mountain; Scissors Crossing. Nov.-Apr.

A. tener Gray var. **titi** (Eastw.)Barneby. [*A. titi* Eastw.] Coastal Dune Locoweed. Scarce, in alkaline areas; coastal, below 20m: Santa Margarita River estuary; Soledad; Silver Strand. Mar.-May.

A. trichopodus (Nutt.)Gray ssp. **leucopsis** (T.& G.)Thorne. [*A. l.* T.& G.; *A. t.* var. *lonchus* (Jones)Barneby] Ocean Locoweed. Frequent, in open areas; coastal, below 80m: e.g. San Onofre; Ysidora; Fallbrook; Escondido; Torrey Pines; Pacific Beach; San Diego; La Mesa; Mission Gorge; Point Loma; Jamul; Otay Valley; Imperial Beach; Barrett Dam. Feb.-June. The type from San Diego. $2n$=11 pairs.

Caesalpinia L.
(Caesalpinioideae)

C. virgata E.M. Fischer. [*Hoffmannseggia microphylla* Torr.] Desert Bird-of-Paradise. Uncommon, in canyons and washes; desert, below 200m: Split Mountain; Carrizo Creek. Mar.-May.

Calliandra Benth.
(Mimosoideae)

C. eriophylla Benth. Fairy Duster. Uncommon, rocky ridges; desert, below 300m: northeastern corner of county; south end of Pinyon Mountain; upper Blair Valley. Jan.-Apr.

Caragana Fabr.
(Faboideae)

***C. arborescens** Lam. Siberia Pea-Shrub. Native of central Asia. Local, naturalized on Palomar Mountain, above 1200m. Mar.-July.

Cercidium Tulasne Palo Verde
(Caesalpinioideae)

C. floridum Benth. Blue Palo Verde. Uncommon, in washes; desert, below 1000m: northeastern corner of county; Palo Verde Canyon. Apr.-May. $2n$=14 pairs.

Cercis L. Redbud
(Caesalpinioideae)

C. occidentalis Torr. ex Gray. [*C. nephrophylla* Greene, the type from "some uncertain mountain district in southwestern California, in San Diego County".] Local, Chaparral; montane, above 1400m: Julian; Chariot Canyon; Banner; Oriflamme Canyon; Cuyamaca Lake; Vallecito Trail; Oasis; Shrine Camp; Storm Canyon; Shaw Canyon. Feb.-Apr. n=7.

Cytisus L. Broom
(Faboideae)

***C. monspessulanus** L. French Broom. Native of the Canary Islands. Local, naturalized at Pine Hills, 1300m. Mar.-May. n=ca. 23.

Dalea Juss. Indigo Bush
(Faboideae)

Flowers 4-5mm long; calyx 3-4mm long; pubescence of whitish hairs, rarely turning rusty on drying . . . . . . *D. mollis*
Flowers 6-8mm long; calyx about as long as corolla; pubescence of inflorescence often turning rusty brown on drying . . . . . . *D. mollissima*

D. mollis Benth. Uncommon, in open, sandy places; desert, below 200m: Borrego Palm Canyon; Borrego Springs; Borrego Valley; The Narrows; Split Mountain; Carrizo Stage Station. Jan.-Apr.

D. mollissima (Rydb.)Munz. [*D. mollis* var. *m.* (Rydb.)Munz] Frequent, in Creosote Bush Scrub; desert, below 300m: e.g. Borrego Palm Canyon; Borrego Valley; northeast corner of county; San Felipe Wash; Kane Spring Road; The Narrows; Vallecito Stage Station; Palm Spring; Carrizo Stage Station. Mar.-May.

Dolichos L.
(Faboideae)

***D. lignosus** Pers. Flowering Bean. Native of Paleotropics. Local, adventive at Point Loma, 80m. Feb.-May.

Glycyrrhiza L. Wild Licorice
(Faboideae)

G. lepidota Pursh var. **glutinosa** (Nutt.)Wats. American Licorice. Rare, known in the county only in the cienega of Sentenac Canyon, 750m. May-June. 2n=16.

Hoffmannseggia Cav.
(Caesalpinioideae)

H. densiflora Benth. ex Gray. [*H. fulcaria* auct. non Cav.] Hog Potato. Uncommon, in alkaline areas; desert, below 1000m: San Felipe Valley; Earthquake Valley; Blair Valley. Apr.-Sep.

Lathyrus L. Pea
(Faboideae)

Leaflets 2; introduced plants . . . . . . *L. tingitanus*
Leaflets at least 4 on some of the leaves; native perennial plants
  Corolla deep rich red, 3-4cm long, the banner reflexed and lying in nearly a straight line with wings and keel . . . . . . *L. splendens*
  Corolla mostly other than deep red, or if so, less than 3cm long
    Banner not reflexed more than 90 degrees
      Flowers whitish to pale flesh-color . . . . . . *L. laetiflorus* ssp. *l.*
      Flowers pink to lavender . . . . . . *L. laetiflorusps* ssp. *barbarae*
    Banner reflexed more than 90 degrees . . . . . . *L. laetiflorus* ssp. *alefeldii*

L. laetiflorus Greene ssp. **alefeldii** (White)Brads. [*L. a.* White; *L. strictus* Nutt. in T. & G., the type from "bushy places around St. Diego".] San Diego Sweetpea, Canyon Pea. Common, in Chaparral and Sage Scrub; cismontane: e.g. De Luz; Fallbrook; Red Mountain; Bonsall; Rainbow; Moosa Canyon; Rincon Grade; Pala; Ramona; Eagles Nest; Volcan Mountain; Banner; Cuyamaca Lake; Laguna Public Camp; Garnet Peak; Anahuac School; Viejas Valley; Guatay; Pine Valley; Clevenger Canyon; Lyons Valley; Lee Valley Dehesa; Barrett; Otay Mountain; Whispering Oaks; Pacific Beach. Mar.-June. Type from San Diego. n=7.

L. laetiflorus ssp. **barbarae** (White)C. L. Hitchc. Santa Barbara Sweetpea. Uncommon, in Chaparral; north coastal and north foothill, below 1000m: Santa Margarita Mountains. Mar.-June.

L. laetiflorus ssp. **laetiflorus**. Uncommon, known from a collection on Banner Grade, 1200m. Apr.-May. 2n=7pairs.

L. splendens Kell. Pride of California. Frequent, in Chaparral; foothill, 400-900m: Jamul; Dulzura; Barrett; Tecate Mountain; Potrero Grade; Campo; Boulevard; Manzanita; White Star; Walker Ranch. Apr.-June. The type from "southern California". n=7.

***L. tingitanus** L. Tangier Pea. Native of Eurasia. Local, adventive in waste places; coastal, below 150m: East San Diego. May-July. n=7.

Lotus L.
(Faboideae)

Stipules expanded, not glandlike
  Stipules green, resembling and almost as large as the 3 leaflets; introduced weed in lawns, etc. . . . . . . *L. corniculatus*
  Stipules membranous or hyaline, not like the several leaflets
    Pods 4-5mm wide; plants of dry places
      Plants inconspicuously strigulose or glabrate . . . . . . *L. crassifolius* var. *c.*
      Plants conspicuously villous . . . . . . *L. crassifolius* var. *otayensis*
    Pods 1.5-2mm wide; plants of wet places . . . . . . *L. oblongifolius*
Stipules reduced to dotlike often dark or reddish glands
  Pods dehiscent, straight or nearly so, abruptly short-beaked
    Plants perennial
      Leaf-rachis 2-5cm long; corolla 1.5-2.4cm long . . . . . . *L. grandiflorus*
      Leaf-rachis not more than 1cm long. . . . . . . *L. rigidus*

Plants annual
Flowers yellow, often aging red
Wings conspicuously longer than the keel; styles hairy at the base of the stigma; seeds not smooth
Leaflets linear-oblong to elliptic, not noticeably succulent
Herbage strigose . . . . *L. strigosus* var. *s.*
Herbage with spreading pubescence . . . . *L. strigosus* var. *hirtella*
Leaflets mostly obovate, quite succulent; herbage canescently tomentulose. . . *L. strigosus* var. *tomentellus*
Wings about as long as or shorter than keel; styles glabrous; seeds smooth
Flowers 1-5 on peduncle; pods 1.5-2mm wide
Corolla 3-5mm long . . . . *L. salsuginosus* ssp. *brevivexillus*
Corolla 6-9mm long . . . . *L. salsuginosus* ssp. *s.*
Flowers solitary, subsessile; pods about 3mm wide
Plant subglabrous to strigose; calyx-teeth about as long as tube; pods 10-15mm long. . . . . *L. subpinnatus*
Plant densely villous-pubescent; calyx-teeth about as long as tube; pods 10-15mm long. . . *L. humistratus*
Flowers cream-white or red or pink
Calyx-teeth shorter than the tube; corolla pinkish or pale salmon, tinged or turning red; pods not deflexed; leaflets mostly 3-10mm long . . . . *L. micranthus*
Calyx-teeth longer than the tube; corolla whitish, tinged with rose; pods deflexed; leaflets mostly 10-15mm long. . . . . *L. purshianus*
Pods indehiscent; often strongly curved, tapering to an elongate beak
Plants mostly annual
Umbels pedunculate; bract usually present, unifoliolate; corolla 5-7mm long; beak of pod glabrous . . . . *L. nuttallianus*
Umbels subsessile; bract lacking; corolla 3-4.5mm long; beak of pod strigose . . . . *L. hamatus*
Plants perennial
Pubescence spreading, woolly-villous
Herbage extremely long villous . . . . *L. heermannii* var. *orbicularis*
Herbage short villous. . . . . *L. heermannii* var. *h.*
Pubescence appressed or almost lacking
Flowers on peduncles mostly more than 5mm long
Plant erect, sparsely strigose; desert edge. . . . . *L. haydonii*
Plants prostrate, densely strigose to silvery
Plant mostly canescent throughout, but scarcely silvery; calyx-teeth about half as long as the tube . . . . *L. nevadensis*
Plant more silvery
Calyx-teeth almost as long as the tube; banner reflexed or at right angles to the wings . . . . *L. nevadensis* var. *davidsonii*
Calyx-teeth about half as long as the tube; banner ascending, at an angle acute to the wings . . . . *L. argophyllus* ssp. *decorus*
Flowers sessile or subsessile, in leaf-axils
Plant plainly grayish or whitish with appressed pubescence . . . . *L. argophyllus* ssp. *d.*
Plant green, subglabrous except at growing tips; body of pod well exserted
Wings shorter than the keel . . . . *L. scoparius* ssp. *brevialatus*
Wings as long as keel. . . . . *L. scoparius* ssp. *s.*

L. argophyllus (Gray)Greene ssp. **argophyllus**. [*Hosackia a.* Gray] Silver-leaf Lotus. Common, on rocky slopes; foothill, montane, and desert, above 300m: e.g. Moosa Canyon; Harrison Grade; Mountain Beauty Mine; Eagles Nest; Mesa Grande; Montezuma View Point; Julian; Oriflamme Mine; Witch Creek; Barona Valley; Foster; Pine Hills; Cuyamaca Lake; Oasis; Buckman Springs; La Posta; Pinyon Mountain; Mason Valley; Dubber; Live Oak Springs; Barrett Lake; Japatul Valley; Tecate Mountain. Jan.-May. Type from "San Isabel [Santa Ysabel] California, on rocks". n=7.

L. argophyllus ssp. **decorus** (Jtn.)Munz. [*Hosackia a.* var. *d.* Jtn.] Satin Lotus. Uncommon, on dry slopes; foothill, 1000-1200m: La Posta; southeast corner of county. Apr.

***L. corniculatus** L. var. **corniculatus**. Native of Europe. Naturalizing in moist, disturbed and cultivated areas; cismontane: Dos Picos County Park Campground. n=12.

L. crassifolius (Benth)Greene. var. **crassifolius**. [*Hosackia c.* Benth.] Buck Lotus. Infrequent, in Chaparral and Montane Coniferous Forest; foothill and montane, above 600m: Palomar Mountain; Fry Creek; Eagles Nest; Cuyamaca Peak; Hauser Creek; Marron Valley; Otay Mountain. May-June. 2n=14.

L. crassifolius var. **otayensis** Moran in Isely. Otay Lotus. Rare, Chaparral; foothill, 300-950m: Otay Mountain. Type from 1.5 miles east of Dog House Junction, Otay Mountains.

L. grandiflorus (Benth.)Greene. var. **grandiflorus**. [*Hosackia g.* Benth.; *Lotus confinis* Greene, the type from "Witch Creek, in the mountains of San Diego Co., Calif., near the United States and Mexican boundary".] Grand Lotus. Occasional, in open Chaparral areas; foothill and montane, 400-1400m: Harrison Grade; Witch Creek; Cuyamaca Lake; Descanso; Barrett Honor Camp; Lawson Peak; Lyons Peak; Potrero; Tecate Mountain; Laguna Mountain. Apr.-July. n=7.

L. hamatus Greene. [*Hosackia micrantha* Nutt.] Grab Lotus. Occasional, on open slopes and burns; cismontane, below 1300m: e.g. O'Neill Lake; Bonsall; Rainbow; Pala; Doane Valley; Henshaw Dam; Borrego Palm Canyon; Anahuac School; San Pasqual; Foster; Poway Junction; Torrey Pines; Mission Hills; Fletcher Hills; Silver Strand; Sloane Ranch; Lee Valley; Otay Mesa; Sweetwater Dam; Alpine. Also lower Box Canyon. Mar.-July. Type said to be from "near Monterey", where the species is not known to occur; probably from San Diego.

L. haydonii (Orcutt)Greene. [*Hosackia h.* Orcutt; *Lotus spencerae* Macbr., the type from stony slopes, Mountain Springs.] Pygmy Lotus. Scarce, on rocky slopes; desert, 700-1100m: Borrego Palm Canyon; southeast slope of Pinyon Mountain; Mountain Springs. Jan.-Apr. Type from a "canyon leading into Colorado Desert, on San Diego to Fort Yuma Stage line".

L. heermannii (Dur. & Hilg.)Greene var. **heermannii**. [*Hosackia tomentosa* ssp. *glabriuscula* (H. & A.)Abrams] Common, on dry, rocky slopes; cismontane, below 1300m: e.g. upper San Onofre Canyon; De Luz; Rainbow; Harrison Grade; Oak Grove; Borrego Palm Canyon; Collins Valley; Mesa Grande; San Clemente Canyon; Mission Valley; Point Loma; Bear Valley; Otay; Tijuana River; El Capitan; Descanso; Barona Valley; Buckman Springs; Pine Valley; Hauser Creek; Jacumba; Live Oak Springs. Mar.-Oct.

L. heermannii var. **orbicularis** (Gray)Isely. [*Hosackia tomentosa* H. & A. var. *eriophorus*] Woolly Lotus. Uncommon, in sandy areas; coastal, below 150m: Carlsbad; Murphy Canyon; Coronado. Also Morgan Hill. Feb.-Sep.

L. humistratus Greene. [*Hosackia brachycarpa* Benth.] Hill Lotus. Infrequent, on rocky slopes; foothill and desert, 600-1400m: Pine Hills; Cuyamaca Lake; San Felipe Valley; Scissors Crossing. May-July. n=6.

L. micranthus Benth. [*Hosackia parviflora* Benth.] Miniature Lotus. Occasional, on grassy slopes; montane, above 1300m: Nellie McQueen Ranch; Mesa Grande; Pine Hills; Cuyamaca Lake. Mar.-May.

L. nevadensis (Wats.)Greene var. **davidsonii** (Greene)Isely. [*Hosackia sulphurea* (Greene)Abrams] Uncommon, on dry slopes, Montane Coniferous Forest; montane, above 1400m: Doane Valley; North Peak. June-Aug.

L. nevadensis var. **nevadensis**. [*Hosackia n.* (Wats.)Parish] Frequent, on dry slopes; montane, above 1400m: Doane Valley; Wyss Ranch; French Valley; Volcan Mountain; Cuyamaca Lake; Cuyamaca Peak; Laguna Public Camp. Also Descanso; Campo. June-Aug. n=7.

L. nuttallianus Greene. [*Hosackia prostrata* Nutt.] Prostrate Lotus. Occasional, on sandy beaches; coastal, below 30m: mouth of San Luis Rey River; Encinitas; Del Mar; La Jolla; Pacific Beach; Ocean Beach; Coronado; Silver Strand; National City; Tijuana River estuary; Boundary Monument No. 258. Mar.-June. Type from "plains near the sea, St. Diego and St. Barbara".

L. oblongifolius (Benth.)Greene. [*Hosackia o.* Benth.] Stream Lotus. Frequent, along streams; foothill and montane, above 600m: e.g. Azalea Creek; lower Doane Valley; Mesa Grande; upper Borrego Palm Canyon; Eagles Nest; Volcan Mountain; Cuyamaca Lake; North Peak; Green Valley; Cuyamaca Peak; El Capitan; Laguna Public Camp; Glen Cliff Campground; Pine Valley; Jamacha. Apr.-Sep.

L. purshianus (Benth.)Clem. & Clem. ssp. **purshianus**. [*Hosackia americana* (Nutt.)Piper] Spanish-Clover. Common, dry fields and other disturbed places; cismontane: e.g. San Onofre Canyon; San Luis Rey River; De Luz; Doane Meadow; Warner Hot Springs; Pine Hills; Witch Creek; Ramona; Mussey Grade; Barona Valley; Lake Hodges; Lake Henshaw; San Pasqual; Kearny Mesa; Jamacha; La Mesa; Sears Ranch; Dulzura; Potrero; Alpine; Pine Valley; Milk Ranch; Laguna Public Camp. May-Oct. n=7.

L. rigidus (Benth.)Greene. [*Hosackia r.* Benth.] Broom Lotus. Frequent, on dry slopes; desert, below 700m: e.g. Lone Palm; Borrego Palm Canyon; Borrego Valley; Yaqui Well; Sentenac Canyon; San Felipe Valley; east of Banner; Box Canyon; Pinyon Mountain; Vallecito Stage Station; Canebrake Canyon; Bow Willow Canyon. Dec.-May. 2n=7 pairs.

L. salsuginosus Greene ssp. **brevivexillus** (Ottley)Munz. Humble Lotus. Occasional, in sandy places; desert, below 900m: Borrego Palm Canyon; The Narrows; northeast corner of county; Wagon Wash; Vallecito Stage Station; Jacumba; McCain Valley. Feb.-May. 2n=7 pairs.

L. salsuginosus ssp. **salsuginosus**. [*Hosackia maritima* Nutt.] Alkali Lotus. Occasional, on dry slopes and in fields; foothill and coastal, below 900m: e.g. San Onofre; Ysidora; Fallbrook; Lake Henshaw; Poway Grade; Torrey Pines State Reserve; Mission Grade; Adobe Falls; San Diego. Mar.-May. n=7.

L. scoparius (Nutt. in T. & G.)Ottley ssp. **brevialatus** (Ottley)Munz. [*Hosackia glabra* ssp. *b.* (Ottley)Abrams] Deerweed. Abundant, on dry slopes and open areas; cismontane and desert, below 1300m: e.g. Rainbow; Rincon; Montezuma View Point; Borrego Palm Canyon; San Felipe; Banner; Anahuac School; Lake Hodges; Escondido; San Marcos; El Capitan Mountain; Poway; Lee Valley; Barrett Junction; Dulzura; Harbison Canyon; Live Oak Springs; Mountain Springs; Mason Valley. Feb.-Aug. 2n=7 pairs.

L. scoparius ssp. **scoparius**. [*Hosackia glabra* (Vogel)Torr.] Coastal Deer Weed. Abundant, on open slopes; Chaparral and Inland Sage Scrub; coastal, below 500m: e.g. San Onofre; San Mateo Creek; Santa Margarita Ranch; De Luz; Fallbrook; Oceanside; Carlsbad; Pala; Vista; Olivenhain; Cardiff; Torrey Pines; Mount Soledad; Kearny Mesa; Point Loma; Chollas Valley; Mission Valley; Telegraph Canyon; Silver Strand; Boundary Monument No. 258. Also Otay Mountain; Barona Valley; Witch Creek; Banner Grade. Mar.-Aug. n=7.

L. strigosus (Nutt. in T. & G.)Greene var. **hirtella** (Greene)Ottley. [*Hosackia s.* var. *h.* (Greene)Hall] Hirsute Lotus. Frequent, dry, disturbed places; foothill and montane, occasional desert, above 700m: e.g. Doane Valley; Eagles Nest; Warner Hot Springs; Culp Valley; Witch Creek; Pine Hills; Sentenac Canyon; Banner; Julian; Cuyamaca Lake; Harper Ranch; Cuyamaca Peak; Inspiration Point; Laguna Lakes; Descanso Creek; Glen Cliff Campground. Mar.-June. 2n=7 pairs.

L. strigosus var. **strigosus**. [*Hosackia s.* Nutt.] Bishop's Lotus. Common, on open slopes and burns; coastal and foothill, infrequent desert, below 900m: e.g. upper San Onofre Canyon; Fallbrook; Lilac; Pala; Rincon; Oak Grove; Warner Pass; Yaqui Well; Pine Hills; Mount Woodson; Barona Valley; Mount Soledad; Pacific Beach; Kearny Mesa; Murphy Canyon; Poway Grade; Silver Strand; Lee Valley; Foster; Campo; Hipass; Buckman Springs; Pine Valley; Oakzanita; Viejas Valley. Mar.-June 2n=7 pairs.

L. strigosus var. **tomentellus** (Greene)Isely. [*Hosackia t.* (Greene)Abrams; *L. t.* Greene] Uncommon, in sandy areas; desert, below 150m: Borrego Springs; Buttes Pass; Kane Spring Road; Vallecito Stage Station;

Palm Spring; Lava Flow Wash; Vallecito Wash; Alta Bisnaga Wash; Ocotillo Wells; East Dolomite Mine Road; Mortero Wash; June Wash. Feb.-May.

L. subpinnatus Lag. [*Hosackia s.* (Lag.)T. & G.] Calf Lotus. Occasional, on grassy slopes; cismontane, below 1300m: San Luis Rey; Mesa Grande; Lake Henshaw; Poway Grade; San Diego; Point Loma; Montezuma Mesa; Lee Valley; Witch Creek. Also Vallecito Stage Station. Mar.-June. n=6.

Lupinus L. Lupine
(Faboideae)

Plants annual
Flowers definitely whorled or mostly so
Keel ciliate on upper margins near the claws
Keel ciliate on lower edges also, near the claws
Plants 0.5-1.5dm tall; peduncles 2-4cm long . . . . *L. concinnus* var. *brevior*
Plants 2-10dm tall; flowers spreading in anthesis, 12-14mm long; peduncles mostly 3-15cm long *L. succulentus*
Keel not ciliate on lower edges
Banner 8-9mm long, 2-6mm wide; flowers suberect at anthesis . . . . *L. ruber*
Banner 12-16mm long, 8-9mm wide; flowers spreading at anthesis
Calyx rather inconspicuously pubescent. . . . . *L. densiflorus* var. *lacteus*
Calyx bushy-villous near the base . . . . *L. densiflorus* ssp. *austrocollium*
Keel ciliate on upper margins toward the apex or not ciliate at all
Keel short, broad, almost straight; reflexed portion of banner much shorter than the straight part *L. polycarpus*
Keel slender with long narrow mostly upturned beak; reflexed portion of banner generally as long as the straight part
Keel not or scarcely ciliate; flowers 5-8mm long . . . . *L. bicolor* ssp. *pipersmithii*
Keel definitely ciliate on upper edge toward tip
Lower calyx-lip deeply cleft, the teeth 1-5mm long; banner obovate, 6-9mm long; plants erect to decumbent . . . . *L. bicolor* ssp. *umbellatus*
Lower calyx-lip entire or with teeth not more than 1mm long
Banner oval, lemon-shaped when flattened, 3.6-6mm long, mucronate; keel with short blunt acumen . . . . *L. bicolor* ssp. *microphyllus*
Banner obovate to oblong; keel with slender acumen
Banner obovate, rounded to truncate at tip
Banner 4.5-7.5mm long, truncate to emarginate at apex; whorls 3-9. . . . *L. bicolor* ssp. *marginatus*
Banner 5.7-8.6mm long, mostly rounded at apex; whorls mostly 1-2 . . . . *L. bicolor* ssp. *umbellatus*
Banner oblong, truncate at apex, sometimes mucronate . . . . *L. bicolor* ssp. *tridentatus*
Flowers not at all whorled
Keel not ciliate
Cotyledons petioled; ovules 2-4
Leaflets linear-spatulate, mostly subglabrous above; plants sparsely spreading-pilose; flowers bluish-purple . . . . *L. concinnus* var. *agardhianus*
Leaflets spatulate-oblanceolate to obovate, amply pubescent above; plants abundantly spreading-pilose; flowers pinkish-white to lavender
Leaflets 1.5-3mm wide . . . . *L. concinnus* ssp. *c.*
Leaflets 4-7mm wide
Flowers 9-11mm long . . . . *L. concinnus* ssp. *optatus*
Flowers 6-8mm long . . . . *L. concinnus* ssp. *orcuttii*
Cotyledons sessile and perfoliate; ovules usually 2 . . . . *L. shockleyi*
Keel ciliate at least on lower margin near the base
Plant covered with long stinging hairs . . . . *L. hirsutissimus*
Plant lacking stinging hairs
Keel stout, with short blunt acumen; plant subglabrous to sparsely strigulose. . . . . *L. truncatus*
Keel with slender acute acumen; plant usually villous
Petals blue, lilac or purple
Flowers blue to lilac, 7-17mm long; leaflets mostly 1.5-3mm wide
Flowers 10-13mm long . . . . *L. sparsiflorus* ssp. *s.*
Flowers 7-11mm long
Stems erect or ascending; plants branched about at and above middle *L. sparsiflorus* ssp. *inopinatus*
Stems more or less decumbent; plants branched at or near base . . . . *L. sparsiflorus* ssp. *pondii*
Flowers pale purplish-pink, 8-10mm long . . . . *L. arizonicus*
Petals pale blue to whitish or yellowish, 6-7mm long. . . . . *L. pallidus*
Plants perennial
Plants shrubby or with definite woody stems of some length above the roots
Banner somewhat pubescent on the back; plants mostly appressed-silky; keel narrow with long slender acumen
Plants appressed-silky; flowers 10-13mm long, or if more, plant strigose; keel narrowed toward the base . . . . *L. albifrons*
Plants with appressed pubescence to tomentose; flowers 10-18mm long; keel not narrowed toward the base
Flowers 10-13mm long . . . . *L. excubitus* ssp. *medius*
Flowers 14-18mm long
Stems scarcely if at all woody, but quite herbaceous . . . . *L. excubitus* ssp. *austromontanus*
Stems woody at base . . . . *L. excubitus* ssp. *hallii*
Banner glabrous on back; keel mostly not markedly slender at the acumen
Herbage somewhat silvery; racemes mostly 1-2dm long above the peduncles. . . . . *L. excubitus* et sspp.
Herbage greenish; racemes mostly 2-4dm long above the peduncles . . . . *L. longifolius*
Plants strictly herbaceous
Leaves glabrous on upper surface or essentially so . . . . *L. latifolius*
Leaves somewhat hairy above
Banner glabrous on back
Keel ciliate on upper edges at least toward tip . . . . *L. excubitus* ssp. *austromontanus*
Keel glabrous on upper edges or practically so
Flowers yellow . . . . *L. adsurgens*

Flowers blue or violet
  Flowers 10-14mm long, the banner obovate; racemes mostly 4-10cm long . . . . *L. adsurgens*
  Flowers 12-18mm long, the banner roundish; racemes mostly 10-25cm long
    Pubescence appressed-silky . . . . *L. formosus* var. *f.*
    Pubescence spreading . . . . *L. formosus* var. *bridgesii*
Banner pubescent on back, the pubescence sometimes almost concealed by the upper lip of the calyx
 . . . . *L. excubitus* ssp. *austromontanus*

L. adsurgens E. Drew. Uncommon, in Chaparral; foothill, 800-1100m: Oak Grove; Green Valley; Live Oak Springs; Middle Peak; Cuyamaca Lake; Jacumba. May.

L. albifrons Benth. ex Lindl. ssp. **eminens** (Greene)D. Dunn ex Thorne. [*L. brittonii* Abrams, the type from Cottonwood Valley.] Occasional, Chaparral; foothill and montane, above 300m: Harrison Grade; Doane Valley; Palomar Observatory; McGee Ranch; Harper Ranch; Shaw Canyon; Corte Madera; Cottonwood Valley; White Star; InKoPah; southeast corner of county; Buckman Spring; Barrett; Lyons Peak; Ramona. Apr.-July.

L. arizonicus (Wats.)Wats. [*L. sparsiflorus* var. *a* (Wats.)C.P. Sm.; *L. arizonicus* var. *barbatulus* (Thornb.)Jtn.; *L. sparsiflorus* var. *b.* Thornb. in C. P. Sm.] Occasional, sandy places; desert, below 600m: Lone Palm; Borrego Palm Canyon; Rockhouse Canyon; Borrego Valley; northeast corner of county; Buttes Pass; Split Mountain; San Felipe; Carrizo Stage Station; Mountain Springs; The Narrows; Yaqui Well; Vallecito Stage Station. Jan.-May.

L. bicolor Lindl. ssp. **marginatus** D. Dunn. Dove Lupine. Frequent, on grassy slopes; foothill and montane, above 800m: Lake Henshaw; Warner Hot Springs; San Felipe Ranch; Banner; Witch Creek; Cuyamaca Lake; Harper Ranch; Laguna Public Camp; Descanso; Pine Hills. May-June.

L. bicolor ssp. **microphyllus** (Wats.)D. Dunn. Frequent, grassy fields and on slopes; cismontane: e.g. San Onofre; Fallbrook; Bonsall; Rincon; Doane Valley; Fry Creek; Warner Ranch; Culp Valley; San Felipe Valley; Cuyamaca Lake; Santa Ysabel; Laguna Public Camp; Oceanside; Rainbow; Pala Grade; San Marcos; Crown Point; Ocean Beach; Kearny Mesa; San Diego; Lee Valley; Campo; Alpine; Jacumba. Apr.-June.

L. bicolor ssp. **pipersmithii** (Heller)D. Dunn. Rare, known in the county only at Banner, 810m. Apr.

L. bicolor ssp. **tridentatus** (Eastw. ex C.P. Sm.)D. Dunn. Infrequent, on open slopes; montane, above 1200m: Lake Henshaw; Warner Ranch; Ramona; Banner; Pine Hills; Cuyamaca Lake; Harper Ranch; Stonewall Mine; Descanso; Laguna Public Camp; Desert View; Descanso. May-June.

L. bicolor ssp. **umbellatus** (Greene)D. Dunn. Occasional, grassy slopes; foothill and coastal, below 1000m: Oceanside; Ramona; Santa Ysabel; Leucadia; La Jolla. Mar.-May.

L. concinnus var. **brevior** (Jeps.)D. Dunn. [*L. b.* (Jeps.)Christian & Dunn] Sentenac Lupine. Infrequent, in sandy places; desert, below 700m: Borrego Palm Canyon; Scissors Crossing; Wagon Wash; lower Box Canyon; Mountain Springs. Also Morena. Apr.-May.

L. concinnus var. **agardhianus** (Heller)C.P.Smith. [*L. a.* Heller] Agardh's Lupine. Frequent, on open or disturbed slopes; coastal and foothill, below 1100m: e.g. De Luz; Rainbow; Borrego Palm Canyon; Banner; Grapevine Spring; Witch Creek; Box Canyon; Foster; Poway; Torrey Pines Mesa; Murphy Canyon; Grossmont; Sweetwater Lake; Lyons Valley; Otay Creek; Hidden Glen; Otay Ranch; Otay Mountain; Barrett; Oakzanita; Espinosa Trail; Jacumba. Mar.-May.

L. concinnus Agardh ssp. **concinnus**. Bajada Lupine. Occasional, in sandy places; foothill, 400-1200m: San Felipe Valley; Witch Creek; Banner; Campo; Hipass; La Posta; Boulevard; Bankhead Spring; Dubber; Jacumba. Apr.-May.

L. concinnus ssp. **optatus** (C.P. Smith)D. Dunn. Frequent, on granitic slopes; foothill and montane, above 400m: e.g. Boucher Hill; Doane Valley; Oak Grove Trail; Hot Springs Mountain; Warner Ranch; Volcan Mountain; Pine Hills; Harper Ranch; Cuyamaca Lake; Stonewall Mine; Green Valley; Laguna Camp; Live Oak Springs; Boulevard; Jacumba; Scissors Crossing; Vallecito Stage Station; Lakeside. Mar.-May.

L. concinnus ssp. **orcuttii** (Wats.)D. Dunn. Infrequent, on rocky slopes, 700-1100m: San Felipe Valley; Culp Valley; Montezuma View Point; Vallecito Valley; Mountain Springs. Mar.-Apr.

L. densiflorus Benth. ssp. **austrocollium** (C. P. Sm.)D. Dunn ex Thorne. Occasional, sandy places; foothill and coastal, below 300m: O'Neill Lake; Moosa Canyon; Miramar Naval Air Station; Ocean Beach; San Diego; Chollas Valley; Mahogany Canyon; Glen Abbey; Otay Mesa; Dehesa; Jamacha; Sweetwater Lake; Dulzura; El Cajon. Apr.-June. Type from near St. Mary's (now Mercy) Hospital, San Diego.

L. densiflorus ssp. **lacteus** (Kell.)Beauchamp. Uncommon, dry places; foothill and montane, above 800m: Warner Hot Springs; Laguna Mountains; Descanso; Coyote Canyon. May-July.

L. excubitus Jones ssp. **austromontanus** (Heller)Beauchamp. Occasional, in Chaparral and openings of Montane Coniferous Forest; montane, above 1200m: Palomar Mountain; Doane Valley; Cuyamaca Lake; Cuyamaca Mountains; Laguna Camp; Garnet Peak; Desert View; Jacumba; Echo Valley. May-July.

L. excubitus ssp. **hallii** (Abrams)D. Dunn ex Thorne. Occasional, on shaded slopes; foothill and montane, 200-1200m: San Onofre Canyon; Fallbrook; Doane Valley: Palomar Mountain; Henshaw Dam; Hot Springs Mountain; Oak Grove; Escondido Grade; Mount Woodson; Cuyamaca Lake; Clevenger Canyon; Ramona; Alpine; Pine Valley. Feb.-June.

L. excubitus ssp. **medius** (Jeps.)Beauchamp. Mountain Springs Lupine. Infrequent, sandy areas, desert 500-1200m: The Potrero; Canebrake Canyon; Rockhouse Canyon; Carrizo Gorge, Table Mountain; McCain Valley. Type from the summit of Mountain Springs Grade, 945m. Apr.

L. formosus Greene var. **bridgesii** (Wats.)Greene. Infrequent, open fields and sandy places; foothill and coastal, below 800m: Lake Hodges; Glen Lonely. Apr.-June.

L. formosus var. **formosus**. Occasional, on open slopes; foothill and montane, above 400m: Doane Valley; Oak Grove; Eagles Nest; Warner Ranch; San Felipe Valley; Banner; Cuyamaca Mountains; Laguna Public Camp; Desert View; Jacumba; Bankhead Spring; Eckener Pass; Descanso; Dehesa; Poway. Apr.-July.

L. hirsutissimus Benth. Stinging Lupine. Common, sandy places; cismontane, but infrequently desert, below 1400m: e.g. San Onofre; De Luz; Moro Hill; Rainbow; Bonsall; Rincon Spring; Pauma; Fry Creek; Lake Henshaw; Volcan Mountain; San Felipe Valley; Sentenac Canyon; Cuyamaca Lake; Witch Creek; Spencer Valley; Ramona; Ballena; Poway Grade; El Capitan Mountain; San Pasqual; Mission Valley; Morena; Otay Mountain; Dehesa; Barrett; Canyon City; Hipass; Jacumba. Feb.-May.

L. latifolius Agardh ssp. **parishii** (C.P. Sm.)Kinney & D. Dunn. Parish's Stream Lupine. Occasional, in wet areas; foothill and montane, above 400m: De Luz; Fallbrook; Castro Canyon; Chimney Flats; Nellie; Mesa Grande; Eagles Nest; Volcan Mountain; Pine Hills; Banner Grade; Cuyamaca Lake; Middle Peak; Oasis; North Peak; El Capitan; Viejas Valley; Wohlford Reservoir; Hidden Glen; Barrett Dam; Jamul; Dehesa; McCain Valley. May-July.

L. longifolius (Wats.)Abrams. Pauma Lupine. Occasional, shaded slopes; coastal and foothill, below 1100m: San Onofre; Santa Margarita Mountains; Ysidora; De Luz; Bonsall; Rainbow; Mesa Grande; Ramona; Vista; Poway Grade; San Pasqual; Barrett Junction. Feb.-June. Type from San Diego.

L. pallidus Bdg. [*L. concinnus* var. *p.* (Bdg.)C.P. Sm.] Pallid Lupine. Uncommon, in sandy areas; desert, below 800m: Borrego Palm Canyon; San Felipe; Vallecito Stage Station; Carrizo Creek. Mar.-Apr.

L. polycarpus Greene. [*L. micranthus* Dougl. in Lindl.] Field Lupine. Uncommon, sandy places; cismontane: Doane Valley; Cuyamaca Lake; Witch Creek; Mesa Grande; Mission Valley. Mar.-May. $2n=48$.

L. ruber Heller. [*L. microcarpus* var. *r.* (Heller)C.P. Sm.] Red-Flowered Lupine. Rare, sandy areas, desert, 700-1200m: Banner; San Felipe; Jacumba. Mar.-June.

L. shockleyi Wats. Desert Lupine. Infrequent, sandy areas, desert, below 200m: Borrego Springs; Oil Well; southwest of Carrizo Stage Station. Mar.-Apr.

L. sparsiflorus Benth. ssp. **inopinatus** (C.P. Sm.)Dziekanowski & D. Dunn. Uncommon, bare slopes; coastal, below 200m: Lake Henshaw; San Diego. Apr. Type from San Diego.

L. sparsiflorus ssp. **sparsiflorus**. Coulter Lupine. Frequent, disturbed areas; foothill and desert, below 800m: San Onofre; Pala Grade; Warner Hot Springs; Borrego Palm Canyon; San Felipe Valley; Scissors Crossing; Sentenac Canyon; San Pasqual; Viejas Valley; Campo. Type may have been from San Diego County, taken by Coulter during his journey to Yuma. Mar.-May.

L. sparsiflorus ssp. **pondii** (Greene)Dziekanowski & D. Dunn. Rare, sandy areas; cismontane and desert, below 500m: Coyote Canyon; Tijuana River Valley. Mar.-May.

L. succulentus Dougl. ex Koch. Arroyo Lupine. Common, clay or grassy slopes; coastal, below 200m: e.g. San Onofre; Vista; Rancho Santa Fe; Solana Beach; Mount Soledad; Kearny Mesa; Poway; Bostonia; Fletcher Hills; Alpine; Sunnyside; Glen Abbey; Otay Ranch; Otay Lake; Adobe Falls; Point Loma; Encanto; Rose Canyon. Jan.-Apr.

L. truncatus Nutt. ex H.& A. Collar Lupine. Common, on grassy slopes; coastal and foothill, below 900m: e.g. San Onofre; Fallbrook; Bonsall; Moosa Canyon; San Dieguito; La Jolla; Ocean Beach; Mahogany Canyon; Silver Strand; Otay Mesa; Dehesa; Otay Mountain; The Willows; Potrero Grade; Barona Indian Reservation; Campo; Hipass. Feb.-May. Type from San Diego.

Marina Liebm.
(Faboideae)

M. parryi (T.& G.)Barneby. [*Dalea p.* T.& G.] Infrequent, rocky flats; desert, below 700m: northeast corner of county; Plum Canyon; Wagon Wash; Sentenac Canyon; Mine Canyon; Agua Caliente Springs; Little Blair Valley; Alta Bisnaga Wash; south end of Pinyon Mountain; Vallecito Stage Station. Jan.-May. $2n=10$ pairs.

Medicago L. Medick
(Faboideae)

Plants perennial; flowers blue . . . *M. sativa*
Plants usually annual; flowers yellow
  Pods sub-reniform, 1-seeded; flowers many, in dense elongate spike-like racemes . . . *M. lupulina*
  Pods spirally coiled; flowers few, not in spikes
    Stipules lanceolate, short-dentate at base; plants pilose . . . *M. minima*
    Stipules pectinate, with linear lateral segments; plants subglabrous or nearly so . . . *M. polymorpha*

***M. lupulina** L. Black Medick. Native of Europe. Common, waste places; cismontane: e.g. Balboa Park; East San Diego; Spencer Valley; Banner; upper Borrego Palm Canyon; Eagles Nest; San Felipe Valley; Cuyamaca Lake; Middle Peak; Pine Valley. Apr.-July. $2n=16$.

***M. minima** (L.)Desr. Hairy Bur-Clover. Native of Eurasia. Rare, moist areas, montane: Wynola; Pine Hills. Mar.-May. $n=8$.

***M. polymorpha** L. [*M. hispida* Gaertn.; *M. p.* var. *brevispina* (Benth.)Heyn.] Bur-Clover. Native of Europe. Common, disturbed areas; cismontane: e.g. San Onofre; San Marcos; Vista; Buena; Cuyamaca Lake; Witch Creek; El Cajon Mountain; The Willows; Glenview; San Pasqual; Santa Ysabel; Anahuac

School; Kearny Mesa; Lemon Grove; Mission Valley; Lee Valley; Paradise Valley; Pacific Beach; Otay Mountain. Feb.-Apr. *n*=7.

***M. sativa** L. Alfalfa. Native of Old World. Common, escaping from cultivation around fields and moist places; cismontane: e.g. San Onofre; Buena; Cuyamaca Mountains; Descanso; Barona Valley; Ramona; Flynn Spring; Chula Vista; Silver Strand; Point Loma. Apr.-Oct. *n*=16, 32, 64.

Melilotus Mill. Sweet-Clover (Faboideae)

Flowers white . . . . *M. albus*
Flowers yellow
  Flowers 2-3mm long, on pedicels less than 1mm long . . . . *M. indicus*
  Flowers 5-7mm long, on pedicels 1.5-2mm long . . . . *M. officinalis*

***M. albus** Medicus. White Sweet Clover. Native of Eurasia. Common, waste places; cismontane: e.g. San Onofre Bluffs; De Luz; Doane Valley; Inaja Indian Reservation; Banner; Borrego Palm Canyon; Descanso; Campo; Silver Strand. Apr.-Oct. *n*=8.

***M. indicus** (L.)All. Indian Sweet Clover. Native of Eurasia. Common, waste places; cismontane: e.g. San Onofre; De Luz; San Luis Rey; Rainbow; Eagles Nest; Witch Creek; Borrego Palm Canyon; Pine Valley; Hipass; White Star; Lee Valley; La Mesa; Mission Valley; Olivenhain; Silver Strand. Apr.-Oct. *n*=8.

***M. officinalis** (L.)Lam. Yellow Sweet Clover. Native of Eurasia. Infrequent, disturbed areas; cismontane: Chimney Creek; Boulevard. Aug.-Sep. *n*=8.

Olneya Gray Desert-Ironwood (Faboideae)

O. tesota Gray. Frequent, seeps, slopes and washes; desert, below 900m: e.g. Grapevine Spring; Yaqui Well; Borrego Springs; The Narrows; Fish Creek. Apr.-May.

Parkinsonia L. Palo Verde (Caesalpinioideae)

***P. aculeata** L. Mexican Palo Verde. Native of Arizona and elsewhere. Infrequent, escaping from cultivation: Borrego Springs; San Diego. May-Sep. *n*=14.

Pickeringia Nutt. Chaparral-Pea (Faboideae)

P. montana Nutt. ssp. **tomentosa** (Abrams)Abrams. [*Xylothermia t.* Abrams] Local, in Chaparral; foothill, 200-1100m: Santa Margarita Mountains; Mount San Miguel; Jamul Mountains; Otay Mountain; Dulzura; Tecate Peak; Skye Valley. May-Aug. 2*n*=14 pairs. The type from "near El Nido, San Diego County".

Prosopis L. (Mimosoideae)

Fruits straight or curved, not spirally curved
  Leaflets more than 5 times as long as broad . . . . *P. glandulosa*
  Leaflets less than 5 times as long as broad . . . . *P. velutina*
Fruits tightly spirally coiled . . . . *P. pubescens*

P. glandulosa Torr. var. **torreyana** (L. Benson)M.C. Jtn. [*P. juliflora* var. *t.* Benson] Honey Mesquite. Common, washes and arroyos; desert and coastal, below 1000m: Warner Hot Springs; Borrego Palm Canyon; Collins Valley; Seventeen Palms; Sentenac Canyon; Vallecito and Carrizo Stage Stations; Dubber; Olivenhain; Mission Gorge; Santee; Sweetwater Lake; Jamacha Junction; Proctor Valley; Dehesa; Jamul Valley; Telegraph Canyon. Mar.-June.

P. pubescens Benth. Screw-Bean Mesquite. Occasional, moist areas; desert, below 1000m: e.g. Sentenac Canyon; Vallecito Stage Station; Bow Willow Canyon; Dubber. Also Tijuana River Valley. May-July.

***P. velutina** Woot. [*P. juliflora* var. *v.* (Woot.)Sarg.] Velvet Mesquite. Native of Arizona and elsewhere. Local, apparently naturalized in the county from its natural range to the east, below 1000m: Grantville; San Clemente Canyon; 19km east of Julian. Apr.-Aug.

Psoralea L. (Faboideae)

Stems prostrate; petioles and peduncles erect, much elongated . . . . *P. orbicularis*
Stems erect
  Corolla ochroleucous or whitish with purple-tipped keel; calyx-teeth subequal
    Leaflets 1-1.5 times as long as wide, broadly ovate; calyx inflated in fruit, the lobes 1-2mm long . . . . . . *P. physodes*
    Leaflets 2-3 times as long as wide, lance-ovate; calyx not inflated in fruit, the lobes 3-5mm long . . . . *P. rigida*
  Corolla purple, about 1cm long; the lower calyx-lobe much longer than others . . . . *P. macrostachya*

P. macrostachya DC. Leather Root. Occasional, wet places; cismontane: San Onofre Bluffs; De Luz; Banner; San Felipe; Volcan Mountain; Barona Valley; El Capitan Mountain; Oasis; Vallecito Creek Canyon; Paso Picacho. May-Aug. *n*=10; 2*n*=11 pairs.

P. orbicularis Lindl. Round-leaf Psoralea. Frequent, in wet places; montane, above 1100m: Thunder Ridge on Palomar Mountain; Henshaw Dam; Witch Creek; Julian; Cuyamaca Dam; Pine Valley; Buckman Spring. June-Sep.

P. physodes Dougl. California Tea. Rare, known in the county at a site 3.5km west of De Luz, 200m. May. n=11.

P. rigida Parish. Parish's Psoralea. Occasional, on dry slopes; foothill and montane, above 700m: Oak Grove; Julian; Cuyamaca Lake; Middle Peak; Shrine Camp; Laguna Public Camp; Pine Valley; Laguna Junction; Boulder Creek. May-July. The type from dry hills, Oak Grove.

Psorothamnus Rydb.
(Faboideae)

Leaves simple, sometimes wanting
  Herbage silvery-silky, the leaves often lacking except on young growth . . . . *P. spinosus*
  Herbage green, subglabrous, the leaves mostly present during vegetative period . . . . *P. schottii*
Leaves pinnate, present during vegetative periods . . . . *P. emoryi*

P. emoryi (Gray)Rydb. [*Dalea e.* Gray] White Dalea. Occasional, sandy flats; desert, below 200m: Borrego Valley; Silent Canyon; Ocotillo Wells; The Narrows; Palm Spring; Carrizo Stage Station. Jan.-May. $2n$=10 pairs.

P. schottii (Torr.)Barneby. [*Dalea s.* Torr.; *D. s.* var. *puberulus* (Parish)Munz] Infrequent, sandy places, rocky flats; desert, below 200m: northeast corner of county; Plum Canyon; Wagon Wash; Sentenac Canyon; Mason Valley; Pinyon Mountain; Vallecito and Carrizo Stage Stations; Yaqui Wells; Mountain Springs. Mar.-May. $2n$=10 pairs.

P. spinosus (Gray)Barneby. [*Dalea s.* Gray] Smoke Tree. Common washes; desert; below 900m: e.g. Borrego Palm Canyon; Borrego Valley; northeast corner of county; The Narrows; Fish Creek; Carrizo Creek; Smoke Tree Wash; Mountain Springs. Nov.-May.

Robinia L. Locust
(Faboideae)

***R. pseudo-acacia** L. Native of eastern United States. Local, persisting around abandoned dwellings and plantings: Palomar Mountain. May-June. n=10.

Senna Mill.
(Caesalpinioideae)

Herbage nearly glabrous; leaflets 4-8mm long; racemes terminal . . . . *S. armata*
Herbage densely white-pubescent; leaflets 1-3cm long; racemes axillary . . . . *S. covesii*

S. armata (Wats.)Irwin & Barneby. [*Cassia armata* Wats.] Frequent, sandy washes and open places; desert, below 200m: e.g. Borrego Palm Canyon; Yaqui Well; Wagon Wash; The Narrows; Palm Spring; Canebrake Canyon. Apr.-May. $2n$=14 pairs.

S. covesii (Gray)Irwin & Barneby. [*Cassia c.* Gray] Uncommon, sandy places; desert, below 800m: northeastern corner of county; Wagon Wash; Sentenac Canyon; Earthquake Valley; upper Blair Valley; Box Canyon; Mason Valley. Mar.-Apr.

Sesbania Adans., corr. Scop.
(Faboideae)

***S. exaltata** (Raf.)Cory. [*S. macrocarpa* Muhl.] Colorado River-Hemp. Native of eastern Colorado Desert and elsewhere. Local, adventive in alkaline areas: Bradbury Ranch in San Pasqual Valley, 120m. Aug. $2n$=12.

Spartium L.

***S. junceum** L. Spanish Broom. Native to Mediterranean Europe. Naturalized at lower Cottonwood Canyon near De Luz; the base of Mount Woodson and at Crest. Mar.-July.

Thermopsis R. Br. False-Lupine
(Faboideae)

T. macrophylla H. & A. ssp. **semota** (Jeps.)Beauchamp. [*T. californica* auct. non Wats.] Endemic to San Diego County. Occasional, in meadows and on grassy slopes; montane, above 1200m: Pine Hills; Wynola; Julian; North Peak; Middle Peak; Cuyamaca Lake; Laguna Camp; Laguna Lakes. Apr.-July. Type from Spencer Valley.

Trifolium L. Clover
(Faboideae)

Heads without an involucre at base of flowers
  Plants annual
    Flowers pedicellate, reflexed in age; calyx mostly glabrous
      Flowers purple, pink to white
        Flowers blood-red to white. . . . . *T. incarnatum*
        Flowers purple; calyx 10-nerved
          Plants villous on peduncles and calyx . . . . *T. bifidum*
          Plants nearly or quite glabrous
            Calyx not ciliolate on the lobes. . . . . *T. gracilentum*
            Calyx ciliolate on the lobes with short flat appendages. . . . . *T. ciliolatum*
      Flowers yellow; calyx 5-nerved . . . . *T. procumbens*
    Flowers sessile, not reflexed in age; calyx densely villous
      Heads sessile, usually in pairs, subtended by the upper leaves and their stipules . . . . *T. macraei*

- Heads peduncled, solitary, not subtended by upper leaves . . . . . *T. albopurpureum*
- Plants perennial
  - Peduncles axillary; flowers on pedicels mostly 4-5mm long, recurved in fruit . . . . . *T. repens*
  - Peduncles terminal or subterminal; flowers sessile or on shorter pedicels . . . . . *T. pratense*

Heads with an involucre at the base of the flowers, this reduced to a mere ring in *T. depauperatum*

- Corolla not inflated in age; involucral bracts united
  - Involucre campanulate to bowl-shaped . . . . . *T. microcephalum*
  - Involucre flat, rotate
    - Plants perennial . . . . . *T. wormskioldii*
    - Plants annual
      - Plants viscid-pubescent and clammy; corolla 12-14mm long, pale with a dark central spot . . . *T. obtusiflorum*
      - Plants glabrous
        - Leaflets obovate to oblanceolate to oblong; involucre with 4-12 lobes, these 3-7-toothed; clayx-teeth simple or one bifid
          - Involucre with 1-4 lobes . . . . . *T. variegatum* var. *pauciflorum*
          - Involucre with 4-12 lobes . . . . . *T. variegatum* var. *v.*
        - Leaflets linear to lance-oblong; involucre unevenly laciniate but not lobed; calyx-teeth often with 2 small lateral teeth or shoulders or simple, entire
          - Calyx-tube 10-nerved . . . . . *T. tridentatum* var. *t.*
          - Calyx tube 20-25-nerved . . . . . *T. tridentatum* var. *aciculare*
- Corolla conspicuously inflated in age; involucral bracts distinct
  - Flowers 12-25mm long
    - Involucral divisions 6-18mm long . . . . . *T. fucatum*
    - Involucral divisions 3-4mm long . . . . . *T. fragiferum*
  - Flowers 5-8mm long; involucral bracts to about 4mm long
    - Involucral bracts oblong, about as long as the calyx . . . . . *T. depauperatum* var. *truncatum*
    - Involucral bracts deeply lobed, the lobes 3-4mm long . . . . . *T. amplectens* var. *.a.*

T. albopurpureum T. & G. Rancheria Clover. Frequent, grassy slopes and meadows; foothill and montane, above 800m: e.g. Palomar Mountain State Park; Fry Creek; Warner Hot Springs; Henshaw Dam; Eagles Nest; Warners Pass; Indian Canyon; Chariot Canyon; Pine Hills; Anahuac School; Ramona; Bernardo; Mesa Grande; Descanso Creek; Springhill; Green Valley; Cuyamaca Lake; Harper Ranch; Descanso; Laguna Camp; Jacumba. Apr.-May. 2*n*=16.

T. amplectens T. & G. Pale Bladder Clover. Uncommon, grassy slopes; coastal, below 200m: Point Loma; La Jolla; Torrey Pines Mesa. Apr.-June.

T. bifidum Gray var. **decipiens** Greene. Pinole Clover. Rare, on grassy slopes: San Diego; Banner. Apr. 2*n*=16.

T. ciliolatum Benth. [*T. ciliatum* Nutt.] Tree Clover. Frequent, on grassy slopes and in fields; foothill, 200-1000m: e.g. upper San Onofre Canyon; De Luz; Fallbrook; Dyche Valley; Eagles Nest; Mesa Grande; Witch Creek; Pine Hills; Springhill; Ramona; Anahuac School; Barona Valley; Hidden Glen; Foster; El Capitan Mountain; Poway; Descanso; San Diego; Sears Ranch. Apr.-June. *n*=8.

T. depauperatum Desv. var. **truncatum** (Greene)Martin ex Isely. [*T. amplectens* var. *truncatum* (Greene)Jeps; *T. a.* var. *stenophyllum* (Nutt.)Jeps.; *T. anodon* Greene, the type from San Diego; *T. decodon* Greene, the type from San Diego.] Common, on grassy slopes; cismontane, below 1400m: e.g. Escondido; Oceanside; Poway; Fletcher Hills Ramona; Miramar Mesa; Mount Soledad; Pacific Beach; Kearny Mesa; Mission Hills; San Diego; Otay Mesa; Sweetwater Valley; Proctor Valley; Inspiration Point; Cuyamaca Lake; Harper Ranch, Henshaw Dam. Mar. May.

***T. fragiferum** L. Strawberry Clover. Native of Europe. Local, established on a hydro-seeded road-cut Torrey Pines Road. Mar.-June.

T. fucatum Lindl. var. **fucatum**. Bull Clover. Rare, on mesas and in fields; coastal, below 150m: San Diego; Chula Vista. Apr.-June. *n*=8.

T. gracilentum T. & G. Pin-Point Clover. Occasional, in grassy places; coastal and foothill, below 1100m: Oceanside, Will Valley; Mesa Grande; Warners Pass; Banner; Harper Ranch; Lakeside; Dulzura; San Ysidro; Silver Strand. Apr.-June.

***T. incarnatum** L. Native of Europe. Uncommon, escape from cultivation: Fallbrook, Carmel Valley. Mar.-May.

T. macraei H.& A. Chile Clover. Rare, known in the county from an old collection from Witch Creek, 800m. Apr. 2*n*=16.

T. microcephalum Pursh. Maiden Clover. Common, on grassy slopes; cismontane: e.g. San Onofre; Moro Hill; Fallbrook; De Luz; Rainbow; Morgan Hill; Fry Creek; Doane Valley; Will Valley; Henshaw Dam; Witch Creek; Pine Hills; Lakeside; Anahuac School; Cuyamaca Lake; Middle Peak; El Cajon Mountain; Foster; Escondido; Mount Soledad; Point Loma; Fletcher Hills; San Ysidro; Otay Lake; Dulzura; Lee Valley. May-June. *n*=8.

T. obtusiflorum Hook. Creek Clover. Frequent, in wet areas; cismontane: e.g. San Mateo; upper San Onofre Canyon; De Luz; Wyss Ranch; Palomar Mountain; Lake Henshaw; Witch Creek; Mussey Grade; Barona Valley; Viejas Valley; Pine Valley; Otay Mountain; Tecate Peak. May-July. 2*n*=16.

***T. pratense** L. Red Clover. Native of Eurasia. Occasional, in disturbed areas: Pine Hills; Laguna Meadows; Dulzura. Apr.-Oct. *n*=7.

***T. procumbens** L. Hop Clover. Native of Europe. Occasional, lawns: San Diego. Sep. *n*=7.

***T. repens** L. White Clover. Native of Europe. Infrequent, in meadows and lawns: Wyss Ranch; Warner Hot Springs. Apr.-Dec. *n*=16.

T. tridentatum Lindl. var. **aciculare** (Nutt.)McDer. Valley Clover. Common, on grassy slopes and fields; cismontane: e.g. De Luz; Fallbrook; Bonsall; Pala; Harrison Grade; Castro Canyon; Wyss Ranch; Banner; Pine Hills; Cuyamaca Lake; Clevenger Canyon; Ramona; Barona Valley; Foster; San Marcos; Del Dios; La Jolla; Point Loma; Lee Valley; San Miguel Mountain; Whispering Oaks; Hipass. Apr.-July.

T. tridentatum var. **tridentatum**. Tomcat Clover. Occasional, grassy slopes and fields; cismontane: e.g. Moro Hill; Poway Grade; Kearny Mesa; Soledad; Dulzura; Tecate Peak; Campo; Laguna Camp; Cuyamaca Lake. Mar.-June. 2*n*=16.

T. variegatum Nutt. in T.& G. var. **pauciflorum** (Nutt.)McDer. Occasional, moist areas; cismontane: Doane Valley; San Clemente Canyon; Lee Valley; Sears Ranch. Mar.-June.

T. variegatum var. **variegatum**. White-Tip Clover. Frequent, moist spots; cismontane: e.g. Morgan Hill; French Valley; Henshaw Dam; Hot Springs Mountain; upper Borrego Palm Canyon; Banner; Cuyamaca Lake; Laguna Public Camp; Campo; Walker Ranch; Escondido; Witch Creek; Point Loma; Mount Soledad; El Cajon Mountains. Mar.-June. 2*n*=16.

T. wormskioldii Lehm. Cow Clover. Common, moist areas; foothill and montane, above 800m: e.g. De Luz; Wyss Ranch; Palomar Observatory; Henshaw Dam; Warner Hot Springs; Eagles Nest; Volcan Mountain; Witch Creek; Pine Hills; San Pasqual; Cuyamaca Lake; Laguna Camp; Mesa Grande; Middle Peak; Descanso; Pine Valley; Campo; Morena Dam; Jacumba. May-July. *n*=16; 2*n*=ca.48.

Ulex L. Gorse
(Faboideae)

*__U. europaeus__ L. Prickly Broom. Native of Europe. Infrequent, naturalized in disturbed areas; foothill and coastal, below 400m: San Diego. Feb.-July. *n*=48.

Vicia L. Vetch
(Faboideae)

Flowers borne near the ends of evident peduncles; stipules lacking nectary patches
 Flowers 4-5mm long
  Calyx-teeth about as long as the tube; pods 8-12mm long . . . . *V. tetrasperma*
  Calyx-teeth about one-quarter as long as tube; pods 20-30mm long
   Stylar hairs encircling and of the same length . . . . *V. ludoviciana*
   Stylar hairs forming an abaxial brush . . . . *V. hassei*
 Flowers 10-18mm long
  Leaflets 4-8 pairs; peduncles 3-9 (-10)-flowered . . . . *V. americana*
  Leaflets mostly 8-12 pairs; peduncles with dense 1-sided, many-flowered racemes; plants annual . . . . . . *V. villosa*
Flowers 1-4 in leaf-axils, sessile or subsessile; stipules with nectary patches
 Leaves without tendrils; leaflets 1-3 pairs . . . . *V. faba*
 Leaves with tendrils; leaflets more than 3 pairs
  Flowers 18-25mm long, purple with violet wings; pods torulose; seeds 5mm broad . . . . *V. sativa*
  Flowers 10-18mm long, uniformly purple; pod plane; seeds 3mm broad . . . . *V. angustifolia*

V. americana Muhl. ex Willd. [*V. a.* var. *linearis* (Nutt.)Wats.; *V. a.* ssp. *oregana* (Nutt.)Abrams; *V. a.* var. *truncata* (Nutt.)Brew.; *V. californica* Greene; *V. californica* Greene] American Vetch. Infrequent, on shrubby slopes; foothill and montane: Fallbrook; Rainbow; Mesa Grande; Palomar Mountain; Cuyamaca Lake; Witch Creek; Volcan Mountain; Garnet Peak; Pine Hills; Anahuac School; Lyons Valley; Corte Madera. Apr.-June.

*__V. angustifolia__ Reichard. Common Vetch. Native of Old World. Uncommon introduction as at Morena Dam. 2*n*=12.

V. ludoviciana Nutt. [*V. exigua* Nutt. in T. & G.] Deerpea Vetch. Common, on grassy slopes; coastal and foothill, below 900m: e.g. De Luz; Moro Hill; Monserrate; Lilac; Rainbow; Pala; Hormiguero Ranch; Witch Creek; Ramona; Foster; San Pasqual; Soledad; Mission Valley; Balboa Park; Imperial Beach; Lee Valley; Pine Valley; San Felipe Valley. Mar.-May.

*__V. faba__ L. Broad-Bean. Native of Eurasia. Local, escaping from cultivation: San Diego; Montezuma Mesa. Apr.-June. 2*n*=12, 14.

V. hassei Wats. [*V. exigua* Nutt. in T.& G. var. *h.* (Wats.)Jeps.] Slender Vetch. Infrequent, in shaded locations; cismontane, below 1000m: Moro Hill; San Miguel Mountain; Lyons Valley. Mar.-Apr.

*__V. sativa__ L. Spring Vetch. Native of Europe. Local, escaping from cultivation: Hidden Glen, 630m. May. 2*n*=12, 14.

*__V. tetrasperma__ (L.)Moench. Slender Tare. Native of Europe. Infrequent, naturalized in disturbed areas: Mount Soledad, 100m. Apr. 2*n*=14.

*__V. villosa__ Roth. [*V. dasycarpa* Tenore] Winter Vetch. Native of Europe. Infrequent, naturalized in disturbed areas; coastal and foothill, below 900m: De Luz Creek; Rancho Santa Fe; Vista Grande; east of Ramona. Apr.-May. 2*n*=14.

FAGACEAE. Oak Family

Quercus L. Oak

Bark dark, not scaly but smooth, or on old trunks irregularly ruptured; stigmas on slender styles; involucral cups with thin closely imbricated scales; nut tomentose on inner surface (Black Oaks)
 Plant deciduous, the leaves usually 1dm or more long, with large bristle-tipped lobes . . . . *Q. kelloggii*
 Plant evergreen, the leaves smaller, coriaceous
  Acorns maturing the second autumn; leaves plane, glabrous beneath . . . . *Q. wislizenii*

Acorns maturing the first autumn; leaves mostly convex on upper surface, with hair beneath, especially in axils of leaves
Leaves somewhat stellate-pubescent beneath . . . . *Q. agrifolia* var. *a.*
Leaves densely stellate beneath . . . . *Q. agrifolia* var. *oxyadenia*
Bark light in color, scaly, furrowed only on large trees; stigmas subsessile; involucral cups usually with tuberculate scales (White Oaks)
Fruit maturing the first season; shell of the nut glabrous on the inner surface
Tree; leaves mostly 3-6cm long; cup enclosing nearly half the nut . . . . *Q. engelmannii*
Shrubs, occasionally somewhat arborescent; leaves mostly 1-2.5 cm long; cup more shallow
Leaf-blades shining and subglabrous above; cismontane . . . . *Q. dumosa*
Leaf-blades stellate-pubescent above and dull; largely desert. . . . . *O. cornelius-mulleri*
Fruit maturing the second season; shell of the nut tomentose on the inner surface
Trees; acorn-cup large, thick-walled, densely tomentose; leaves mostly 2-5cm long, the lateral veins not prominent or parallel . . . . *Q. chrysolepis*
Plant shrubby; the cup with rather thin walls; branchlets rigid; leaves spinose-toothed; mostly desert-edge . . . . *Q. dunnii*

Q. agrifolia Nees var. **agrifolia**. Coast Live Oak, Encina. Common, in valleys and on slopes, an indicator of Southern Oak Woodland; cismontane, below 1200m: e.g. De Luz; Rincon; Oak Grove; Julian; Cuyamaca Mountains; Tecolote Canyon; El Monte County Park; Alpine; Campo; Live Oak Springs. Mar.-Apr. n=12.

Q. agrifolia var. **oxyadenia** (Torr.)J.T. Howell. Interior Live Oak. Common, on slopes in Chaparral and Southern Oak Woodland; foothill, 600-1200m: e.g. Doane Valley; Banner; San Felipe Valley; Live Oak Springs; Descanso; Dehesa; Boulevard; Campo; Jacumba; The Willows; Dulzura Creek. Mar.-Apr. Type from "Santo Isabelle (Santa Ysabel)".

Q. chrysolepis Liebm. Canyon Live Oak. Common, on slopes and in canyons; montane, above 900m: e.g. Agua Tibia; Palomar, Hot Springs, Volcan, Cuyamaca, and Laguna Mountains; Banner Grade; Morena Lake. Apr.-May. n=12.

Q. cornelius-mulleri Nixon & Steele. [*Q. acutidens* Torr., the type probably from east of San Luis Rey, near Warner Springs] Desert Scrub Oak. Occasional, in Chaparral on rocky slopes; desert and foothill, above 700m: Montezuma Valley; San Felipe; Banner; Oriflamme Canyon; Desert View; Manzanita Indian Reservation; McCain Valley; Mountain Springs. Mar.-May. Type from McCain Valley.

Q. dumosa Nutt. Scrub Oak. Prevalent, in Chaparral; cismontane: e.g. Santa Margarita Mountains; Fallbrook; Lost Valley; Palomar, Hot Springs, Volcan, Cuyamaca, Laguna, and Corte Madera Mountains; Buckman Springs; Boulevard; Echo Valley; Campo; Otay Mountain; Point Loma; Del Mar; Bernardo. Apr.-July. n=12.

Q. dunnii Kell. [*Q. palmeri* Engelm., the type from "mountains 80 miles east of San Diego".] Palmer's Oak. Infrequent, on rocky slopes; desert, 700-1300m: Montezuma View Point; Bankhead Springs; Boulevard; Jacumba. Apr.-May.

Q. engelmannii Greene. Mesa Blue Oak. Frequent, in canyons and on open slopes; foothill and coastal, below 1300m: e.g. Santa Margarita Mountains; Fallbrook; Mesa Grande; Volcan Mountains; San Felipe; Lost Valley; Banner; Descanso; Bernardo; Dulzura; El Nido; Alpine; Buckman Spring. Apr.-May. n=12.

Q. × ganderi C.B. Wolf. (*Q. agrifolia* var. *oxyadenia* × *Q. kelloggii*.) Gander Oak. Local, on wooded slopes about Mesa Grande, 900-1000m: Volcan Indian School; Morretti Ranch; Mesa Grande. Mar.-Apr. Type from about 300 yards north of the entrance to the Volcan Indian School on California State Highway No. 79 between Santa Ysabel and Lake Henshaw, at 3100 feet elevation, in decomposed granite soil, near an occasional stream just west of the highway.

Q. × grandidentata Ewan. (*Q. engelmannii* × *Q. dumosa*). Infrequent, on open slopes; foothill and montane, above 400m: San Felipe; Echo Valley. Apr.-May.

Q. kelloggii Newb. California Black Oak. Prevalent, in Montane Coniferous Forest; montane, above 1050m: e.g. Palomar, Hot Springs, Volcan, Cuyamaca, and Laguna Mountains. Apr.-May. n=12.

Q. × morehus Kell. (*Q. kelloggii* × *Q. wislizenii* var. *frutescens*). Oracle Oak. Uncommon, in Montane Coniferous Forest; montane, above 1000m: Palomar Mountain; Santa Ysabel; Hot Springs Mountain; Monument Peak; Noble Canyon; Guatay Mountain. Apr.-May.

Q. wislizenii A. DC. var. **frutescens** Engelm. Scrub Live Oak. Occasional, in Chaparral; foothill and montane, above 1000m: Agua Tibia, Palomar, Hot Springs, Cuyamaca, and Laguna Mountains; Boulevard. Mar.-May.

FOUQUIERIACEAE. Ocotillo Family

Fouquieria H.B.K. Ocotillo

F. splendens Engelm. Prevalent, on rocky slopes and alluvial fans; desert, below 900m: e.g. Borrego Valley; Borrego Palm Canyon; Sentenac Canyon; Fish Creek; Pinyon Mountain; Box Canyon; Vallecito Stage Station; Bow Willow; Dos Cabezas; Mountain Springs. Mar.-July. $2n$=12 pairs.

FRANKENIACEAE.

Frankenia L.

Leaves 5-15mm long, linear-oblanceolate to obovate, somewhat expanded; style 3-cleft . . . . *F. salina*
Leaves 2-4mm long, revolute and terete; style 2-cleft . . . . *F. palmeri*

F. palmeri Wats. Yerba Reuma. Rare, upper Coastal Salt Marsh; coastal, below 10m: mouth of the Sweetwater River; E Street Marsh on Gunpowder Point in San Diego Bay. May-June.

F. salina Jtn. [*F. grandifolia* Cham. & Schlecht.; *F. g.* var. *campestris* Gray] Alkali-Heath. Abundant, in Coastal Salt Marsh and other alkaline spots; coastal, and occasionally foothill, below 800m: e.g. San Mateo Creek; Lake Guajome; Oak Grove; San Marcos Creek; Encinitas; Del Mar; La Jolla; San Diego Bay; Bonita; Lakeside; National City; Imperial Beach; Tijuana River estuary. June-Oct.

FUMARIACEAE

KEY TO GENERA

Outer petals both saccate or spurred at base . . . . *Dicentra*
Outer petals with one spurred, the other not . . . . *Fumaria*

Dicentra Bernh.

D. chrysantha (H.& A.)Walp. Golden Ear-Drops. Occasional, along roadways, firebreaks, and other disturbed areas; foothill and montane, above 300m: e.g. Poway Grade; Bear Valley; Banner Grade; Laguna Junction; East Mesa, Cuyamaca Rancho State Park; Pine Valley; Buckman Spring; Echo Valley; Campo; Lawson Peak; Potrero Grade; Otay Mountain; Cowles Mountain; San Miguel Mountain; north of Alpine; Deer Spring & I-15; Clevenger Canyon. Apr.-Sep. n=12.

Fumaria L. Fumitory

Flowers purple, 5-7mm long; nutlet depressed-globose . . . . *F. officinalis*
Flowers cream with inner petals purple at tip, 3-4mm long; nutlet apiculate . . . . *F. parviflora*

***F. officinalis** L. Native of Europe. Rare weed as at Pacific Beach, 10m. Apr.-July. n=16

***F. parviflora** Lam. Native of Europe. Rare, agricultural weed at Harborside, Chula Vista. Mar.-May. 2n=28

GARRYACEAE. Silk-Tassel Family

Garrya Dougl. Silk-Tassel Bush

Lower leaf surfaces densely covered with strongly curled, woolly or villous trichomes; ovary densely pubescent . . . . *G. veatchii*
Lower leaf surfaces glabrous, slightly or densely covered with straight or somewhat wavy upward appressed trichomes; ovary glabrous, sparsely or densely pubescent
Lower leaf surfaces glabrous or very sparsely covered with pilose trichomes when young . . . . *G. fremontii*
Lower leaf surfaces more or less covered with straight or wavy upwardly appressed trichomes
Lower leaf surfaces glabrate or sparsely to moderately covered with coarse, wavy, more or less appressed and ascending trichomes . . . . *G. flavescens* ssp. *pallida*
Lower leaf surfaces densely covered with fine, straight, long, upwardly appressed trichomes . *G. flavescens* ssp. *f.*

G. flavescens Wats. ssp. **flavescens** Infrequent, in Chaparral; montane and transmontane, above 1000m: Deer Flat, Palomar Mountain; Middle Peak, Cuyamaca Mountains; Inspiration Point; Harper Ranch; Oasis Spring; Monument Peak; Laguna Camp; Vallecito Canyon. Mar.-May.

G. flavescens ssp. **pallida** (Eastw.)Dahling. Occasional, in Chaparral; montane and transmontane, above 1200m: Oak Grove Trail on Aguanga Mountain; Hot Springs Mountain; Montezuma Valley; San Felipe; Julian; Harper Ranch; Middle Peak, Cuyamaca Mountains. Feb.-Apr.

G. fremontii Torr. Rare, in Chaparral; foothill and montane, above 1000m: west of Jacumba; Oasis Spring; Monument Peak. Jan-Apr.

G. veatchii Kell. [*G. v.* var. *palmeri* (Wats.)Eastw., the type from Milguatay (Campo Valley), 60 miles from San Diego, on the Fort Yuma road.] Infrequent, in Chaparral; foothill, 300-1200m: Dodge Valley; Mesa Grande; Warner Springs;El Capitan Mountain; east of Encinitas; Mount Woodson; Lyons Peak; Descanso Junction; Japatul Valley; La Posta; Boulevard; Clover Flats; Campo; Morena Dam; Buckman Spring; Bankhead Springs. Feb.-Apr.

GENTIANACEAE. Gentian Family

KEY TO GENERA

Corolla rotate and with conspicuous glands on upper surface; flowers 4-merous; in a terminal panicle . . . . . . *Swertia*
Corolla campanulate to funnelform or salverform, without glands on upper surface
Anthers coiled or spirally twisted after anthesis; flowers mostly red or pink . . . . *Centaurium*
Anthers not coiled or twisted after anthesis . . . . *Eustoma*

Centaurium Hill. Centaury

Corolla-lobes less than half as long as the tube; anthers 1.5-2.5mm long, oblong . . . . *C. exaltatum*
Corolla-lobes more than half as long as tube; anthers 3.5mm long, linear . . . . *C. venustum*

C. exaltatum (Griseb.)W. Wight. Infrequent, in seepage areas and along streams; foothill, 300-1100m: Pine Valley; Cuyamaca Mountains; Lake Hodges; San Pasqual; Barrett Junction; Harbison Canyon; Jamul; McCain Valley; Carrizo Creek. May-Aug. n=40

C. venustum (Gray)Rob. Canchalagua. Frequent, in grassy areas and on burns; coastal and foothill, below 1000m: e.g. De Luz; Fallbrook; Vista; Bonsall; Pala; Palomar Mountain; Mesa Grande; Barona Valley; Hauser Creek; Alpine; Espinosa Trail; Japatul Valley; Otay Mountain; Point Loma; Kearny Mesa; Rancho Santa Fe; Torrey Pines. Also Indian Canyon and Mason Valley. Mar.-July. n=17

Eustoma Salisb. Catchfly Gentian

E. exaltatum (L.)Griseb. Alkali Chalice. Infrequent, seepage areas and along water courses; desert, below 400m: San Felipe; Vallecito and Carrizo Stage Stations; Mountain Palm Springs. Also Fallbrook. Mar.-Sep.

Swertia L. Green Gentian

S. parryi (Torr.)Kuntze. [*Frasera parryi* Torr.] Deer's Ears. Occasional, in open areas of Chaparral and Montane Coniferous Forest; foothill and montane, above 300m: Encinitas; Rincon Grade; Palomar Mountain; Oak Grove; San Felipe; Ballena; Descanso; El Capitan; Pine Valley; Campo; Hipass; Jacumba; Corte Madera; Monument Peak; Pine Hills. Apr.-July. Type from mountains east of San Diego.

GERANIACEAE. Geranium Family

KEY TO GENERA

Flowers without nectar-spur, and with glands alternating with the petals; petals alike
- Leaves palmately veined or divided; stamens all bearing anthers . . . *Geranium*
- Leaves pinnately veined or divided; stamens having outer filaments without anthers . . . *Erodium*

Flowers with nectar-spur at base of calyx (discovered by sectioning pedicel) and without glands; upper petals usually larger . . . *Pelargonium*

Erodium L'Her. Filaree

Leaves somewhat cordate at base, usually 3-more-lobed
- Petals unequal, much exceeding sepals; pedicels strigulose, not glandular; deserts . . . *E. texanum*
- Petals equal, rarely longer than sepals; cismontane
  - Leaves deeply 3-5-parted; pedicels retrorse-pubescent, not glandular; carpel-bodies 5-6mm long . . . *E. cygnorum*
  - Leaves shallowly 5-7-lobed; pedicels glandular-pilose; carpel-bodies 8-10mm long . . . *E. macrophyllum*

Leaves not cordate at base
- Leaves simple, often lobed or divided; style-column 5-12cm long
  - Style-column 5-9cm long; concavities at top of fruit subtended by a single fold, the upper part of the carpel-body pubescent; sepals with a short green tip . . . *E. brachycarpum*
  - Style-column 9-12cm long; concavities at top of fruit subtended by 2 folds, the upper part of the carpel-body glabrous; sepals with a prominent reddish mucro . . . *E. botrys*
- Leaves pinnate; style-column 2.5-4cm long
  - Leaves broad, coarsely toothed or serrate; sepal-tips not setose; claws of petals glabrous . . . *E. moschatum*
  - Leaves pinnately lobed or divided; sepal-tips setose; claws of petals ciliate . . . *E. cicutarium*

***E. botrys** (Cav.)Bertol. Long-beak Filaree. Native of Europe. Abundant, in open and disturbed areas; coastal and foothill, below 1100m: Palomar Mountain; San Pasqual; San Marcos; Lakeside; Spencer Valley; Dulzura; Point Loma; Mahogany Canyon; Kearny Mesa; El Cajon Valley; Lake Hodges. Mar.-May. n=20.

***E. brachycarpum** (Godr.)Thell. [*E. obtusiplicatum* (Maire, Weiller, & Wilcz.)J.T. Howell.] Short-beak Filaree. Native of North Africa. Occasional, grassy places; foothill and coastal, below 900m: e.g. Lake Henshaw Resort; Escondido; Buena; San Pasqual; Ramona; Poway Grade; Kearny Mesa; Lakeside; Fletcher Hills; Dulzura; San Miguel Mountain; Proctor Valley; Point Loma; Mount Soledad. Apr.-Aug. $2n$=40.

***E. cicutarium** (L.)L'Her. Red-stem Filaree. Native of Europe. Abundant, open and disturbed places; cismontane and desert: e.g. San Onofre, San Luis Rey; Pauma School; Palomar, Hot Springs, Cuyamaca, and Laguna Mountains; Spencer Valley; Pine Valley; El Capitan; Pacific Beach; Kearny Mesa; Silver Strand; Borrego Valley; upper Blair Valley; Yaqui Pass; Vallecito and Carrizo Stage Stations. Feb.-May. $2n$=40.

***E. cygnorum** Nees. Swan Filaree. Native of Australia. Local, naturalized on "mesa near San Diego". Mar.-May.

E. macrophyllum H.& A. var. **macrophyllum**. Large-leaf Filaree. Rare, open places; coastal, below 300m: El Capitan Dam; San Diego; calcareous clay at old cement factory quarry, Jamul Ranch. Mar.-May.

***E. moschatum** (L.)L'Her. White-stem Filaree. Native of Europe. Prevalent, disturbed and open areas; foothill and coastal, below 600m: e.g. San Onofre Bluffs; Vista; San Luis Rey; Palomar; San Pasqual; San Diego; Lake Murray; Roseville; Barrett Junction; Dulzura. Feb.-May. $2n$=20.

E. texanum Gray. Desert Filaree. Uncommon, sandy places; desert, below 900m: Coyote Creek; Borrego Palm Canyon; The Narrows; Yaqui Pass; Dry Lake School; Jade, east of Jacumba. Also San Diego. Mar.-May. $2n$=20.

Geranium L. Cranesbill

Petals mostly 3-8mm long; mature style-column less than 2 cm long; annuals or perennials, mostly introduced
- Sepals prominently awned or subulate-tipped; seeds reticulate
  - Plants perennial from a carrotlike root; no hairs with gland-tips . . . *G. retrorsum*
  - Plants annual; some hairs of pedicels gland-tipped
    - Carpel-bodies with short stiff subequal spreading hairs; lobes of upper leaves acute; pits of seed square or rounded . . . *G. dissectum*
    - Carpel-bodies with long unequal ascending hairs; lobes of leaves obtuse; pits of seed elongate *G. carolinianum*
- Sepals awnless, at most with callous tips; seeds smooth . . . *G. molle*

Petals 10-20mm long; mature style-column 2-5cm long; native perennials
- Style-branches 3-4mm long; petals white or whitish; pedicels usually with purplish glands . . . *G. richardsonii*
- Style-branches 4.5-9mm long; petals pink or lavender; pedicels with yellowish glands . . . *G. californicum*

G. californicum Jones & Jones. California Geranium. Rare, in shaded places of the Montane Coniferous Forest; montane, above 1500m: Laguna Mountains. June-July.

***G. carolinianum** L. Carolina Geranium. Native of eastern North America. Occasional, in grassy places; cismontane, below 1500m: De Luz; Fallbrook; San Luis Rey; Rainbow; Palomar Mountain; Poway Grade; Featherstone Creek; Murray Dam; Rolando; Mount Soledad; Balboa Park. Feb.-Apr. n-26.

***G. dissectum** L. Cut-leaf Geranium. Native of Europe. Local, waste places: Kearny Mesa, 120m. Apr. $2n=22$.

***G. molle** L. Dove-foot Geranium. Native of Europe. Infrequent, lawns and moist places; foothill and coastal, below 800m: Lake Henshaw; Kearny Mesa; San Marcos. May-July. $n=13$.

***G. retrorsum** L'Her. [*G. glabratum* auct. non Small] Traveling Geranium. Native of Australia. Local, adventive at Nellie, 1500m. May-Sep.

***G. richardsonii** Fisch. & Trautv. Richardson Geranium. Native of western United States but not of San Diego County. Infrequent, escaping from cultivation; cismontane: Dehesa, 150m. July-Aug. $n=26$.

Pelargonium L'Her. Geranium

Stems fleshy at least when young; leaves scarcely if at all lobed .... *P. zonale*
Stems not fleshy, shallowly several-lobed .... *P. grossularioides*

***P. grossularioides** (L.)Ait. Coconut-scented Geranium. Native of South Africa. Local, escaping from cultivation; coastal: Fallbrook, 220m. Mar.-May.

***P. zonale** (L.)Ait. Zonal Geranium. Native of South Africa. Local, escaping from cultivation; coastal: Oceanside. Mar.-May.

GROSSULARIACEAE. Currant Family

Ribes L. Currant; Gooseberry

Nodal spines lacking; berry spineless; flowers usually several to many in raceme
  Plants evergreen .... *R. viburnifolium*
  Plants deciduous
    Flowers white or greenish-white
      Free portion of floral-tube 2.5-3 times as long as broad; berry red .... *R. cereum*
      Free portion of floral-tube 1-2 times as long as broad; berry blue or black .... *R. indecorum*
    Flowers pink or red or purplish
      Sepals erect; free part of floral-tube short, bowl-shaped .... *R. nevadense*
      Sepals spreading or recurved; free part of floral-tube cylindric to urceolate .... *R. canthariforme*
Nodal spines present; berry often spiny; flowers mostly 1-few
  Sepals 4, red, erect, about 1cm long; petals as long; stamens 2-3 times as long .... *R. speciosum*
  Sepals 5, mostly not erect; petals shorter than sepals
    Ovary and berry without spines or prickles .... *R. quercetorum*
    Ovary and berry with spines or bristles
      Ovary and berry usually with all or nearly all of the bristles gland-tipped; leaves glandular beneath *R. amarum*
      Ovary and berry with eglandular spines; leaves not glandular beneath .... *R. roezlii*

R. amarum McClat. Bitter Gooseberry. Uncommon, in shaded areas of Montane Coniferous Forest; north montane, above 1200m: Palomar Mountains. Feb.-Apr.

R. canthariforme Wiggins. Moreno Currant. Endemic to San Diego County. Local, in the shade of large rocks in the Chaparral; foothill, 500-1200m: Barona Valley; El Cajon Mountain; Descanso Junction; Lyons Peak; Corte Madera; near Morena Dam. Feb.-June. Type in Chaparral on a northeast slope near Morena Dam.

R. cereum Dougl. Squaw Currant. Rare, known in the county only from a collection on Pinyon Mountain. $n=8$.

R. indecorum Eastw. Winter Currant. Frequent, in Chaparral; foothill, 200-1300m: e.g. Rainbow; Fallbrook; Julian; Oriflamme Canyon; Mesa Grande; Poway; Olivenhain; San Diego; Campo; Jamul; Otay Mountain. Also Monument Peak. Type from Cajon Heights, near San Diego.

R. nevadense Kell. Sierra Currant. Infrequent, along streams; montane, above 1300m: Palomar Mountain; Cutca Valley; Eagles Nest; Cuyamaca Mountains. Apr.-June.

R. quercetorum Greene. Rock Gooseberry. Occasional, on dry slopes in Chaparral and Southern Oak Woodland; foothill, 700-1300m: Warner Ranch; Montezuma View Point; Banner; Grapevine Canyon; Angelina Spring; La Posta; Jacumba; Eckener Pass. Also Otay Mountain. Mar.-Apr.

R. roezlii Regel. Sierra Gooseberry. Infrequent, in open areas of Montane Coniferous Forest; montane, above 1200m: Palomar Mountain; Castro Canyon; Hot Springs Mountain; Cuyamaca Mountains. May-July.

R. speciosum Pursh. Fuchsia-Flower Gooseberry. Frequent, on shaded slopes; coastal and foothill, below 700m: e.g. Santa Margarita River estuary; Fallbrook; Poway; Del Dios; Carroll Canyon; Murphy Canyon; Mission Valley. Feb.-June.

***R. viburnifolium** Gray. Island Currant. Native to Santa Catalina Island and northwestern Baja California. Persisting from a planting in San Clemente Canyon, San Diego, 100m. Mar.-May. $n=8$.

HALORAGACEAE. Water-Milfoil Family

Myriophyllum L. Water-Milfoil

M. exalbescens Fern. American Milfoil. Rare, in shallow fresh water bodies; cismontane: Lilac; Lake Murray. June-Sep. $2n=14$.

HYDROPHYLLACEAE. Waterleaf Family

Styles mostly united
  Perennial herbs
    Leaves entire to toothed, but not deeply divided or lobed
      Stamens unequally inserted, unequal in length . . . . . *Nama*
      Stamens equally inserted, subequal in length . . . . . *Phacelia*
    Leaves deeply divided or lobed, largely cauline . . . . . *Phacelia*
  Annual herbs
    Herbage viscid and scented; ovules borne on both faces of the placentae; calyx lacking auricles between the lobes . . . . . *Eucrypta*
    Herbage not viscid or scented; ovules borne only on the front of the placentae
      Ovary 1-celled, with broad placentae; inflorescence not a clearly marked cyme; calyx usually with evident auricles between lobes
        Stems recurved-prickly; capsules prickly or bristly; seeds without a cucullus or hood-like process on the seed . . . . . *Pholistoma*
        Stems usually not recurved-prickly; capsules pubescent; seeds with a terminal cucullus . . . . . *Nemophila*
      Ovary falsely 2-celled by intrusion of narrow placentae; inflorescence usually an evident coiled cyme
        Stamens equally inserted, and if unequal in length, the seeds corrugate
          Corolla mostly deciduous, blue to purplish or white, if persistent, the flowers erect . . . . . *Phacelia*
          Corolla persistent, yellowish or reddish, pendulous . . . . . *Emmenanthe*
        Stamens unequally inserted and unequal in length; seeds ovoid, reticulate, often pitted . . . . . *Nama*
Styles separate to base
  Low annuals; seeds pitted or alveolate . . . . . *Nama*
  Perennial or shrubby plants
    Herbaceous plants; flowers in an elongate thyrsoid panicle or cymes; seeds longitudinally striate . . . . . *Turricula*
    Shrubby plants; flowers not paniculate; seeds transversely corrugate or winged
      Leaves to 1.5dm long . . . . . *Eriodictyon*
      Leaves to 4.5dm long . . . . . *Wigandia*

Emmenanthe Benth.

E. penduliflora Benth. Whispering Bells. Frequent, on open slopes and burns; coastal, foothill, and desert, below 1000m: e.g. Moosa Canyon; Lilac; Pala; Eagles Nest; Chihuahua Valley; Borrego Valley; Hellhole Canyon; northeast corner of county; Coyote Creek; Yaqui Well; San Felipe; lower Box Canyon; Banner; Vallecito Stage Station; Springhill; Buckman Spring; Boulevard; southeast corner of county; InKoPah County Park; Mountain Springs; Tecate Peak; Dulzura; Otay Mountain; Foster; East San Diego; Ramona; Mission Valley; San Ysidro. Apr.-June. *n*=18.

Eriodictyon Benth. Yerba Santa

Upper surface of leaves and the young stems glabrous or subglabrous, glutinous . . . . . *E. trichocalyx*
Upper surfaces of leaves and young stems pubescent to tomentose . . . . . *E. crassifolium*

E. crassifolium Benth. Occasional, in dry areas; foothill and coastal, below 900m: e.g. San Luis Rey; Rainbow; Pala; Oak Grove; Los Tules; Encinitas; Mussey Grade; Barona Valley; Kearny Mesa; Poway Grade; Torrey Pines; Murphy Canyon; Point Loma; Chollas Valley; San Miguel Mountain; Lyons Peak; El Capitan; Hidden Glen; Marron Valley; Lee Valley. Apr.-June. Type from San Diego. *n*=18.

E. trichocalyx Heller ssp. **lanatum** (Brand)Munz. Occasional, on rocky slopes; desert and south foothill: Otay Mountain; Sweetwater Valley; Dulzura; Tecate Peak; Potrero; Campo; Boulevard; Jacumba; Laguna Mountains; Oriflamme Canyon; Canebrake Canyon; Vallecito Canyon; Montezuma View Point; San Felipe; upper Borrego Palm Canyon. Mar.-July. Type from between Campo and Jacumba.

Eucrypta Nutt.

Leaves 2-3-pinnatifid, mostly opposite; mature calyx stellate-rotate beneath the capsule
  Leaves 13-9-lobed . . . . . *E. chrysanthemifolia* var. *c.*
  Leaves 7-9 lobed . . . . . *E. chrysanthemifolia* var. *bipinnatifida*
Leaves 1-pinnate, alternate, the lobes mostly entire or few-toothed; mature calyx erect and enclosing capsule . . . . . *E. micrantha*

E. chrysanthemifolia (Benth.)Greene var. **bipinnatifida** (Torr.)Const. Infrequent, in rocky areas; desert, below 700m: Lone Palm; Borrego Palm Canyon; San Felipe Valley; Box Canyon. Also Hipass; Whispering Oaks. Mar.-Apr.

E. chrysanthemifolia var. **chrysanthemifolia**. Frequent, in shaded places; coastal and foothill, below 900m: e.g. De Luz; Fallbrook; Lilac; Pala; Moosa Canyon; Red Mountain Grade; Bonsall; Springhill; Miramar; Poway Grade; Mount Soledad; Mission Valley; Point Loma; Chollas Heights; Lyons Valley; Foster; Potrero; Campo; Jacumba; Chariot Canyon; Montezuma View Point. Feb.-June. *n*=10.

E. micrantha (Torr.)Heller. Infrequent, in shaded areas; desert, below 700m: northeast corner of county; Borrego Valley; Mason Valley; Vallecito Stage Station; Jade, east of Jacumba. Mar.-Apr. *n*=6, 12.

Nama L.

Calyx divided ca. 3/4 its length, the tubular base adnate to the somewhat inferior ovary; calyx-lobes indurate and recurved in fruit . . . . . *N. stenocarpum*
Calyx divided nearly to base, the ovary superior; calyx-lobes mostly erect and not indurate
  Stems quite erect; adnate portion of filaments not winged; seeds many, dark brown
    Leaves short-hispid, plane to somewhat revolute . . . . . *N. hispidum* var. *spathulatum*
    Leaves soft-hirsute as well as hispid, strongly revolute . . . . . *N. hispidum* var. *revolutum*
  Stems prostrate; adnate portion of filaments with narrow margin; seeds 15-25, yellowish-brown . . . . . *N. demissum*

N. demissum Gray var. **deserti** Brand. Purple Mat. Occasional, in sandy areas; desert, below 1000m: Lone Palm; Borrego Valley; Borrego Palm Canyon; Montezuma View Point; Banner Grade; San Felipe Wash; Mason Valley; Vallecito Stage Station; Mountain Springs; southeast corner of county; Boulevard. Mar.-May. *n*=7.

N. hispidum Gray var. **revolutum** Jeps. Rare, in sandy areas; desert, below 300m: Borrego Valley; Split Mountain. Mar.

N. hispidum var. **spathulatum** (Torr.)C.L. Hitchc. Rare, in sandy areas; desert, below 300m: Dry Lake School; Mountain Palm Springs. Mar.-Apr.

N. stenocarpum Gray. Local, on muddy shore of ponds and lakes; coastal, below 300m: San Luis Rey; Sweetwater Lake; Rickey Dam; Sweetwater Valley. Jan.-June. *n*=7.

Nemophila Nutt. ex Barton

Corolla 0.7-4cm broad
  Corolla mostly deep blue
    Lower leaves with 5-7 mostly entire lobes . . . . *N. menziesii* ssp. *integrifolia*
    Lower leaves with 9-11 rounded or oblong divisions . . . . *N. menziesii* ssp. *m.*
  Corolla light blue . . . . *N. rotata*
Corolla 0.3-0.6cm broad, mostly paler
  Leaves oblong or oval, deeply divided, weakly cuneate to truncate at base . . . . *N. pedunculata*
  Leaves spatulate, shallowly lobed, strongly cuneate at base . . . . *N. spatulata*

N. menziesii H. & A. ssp. **integrifolia** (Parish)Munz. Frequent, in shady places, Montane Coniferous Forest and Southern Oak Woodland; foothill and montane, above 600m: e.g. Palomar Mountain; Lake Henshaw; Eagles Nest; Mesa Grande; Julian; Cuyamaca Mountains; Laguna Mountains; Descanso; Corte Madera; Buckman Spring; Campo. Apr.-June. *n*=9.

N. menziesii ssp. **menziesii**. Baby Blue Eyes. Occasional, in open areas; foothill and montane, above 800m: Palomar Mountain; Oak Grove; Chihuahua Valley; Warner Ranch; Cañada Buena Vista; Montezuma View Point; Culp Valley; Banner Grade; Foster. Apr.-May.

N. pedunculata Dougl. ex Benth. Rare, known in the county only from old collection from Witch Creek, 810m. May. *n*=9.

N. rotata Eastw. Occasional, in shady places; coastal and foothill, below 1000m: Fallbrook; Bonsall; Ramona; Point Loma; San Diego; Lyons Valley; Potrero Grade; Campo; Buckman Spring. Mar.-May. Type from near San Diego.

N. spatulata Cov. Scarce, in shaded areas of Chaparral; foothill, 800-1100m: Echo Valley; Campo; Cameron Corners. May-July. *n*=9.

Phacelia Juss.

Plants perennial or biennial
  Leaves pinnate into toothed or pinnatifid lobes; calyx-lobes mostly spatulate; corolla 6-8mm long
    Stems pubescent and setose-hispid; lacking gland-tipped hairs . . . . *P. ramosissima* var. *austrolitoralis*
    Stems glandular-pubescent and hispid, with basal pustules . . . . *P. suffrutescens*
  Leaves entire to pinnate into entire lobes or leaflets; calyx-lobes mostly not widened upward
    Inflorescence mostly elongate, narrow, dense, consisting of many short lateral cymes along a central axis; stems mostly simple . . . . *P. imbricata* ssp. *bernardina*
    Inflorescence usually shorter, more open, widely branched; stems often branched except in reduced forms . . . . *P. imbricata* ssp. *patula*
Plants annual
  Seeds not transversely corrugated, or if so, the ovules only 2 to a placenta
    Leaves pinnately toothed to compound, their divisions further toothed or divided
      Seeds terete or angled, usually pitted, but not excavated on either side of a salient ridge; plants not markedly viscid or ill-scented
        Corolla lavender or white; calyx-lobes linear or nearly so, up to 1cm long in fruit, often conspicuously clawed and loosely enveloping the capsule
          Corolla open-campanulate, 8-15mm long
            Herbage heavily glandular-pubescent, viscid . . . . *P. ixodes*
            Herbage less densely glandular-viscid . . . . *P. cicutaria*
          Corolla tubular-campanulate, 4-7mm long, not much exceeding calyx . . . . *P. cryptantha*
        Corolla blue or bluish; calyx-lobes linear to obovate, less than 1cm long and closely investing the capsule
          Calyx-lobes 4-5mm long, mostly oblanceolate to obovate, corolla promptly deciduous; capsule hairy in whole upper half; stamens mostly not long-exserted
            Seeds 2-4, pitted . . . . *P. distans*
            Seeds 4, papillate . . . . *P. umbrosa*
          Calyx-lobes linear to lance-linear, 6-8mm long; corolla tardily deciduous; capsules hairy only at tip; stamens long-exserted . . . . *P. tanacetifolia*
      Seeds flattened, excavated on each side of a salient ridge; plants very viscid, ill-scented
        Pedicels shorter than calyx; lobes of leaf-blades mostly oblong
          Corolla 6-10mm long, open-campanulate . . . . *P. crenulata*
          Corolla 4-5mm long, tubular-campanulate . . . . *P. minutiflora*
        Pedicels at least as long as calyx; lobes of leaf-blades almost round . . . . *P. pedicellata*
    Leaves entire or pinnate, but then with entire divisions
      Corolla-scales present, attached to tube at base of each filament
        Leaves mostly entire, or if pinnate, plants tending to be montane; seeds ca. 1mm long . . . . *P. curvipes*
        Leaves mostly pinnate to pinnatifid
          Corolla 9-15mm long; Montane Coniferous Forest . . . . *P. davidsonii*
          Corolla 10-12mm long; coastal . . . . *P. douglasii*

Corolla-scales lacking, the filament-bases sometimes dilated or winged
Filaments with a dilation at base; style parted above the middle
Corolla open-campanulate, the tube ca. as long as the limb . . . . . *P. parryi*
Corolla tubular-campanulate or campanulate-funnelform, the tube ca. twice as long as the limb
Corolla purple, slightly constricted at throat; cismontane . . . . . *P. minor*
Corolla deep blue, not constricted at throat; deserts . . . . . *P. campanularia*
Filaments without a basal dilation; style parted to below the middle
Cymes lax in fruit, the pedicels becoming 1-2cm long; corolla 8-18mm long, blue with purpllsh or whitish center; filaments pubescent . . . . . *P. viscida*
Cymes dense, the pedicels less than 1cm long; corolla 12-30mm long, the center colored like the periphery; filaments glabrous . . . . . *P. grandiflora*
Seeds transversely corrugated; ovules many
Leaves entire to crenate-dentate, not deeply lobed; deserts . . . . . *P. neglecta*
Leaves pinnate to bipinnatifid
Corolla 2.5-6.5mm long, not longer than calyx
Calyx linear to narrow-oblanceolate; corolla white with yellowish tube . . . . . *P. ivesiana*
Calyx-lobes spatulate; corolla lavender with yellowish tube . . . . . *P. affinis*
Corolla mostly 7-15mm long, at least twice as long as calyx . . . . . *P. brachyloba*

P. affinis Gray. Occasional, on rocky slopes; desert and foothill, above 800m: Montezuma View Point; Grapevine Spring; San Felipe Valley; east of Banner; Buckman Spring; Boulevard; Jacumba; Kitchen Creek. Apr.-May. *n*=11.

P. brachyloba (Benth.)Gray. [*P. leucantha* Lemmon ex Greene, the type from Del Mar.] Common, in open areas and on burns; foothill and montane, above 700m: e.g. Palomar Mountain; Oak Grove; Warner Ranch; Eagles Nest; San Felipe Valley; Banner; Bear Valley; Hauser Creek; Alpine; Green Valley; Laguna Junction; Buckman Spring; Monument Peak; Campo; Morena Dam; Boulevard; Jacumba. Also Del Mar. Apr.-June. *n*=12.

P. campanularia Gray ssp. **campanularia**. Infrequent, in sandy areas; desert, below 900m: Borrego Palm Canyon; Grapevine Spring; Sentenac Canyon; Box Canyon; Mountain Springs; southeast corner of county; Dubber; Jacumba; Hipass. Apr.-May. *n*=11.

P. cicutaria Greene ssp. **hispida** (Gray)Beauchamp ex Thorne. Caterpillar Phacelia. Frequent, in Chaparral and Coastal Sage Scrub; coastal, foothill, and desert, below 1200m: e.g. San Onofre Canyon; De Luz; Monserrate; Bear Valley; Pala Grade; Lake Henshaw; Mesa Grande; Banner; Chariot Canyon; Vallecito Canyon; Mount Woodson; Viejas Valley; Pine Valley; Buckman Spring; Jacumba; Box Canyon; Barrett; Dulzura; Jamul; Lyons Valley. Jan.-Mar. *n*=11.

P. crenulata Torr. var. **ambigua** (Jones)Macbr. Occasional, in sandy areas; desert, below 500m: northeast corner of county; Silent Canyon; Benson Dunes; Dry Lake School; Borrego Mud Hills; Yaqui Pass; Yaqui Well; Vallecito and Carrizo Stage Stations; Lava Flow Wash; north Mortero Wash; Ocotillo Wells; East Dolomite Mine Road; June Wash. Jan.-Mar. *n*=11.

P. cryptantha Greene. [*P. eremica* Jeps., the type from rocky canyon walls, Collins Valley.] Scarce, on rocky slopes; desert, below 500m: Borrego Palm Canyon; Indian Canyon; Box Canyon. Apr. *n*=11.

P. curvipes Torr. ex Wats. Infrequent, on dry slopes; montane, above 1000m: Oak Grove Trail; Volcan, Cuyamaca, and Laguna Mountains; Pine Valley. Apr.-May.

P. davidsonii Gray. [*P. aldersonii* Greene, the type from Witch Creek.] Local, on open slopes; montane, above 1500m: Laguna Public Camp; Little Laguna Lake; Garnet Peak; Monument Peak. Also Witch Creek. May-June. *n*=10.

P. distans Benth. Wild-Heliotrope. Common, in open fields and on slopes; coastal, foothill, and desert, below 1200m: e.g. San Onofre; Oceanside; Fallbrook; Doane Valley; Mesa Grande; Oak Grove; Warner Ranch; Eagles Nest; Montezuma View Point; Borrego Valley; Borrego Palm Canyon; Rockhouse Canyon; Hellhole Canyon; Sentenac Canyon; Banner; Box Canyon; Carrizo Stage Station; Dos Cabezas; Jacumba; Campo; Hipass; Hidden Glen; Descanso; Foster; San Diego; San Ysidro; Silver Strand; Mount Soledad; Torrey Pines; Miramar. Mar.-Aug. *n*=11.

P. douglasii (Benth)Torr. ssp. **cryptantha** (Brand)Beauchamp. Local, in sandy areas of Coastal Sage Scrub; coastal, below 30m: Crown Point; San Diego River bed; Old Town; Silver Strand; Border Field State Park. Mar.-Apr. Type from San Diego.

P. grandiflora (Benth.)Gray. [*Eutoca speciosa* Nutt., the type from "near St. Diego".] Occasional, on open slopes and burns; coastal and foothill, below 1000m: e.g. Fallbrook; Santa Margarita Mountains; Valley Center; Cole Grade; Bonsall; Lilac; Poway Grade; Escondido; Del Dios; Torrey Pines; Point Loma; San Diego; Jamul; Otay Mountain. May-June. *n*=11.

P. imbricata Greene ssp. **bernardina** (Greene)Heckard. [*P. californica* var. *b.* (Greene)Jeps.] Scarce, on rocky slopes; montane, above 1100m: Pauma Creek; Harper Ranch. Apr.-May. *n*=11.

P. imbricata ssp. **patula** (Brand)Heckard. [*P. californica* var. *p.* (Brand)Jeps.] Frequent, in rocky places in Chaparral and Montane Coniferous Forest; montane and foothill, above 900m: e.g. Palomar, Hot Springs, Volcan, Cuyamaca, and Laguna Mountains; Julian; Descanso; Wildwood Glen; Pine Hills; Pine Valley; Oakzanita; Campo. June-July. Type from Stonewall Mine, 4600 feet altitude in the Cuyamaca Mountains. *n*=11.

P. ivesiana Torr. Rare, in sandy areas; north desert, below 300m: Borrego Springs. Apr. *n*=11.

P. ixodes Kell. Rare, on coastal bluffs at Border Field State Park.

P. minor (Harv.)Thell. [*P. whitlavia* var. *jonesii* forma *gracillima* Brand, the type from Fallbrook.] Wild Canterbury-Bell. Frequent, in disturbed areas and burns; coastal and foothill, below 1100m: e.g. upper San Onofre Canyon; De Luz; Rainbow; Fallbrook; Palomar Mountain; Oak Grove; Warner Ranch; Chihuahua Valley; Eagles Nest; Lone Palm; Borrego Palm Canyon; Scissors Crossing; Hipass; InKoPah County Park. Mar.-May. *n*=11.

P. minutiflora J. Voss. Scarce, in sandy areas; desert, below 300m: Borrego Valley; Borrego Springs. Mar.-Apr. *n*=11.

P. neglecta Jones. Scarce, on desert pavement; desert, below 800m: Ocotillo Wells; valley south of Coyote Mountain. Mar.-May. *n*=11.

P. parryi Torr. Common, on open slopes and burns; cismontane: e.g. Fallbrook; Bonsall; Moosa Canyon; Pala; San Pasqual; Harrison Grade; Montezuma View Point; Anahuac School; Cuyamaca Peak; El Capitan; Barona Valley; Springhill; Mount Helix; Mission Valley; Tecate Peak; Lyons Valley; Potrero Grade; Campo; Buckman Spring; Sweetwater Valley. Mar.-June. Type from mountains east of San Diego. *n*=11.

P. pedicellata Gray. Infrequent, in sandy and rocky areas; desert, below 800m: Borrego Palm Canyon; Borrego Valley; The Narrows; Fish Creek; Sentenac Canyon; Lava Flow Wash; north Mortero Wash; Coyote Mountain; Mortero Wash; June Wash; Inner Pasture Outlet Canyon. Mar.-May. *n*=11.

P. ramosissima Dougl. ex Lehm. var. **austrolitoralis** Munz. Scarce, on sandy beaches; coastal, below 30m: Carlsbad; San Marcos Creek. Feb.-June.

P. suffrutescens (Parry)Const. [*P. ramosissima* var. *suffrutescens* (Parry)Parry; *P. polystachya* Greene, the type from Witch Creek.] Frequent, on rocky slopes and canyons; foothill and montane, above 400m: e.g. Fallbrook; Rainbow; Pauma; Palomar Mountain; Borrego Palm Canyon; Hot Springs Mountain; Witch Creek; Ramona; Barona Valley; Lakeside; Campo. May-Aug. *n*=10.

P. tanacetifolia Benth. Rare, known in the county only from Coyote Creek, 300m. Apr. *n*=11.

P. umbrosa Greene. Rare, an obscure collection labelled "southwest Colorado Desert, Orcutt", perhaps from the desert area of San Diego County. The taxon is poorly known.

P. viscida (Benth.)Torr. Rare, known in the county only from a collection from San Pasqual Valley, 100m. Mar.-June. *n*=11.

Pholistoma Lilja ex Lindbl.

Calyx with auricles between the lobes, enveloping the capsule; stems with retrorse prickles on the angles
  Petioles broadly winged and auriculate-clasping; corolla 1-3cm broad . . . . *P. auritum*
  Petioles narrowly winged, scarcely auriculate-clasping; corolla 0.6-1cm broad . . . . *P. racemosum*
Calyx without auricles, stellate-rotate under the capsule; stems slightly scabrous, without retrorse prickles . . . . *P. membranaceum*

P. auritum (Lindl.)Lilja ex Lindl. Fiesta-Flower. Frequent, on shaded slopes; coastal and foothill, below 600m: e.g. San Onofre; De Luz; Fallbrook; Bonsall; Palomar Mountain; San Marcos Creek; Foster; Mission Canyon; Kearny Mesa; San Diego. Feb.-Apr. *n*=9.

P. membranaceum (Benth.)Const. Occasional, on shaded slopes; cismontane and desert, below 1000m: Bernardo; Ocean Beach; Silver Strand; La Jolla; San Ysidro; Jacumba; Mountain Springs; San Felipe Valley; Sentenac Canyon; Borrego Palm Canyon. Feb.-May. *n*=9.

P. racemosum (Nutt.)Const. Frequent, on shaded slopes; south coastal, below 400m: San Pasqual; Miramar; Pacific Beach; Mission Hills; Mission Gorge; Barrett Dam; Sloane Ranch; Dehesa; Emerald Hills; San Diego; Murray Dam; Otay Ranch. Feb.-May. Type from San Diego. *n*=9.

Turricula Macbr.

T. parryi (Gray)Macbr. [*Eriodictyon p.* (Gray)Greene] Sticky Nama. Occasional, in open areas and burns; foothill and montane, above 1000m: Palomar Mountain; Cuca Grade; Rincon; Lost Valley; Hauser Creek; Banner Grade; San Felipe Valley; Oriflamme Canyon; Cuyamaca Peak; Laguna Mountains; Laguna Junction; Pine Valley; Espinosa Trail. Also Mount Soledad. May-Sep. *n*=13.

Wigandia H.B.K.

***W. caracasana** H.B.K. var. **macrophylla** Brand. Native of Mexico. Uncommonly persisting from cultivation; coastal, below 300m: San Diego State University; North Park. June-Sept.

HYPERICACEAE
(Guttiferae)

Hypericum L. St. John's Wort

Stems mostly prostrate; leaves 4-12mm long; petals 2-3mm long . . . . *H. anagalloides*
Stems not prostrate; leaves larger; petals longer
  Sepals linear-deltoid, acuminate; stems with sterile shoots in lower leaf-axils . . . . *H. perforatum*
  Sepals ovate to obovate, acute to obtuse; stems lacking sterile shoots in lower axils
    Leaves 1-4cm long; herbs to 7dm tall . . . . *H. formosum*
    Leaves 5-7cm long; shrubs to 5m tall . . . . *H. canariense*

H. anagalloides Cham. & Schlecht. Tinker's Penny. Occasional, in wet places of the Montane Coniferous Forest; montane, above 900m: Doane Valley; Wyss Ranch; Hot Springs Mountain; Cuyamaca Dam; Mother Grundy Peak; Dulzura; Tecate Peak. July-Sep.

***H. canariense** L. Native of Canary Islands. Uncommon, escape from cultivation; coastal, below 200m: La Mesa; Balboa Park. June-Aug.

H. formosum H.B.K. ssp. **scouleri** (Hook.)Thorne. Frequent, in wet areas, Montane Coniferous Forest; montane, above 900m: Wyss Ranch; Harrison Grade; Mesa Grande; Boulder Creek; Cuyamaca Lake; San Felipe Valley; Witch Creek; Cuyamaca Peak; Paso Picacho; Green Valley; Laguna Public Camp; Guatay Mountain. June-Aug.

***H. perforatum** L. Klamath Weed. Native of Europe. Uncommon, open fields; montane, above 1000m: Spencer Valley; Wynola. June. $2n=32$.

JUGLANDACEAE. Walnut Family

Juglans L. Walnut

J. californica Wats. Scarce, on open slopes; north foothill, below 600m: Cottonwood Canyon; De Luz; Bonsall. Also spreading near Jamacha Junction and Sloane Canyon where it was probably planted. Apr.-May.

KRAMERIACEAE. Krameria Family

Krameria L.

Barbs of spines of fruits in an umbrella-like group at apex . . . . . *K. grayi*
Barbs of spines of fruits scattered along whole upper portion . . . . . *K. parvifolia*

K. grayi Rose & Painter. White Ratany. Frequent, sandy areas; desert, below 1000m: Borrego Valley; San Felipe Valley; The Narrows; Fish Creek; Lava Flow Wash; Plum Canyon; Ocotillo Wells; Sentenac Canyon; Blair Valley; Vallecito and Carrizo Stage Stations; Mountain Springs. Mar.-May. Known host: *Ambrosia dumosa*.

K. parvifolia Benth. var. **imparata** Macbr. Pima Ratany. Frequent, on sandy flats and slopes; desert, below 1000m: Borrego Valley; Borrego Palm Canyon; Silent Canyon; San Felipe Valley; Fish Creek; Pinyon Mountain; Vallecito Stage Station; Mountain Springs; Jacumba. Mar.-May,. Known host: *Ambrosia dumosa*. Type from Mountain Springs.

LAMIACEAE. Mint Family
(Labiatae)

KEY TO GENERA

Ovary of 4 united nutlets; style not basal; nutlets laterally attached. . . . . *Trichostema*
Ovary of 4 distinct or nearly distinct nutlets; style basal; nutlets basally attached
  Calyx 2-lipped, the lips entire
    Shrub; calyx inflated in fruit, not crested or gibbous on the back . . . . . *Salazaria*
    Herbs; calyx with a gibbous or helmet-like crest on the back . . . . . *Scutellaria*
  Calyx regular or 2-lipped, the lips not entire
    Corolla strongly 2-lipped
      Stamens included in the corolla-tube; calyx with 10 spinescent hooked teeth at the tip . . . . . *Marrubium*
      Stamens exserted from the corolla or included in throat; calyx-teeth not hooked at tip
        Stamens declined and enveloped by the lower lip of the corolla; scurfy-tomentose desert shrub . . . . . *Hyptis*
        Stamens ascending, not declined and enveloped by the lower lip
          Upper lip of corolla concave
            Fertile stamens 4
              Calyx 2-lipped, closed in fruit, the upper lip truncate; flowers in dense spike, purple; low herb *Prunella*
              Calyx 5-toothed, not closed in fruit; if 2-lipped, the upper lip not truncate
                Plants shrubby; flowers 2.5-4cm long; leaves coarsely serrate . . . . . *Leonotis*
                Plants herbaceous; flowers shorter
                  Calyx-teeth not spine-tipped; corolla-tube without a hairy ring within; nutlets sharply 3-sided; plants annual . . . . . *Lamium*
                  Calyx-teeth spine-tipped; corolla-tube with hairy ring within; nutlets ovoid; plants mostly perennial . . . . . *Stachys*
            Fertile stamens 2
              Pollen-sacs in each anther 2, with very short connective between; plants annual with spines on bracts and calyx-teeth . . . . . *Acanthomintha*
              Pollen-sacs in each anther 1 or 2, at the ends of elongate arms of connective jointed to the filaments; plants shrubs to annuals, usually not spiny on bracts of calyx-teeth . . . . . *Salvia*
          Upper lip of corolla plane
            Flowers solitary, or few in the leaf-axils
              Corolla 2.5-4cm long; leaves ovate or deltoid . . . . . *Lepechinia*
              Corolla 1-1.3cm long . . . . . *Satureja*
            Flowers in dense heads
              Plants annual, with round to spatulate leaves not more than 2cm long. . . . . *Pogogyne*
              Plants perennial, with ovate to lance-ovate leaves mostly 2-8cm long; calyx naked in the throat; bracts green . . . . . *Pycnanthemum*
    Corolla regular or nearly so, the lobes subequal
      Flowers in terminal heads; stamens 4 . . . . . *Monardella*
      Flowers in axillary whorls which may be aggregated in terminal spikes . . . . . *Mentha*

Acanthomintha Gray

A. ilicifolia (Gray)Gray. San Diego Thorn-mint. Uncommon, clay soils of mesas and valleys; coastal, below 300m: Encinitas; San Marcos; Bernardo; La Costa; University Heights; Alvarado Canyon; San Diego; Spring Valley; Rice Canyon; McGinty Mountain; Viejas Mountain; Sweetwater and Otay Lakes; Poser Mountain. Apr.-May. Type from San Diego.

Hyptis Jacq.

H. emoryi Torr. Desert-Lavender. Prevalent, in washes and other sandy places; desert, below 600m: e.g. Coyote Canyon; Borrego Palm Canyon; Yaqui Well; Campbell Ranch; Dos Cabezas. Jan.-Apr.

Lamium L. Henbit

***L. amplexicaule** L. Native of Europe. Infrequent, naturalized in waste places; cismontane: Palomar Mountain; Santee; Lakeside; Mission Valley; Morrettis. Mar.-Aug. $2n=18$.

Leonotis R. Br. Lion's-Ear

***L. leonurus** (L.)R. Br. Native of South Africa. Infrequent, escaping from cultivation; coastal, below 200m: Cardiff; San Diego. July-Sep.

Lepechinia Willd. Pitcher Sage

Calyx-teeth deltoid, mostly shorter than the mature tube . . . . . *L. cardiophylla*
Calyx-teeth lanceolate or acicular, equal to or longer than the mature tube; nutlets glabrous . . . . . *L. ganderi*

L. cardiophylla Epl. Rare, known in the county from near Iron Mountain, 500-900m. Apr.-June.

L. ganderi Epl. Local, in Chaparral; foothill, 500-1060m: San Miguel Mountain; Otay Mountain; Jamul Mountains. June-July. Type from Otay Mountain, 3000 feet. $n=16$.

Marrubium L. Horehound

***M. vulgare** L. Native of Europe. Common, weed in disturbed areas; cismontane and desert: e.g. San Onofre; San Marcos; Banner; Poway; Montezuma Mesa; Bonita; National City. Mar.-Aug. $2n=34$.

Mentha L. Mint

Flower-whorls distant and in leaf-axils, mostly exceeded by the subtending leaves
  Leaves 1-2cm long, grayish; calyx-teeth dissimilar, the 2 lower lanceolate-subulate . . . . . *M. pulegium*
  Leaves 2-7cm long, light green; calyx-teeth quite alike, subequal . . . . . *M. arvensis*
Flower-whorls crowded, usually in terminal spikes, or some of the lower more distant
  Plants glabrous or nearly so . . . . . *M. spicata*
  Plants villous or canescent . . . . . *M. pulegium*

M. arvensis L. Tule Mint. Occasional, in wet places; cismontane: Cuyamaca Lake; Laguna Mountains; Lake Murray. July-Oct. $2n=12, 54, 60, 64, 72, 92$.

***M. pulegium** L. Pennyroyal. Native of Europe. Local, naturalized at Mesa Grande, 950m. June-Sep.

***M. spicata** L. Spearmint. Native of Europe. Occasional, naturalized in moist places; cismontane: Lake Henshaw; Julian; Palomar Mountain; Lake Murray. July-Oct. $2n=36, 48$.

Monardella Benth.

Flowers in rather loose heads; limb of the corolla 1/5-1/2 the length of the tube; calyx 10-25mm long
  Corolla 35-45mm long, scarlet or yellowish; calyx 20-25mm long
    Corolla yellow, limb 6-10mm long . . . . . *M. macrantha* ssp. *hallii*
    Corolla scarlet to yellow, limb 5-6mm long . . . . . *M. macrantha* ssp. *m.*
  Corolla 25-30mm long, pale to pinkish; calyx 12-15mm long
    Stems with a short retrorse pubescence
      Leaf-blades subglabrous above . . . . . *M. nana* ssp. *n.*
      Leaf-blades cinereous above . . . . . *M. nana* ssp. *tenuiflora*
    Stems villous with spreading hairs . . . . . *M. nana* ssp. *leptosiphon*
Flowers in dense heads; limb of corolla 1/2-2/3 the length of the tube; calyx 5-10mm long
  Plants perennial
    Leaves with well developed tomentum on under surface
      Leaves green and glabrous above . . . . . *M. hypoleuca* ssp. *h.*
      Leaves pubescent to lanate above . . . . . *M. hypoleuca* ssp. *lanata*
    Leaves lacking a felt-like tomentum beneath
      Leaves silvery-gray . . . . . *M. linoides* ssp. *l.*
      Leaves greenish . . . . . *M. linoides* ssp. *viminea*
  Plants annual
    Flower heads 1-2cm broad . . . . . *M. lanceolata* var. *microcephala*
    Flower heads 1.5-3 cm broad . . . . . *M. lanceolata* var. *l.*

M. hypoleuca Gray ssp. **hypoleuca**. Rare, on dry slopes in Chaparral; montane, above 1200m: Palomar Mountain. July-Sep.

M. hypoleuca ssp. **lanata** (Abrams)Munz. [*M. l.* Abrams] Local, in Chaparral; foothill, 300-1000m: Viejas Grade; Alpine; Lawson Peak; Mount Woodson; Otay Mountain; Tecate Peak; Potrero Mountain. June-July. Type from Descanso Grade, near the top, "Cuiamaca" [Cuyamaca] Mountains.

M. lanceolata Gray var. **lanceolata**. [*M. sanguinea* Greene, the type from near Julian.] Occasional, in dry places; foothill and montane, above 300m: Palomar Mountain; Lake Henshaw; Cuyamaca Mountains; Julian; Boulder Creek; Viejas Valley; Potrero. May-Aug.

M. lanceolata var. **microcephala** Gray. Occasional, in dry places; foothill, 300-1200m: Palomar Mountain; Santa Ysabel; Cold Spring; Pine Valley; Buckman Spring; Campo; Potrero. June-Sep. Type from "Potrero, near San Diego".

M. linoides Gray ssp. **linoides**. Uncommon, in dry places and rocky slopes; montane, above 900m: Oriflamme and Vallecito Canyons; Desert View; Cuyapaipe Indian Reservation; Bankhead Springs. June-Sep. Type from "mountains east of San Diego [Oriflamme Mine, Cuyamaca Mountains]".

M. linoides ssp. **viminea** (Greene)Abrams. [*M. v.* Greene] Endemic to San Diego County. Infrequent, in canyons; coastal, below 200m: Poway; Mira Mesa; Miramar; Murphy Canyon; Sycamore Canyon; San Diego. June-Aug. Type from mountains of San Diego County.

M. macrantha Gray ssp. **hallii** (Abrams)Abrams. Infrequent, on floor of Montane Coniferous Forest; montane, above 1200m: Palomar Mountain; Cuyamaca Mountains. June. Type from Palomar Mountain.

M. macrantha ssp. **macrantha**. Scarlet Monardella. Occasional, in shaded areas, Montane Coniferous Forest; montane, above 1200m: Palomar Mountain; Julian; Cuyamaca and Laguna Mountains. Also below Barrett Dam. Apr.-Aug. Type from "Cuiamaca Mountains and near Julian City".

M. nana Gray ssp. **leptosiphon** (Torr.)Abrams. Local, on rocky slopes; montane, above 1200m: Palomar Mountain; Aguanga Ridge; Banner Grade; east of Julian; San Felipe; Laguna Camp. June-July. Type from San Felipe.

M. nana ssp. **nana**. Uncommon, on rocky slopes; montane, above 1200m: Julian; Oriflamme Canyon; Laguna Mountains; Thing Valley; McCain Valley. June-July. Type from Mountains behind San Diego.

M. nana ssp. **tenuiflora** (Wats.)Abrams. Rare, known in the county only from a collection from near Warner Ranch, 940m. May.

Pogogyne Benth.

Flower-bracts and calyx-lobes conspicuously hirsute, bristly-ciliate . . . . . *P. abramsii*
Flower-bracts and calyx-lobes subglabrous . . . . . *P. nudiuscula*

P. abramsii J.T. Howell. San Diego Mesa-mint. Endemic to San Diego County. Occasional, in vernal pools; coastal, 150m: Peñasquitos; Miramar; San Diego. Apr.-June. Type from mesa north of San Diego.

P. nudiuscula Gray. Loma Alta Mesa-mint. Uncommon, in vernal pools; coastal, 150m: San Diego; Otay Mesa. May-June. Type from near San Diego.

Prunella L. Selfheal

P. vulgaris L. ssp. **lanceolata** (Barton)Hult. Uncommon, in moist areas; montane, above 1200m: Palomar Mountain. May-Sep. n=14.

Pycnanthemum Michx. Mountain-Mint

P. californicum Torr. Occasional, in wet places; montane, above 1300m: Palomar, Cuyamaca, and Laguna Mountains; Milk Ranch. June-Sep. $2n$=40.

Salazaria Torr. Bladder-Sage

S. mexicana Torr. Rare, known in the county only at Mountain Springs, 1000m. Mar.-June. $2n$=50.

Salvia L. Sage

Both branches of the stamen-connective evident and bearing fertile anthers; leaves pinnatifid or spinose-sinuate or -lobed
  Corolla lavender, 25-35mm long, the lower lip fimbriate; leaves spiny, tomentose . . . . . *S. carduacea*
  Corolla blue, 12-16mm long, the lower lip not fimbriate; leaves not spiny or tomentose . . . . . *S. columbariae*
Lower end of stamen-connective obsolete or forming only a tooth, rarely bearing an anther
  Plants creeping perennial herbs . . . . . *S. sonomensis*
  Plants shrubby
    Inflorescence thyrsoid-paniculate with lax few-flowered glomerules; corolla-tube shorter than lower lip; upper lip very short
      Panicles 5-15cm long . . . . . *S. apiana* var. *a.*
      Panicles condensed to ca. 4cm long . . . . . *S. apiana* var. *compacta*
    Inflorescence with dense glomerules in an interrupted spike, the spikes occasionally branched; corolla-tube usually longer than lower lip; upper lip well developed
      Leaf-blades entire, mostly broader in upper half, not rugose; flower-bracts membranous, usually colored . . . . . *S. pachyphylla*
      Leaf-blades crenulate to crenate, broader in their lower half, rugose; bracts often not colored
        Stamens not or scarcely exceeding the upper lip of the corolla, lying close under it; corolla 8-12mm long
          Leaf-blades oblong-elliptic, mostly 3-6cm long, ashy beneath; corolla mostly pale blue to whitish *S. mellifera*
          Leaf-blades oblong-obovate, mostly 1-2cm long, ashy on both surfaces; corolla clear blue . . . . . *S. munzii*
        Stamens well exserted from upper corolla-lip and not lying close under it; corolla 15-25mm long
          Calyx-teeth not tipped with long spines
            Leaves green and rugose on upper surface, not ashy beneath; flower-bracts purplish-green, spinulose-tipped; corolla 20-25mm long . . . . . *S. eremostachya*
            Leaves rugose, but tomentose beneath; corolla ca. 2cm long . . . . . *S. clevelandii*
          Calyx-teeth tipped with long spines; leaves whitish; corolla white, 2cm long . . . . . *S. vaseyi*
    Inflorescence thyrsoid-paniculate with lax few-flowered glomerules; corolla-tube shorter than lower lip; upper lip very short
      Panicles 5-15cm long . . . . . *S. apiana* var. *a.*
      Panicles condensed to ca. 4cm long . . . . . *S. apiana* var. *compacta*

S. apiana Jeps. var. **apiana**. [*Audibertia polystachya* Benth. non *Salvia p.* Ortega] White Sage. Abundant, on open slopes; cismontane: e.g. San Onofre Canyon; Palomar and Hot Springs Mountains; Pine Hills; Lake Hodges; Poway; Moosa Canyon; Jacumba; Dulzura; San Miguel Mountain. Mar.-July. n=15, 16.

S. apiana var. **compacta** Munz. Frequent, in valleys and on slopes; desert, below 120m: Culp Valley; Borrego Palm Canyon; The Narrows; Scissors Crossing; Vallecito Stage Station; Rockhouse Canyon; Jacumba; Mountain Springs. Also San Ysidro. Apr.-July.

S. × bernardina Parish ex Greene. (*S. columbariae* × *S. mellifera*.) Infrequent, on disturbed slopes; foothill, 200-1000m: Lost Valley; Fallbrook; Jamul. Apr.-June.

S. carduacea Benth. Thistle Sage. Occasional, in sandy areas; desert and foothill, below 1250m: San Felipe Valley; eastern base of Laguna Mountains; Jacumba; Cottonwood Valley; Potrero; San Pasqual; Chollas Valley; Bear Valley. Mar.-June. *n*=12, 16.

S. clevelandii (Gray)Greene. [*Audibertia c.* Gray] Fragrant Sage. Local, in Chaparral; coastal and foothill, below 1000m: e.g. Tenaja area; Henshaw Grade; Pala Creek; San Marcos Mountains; Witch Creek; Potrero; Torrey Pines State Reserve; Harbison Canyon; Flynn Springs; Viejas Grade; Tecate; Cottonwood Grade; Jamul Mountains; Potrero Peak; McGinty Mountain; Type from mountains behind San Diego, California, at an elevation of about 2200 feet, i.e. Ballena *fide* Cleveland.

S. columbariae Benth. ssp. **columbariae**. Chia. Common, in open areas; cismontane and desert, below 1250m: e.g. Fallbrook; Warner Hot Springs; Scissors Crossing; Borrego Valley; Mount Laguna; Jacumba; Campo; Point Loma. Mar.-June. *n*=8.

S. eremostachya Jeps. Sunny Sage. Infrequent, on dry, rocky slopes; desert, 300-1000m: Indian Canyon; Yucca Valley; Rockhouse Canyon. Mar.-May. Type from Indian Canyon, Collins Valley. 2*n*=30.

S. mellifera Greene. [*Audibertia stachyoides* Benth. non *Salvia s.* Kunth] Black Sage. Prevalent, in Chaparral and Sage Scrub; foothill and coastal, below 500m: e.g. San Onofre; Escondido; Moosa Canyon; Ramona; Torrey Pines; Kearny Mesa; Harbison Canyon; Jamul. Apr.-July. *n*=16.

S. munzii Epl. [*S. mellifera* var. *jonesii* Munz] Munz Sage. Frequent, in Sage Scrub; south foothill and coastal, below 500m: Dictionary Hill; San Miguel, Jamul, and Otay Mountains; Proctor Valley; Lower Otay Lake. Feb.-Apr. 2*n*=30.

S. pachyphylla Epl. ex Munz. Rose Sage. Rare, known in the county only at Mountain Springs, 1000m; but to be expected in northern high desert areas of the county. July-Sep. 2*n*=30.

S. × palmeri (Gray)Greene. (*S. apiana* × *S. clevelandii*.) [*Audibertia p.* Gray] Palmer's Sage. Infrequent, in disturbed places; montane and foothill, above 300m: Ramona; Palomar Mountain; Pala; Cuyamaca Mountains; Type from near Tighes Ranch in the mountains northeast of San Diego.

S. sonomensis Greene. [*Audibertia humilis* Benth. non *Salvia h.* Benth.] Creeping Sage. Local, in Chaparral understory; foothill and montane, above 1100m: Cutca Valley; Oriflamme Canyon; Rattlesnake Valley; Cuyamaca Mountains; Hauser Creek; Guatay Mountain; Otay Mountain; Lawson Peak; Sunrise Highway; Corte Madera; Los Pinos Mountain. May-June. 2*n*=30.

S. vaseyi (Porter)Parish. [*Audibertia v.* Porter] Wand Sage. Infrequent, on rocky slopes; desert, below 800m: Montezuma Grade; Borrego Palm Canyon; Yaqui Well; Carrizo Gorge; Mountain Springs; north Mortero Wash; south of Lava Flow Wash. Apr.-June. Type from Mountain Springs.

Satureja L.

S. chandleri (Bdg.)Druce. San Miguel Savory. Occasional, in Chaparral; foothill, 500-750m: De Luz Heights; Barona Valley; San Miguel, McGinty, Jamul, and Otay Mountains. Mar.-May. Type from Mount San Miguel.

Scutellaria L. Skullcap

Corolla 12-33 mm long; flowers solitary in axils of leaves
  Petioles of all but lowest leaves 1-3mm long; blades truncate-cordate at base. . . . . *S. bolanderi*
  Petioles 5-30mm long; blades of all but lowest leaves not truncate-cordate
    Stems with short curved hairs; corolla 24-29mm long . . . . *S. austinae*
    Stems loosely villous with spreading hairs; corolla 15-20mm long . . . . *S. tuberosa*

S. austinae Eastw. [*S. linearifolia* Eastw., the type from "San Diego".] Austin's Skullcap. Rare, reported in the county only from an old collection whose origin must be considered doubtful. May-July.

S. bolanderi Gray ssp. **austromontana** Epl. Uncommon, in wet places; foothill and montane, above 600m: Palomar Mountain; Lake Henshaw; Japatul; Morena; Green Valley; Viejas Valley; below Cuyamaca Dam. June-July. Type from Carrizo Creek, near Lake Henshaw.

S. tuberosa Benth. ssp. **australis** Epl. Danny's Skullcap. Uncommon, on moist, shaded slopes; foothill and montane, 200-1200m: Camp Pendleton; Rattlesnake Valley; Alpine; Mount Woodson; Japatul Valley; Lyons Valley; Noble Canyon; Otay Mountain; Rice Canyon; Potrero Grade. Apr.-May.

Stachys L. Hedge-Nettle

Plants 6-10dm high; leaves with sub-cordate base. . . . . *S. rigida* ssp.*r.*
Plants 1.5-4dm high; leaves ovate . . . . *S. rigida* ssp. *quercetorum*

S. rigida Nutt. ex Benth. ssp. **quercetorum** (Heller)Epl. Infrequent, in moist areas; cismontane: Camp Pendleton; Oak Grove; Montezuma Valley; Laguna Lakes; Campo; La Jolla; Torrey Pines State Reserve. Apr.-Sep.

S. rigida ssp. **rigida**. Occasional, in moist areas; cismontane: Palomar, Cuyamaca, and Laguna Mountains; Mission Valley; Jamul Valley; Otay. Apr.-Sep.

Trichostema L. Bluecurls

Plants annual, with pale pubescence
  Stamens 10-20mm long; corolla-tube 5-10mm long . . . . *T. lanceolatum*
  Stamens 2-6mm long
    Calyx-lobes ca. twice as long as the tube, approximately equal to each other in width; stamens 3-6mm long . . . . *T. austromontanum*
    Calyx-lobes 1-2 times as long as the tube, the uppermost more slender than the other 4; stamens 2-3mm long . . . . *T. micranthum*
Plants low shrubs, with purplish wool in the inflorescence
  Corolla-tube 9-14mm long; calyx ca. 8mm long, with lanceolate lobes; filaments exserted 2.5-3cm . . . . . *T. lanatum*
  Corolla-tube 4-7mm long; calyx ca. 5mm long, wtih triangular lobes; filaments exserted 1-1.5cm . . . . *T. parishii*

T. austromontanum Lewis. San Jacinto Blue-Curls. Scarce, in grassy areas; foothill and montane, above 900m: Palomar Mountain; Laguna Mountains; Japatul Valley; east of Stonewall Peak; Cuyamaca Peak; North Peak. June-Sep. n=14.

T. lanatum Benth. Woolly Blue-Curls. Occasional, in Chaparral and Coastal Sage Scrub; foothill and coastal, below 700m: Carlsbad; Del Mar Heights; Mission Gorge; Kearny Mesa; Otay Mountain. May-Aug. n=10.

T. lanceolatum Benth. Vinegar Weed. Frequent, in disturbed areas; cismontane, below 1100m: e.g. Camp Pendleton; De Luz; Warner Hot Springs; Santa Ysabel; Descanso; Lake Hodges; Sweetwater Lake. Aug.-Oct. n=7.

T. micranthum Gray. Small-flowered Blue-curls. Rare, known in the county from a collection on Palomar Mountain. Aug. n=7.

T. parishii Vasey. Mountain Blue-Curls. Frequent, in Chaparral; foothill and montane, above 500m: e.g. Palomar Mountain; Oak Grove; Warner Hot Springs; Banner Queen; Laguna Junction; Alpine; McCain Valley; Jacumba; Campo; San Miguel Mountain. May-June. Type from San Diego County. n=10.

LAURACEAE. Laurel Family

Umbellularia Nutt. California-Bay

U. californica (H. & A.)Nutt. Local, on shaded slopes; foothill and montane, above 4500m: Roblar Grade and Cottonwood Canyon, Santa Margarita Mountains; Oriflamme Canyon; Garnet Peak; Noble Canyon; Shaw Canyon; Oasis Spring. Dec.-May. 2n=24.

LENNOACEAE.

Pholisma Nutt. ex Hook.

P. arenarium Nutt. ex Hook. [*P. paniculatum* Templeton] Sand Plant. Uncommon, in sandy areas; coastal, foothill, and desert, below 1000m: La Costa; Poway; Miramar Naval Air Station; Point Loma; Mission Hills; Mission Valley; Mountain Springs; The Narrows; Sentenac Canyon; Borrego Springs; Mortero Wash; Alta Bisnaga Wash; canyon from Inner Pasture; Yaqui Wells; Font's Point; Agua Caliente. Apr.-Oct. Known hosts: *Croton*, *Eriodictyon*; *Isocoma*, *Chrysothamnus*, *Hymenoclea*, *Ambrosia*. Type from "Monterey and San Diego"; however, since the plant has never again been collected near Monterey, the type is assumed to be from San Diego. n=18.

LIMNANTHACEAE. False Mermaid Family.

Limnanthes R. Br. Meadow-Foam

L. gracillis Howell ssp. **parishii** (Jeps.)Beauchamp. [*L. douglasii* auct. non R. Br.] Cuyamaca Meadow-Foam. Occasional, in wet places and along lake shores; montane, above 1300m: Mendenhall Valley; Palomar Mountain; Cuyamaca and Laguna Mountains; Cuyamaca Lake; Pine Hills; Coleman Creek; Stonewall Creek; Laguna Public Camp. Apr.-June. Type from Stonewall Mine, Cuyamaca Mountains. n=5.

LINACEAE. Flax Family

KEY TO GENERA

Petals blue, 10-20mm long; styles 5; capsules 5-10mm long . . . . *Linum*
Petals white to pink, 2-3mm long; styles 3; capsules 1-2mm long. . . . . *Hesperolinon*

Hesperolinon Small

H. micranthum (Gray)Small. [*Linum m.* Gray] Thread-stem Dwarf-Flax. Occasional, on open slopes; coastal and foothill, below 1100m: e.g. upper San Onofre Canyon; Hot Springs Mountain; Inspiration Point; Pine Hills; Descanso; Alpine; Barona Valley; El Cajon Mountain; Tecate Peak; San Miguel Mountain; San Diego. May-July. 2n=18 pairs.

Linum L. Flax

L. lewisii Pursh. Frequent, in Montane Coniferous Forest; montane, above 1050m: Hot Springs Mountain; Cuyamaca Lake; Laguna Mountains; Pine Valley; Descanso. May-Sep.

LOASACEAE. Stick-Leaf Family

KEY TO GENERA

Stamens 5, capsules indehiscent, 1-seeded . . . . . *Petalonyx*
Stamens many, capsules dehiscent by apical valves, few to many seeds
  Placentae 5, seeds minute, flowers green . . . . . *Eucnide*
  Placentae 3, seeds 1-4mm long; flowers white to golden yellow . . . . . *Mentzelia*

Eucnide Zucc.

Eucnide rupestris (Baill.)Thompson & Ernst. Rock Nettle. Rare, on talus slopes and washes; desert, 300m: Indian Gorge; Mountain Palm Springs.

Mentzelia L. Blazing Star

Petals 5 and clearly distinct from the outer stamens
  Outer stamens with narrow filaments
    Perennial; petals lanceolate, longer than 4cm . . . . . *M. laevicaulis*
    Annual; petals shorter than 2.5cm, broadly oval or obovate
      Seeds (and maturing ovules) angular, prismatic, regularly grooved along 3 sides; capsules not narrowed at base
        Petals ovate, apex acute, leaves deeply and irregularly lobed . . . . . *M. affinis*
        Petals broadly oval with rounded tips, or obovate with retuse tips; leaves entire or dentate
          Leaves, at least upper, orbicular; capsules less than 1cm . . . . . *M. dispersa*
          Leaves entire or dentate, linear to lanceolate; capsules longer than 1cm . . . . . *M. pinetorum*
      Seeds not regularly grooved, irregularly angled or round; capsules narrowed toward base
        Bracts lobed; all capsules erect
          Petals rounded to retuse at apex, 2-20mm long; capsules 10-22mm long
            Bracts without white area at base . . . . . *M. albicaulis*
            Bracts with white area at base
              Petals 0.7-1.7cm long; styles 0.6-1.1cm long
                Stems greenish; leaves entire to pinnatifid; petals 12-20mm long . . . . . *M. gracilenta*
                Stems whitish; leaves mostly pectinately pinnatifid; petals 6-10mm long . . . . . *M. pectinata*
              Petals mostly 2-7mm long; styles 2-7mm long
                Flowers terminal, not solitary in axils; bracts concealing fruit . . . . . *M. congesta*
                Flowers and bracts not as above
                  First formed capsules 0.9-1.7cm long; cauline leaves often entire, linear . . . . . *M. montana*
                  First formed capsules 1.4-3cm long; leaves never entire . . . . . *M. veatchiana*
          Petals abruptly acuminate to acute with definite mucro and 20-24mm long; capsules 25-40mm long . . . . . *M. lindleyi*
        Bracts usually entire; oldest capsules recurved
          At least the oldest capsules recurved more than 90 degrees
            Seeds uniformly light in color, not tessellate
              Seeds rounded or sharply but irregularly angled, surface smooth; leaves with short rounded lobes; on fine sandy soils . . . . . *M. desertorum*
              Seeds rounded, surface with slight, pointed papillae; leaves with long, pointed lobes . . . . . *M. obscura*
            Seeds moderately to heavily tessellate . . . . . *M. nitens*
          None of the capsules recurved more than 90 degrees . . . . . *M. albicaulis*
  Outer stamens with broad filaments
    Desert annuals; bracts laciniate; seeds round or obovate with 1 or 2 grooves on both faces, surface tuberculate
      Bracts scarious with conspicuous white area and narrow green margin . . . . . *M. involucrata*
      Bracts green without white area . . . . . *M. hirsutissima*
    Coastal annual; bracts ovate, entire, cupped; seeds prismatic, surface smooth . . . . . *M. micrantha*
Petals more than 5, or if appearing to be 5, then these grading through a whorl of petal-like staminodia to stamens with broad filaments . . . . . *M. multiflora*

M. affinis Greene. Hydra Stick-Leaf. Occasional, in open areas; cismontane and desert: northeast corner of county; Montezuma View Point; Coyote Creek; Palomar Mountain; Mount Soledad; Point Loma; San Ysidro; Box Canyon. Mar.-May. *n*=9.

M. albicaulis (Dougl. ex Hook.)T.& G. [*M. albicaulis* var. *gracilis* (Rydb.)Darlington] Kuha. Occasional, in sandy areas; foothill and desert, below 1250m: Warner Hot Springs; Borrego Palm Canyon; Vallecito Stage Station; The Narrows; Sentenac Canyon; Campo; Jacumba; Table Mountain. Mar.-July. *n*=36.

M. congesta T. & G. Ventana Stick-Leaf. Local, on disturbed or open slopes, montane, above 1200m: East Grade; Palomar Mountain. June-July. *n*=9.

M. desertorum (A. Davids.)Thomps. & Robts. Desert Stick-Leaf. Infrequent, in sandy areas; desert, below 500m: northeast corner of county; Borrego Palm Canyon. Apr.

M. dispersa Wats. Nada Stick-Leaf. Rare, on open slopes; montane, above 1200m: Lost Valley. May-Aug.

M. gracilenta T.& G. Slender Stick-Leaf. Scarce, on rocky slopes; cismontane, above 500m: Palomar Mountain; Campo. Apr.-May. *n*=18.

M. hirsutissima Wats. var. **stenophylla** (U. & G.)Jtn. [*M. peirsonii* Jeps.; *M. tricuspis* var. *brevicornuta* auct.] Hairy Stick-Leaf. Uncommon, in rocky areas; desert, below 1000m: Borrego Palm Canyon; Box Canyon; Mountain Springs; southeast of Agua Caliente; Canebrake Canyon; Rockhouse Canyon. Mar.-Apr.

M. involucrata Wats. Sand Blazing Star. Occasional, on rocky slopes; desert, below 700m: e.g. northeast corner of county; Palm Wash; Borrego Palm Canyon; Butler Canyon; San Felipe Creek; Yaqui Well; Carrizo Stage Station; Mountain Springs; Lava Flow Wash; Alta Bisnaga Wash; Mortero Wash; June Wash. Jan.-May. *n*=9.

M. laevicaulis (Dougl. ex Hook.)T.& G. Giant Blazing Star. Rare, known in the county only from an old collection from "Southwestern Colorado Desert, San Diego County". *n*=11.

***M. lindleyi** T.& G. Lindley's Blazing Star. Native of central California. Local, escaping from cultivation in Chollas Heights and Pine Valley. May. 2*n*=36.

M. micrantha (H.&. A.)T.& G. Small-flowered Stick-Leaf. Rare, in Chaparral and burns; cismontane, below 1100m: Cottonwood Valley; Otay Mountain. Apr.-June. *n*=9.

M. montana (A. Davids.)A. Davids. ssp. **montana**. Montane Mentzelia. Occasional, in open and disturbed areas; montane, above 900m: Palomar Mountain State Park; Chihuahua Valley; Lost Valley; Harper Ranch; Hipass; Boulevard. May-June. *n*=18.

M. nitens Greene. Venus Blazing Star. Rare, known in the county only at Indian Canyon, 350m. Mar. *n*=9.

M. multiflora (Nutt.)Gray ssp. **longiloba** (Darl.)Thompson & Zavortink. [*M. l.* Darl.] Yerba Amarilla. Rare, known in the county only at San Felipe Canyon, 550m. Mar. *n*=9.

M. obscura Thomps. & Robts. Obscure Blazing Star. Uncommon, in sandy areas; desert: Carrizo Stage Station. Apr.

M. pectinata Kell. San Joaquin Blazing Star. Rare, known in the county only on Palomar Mountain, above 1300m. May.

M. pinetorum Heller. Pine Stick-Leaf. Infrequent, in open areas, montane, above 1100m: Fry Creek, Palomar State Park; Eagles Nest; Harper Ranch. May-June.

M. veatchiana Kell. [*M. gracilenta* var. *v.* (Kell.)Jeps.] Veatch Stick-Leaf. Occasional, in open and disturbed areas; transmontane, above 900m: Montezuma View Point; Hot Springs Mountain; Laguna Summit; Table Mountain; InKoPah County Park. Mar.-May. *n*=27.

Petalonyx Gray   Sandpaper Plant

Petals distinct . . . . *P. linearis*
Petals connivent, claws forming a tube . . . . *P. thurberi*

P. linearis Greene. Rare, known in the county only from Borrego Palm Canyon, 200m; canyon from Inner Pasture, 400m; northwest corner Borrego Valley; unnamed wash north of June Wash. Mar.-May. *n*=23.

P. thurberi Gray ssp. **thurberi**. Occasional, in sandy areas; desert, below 1300m: Fish Creek; San Felipe Valley; The Narrows; Mason Valley; Vallecito Stage Station; Mountain Springs. May-July. *n*=23.

LYTHRACEAE.   Loosestrife Family

KEY TO GENERA

Flower-tube short, campanulate to globular; leaves opposite; petals 4 . . . . *Ammannia*
Flower-tube cylindric; petals usually 6; leaves mainly alternate . . . . *Lythrum*

Ammannia L.

A. coccinea Rottb. Valley Red-Stem. Occasional, on lake shores and other moist areas; coastal, below 300m: Lakeside; Santee; San Pasqual; Lake Hodges; Lake Murray; Sweetwater Lake; Lower Otay Lake. May-Oct.

Lythrum L.

Flowers sessile or nearly so . . . . *L. hyssopifolia*
Flowers short-pedicelled . . . . *L. californicum*

L. californicum T.& G. California Loosestrife. Frequent, on lake shores and other moist areas; cismontane: Sandia Creek; Palomar Mountain; Lake Henshaw; Peña Spring; Cuyamaca Mountains; Buckman Spring; Poway; San Pasqual; Sweetwater Lake; Otay River. Apr.-Oct.

L. hyssopifolium L. Grass Poly. Common, in vernal pools and other moist places; cismontane: De Luz; Palomar Mountain; Cuyamaca Lake; Hatfield Creek; Pine Valley; Lake Hodges; Mira Mesa; National City; Otay; Dulzura. Apr.-Oct. 2*n*=20.

MALVACEAE.   Mallow Family

KEY TO GENERA

Fruit a capsule, the carpels not separating from each other or the axis; flowers solitary in upper axils; ovules several in each carpel . . . . *Hibiscus*
Fruit of several carpels, usually separating from each other and the axis when mature
  Style-branches terminating in capitate or truncate stigmas
    Involucel of bractlets below calyx wanting
      Carpels not sharply differentiated into a reticulate basal and a wider smooth apical portion; mostly herbaceous; petals yellow or orange; carpels 2-9-seeded
        Carpels leathery or chartaceous, not inflated; fruit hemispheric-truncate, apically umbilicate . . . . . *Abutilon*
        Carpels membranaceous, strongly inflated; fruit subglobose, not umbilicate apically . . . . *Herissantia*
      Carpels sharply differentiated into a reticulate basal and a wider winged smooth apical portion; rather woody . . . . *Horsfordia*
    Involucel of 1 to several bractlets below calyx present
      Carpels sharply differentiated into a reticulate basal and a wider winged smooth apical portion; plants herbaceous or suffruticose . . . . *Sphaeralcea*

Carpels not so differentiated
Ovules 2 or more per carpel; carpels hirsute, 4mm high, reniform, 2-celled . . . . *Modiola*
Ovules solitary; carpels not hirsute
Plants not lepidote (beset with small scurfy scales); flowers white, pink, lavender, or purplish
Carpels indehiscent, rugose on the sides; annuals . . . . *Eremalche*
Carpels splitting into halves at maturity, not rugose; shrubby . . . . *Malacothamnus*
Plants lepidote; petals yellowish . . . . *Malvella*
Style-branches filiform, longitudinally stigmatic on the inner side
Involucel of 3-9 connate bractlets; axis of fruit surpassing the carpels, forming a cone or projection in the center . . . . *Lavatera*
Involucel of 1-3 distinct bractlets or lacking
Bractlets of involucel 3; weedy introduced annuals . . . . *Malva*
Bractlets 0 or 1; native annuals or perennials . . . . *Sidalcea*

Abutilon Mill. Indian Mallow

Plants annual; leaves often 10-20cm wide; carpels more than 10, with long divergent awns . . . . *A. theophrasti*
Plants perennial; leaves 1-5cm wide; carpels usually fewer . . . . *A. palmeri*

A. palmeri Gray. Palmer's Abutilon. Infrequent, on rocky slopes; desert, 600-1200m: Shaw Canyon; Potrero Canyon; Inner Pasture; hills south of Vallecito Stage Station. Apr.-May.

***A. theophrasti** Medic. Chingma Velvet Leaf. Native of southern Asia. Local, escaping from cultivation; coastal, below 300m: Singing Hills; San Diego. July-Aug. $2n$=42.

Eremalche Greene

Leaves crenate; petals 2-3cm long, each with a dark spot . . . . *E. rotundifolia*
Leaves 5-7-lobed; petals not spotted . . . . *E. exilis*

E. exilis (Gray)Greene. [*Malvastrum e.* Gray] Trailing Mallow. Frequent, in sandy places; desert, below 1000m: e.g. Benson Dry Lake; Borrego Valley; Vallecito Stage Station; Jacumba. Also Pala. Mar.-June. n=10.

E. rotundifolia (Gray)Greene. [*Malvastrum r.* Gray] Desert Fivespot. Occasional, in sandy washes and mesas; desert, below 150m: Borrego Valley; The Narrows; Ocotillo Wells; northeast corner of county; Mortero Wash; Vallecito Wash; Jojoba Wash. Mar.-Apr. n=10.

Herissantia Medic.

H. crispa (L.)Brizicky. [*Bogenhardia c.* (L.)Kearney] Curly Abutilon. Rare, known in the county only at Mountain Springs Grade, 700m. Apr. n=7.

Hibiscus L. Rose-Mallow

H. denudatus Benth. Rock Hibiscus. Occasional, on rocky slopes; desert, below 700m: Fish Creek; Mountain Palm Springs; Vallecito Stage Station; Borrego Palm Canyon. Nov.-Dec. and Mar.-Apr.

Horsfordia Gray

H. newberryi (Wats.)Gray. Yellow Feltplant. Infrequent, on rocky slopes; desert, below 800m: Fish Creek; Vallecito Stage Station; Borrego Valley; northeast corner of county; Mortero Wash; Inner Pasture Outlet Canyon; Volcanic Hills; south Lava Flow Wash. Nov.-Dec. and Mar.-Apr. n=15.

Lavatera L. Tree-Mallow

Involucral lobes lanceolate; petals 2.5-4.5cm long . . . . *L. assurgentiflora*
Involucral-lobes rounded-ovate; petals 1-2cm long . . . . *L. cretica*

***L. assurgentiflora** Kell. [*Saviniona suspensa* Greene, the type from San Diego.] Malva Rosa. Native of California Channel Islands. Occasional, escaping from cultivation; coastal, below 200m: Encinitas; San Dieguito Valley; Sorrento Valley; Mission Valley; National City; Tijuana River Valley. Mar.-Nov.

***L. cretica** L. Native of Mediterranean. Rare, escape at National City, 60m . Feb.-July.

Malacothamnus Greene

Calyx-lobes mostly 4-8mm wide at base; bractlets subtending the calyx ovate to broadly lanceolate; leaves cordate at base . . . . *M. aboriginum*
Calyx-lobes mostly not more than 3mm wide at base; bractlets narrowly lanceolate to filiform
Hairs of calyx long, simple to few-branched (the arms to 2mm or more long) . . . . *M. densiflorus* var. *viscidus*
Hairs of calyx short (to 1mm long) and many-armed
Stems appressed-pubescent, or if pubescence loose, then the calyx-lobes abruptly acuminate
Calyx-lobes abruptly acuminate . . . . *M. densiflorus* var. *d.*
Calyx-lobes acutish; leaves thinnish, mostly cordate at base, distinctly but shallowly lobed; inflorescence paniculate
Inflorescence of interrupted spike of headlike cluster . . . . *M. fasciculatus* ssp. *f.*;
Inflorescence open-paniculate, the lower branches often long and loosely flowered . . . . *M. fasciculatus* ssp. *laxiflorus*

M. aboriginum (Rob.)Greene. Indian Valley Bushmallow. Local, in rocky places of Laguna Mountains; montane, above 1550m: Black Mountain, Lusardi; Monument Peak; Shrine Camp; Garnet Peak; Desert View. May.-Aug.

M. densiflorus (Wats.)Greene var. **densiflorus**. [*Malvastrum d.* Wats.] Many-flowered Bushmallow. Occasional, in Chaparral; foothill, 300-1300m: Banner Grade; Barrett Reservoir; Laguna Mountains; El Capitan; Campo. Apr.-July.

M. densiflorus var. **viscidus** (Abrams)Kearney. [*Malvastrum d.* var. *v.* (Abrams)Estes.] San Diego Bushmallow. Occasional, in Chaparral; foothill, 150-1300m: San Felipe Valley; Chariot Canyon; Dehesa; Jamul; Dulzura Grade; Japatul Valley; San Miguel and Otay Mountains. Type from dry hillsides near El Nido (south side of Lower Otay Lake).

M. fasciculatus (Nutt.)Greene ssp. **fasciculatus**. [*Malvastrum f.* (Nutt.)Greene] Mesa Bushmallow. Common, in Chaparral and Coastal Sage Scrub; coastal and foothill, below 500m: e.g. Fallbrook; San Vicente Reservoir; Point Loma; Chollas Valley; San Ysidro. Apr.-July. Type from "St. Barbara"; however, the specimen is like the form occurring at San Diego, where Nuttall probably obtained his specimen, as the form has not since been found at Santa Barbara. n=17.

M. fasciculatus ssp. **laxiflorus** (Gray)Thorne. Lax-flowered Bushmallow. Rare, known in the county only on the seaward slopes of San Onofre Mountain, 200m. May-Aug. n=17.

Malva L. Mallow

Bractlets of involucel ovate to oblong . . . . *M. nicaeensis*
Bractlets linear to nearly so . . . . *M. parviflora*

***M. nicaeensis** All. Bull Mallow. Native of Eurasia. Uncommon, weed in cultivated and disturbed areas: Otay. May-Oct. n=21.

***M. parviflora** L. Cheeseweed. Native of Eurasia. Prevalent, weed in disturbed areas; cismontane and desert: e.g. San Onofre; Moosa Canyon; Borrego Springs; Barona Valley; National City; San Diego; Otay. Jan.-Dec. 2n=42.

Malvella Jaub. & Spach

M. leprosa (Ortega)Krapov. [*Sida hederacea* (Dougl.)Torr.] Alkali Mallow. Frequent, in moist or saline habitats; foothill and coastal, below 500m: e.g. Moosa Canyon; Ramona; San Diego; Bonita; San Ysidro. Mar.-Oct. 2n=22.

Modiola Moench

***M. caroliniana** (L.)D. Don. Wheel Mallow. Native of Neotropics. Uncommon, weed in cultivated areas; coastal, below 150m: Encanto. Apr.-Sep.

Sidalcea Gray Checkers

S. malvaeflora (DC.)Gray ex Benth. ssp. **sparsifolia** C.L. Hitchc. Checker-bloom. Frequent, on shaded slopes and meadows; cismontane: e.g. Rainbow; Santa Margarita Mountains; Palomar Mountain; Mesa Grande; Cuyamaca and Laguna Mountains; El Cajon Mountain; La Jolla; Otay Mountain. Mar.-Aug.

Sphaeralcea St. Hil. Globemallow

Plants annual or biennial; petals orange; the reticulate basal part of the carpel forming 2/3 or more of the whole and conspicuously wider than the unreticulate part . . . . *S. orcuttii*
Plants perennial; petals red (grenadine) to white, pink or lavender; the reticulate part of the carpel forming less than 2/3 of the whole and not conspicuously wider then the unreticulate part
  Leaves not more than 1/3 as wide as long, pubescence yellowish . . . . *S. angustifolia*
  Leaves at least half as wide as long
    Stems usually woody above base and more than 5dm high; inflorescence usually an open long-branched panicle
      Petals grenadine . . . . *S. ambigua* ssp. *a.*
      Petals pink or lavender, often drying violet . . . . *S. ambigua* ssp. *rosacea*
    Stems usually herbaceous above the crown and less than 5dm high; inflorescence racemiform or narrow-thyrsoid . . . . *S. ambigua* ssp. *rugosa*

S. ambigua Gray ssp. **ambigua**. [*S. a.* var. *keckii* Munz] Grenadine Desert Mallow. Common, in canyons and on rocky slopes; desert, below 1300m: Chihuahua Valley; Borrego Valley; San Felipe Valley; Mountain Springs Grade; Jacumba; Boulevard. Also Tijuana Hills. Mar.-July. n=10, 15.

S. ambigua ssp. **rosacea** (M.& J.)Kearney. [*S. r.* M. & J.] Rose Desert Mallow. Uncommon, on rocky slopes; desert, below 1300m: Table Mountain; Box Canyon; Dos Cabezas; Mountain Springs; InKoPah Park. Mar.-Sep. n=15.

S. ambigua ssp. **rugosa** Kearney. Thick-leaf Desert Mallow. Infrequent, northeast corner of county, in Chaparral and rocky slopes; foothill and desert, below 1400m: Warner Springs; Borrego Valley; Box Canyon; Jacumba; Oak Grove; Border Field State Park. Mar.-Sep. n=10.

S. angustifolia (Cav.)G. Don ssp. **cuspidata** (Gray)Kearney. Narrow-leaf Globemallow. Rare, known in the county only near Ocotillo Wells, 50m. June-Sep. n=5, 10.

S. orcuttii Rose. Orcutt's Globemallow. Rare, known in the county only from Carrizo Creek, below 150m. Mar.-May. n=5.

MARTYNIACEAE. Martynia Family

Proboscidea Keller in Schmid Unicorn-Plant

Leaves averaging 5cm or less wide; flowers yellowish or copper-colored . . . . *P. althaeifolia*
Leaves averaging 10cm or more across; flowers deep reddish-purple to pinkish, or at least with purplish blotches . . . . *P. lousianica*

P. althaeifolia (Benth.)Dcne. [*Martynia a.* Benth.] Desert Unicorn-Plant. Infrequent, in sandy areas; desert, below 1000m: Vallecito Stage Station; Carrizo Wash; Sweeny Pass; Egg Mountain; Agua Caliente Springs; Campbell Grade; Canebrake Canyon; Bow Willow. July-Sep.

***P. louisianica** (Mill.)Thell. [*Martynia l.* Mill.; *P. parviflora* (Woot.)Woot. & Standl.] Unicorn Plant. Native of southeastern United States. Uncommon, in disturbed places; coastal, below 150m: north of Batiquitos Lagoon; San Diego; Otay Mesa. June-Aug. n=15.

MELIACEAE.

Melia L.

***M. azedarach** L. var. **umbraculiformis** Berckmans. Umbrella Tree. Native of southwest Asia. Infrequent, persisting around abandoned dwellings or becoming naturalized in drainage areas; coastal, below 200m: Sorrento Valley; Sweetwater Valley; Otay Valley; El Cajon Valley. Mar.-May. 2n=28.

MYOPORACEAE. Myoporum Family

Myoporum Banks

***M. laetum** Forst. f. Ngaio. Native of New Zealand. Local, escaping from cultivation in moist areas; coastal, below 200m: Las Flores; Encinitas; Del Mar; La Jolla; Imperial Beach. Jan.-Aug.

MYRTACEAE. Myrtle Family

Eucalyptus L'Herit

Large plantations of *Eucalyptus* are to be found in the western part of the county. These were planted, for the most part, as a source of wood for railroad ties; however the wood proved to be of improper quality and the plantations have persisted unattended.

Each filament with an anther
  Capsules less than 1cm in diameter
    Anthers oblong, the dorsal gland sometimes showing in front
      Filaments generally attached beneath the gland halfway or less; leaves lemon scented; buds ovoid, operculum hemispherical, scarcely acute; pedicels slender . . . . *E. citriodora*
      Filaments attached near base of the cells so as to appear adnate . . . . *E. cladocalyx*
    Anthers versatile, obovate, the distinct cells opening in longitudinal slits for their entire length; filaments tapering to a sharp point
      Operculum conical to horn-shaped, 2-3 times longer than the calyx-tube . . . . *E. umbellata*
      Operculum short, conical to rostrate, 1-3 times longer than the calyx-tube . . . . *E. camaldulensis*
  Capsule over 1cm in diameter, verrucose, powdery-white . . . . *E. globulus*
Many filaments without anthers
  Anthers adnate, erect on the filament, the cells distinct, opening in ovate slits; gland absent. . . . . *E. sideroxylon*
  Anthers oblique on the filaments, opening by a terminal pore. . . . . *E. polyathemos*

***E. camaldulensis** Dehnhardt. Murray Red Gum. Native of Australia. Common, persisting in planted groves and escaping from cultivation; coastal, below 200m: Rancho Santa Fe; Torrey Pines; Jamul; Otay. Jan.-Apr.

***E. citriodora** Hook. Lemon-scented Spotted Gum. Native of Australia. Infrequent, escaping from cultivation; coastal, below 200m: San Diego. Jan.-Apr.

***E. cladocalyx** F. v. Muell. Sugar Gum. Native of Australia. Uncommon, persisting in planted groves and escaping from cultivation; coastal, below 200m: Miramar Mesa; Otay; Proctor Valley. Feb.-May.

***E. globulus** Labill. Tasmanian Blue Gum. Native of Australia; Common, persisting in planted groves, around abandoned dwellings or escaping from cultivation; coastal, below 200m: Rancho Santa Fe; Kearny Mesa; San Diego. Dec.-May. 2n=22.

***E. polyanthemos** Schauer. Australian-Beech. Native of Australia. Scarce, persisting in planted groves and escaping from cultivation; coastal, below 200m: University of California, San Diego campus. Dec.-June.

***E. sideroxylon** A. Cunn. ex Benth. Red Ironbark. Native of Australia. Infrequent, escaping from cultivation; coastal, below 200m: San Diego State University. Jan.-Apr.

***E. umbellata** (Gaertn.)Domin. [*E. tereticornis* Smith] Forest Red Gum. Native of Australia. Infrequent, escaping from cultivation; coastal, below 200m: Sweetwater Valley. Apr.-July.

NELUMBONACEAE.

Nelumbo Adans.

***N. lutea** (Willd.)Pers. [*N. pentapetala* (Walt.)Fern.] American Water-lily. Native of eastern North America. Known in the county only at Middle Lake on Corte Madera.

NYCTAGINACEAE. Four-O'Clock Family

KEY TO GENERA

Stigma linear; fruit usually winged; perianth salverform . . . . *Abronia*
Stigma subglobose; fruit rarely winged; perianth usually funnelform or campanulate
  Fruit lenticular, the margins inrolled, dentate, the dorsal surface with 2 rows of stipitate glands; involucres 3-parted . . . . *Allionia*

Fruit not lenticular or with dentate margins or glands
Involucre of distinct bracts, 1 at base of each flower . . . . *Boerhavia*
Involucre of united bracts, resembling a gamosepalous calyx . . . . *Mirabilis*

Abronia Juss. Sand-Verbena

Perianth white to rose or purple-red
Flowers deep dark red; fruit 12-20mm broad . . . . *A. maritima*
Flowers white to rose; fruit often narrower
Stems and perianth densely long-villous; body of fruit rugose-veined, the veins coarse and extending into the wings
Perianth 12-16mm long . . . . *A. villosa* var. *v.*
Perianth 16-26mm long . . . . *A. villosa* var. *aurita*
Stems and perianth glabrous to short-villous; body of fruit not rugose-veined . . . . *A. umbellata*

A. maritima Nutt. ex Wats. Red Sand-Verbena. Prevalent, on sandy beaches; coastal, below 10m: e.g. Agua Hedionda; Cardiff; Solana Beach; Torrey Pines Beach; Mission Beach; Ocean Beach; Silver Strand; Boundary Monument No. 258. Feb.-Oct.

A. umbellata Lam. [*A. u.* var. *platyphylla* (Standley)Munz, the type from Del Mar; *A. u.* var. *minor* (Standley)Munz] Beach Sand-Verbena. Prevalent, on sandy beaches; coastal, below 10m: e.g. Oceanside; Encinitas; Torrey Pines Beach; Mission Beach; Silver Strand; Boundary Monument No. 258. Jan.-Dec.

A. villosa Wats. var. **aurita** (Abrams)Jeps. Chaparral Sand-Verbena. Uncommon, in sandy creek beds; foothill, below 800m: San Felipe; Santa Margarita River. Mar.-Aug.

A. villosa var. **villosa**. Desert Sand-Verbena. Common, in sandy areas; desert, below 1000m: e.g. Coyote Creek; Borrego Palm Canyon; Borrego Valley; Benson Dry Lake; Mason Valley; Agua Caliente Hot Springs; Carrizo Stage Station. Jan.-July.

Allionia L.

A. incarnata L. Trailing Windmills. Frequent, in sandy areas; desert, below 1000m: e.g. Coyote Creek; Borrego Valley; Sentenac Canyon; Fish Creek; Vallecito Stage Station; Carrizo Stage Station; Bow Willow. Apr.-Sep.

Boerhavia L.

Plants perennial; fruit pubescent . . . . *B. diffusa*
Plants annual; fruit glabrous
Ultimate branches of inflorescence with flowers in racemose spikes
Bracts persistent, as long as the mature fruit, which is 4-angled; stamens 3-4 . . . . *B. wrightii*
Bracts deciduous, shorter; fruit 5-angled; stamens 1-2 . . . . *B. coulteri*
Ultimate branches of inflorescence with flowers in umbels or solitary on slender pedicels
Flowers in umbels; fruit ca. 3 times as long as thick . . . . *B. intermedia*
Flowers mostly solitary, forming cymes; fruit ca. twice as long as thick . . . . *B. triquetra*

B. coulteri (Hook. f.)Wats. Coulter Ringstem. Rare, known in the county only at San Felipe Valley, 800m. Sep.-Nov.

B. diffusa L. [*B. coccinea* auct. non Mill.] Red Ringstem. Frequent, in sandy places; desert and foothill, below 1000m: Warner Hot Springs; Warner Ranch; San Felipe Valley; Borrego Palm Canyon; Vallecito Canyon; Bow Willow. Also Balboa Park. Apr.-July.

B. intermedia Jones. [*B. erecta* var. *i.* (Jones)Kearn. & Peebl.] Five-wing Ringstem. Rare, known in the county from Sentenac Canyon, 650m; Blair Valley; Little Blair Valley. Aug.-Oct.

B. triquetra Wats. Slender Boerhavia. Rare, known in the county from Scissors Crossing, 675m. Sept.-Dec.

B. wrightii Gray. Wright Ringstem. Scarce, in sandy areas; desert, below 500m: The Narrows; east of Arroyo Salado; Vallecito Stage Station; Sweeny Pass. Aug.-Dec.

Mirabilis L. Four-O'Clock

Involucre 1-flowered; perianth campanulate
Perianth 3-5cm long; fruit 5-angled . . . . *M. jalapa*
Perianth 1-1.5cm long; fruit smooth
Involucral lobes lanceolate, longer than the tube; involucre 11-13mm long; leaf-blades 3-4cm long . . . *M. tenuiloba*
Involucral lobes subovate, not exceeding tube; involucre 5-9mm long; leaf-blades mostly less than 2.5cm long
Flowers white to pale pink; leaves viscid-villous
Stems villous-viscid
Fruit ovoid, elongate . . . . *M. bigelovii* var. *b.*
Fruit subglobose . . . . *M. bigelovii* var. *aspera*
Stems subglabrous below, scabrous-puberulent above with short retrorse hairs . . . . *M. bigelovii* var. *retrorsa*
Flowers rose to red; leaves subglabrous to rough-pubescent . . . . *M. californica*
Involucre 3-10-flowered, 2.5-4cm long; perianth funnelform . . . . *M. multiflora*

M. bigelovii Gray var. **aspera** (Greene)Munz. Desert Wishbone Plant. Uncommon, in sandy places; desert, above 700m: San Felipe; Chariot Canyon; Mountain Springs. Oct.-Nov.

M. bigelovii var. **bigelovii**. Occasional, in sandy areas; desert, below 1000m: e.g. Borrego Palm Canyon; Borrego Valley; Fish Creek; Box Canyon; Sentenac Canyon; Jacumba. Mar.-Apr.

M. bigelovii var. **retrorsa** (Heller)Munz. Frequent, in sandy areas; desert, below 1000m: e.g. Borrego Palm Canyon; Borrego Valley; Sentenac Canyon; Vallecito Stage Station; Pinyon Mountain; Oriflamme Canyon; Mountain Springs; Jacumba. Mar.-Apr.

M. californica Gray. var. **californica**. [*M. laevis* auct. non (Benth.)Curran] Coastal Wishbone Plant. Common, on dry slopes; coastal and foothill, below 700m: e.g. San Onofre; Fallbrook; De Luz; Pala; Bonsall; San Pasqual; Kearny Mesa; Poway Grade; Alpine; Mount Helix; Lee Valley; Dulzura; Point Loma; Silver Strand; San Ysidro. Also Hellhole Canyon; Mountain Springs; Potrero. Dec.-June. Type from "dry hills, San Diego".

M. multiflora (Torr.)Gray in Torr. var. **pubescens** Wats. [*M. froebelii* (Behr)Greene; *Oxybaphus f.* Behr, type from around Warner's Ranch; *M. froebelii* var. *glabrata* (Standl.)Jeps.] Froebel's Four-OClock. Occasional, in sandy areas and on rocky slopes; desert and foothill, above 500m: e.g. Oak Grove; Warner Hot Springs; Chihuahua Valley; Hot Springs Mountain; Warner Ranch; Banner; Oriflamme Canyon; Mount Laguna; Boulevard. Apr.-Aug. $2n=66$.

***M. jalapa** L. Four-OClock. Native of Neotropics. Infrequent, escaping from cultivation; coastal, below 200m: National City. June-Aug.

M. tenuiloba Wats. Long-lobed Wishbone Plant. Infrequent, in canyons and washes; desert, below 800m: Borrego Palm Canyon; Borrego Valley; The Narrows; Mount Tule; Carrizo Gorge; Mountain Springs; Jacumba; Lava Flow Wash; N. Mortero Wash; Inner Pasture Outlet Canyon. Mar.-May.

OLEACEAE. Olive Family

KEY TO GENERA

Fruit a samara; leaves largely pinnately compound . . . . . *Fraxinus*
Fruit a capsule; leaves mostly alternate; small shrubs . . . . . *Menodora*

Fraxinus L. Ash

Corolla present, consisting of 2 white petals; flowers bisexual; style obscurely lobed; samara flat, broadly wing-margined to base . . . . . *F. dipetala*
Corolla absent; style conspicuously 2-lobed
  Leaflets 5-7; branchlets terete; body of samara subterete and narrowly winged along the side for 1/2 the way *F. velutina*
  Leaflets 3, sometimes 1 or 2, rarely petiolate . . . . . *F. jonesii*

F. dipetala H.& A. Rare, in canyons and along streambeds; foothill, 200-1000m: Sloane Ranch, below Loveland Dam. Apr.-May. $2n=23$ pairs.

F. jonesii Lingel. [*F. trifoliolata* (Torr.)Lewis & Epl. non W. W. Smith] Rare, known in the county only from a collection between Lyons Valley and Lawson Valley, 620m. Apr.-May.

F. velutina Torr. var. **coriacea** (Wats.)Rehd. [*F. oregana* Nutt.] Infrequent, in canyons and along streambeds; foothill, 300-1500m: Lost Valley; Eagles Nest; Cañada Agua Caliente; Indian Canyon; Spencer Valley; Noble Canyon; McCain Ranch; Walker Ranch; Olivenhain; Barrett; Marron Valley. Mar.-Apr.

Menodora Humb. & Bonpl.

M. scoparia Engelm. ex Gray. Broom Twinberry. Uncommon, on rocky slopes; desert, 600-1200m: San Felipe; Canebrake Canyon; Dubber; Jacumba; InKoPah; Mountain Springs; Table Mountain. Mar.-May. $2n=22$.

ONAGRACEAE. Evening-Primrose Family

KEY TO GENERA

Sepals persistent; petals 5 or 0; stamens 10 or 3-6 . . . . . *Ludwigia*
Sepals deciduous after anthesis
  Seeds with tuft of hairs (coma) at one end . . . . . *Epilobium*
  Seeds without a coma
    Fruit a dehiscent capsule (although sometimes tardily so)
      Ovary 4-loculed; floral-tube prolonged beyond the ovary
        Stigma capitate, entire, sporogenous tissue continuous in the anther-locules . . . . . *Camissonia*
        Stigma 4-lobed
          Petals 2-lobed, small; pollen in tetrads . . . . . *Boisduvalia*
          Petals entire, usually larger; pollen not in tetrads
            Anthers erect, attached near the base, innate; sporogenous tissue in the anther-locules divided into packets by sterile tissue . . . . . *Clarkia*
            Anthers versatile, attached near their middle; sporogenous tissue continuous . . . . . *Oenothera*
      Ovary 2-loculed; floral-tube not prolonged beyond the ovary; flowers minute; stem branches capillary . . . . . *Gayophytum*
    Fruit indehiscent, nutlike and hard . . . . . *Gaura*

Boisduvalia Spach

Capsule with a tough wall, the seeds in 2 rows in each locule; plant often subglabrous . . . . . *B. glabella*
Capsule with a pliable wall, the seeds in a single row in each locule; plants not subglabrous . . . . . *B. densiflora*

B. densiflora (Lindl.)Wats. [*B. densiflora* var. *imbricata* Greene] Dense-flowered Boisduvalia. Frequent, in wet places; foothill and montane, above 900m: e.g. upper Doane Valley; Lake Henshaw; Morretti Junction; Mesa Grande; Witch Creek; Pine Hills; Cuyamaca Lake; Middle Peak; Paso Picacho; Pine Valley; Japatul Valley. May-Aug. $n=10$.

B. glabella (Nutt.)Walp. Smooth Boisduvalia. Infrequent, in vernal pools and other ephemeral ponds; cismontane: Palomar Mountain; National City; Otay Mesa; San Ysidro. May-Aug. $n=15$.

Camissonia Link

Ovary with a long slender sterile projection below the floral-tube; the plants acaulescent or nearly so; annual
- Petals 5-18mm long; style 3-5.5mm long; plants pilose . . . *C. graciliflora*
- Petals 2-3.5mm long; style 1-2.2mm long; plants strigose . . . *C. palmeri*

Ovary lacking a sterile projection; the plants only occasionally appearing acaulescent, and then when immature
- Capsule on well-defined pedicel, not coiled or contorted; seeds in 2 rows in each locule
  - Floral-tube 0.4-0.8mm long; leaves basal or cauline, often pinnately divided or lobed; pollen shed singly
    - Lower parts of plant villous with spreading hairs; sepals with free caudate projections arising below the apices . . . *C. claviflormis* ssp.*peirsonii*
    - Lower parts of plant strigulose or glabrous; sepals entire or with caudate projections *C. claviformis* ssp. *yumae*
  - Floral-tube 4.5-40mm long; leaves all cauline, simple, usually cordate-orbicular; pollen shed in tetrads . . . *C. cardiophylla*
- Capsule sessile, often coiled or contorted; seeds in 1 row in each locule
  - Petals white, the flowers vespertine
    - Capsule not thickened near the base; plants lacking flowers near lower nodes
      - Petals 3.5-7mm long; stigma held well above the anthers at anthesis . . . *C. refracta*
      - Petals 1.8-3mm long; stigma surrounded by the anthers at anthesis . . . *C. chamaenerioides*
    - Capsule thickened near the base; plants with or without flowers at lower nodes . . . *C. boothii*
  - Petals yellow; the flowers matinal
    - Leaves pinnately lobed; petals usually with a pattern of minute maroon flecks near the base; seeds often with purple dots; capsule often sharply reflexed at maturity . . . *C. californica*
    - Leaves subentire; petals unspotted or with 1 or 2 bright red spots near the base; seeds unspotted; capsule not reflexed
      - Capsule quadrangular, at least when dry; plants usually with flowers at the basal nodes; leaves mostly lanceolate to ovate; seeds dull, flattened, usually over 1mm long
        - Plants perennial, often somewhat woody near the base . . . *C. cheiranthifolia*
        - Plants annual, except occasionally in *C. bistorta*
          - Stigma evidently held above the anthers at anthesis, the petals mostly 7-15mm long . . . *C. bistorta*
          - Stigma surrounded by at least the anthers of the longer set of stamens at anthesis, the flowers often smaller
            - Pollen 25-100 percent 4- or 5-spored . . . *C. robusta*
            - Pollen up to 5 percent 4-spored
              - Capsule markedly quadrangular in transection . . . *C. lewisii*
              - Capsule not markedly quadrangular in transection, at least in living material
                - Upper leaves evidently petiolate, attenuate at the base; capsule usually contorted; plants subglabrous to strigulose, often reddish; cismontane . . . *C. ignota*
                - Upper leaves usually not evidently petiolate, rounded or truncate at the base; capsule straight to coiled; plants strigulose to villous
                  - Plants conspicuously gray-strigose, of the deserts, the branches usually decumbent
                    - Petals 6.5-13mm long, each red-dotted at base . . . *C. pallida* ssp. *hallii*
                    - Petals 3.5-6mm long, rarely red-dotted . . . *C. pallida* ssp. *p.*
                  - Plants not conspicuously gray-strigose, barely reaching the desert-edge
                    - Capsule 0.75-0.9mm thick; upper leaves ovate; plant erect . . . *C. hirtella*
                    - Capsule 0.9-1.8mm thick; upper leaves lanceolate to lance-ovate; plants erect or decumbent
                      - Plants with decumbent branches from the basal rosette; upper leaves narrowly lanceolate; inflorescence usually without short glandular hairs . . . *C. micrantha*
                      - Plants erect; upper leaves lanceolate to lance-ovate; inflorescence with short glandular hairs
                        - Petals 5-10mm long; style 4.5-7.5mm long; plants gray-villous . . . *C. confusa*
                        - Petals 1.5-4mm long; style 2-3.5mm long; plant usually not gray-villous . . *C. intermedia*
      - Capsule subterete; plants lacking flowers in the basal nodes; leaves mostly linear or very narrowly elliptic; seeds shining, often under 1mm long
        - Sepals reflexed singly; plants often with conspicuous spreading pubescence . . . *C. pusilla*
        - Sepals usually reflexed in pairs, remaining attached
          - Stigma held above the anthers at anthesis; petals 4-15mm long . . . *C. campestris*
          - Stigma surrounded by the anthers at anthesis; petals 2-4mm long . . . *C. strigulosa*

C. bistorta (Nutt. ex T.& G.)Raven. [*Oenothera b.* Nutt.; *O. heterophylla* Nutt. ex H.& A. the type from San Diego; *O. b.* var. *veitchiana* Hook.] Southern Sun Cup. Common, on open slopes and burns; cismontane, below 1400m: e.g. Fallbrook; Rincon; Oak Grove; Warner Hot Springs; Julian School; Anahuac; Oakzanita; Campo; Lyons Valley; Jamacha; San Diego; Point Loma; Silver Strand; Kearny Mesa; Lake Hodges; Encinitas. Mar.-June. Type from "St. Diego". n=7.

C. boothii (Dougl. in Hook.)Raven ssp. **condensata** (Munz)Raven. [*Oenothera decorticans* var. *c.* Munz] Desert Lantern. Frequent, on rocky slopes and in sandy washes; desert, below 300m: e.g. Lone Palm; Borrego Palm Canyon; northeast corner of county; Oil Well; Benson Dunes; Dry Lake School; The Narrows; Vallecito and Carrizo Stage Stations. Jan.-June. 2n=7 pairs.

C. californica (Nutt. ex T.& G.)Raven. [*Oenothera leptocarpa* Greene] False-Mustard. Frequent, in open and disturbed places; cismontane and desert, below 1400m: e.g. Fallbrook; Lilac; Rincon; Lake Wohlford; San Pasqual; Springhill; Foster; Viejas Grade; El Capitan Mountain; Dehesa; Dulzura; Campo; Mountain Springs; southeast corner of county; Carrizo Stage Station; Vallecito Mountain; Box Canyon; The Narrows; Borrego Palm Canyon; Lone Palm; northeast corner of county. Apr.-May. Type from "bushy plains near St. Diego". n=7, 14.

C. campestris (Greene)Raven. Field Evening-Primrose. Rare, known in the county only from a collection near Campo. Mar.-May. n=7.

C. cardiophylla (Torr.)Raven. [*Oenothera c.* Torr.] Heart-leaf Evening-Primrose. Occasional, on alluvial fans and rocky slopes; desert, below 600m: Borrego Grade; Borrego Springs; The Narrows; Yaqui Pass; Fish Creek; Carrizo Stage Station. Mar.-May. n=7.

C. chamaenerioides (Gray)Raven. [*Oenothera c.* Gray in Wats.] Willow-herb Evening-Primrose. Infrequent, on slopes and sandy flats; desert, below 500m: northeast corner of county; Borrego Valley; Yaqui Well; San Felipe Canyon; Mountain Palm Springs; Vallecito Stage Station. Mar.-June. *n*=7.

C. cheiranthifolia (Hornem. ex Spreng.)Raimann in Eng. & Prantl. ssp. **suffruticosa** (Wats.)Raven. [*Oenothera viridescens* Hook, the type from either San Diego or Santa Barbara; *Holostigma bottae* Spach, the type from San Diego; *O. spiralis* var. *linearis* Jeps., the type from Sunnyside; *O. c.* var. *s.* Wats.] Beach Evening-Primrose. Common, on sandy beaches and elsewhere; coastal, below 100m: e.g. Fallbrook; Oceanside; Carlsbad; Solana Beach; Torrey Pines Beach; Mission Bay; Sunnyside; Silver Strand; Coronado; Initial Boundary Monument. Apr.-Aug. *n*=7.

C. claviformis (Torr. & Frem.)Raven ssp. **peirsonii** (Munz)Raven. [*Oenothera c.* var. *p.* Munz] Peirson's Evening-Primrose. Common, on sandy flats and in washes; desert, below 900m: e.g. Lone Palm; Borrego Palm Canyon; Oil Well; Kane Spring Road; The Narrows; Canebrake Canyon; Carrizo Stage Station; Mountain Palm Springs; Sweeney Pass; Jacumba. Mar.-May. *n*=7.

C. claviformis ssp. **yumae** (Raven)Raven. Yuma Evening-Primrose. Rare, known in the county only at Yaqui Well, 500m. *n*=7.

C. confusa Raven. Occasional, on open slopes; foothill, 300-900m: Lake Henshaw; Henshaw Grade; Warner Ranch; Santa Ysabel; Banner; Cottonwood Valley. Apr.-May. *n*=14.

C. graciliflora (H.& A.)Raven. Rare, known in the county only from an old collection at Santa Ysabel, 1000m. Mar.-May. *n*=7.

C. hirtella (Greene)Raven. [*Oenothera micrantha* var. *jonesii* (Levl.)Munz] Field Sun Cup. Common, on dry slopes; foothill, 200-1300m: e.g. Rainbow; Pauma; Chihuahua Valley; Oak Grove; Eagles Nest; San Felipe Valley; Banner; Santa Ysabel; Lake Wohlford; Oriflamme Canyon; Green Valley; Cuyamaca Lake; Pine Valley; Jacumba; Campo; El Capitan Mountain; Otay Mountain; Tecate Peak; Mount Soledad. Mar.-May. *n*=7.

C. ignota (Jeps.)Raven. [*Oenothera micrantha* var. *i.* Jeps.] Frequent, on dry slopes; foothill, 200-1200m: e.g. Fallbrook; Moosa Canyon; Rincon Springs; Del Dios; Foster; San Miguel Mountain; Sloane Ranch; Harbison Canyon; Banner; Campo; Jacumba. Mar.-May. *n*=7.

C. intermedia Raven. Infrequent, on open slopes; foothill, 100-300m: Fallbrook; Red Mountain Grade; Rainbow; Jamul; Lee Valley. Apr.-May. *n*=14.

C. lewisii Raven. Occasional, in grassy places; coastal, below 250m: Fallbrook; San Luis Rey; Agua Hedionda; San Pasqual; Bird Rock; Crown Point; Ocean Beach; Point Loma; Balboa Park; Silver Strand; Oneonta; Bonita; Jamul. Mar.-Apr. *n*=7.

C. micrantha (Hornem. ex Spreng.)Raven. [*Oenothera m.* Hornem.] Uncommon, in sandy areas; coastal, below 150m: Fallbrook; Moosa Canyon; Lilac; Mount Soledad. Mar.-May. *n*=7.

C. pallida (Abrams)Raven ssp. **hallii** (Davids.)Raven. [*Oenothera bistorta* var. *h.* (Davids.)Jeps.] Rare, known in the county only from a collection at Warner Hot Springs, 950m. Apr. *n*=7.

C. pallida ssp. **pallida**. [*Oenothera micrantha* var. *exfoliata* (A. Nels.)Munz, the type from Colorado Desert, San Diego County; *O. autrani* Levl., the type from San Diego.] Frequent, in sandy areas; desert, below 1000m: e.g. Borrego Springs; Borrego Palm Canyon; Hellhole Canyon; San Felipe Ranch; Montezuma Valley; Mason Valley; Scissors Crossing; Jacumba; Vallecito Valley; Boulevard. Mar.-Apr. *n*=7.

C. palmeri (Wats.)Raven. Rare, known in the county only from a collection east of Jacumba. Apr.-May. *n*=7.

C. pusilla Raven. Rare, known in the county only from San Felipe Wash, 300m. Apr.

C. refracta (Wats.)Raven. [*Oenothera r.* Wats.] Occasional, on slopes; desert, below 500m: Banner Grade; Vallecito Stage Station; Canebrake Canyon. Mar.-May. *n*=7.

C. robusta Raven. Frequent, on open slopes; coastal and foothill, below 350m: e.g. Moosa Canyon; Pala; Lilac; Springhill; Del Dios; Escondido; San Pasqual Grade; Japatul Valley; Kearny Mesa; Point Loma; Balboa Park; Harbison Canyon; Dulzura; Tecate Junction. Mar.-May. *n*=21.

C. strigulosa (F.& M.)Raven. [*Oenothera contorta* var. *epilobioides* (Greene)Munz] Common, in grassy areas; cismontane: e.g. Fallbrook; Pala; San Luis Rey; Palomar Mountain; Lake Henshaw; Oak Grove; Warner Hot Springs; Eagles Nest; Montezuma View Point; San Felipe; Banner; Julian; Harper Ranch; Milk Ranch; Garnet Peak; Shrine Camp; Buckman Spring; Morena Dam; Boulevard; Jacumba; Campo; Descanso; The Willows; El Cajon; Ramona; San Miguel Mountain. May-June. *n*=14.

Clarkia Pursh

Petal constricted at base into a definite claw which is expanded into a pair of lateral teeth . . . . *C. rhomboidea*
Petals not clawed, or if so the claw entire
  Petals white to cream, fading pink, not flecked with purple, mostly less then 10mm long . . . . *C. epilobioides*
  Petals colored, or if nearly white, flecked with purple, mostly more than 10mm long
    Buds, at least the older ones, pendulous or deflexed
      Claw of the petals slender, equal to or longer than the limb . . . . *C. unguiculata*
      Claw of the petals short or lacking
        Sepals 6-10mm long; petals usually not more than 12mm long
          Rachis of inflorescence reflexed at tip, becoming straight as buds mature; petals pink, often with reddish-purple flecks near the base . . . . *C. similis*
          Rachis of inflorescence straight, the buds deflexed; petals pink to rose-lavender . . . . *C. delicata*
        Sepals mostly 10-20mm long; petals mostly 10-30mm long . . . . *C. deflexa*

Buds erect
Petals 5-15mm long, often with a wedge- or shield-shaped dark spot at or above the middle; stamens equaling the stigma . . . . . . *C. purpurea* ssp. *quadrivulnera*
Petals 15-25mm long, often with a dark blotch in center; stamens shorter than stigma *C. purpurea* ssp. *viminea*

C. deflexa (Jeps.)Lewis & Lewis. Punchbowl Clarkia. Rare, known in the county only from a collection from Franklin Canyon, Santa Margarita Mountains, below 800m. Apr.-June. *n*=9.

C. delicata (Abrams)Nels. & Macbr. Campo Clarkia. Occasional, in shaded areas of Southern Oak Woodland; foothill, 150-1300m: Palomar Mountain; Mesa Grande; Ramona; Bandy Canyon; Barona Valley; Foster; El Capitan Reservoir; Flinn Springs; Harbison Canyon; Lyons Valley; Bratton Valley; Morena Dam; Campo; Potrero; Chariot Canyon. Apr.-May. Type from "shady slopes between Potrero and Campo. *n*=18.

C. epilobioides (Nutt.)Nels. & Macbr. [*Godetia e.* (Nutt.)Wats. in Brewer & Wats.] Canyon Godetia. Occasional, on grassy, shaded slopes; coastal and foothill, below 400m: San Onofre Canyon; Fallbrook; Moro Hill; Bonsall; Rainbow; Lilac; Springhill; Foster; Mission Hills; Harbison Canyon; El Capitan Reservoir; near Barrett Dam; Otay Mountain; Tecate Peak. Mar.-May. Type from "St. Diego". *n*=9.

C. purpurea (Curt.)Nels. & Macbr. ssp. **quadrivulnera** (Dougl. in Lindl.)Lewis & Lewis. [*Godetia q.* (Dougl.)Spach; *G. p.* var. *parviflora* (Wats.)C. L. Hitchc.] Four-spot Clarkia. Frequent, in grassy places; cismontane: e.g. San Onofre Canyon; Bonsall; Lilac; Pala; Love Valley; Henshaw Dam; Ramona; Descanso; Barona Valley; Pine Hills; Cuyamaca Lake; Cuyamaca Peak; Harper Ranch; Garnet Peak; Chariot Canyon; Corte Madera; Lyons Valley; Proctor Valley; Otay Mesa; Lakeside; Alpine; Miramar Mesa; Kensington Park. Apr.-July. *n*=26.

C. purpurea ssp. **viminea** (Dougl. in Lindl.)Lewis & Lewis. Large Clarkia. Uncommon, on grassy slopes; cismontane: Cuyamaca Lake; El Cajon. May-July. *n*=26.

C. rhomboidea Dougl. ex Hook. Diamond Clarkia. Occasional, on dry or shaded slopes; foothill and montane, above 300m: Moosa Canyon; Fry Creek; Eagles Nest; Mesa Grande; Cuyamaca Lake; Pine Hills; Garnet Peak; Descanso; El Cajon Mountain; Otay Mountain; Banner Grade. May-July. *n*=12.

C. similis Lewis & Ernst. Canyon Clarkia. Infrequent, on shaded slopes; coastal and foothill, below 1100m: De Luz; Fallbrook; Warner Hot Springs; San Felipe Creek; Mesa Grande; Banner Grade; Ramona; Barona Valley; Dehesa; Palos Verde near Alpine; Descanso; Potrero Grade; Barrett; Campo; Morena Dam; Tijuana River Valley. Apr.-May. Type from 7.6 miles west of Ramona. *n*=17.

C. unguiculata Lindl. Elegant Clarkia. Scarce, on shaded slopes; coastal, below 300m: Camp Pendleton; Bonsall; San Luis Rey River at Highway 395; Gopher Canyon. May-June. *n*=9.

Epilobium L. Willow-Herb

Floral-tube lacking to less than 1cm long, without a ring of internal scales; flowers not scarlet
Floral-tube not prolonged above the ovary; flowers slightly irregular, the petals 1-2cm long, entire, spreading . . . . . . *E. angustifolium*
Floral-tube prolonged above the ovary; flowers regular, mostly smaller
Plants annual; stems with exfoliating epidermis; of dry situations . . . . . . *E. paniculatum*
Plants perennial, sometimes blooming the first year; epidermis not exfoliating from the stems; mostly of wet situations
Inflorescence glandular-pubescent . . . . . . *E. adenocaulon* var. *a.*
Hairs of inflorescence not gland-tipped
Herbage green, the stems and leaves scarcely pubescent, the inflorescence strigulose . . . . . . *E. ciliatum*
Herbage grayish, the stems and leaves densely soft-pubescent . . . . . . *E. adenocaulon* var. *holosericeum*

Floral-tube 2-3cm long, funnelform, with a row of 8 scales within at about one-half its length; flowers scarlet
Plants herbaceous throughout; leaves broadly lanceolate to ovate, the principal ones more than 6mm wide . . . . . . *E. canum* ssp. *latifolium*
Plants suffrutescent at base; leaves linear to lanceolate, mostly less than 6mm wide
Stems slightly woody; leaves mostly 3-5mm wide, pilose, not canescent . . . . . . *E. canum* ssp. *mexicanum*
Stems obviously woody at the base; leaves 2-3.5mm wide, densely tomentose-canescent . . . . . . *E. canum* ssp. *angustifolium*

E. adenocaulon Hausskn. var. **adenocaulon**. Uncommon, in wet places; foothill and montane, above 1000m: Palomar Mountain; Henshaw Dam. July-Sep.

E. adenocaulon var. **holosericeum** (Trel.)Munz. [*E. californicum* var. *h.* (Trel.)Jeps.] Occasional, in wet places; coastal and foothill, below 1000m: San Diego; El Cajon; Clover Flats. July-Sep.

E. angustifolium ssp. **circumvagum** Mosquin. [*Chamaenerion a.* (L.)Scop.] Red Fireweed. Rare, known in the county only from a collection in the Cuyamaca Mountains, above 1400m. July-Sep. 2*n*=36.

E. canum (Greene)Raven ssp. **angustifolium** (Keck)Raven. [*Zauschneria californica* Presl ssp. *californica*; *Z. c.* ssp. *angustifolia* Keck] California-Fuchsia. Frequent, on shaded slopes; coastal and foothill, below 900m: e.g. De Luz; Fernbrook; Poway Grade; El Cajon; Alvarado Canyon; Mission Valley; National City; Potrero Grade; Japatul Valley; Canyon City; Jacumba. Aug.-Oct. *n*=30.

E. canum ssp. **latifolium** (Hook.)Raven. [*Zauschneria californica* ssp. *latifolia* (Hook.)Keck; *Z. l.* var. *johnstonii* Hilend.] Frequent, in rocky places; foothill, montane, and desert, above 400m: e.g. Doane Valley; Eagles Nest; Sheep Canyon; Borrego Palm Canyon; San Felipe; Pine Hills; North Peak; Cuyamaca Peak; Paso Picacho; Pine Valley; Live Oak Springs; Descanso Junction; Laguna Junction; Jacumba; Dubber; Bow Willow Canyon. Aug.-Dec. *n*=30.

E. canum ssp. **mexicanum** (Presl.)Raven. [*Zauschneria californica* ssp. *mexicana* (Presl)Raven] Occasional, on shaded slopes; coastal and foothill, below 700m: e.g. O'Neill Lake; Ballena; Chollas Valley. Aug.-Oct. *n*=30.

E. ciliatum Raf. ssp. **ciliatum**. [*E. adenocaulon* var. *parishii* (Trel.)Munz; *E. californicum* Hausskn.; *E. palmeri* Levl., the type from Cuiamaca (Cuyamaca) Mountains.] Occasional, in wet areas; foothill and montane, above 900m: near Beauty Mountain; Cuyamaca Lake; San Diego River Falls; Cold Spring; Noble Mine; Campo. July-Sep. *n*=18.

E. paniculatum Nutt. ex T. & G. Summer Cotton-Weed. Frequent, in dry areas; foothill and montane, above 900m: e.g. Doane Valley; Warner Hot Springs; Los Terranitos; Descanso Junction; Inaja Indian Reservation; Middle Peak; Cuyamaca Peak; Pine Valley; Mount Laguna Picnic Grounds. June-Sep. *n*=12.

Gaura L.

Floral-tube 2.5-4mm long; filaments 7-10mm long; fruit 5-9mm long, 1-1.5mm wide . . . . *G. sinuata*
Floral-tube 5-12mm long; filaments 3-6mm long; fruit 5-11mm long, 2-3mm wide . . . . *G. drummondii*

***G. drummondii** (Spach)T.& G. [*G. odorata* auct. non Sesse ex Lag.] Drummond's Gaura. Native of Texas and Mexico. Infrequent, weed in cultivated and disturbed areas; foothill and coastal, below 500m: San Marcos; Escondido; Fletcher Hills; Otay Mesa. May-Sep. *n*=7, 14.

***G. sinuata** Nutt. ex Ser. in DC. Butterfly-Weed. Native of central United States. Infrequent, weed in cultivated and disturbed areas; foothill and coastal, below 1000m: Escondido; San Marcos; Ramona; Lake Hodges; Carlsbad; Santa Ysabel; Jamul. June-Sep. *n*=14.

Gayophytum Juss.

Plants flowering from near the base, the first flowers at 1-4 nodes above the cotyledons *G. diffusum* ssp. *parviflorum*
Plants not flowering near the base, but only several to many nodes above the cotyledons
  Capsule irregularly lumpy by failure of part of the ovules to develop; a high percentage of pollen grains empty . . . . *G. heterozygum*
  Capsule mostly regular in outline, entire or torulose, all of the ovules developing; pollen grains almost all good
    Seeds 1-5 (-6) in a capsule, each about 1.5mm long; capsule about twice as long as pedicels; petals usually 1.5-2mm long . . . . *G. oligospermum*
    Seeds usually 6-10 or more in a capsule, each about 1-1.2mm long; petals mostly 1.5-7mm long . . . . *G. diffusum* ssp. *parviflorum*

G. diffusum T. & G. ssp. **parviflorum** Lewis & Szweykowski. [*G. nuttallii* auct. non T.& G.; *G. lasiospermum* Greene, the type from near Julian.] Hairy-leaf Gayophytum. Occasional, in open areas of Montane Coniferous Forest; montane, above 1100m: Palomar Mountain; Julian; Cuyamaca Peak; Arroyo Seco; Cold Spring; Pine Valley. June-Aug. *n*=7, 14.

G. heterozygum Lewis & Szweykowski. Zigzag Gayophytum. Uncommon, in open areas of Montane Coniferous Forest; montane, above 1000m: Palomar and Cuyamaca Mountains. June-Oct. *n*=7.

G. oligospermum Lewis & Szweykowski. Few-seed Gayophytum. Occasional, in open areas of Montane Coniferous Forest; montane, above 1300m: Cuyamaca Mountains; Cuyamaca Peak; Japacha. June-Sep. *n*=7.

Ludwigia L.

Stamens in 2 series, mostly 8 or 10 in number; petals 10-20mm long
  Flowering stems usually floating or creeping; leaves oblong, 1-10cm long; bracteoles at base of ovary deltoid; capsule mostly 2-3mm thick . . . . *L. peploides*
  Flowering stems usually erect; leaves lanceolate, mostly 5-10cm long; bracteoles lanceolate; capsule 3-4mm thick . . . . *L. uruguayensis*
Stamens in 1 series, 4-5 in number; petals none or small and quickly shed . . . . *L. palustris*

L. palustris (L.)Ell. var. **pacifica** Fern. & Griscom. Pacific Marsh-Purslane. Uncommon, in ponds and slow streams; foothill, 300-900m: Boulder Creek; Echo Valley. June-Sep.

L. peploides (H.B.K.)Raven ssp. **peploides**. [*Jussiaea repens* var. *p.* (H.B.K.)Griseb.] Willow Water-Weed. Infrequent, in ponds and other moist places; coastal, below 100m: Mission San Luis Rey; Agua Hedionda. May-Oct.

***L. uruguayensis** (Camb. in St. Hil.)Hara. [*Jussiaea u.* Camb.] Uruguay Marsh Purslane. Native of South America. Rare, adventive in wet areas; coastal, below 200m: near intersection of Federal and Euclid Boulevards, 40m. Aug.

Oenothera L. Evening-Primrose

Capsule terete or round-angled (winged above in *Oe. speciosa*); stems well developed and bearing leaves and flowers
  Flowers yellow, the buds erect
    Capsule angled, tapering from base toward apex; seeds prismatic-angled . . . . *Oe. elata*
    Capsule terete, not narrowed in upper part; seeds not angled . . . . *Oe. laciniata*
  Flowers white to rose, the buds often nodding
    Capsule sterile and slender in lower part, thicker, fertile and winged in upper; seeds in more than 2 rows in each locule; floral-tube 0.4-2cm long . . . . *Oe. speciosa*
    Capsule cylindric, sessile, not sterile in lower part; seeds in 1 row in a locule; floral-tube 2-4cm long
      Plants annual or surviving longer from a deep taproot; basal leaves rhombic, the blades 2-10cm long; capsule usually woody; plant with exfoliating epidermis . . . . *Oe. deltoides*
      Plants perennial, largely from running underground rootstocks; basal leaves tending to be smaller and more narrow; capsule usually not woody . . . . *Oe. californica*
Capsule crested or tubercled; plants acaulescent or nearly so, cespitose . . . . *Oe. caespitosa* var. *marginata*

Oe. caespitosa Nutt. var. **marginata** (Nutt.)Munz. Uncommon, in sandy areas, northern desert areas, above 900m: Culp Valley. Apr.

Oe. californica (Wats.)Wats. Common, in dry, sandy places; coastal and foothill, below 1100m: e.g. Santa Margarita River ford; Sweetwater Valley; Boulevard; Campo; La Posta; Buckman Spring; Mountain Springs; Walker Ranch; Mason Valley; Grapevine Spring; Montezuma View Point. Apr.-June. *n*=14.

Oe. deltoides Torr. & Frem. ssp. **deltoides**. [*Oe. d.* var. *cineracea* (Jeps.)Munz, the type from Borrego Springs] Dune Evening Primrose. Frequent, in sandy areas, desert, below 600m: Borrego Valley; Yaqui Wells; Ocotillo Wells; San Felipe Wash; Carrizo Stage Station. Mar.-May. *n*=7.

Oe. elata H.B.K. ssp. **hirsutissima** (Gray ex Wats.)Dietrich. [*Oe. hookeri* T.& G. ssp. *grisea* (Bartlett)Munz; *Oe. hookeri* ssp. *venusta* (Bartlett)Munz] Great Marsh Evening-Primrose. Occasional, in marshy areas; cismontane, below 1500m: Murray Dam; Palomar Mountain; Cardiff; Poway Junction; Julian; Ramona; Lake Henshaw; Inaja Indian Reservation; Paso Picacho; Ballena; San Diego River Falls; Dehesa; Jamul; Buckman Spring; Clover Flats. June-Sep. *n*=7.

***Oe. laciniata** Hill. Native of southeast United States.. Naturalized in San Pasqual Valley. May. *n*=7.

***Oe. speciosa** Nutt. Native of south-central United States. Naturalized in disturbed areas; cismontane: National City; Dos Picos County Park Campground.

OROBANCHACEAE. Broom-Rape Family

Orobanche L. Broom-Rape

Flowers solitary on long naked scapose pedicels that are longer than the corolla, without bracts at base of calyx; corolla-lobes orbicular . . . . *O. fasciculata*
Flowers densely spicate, sessile or on pedicels shorter than the corolla, with 2 bracts at the base of the calyx; corolla-lobes tapered to the pointed or rounded apex
  Stem arising from an enlarged tuberlike thickening; inflorescence parts glandular-puberulent . . . . *O. bulbosa*
  Stem not having an enlarged tuberlike base
    Corolla-lips, especially margins, villous with predominantly nonglandular hairs; calyx-lobes often dark purplish . . . . *O. cooperi*
    Corolla-lips with scattered short, glandular hairs; calyx-lobes pale to brownish, sometimes with the tips suffused with lavender or red
      Corolla 25-50mm long, the lips 10-14mm long and widely spreading . . . . *O. californica*
      Corolla 15-25mm long, the lips 4-9mm long, nearly erect to moderately spreading
        Calyx-lobes 7-11mm long, purple . . . . *O. parishii* ssp. *brachyloba*
        Calyx-lobes 10-16mm long; pinkish within . . . . *O. parishii* ssp. *p.*

O. bulbosa (Gray)G. Beck. Chaparral Broom-Rape. Uncommon, in Chaparral; coastal and foothill, below 1000m: Paradise Mountain; Rancho Santa Fe; Witch Creek; San Diego; Otay Mountain; Japatul Valley; Lawson Valley; McCain Valley. Apr.-May. Known host: *Adenostoma*. 2*n*=48.

O. californica Cham. & Schlecht. ssp. **feudgei** (Munz)Heckard. [*O. grayana* var. *f.* Munz] Sagebrush Broom-Rape. Uncommon, in Chaparral and Sagebrush Scrub; foothill, 800-1300m: Dodge Valley; Boulevard; Tierra del Sol; near Jacumba. May-July. Known host: *Artemisia tridentata*. 2*n*=48.

O. cooperi (Gray)Heller. [*O. ludoviciana* var. *c.* (Gray)G. Beck; *O. ludoviciana* Nutt. var. *latiloba* Munz] Desert Broom-Rape. Rare, in sandy areas; desert, below 150m: northeast corner of county; San Felipe Wash; Mason Valley; Pinyon Mountains; Borrego Springs; Vallecito Wash; Alta Bisnaga Wash; 13 miles southeast of Vallecito Stage Station; Ocotillo Wells; Lava Flow Wash; Mortero Wash; Mine Canyon. Jan.-May. Known host: *Ambrosia dumosa*. 2*n*=48, 72, 96.

O. fasciculata Nutt. [*O. f.* var. *franciscana* Achey] Pine Broom-Rape. Occasional, in Montane Coniferous Forest; montane, above 1050m: Boucher Hill; Doane Valley; Eagles Nest; Harper Ranch; Middle Peak; Cuyamaca Peak; Pinyon Point; Laguna Camp; Cuyapaipe Spring; Pine Valley; McCain Valley. May-June. Known hosts: *Eriogonum*, *Eriodictyon*, *Artemisia tridentata*.

O. parishii (Jeps.)Heckard ssp. **brachyloba** Heckard. Beach Broom-Rape. Rare, on sandy beaches; coastal, below 10m: Torrey Pines State Reserve; Point Loma; Silver Strand. May-Sep. Known hosts: *Atriplex californica*; *Isocoma veneta*. 2*n*=96.

O. parishii ssp. **parishii**. [*O. californica* var. *p.* Jeps.] Parish's Broom-Rape. Infrequent, in Chaparral; foothill and montane, above 600m: Warner Hot Springs; Eagles Nest; Julian; Pine Hills; Witch Creek; Ramona; Laguna Observatories; Cameron. Mar.-Sep. Known hosts: *Adenostoma*, *Arctostaphylos*; *Eriodictyon*, *Corethrogyne*. 2*n*=48.

OXALIDACEAE. Wood-Sorrel Family

Oxalis L. Wood-Sorrel

Plants acaulescent, with underground bulblets; petals 20mm long . . . . *O. cernua*
Plants caulescent, without bulblets; petals 8-16mm long
  Stems creeping from a slender taproot and rooting at the nodes . . . . *O. corniculata*
  Stems erect or decumbent, not rooting at nodes
    Stems glabrous or appressed-puberulent with very short crisped hairs 0.3-0.5mm long; pedicels to 4 times longer than their mature capsule; capsule stigulose . . . . *O. albicans* ssp. *californica*
    Stems with straight hairs appressed upward, to 1.2mm long; pedicels to 1.6 times longer than mature capsule . . . . *O. albicans* ssp. *pilosa*

O. albicans H.B.K. ssp. **californica** (Abrams)Eiten. [*O. californica* (Abrams)Kunth.] California Wood-Sorrel. Occasional, on shaded slopes; coastal and foothill, below 500m: San Onofre; Fallbrook; Pala;

Moosa Canyon; Poway Grade; El Capitan Mountain; Morena; Dehesa; Lee Valley; Barrett; Jamul Mountains; Otay Mountain; Emerald Hills. Mar.-May. Type from dry hillsides, Onofre (San Onofre) Mountains.

O. albicans ssp. **pilosa** (Nutt. ex T.& G.)Eiten. [*O. californica* ssp. *pilosa* (Nutt.)Lourt.; *O. p.* Nutt.] Scarce, on shaded slopes; coastal and foothill, below 300m: San Luis Rey; Pala; 11 km south of Escondido; San Diego. Apr.-Nov.

***O. corniculata** L. ssp. **corniculata**. Yellow Sorrel. Native of Europe. Common, weed in lawns and disturbed areas; coastal, below 200m: Balboa Park; East San Diego; National City. Mar.-Nov. 2*n*=24, 36, 42, 48.

***O. cernua** Thunb. [*O. pes-caprae* L.] Bermuda-Buttercup. Native of South Africa. Occasional, naturalized or escaping from cultivation in moist places; coastal, below 300m: Las Flores; Del Mar Heights; Point Loma; Balboa Park; Dehesa; Silver Strand; Initial Boundary Monument. Nov.-Mar.

PAEONIACEAE. Peony Family

Paeonia L. Peony

P. californica Nutt. ex T. & G. California Peony. Frequent, in Chaparral, Coastal Sage Scrub, and Southern Oak Woodland; coastal and foothill, below 1300m: e.g. estuary of Santa Margarita River; Hormiguero Ranch; Poway Grade; Vista; San Pasqual; Spencer Valley; Pine Hills; Lakeside; Echo Valley; Buckman Spring; La Posta; Campo; Dehesa; Otay Lake; El Capitan Mountain; Sweetwater Valley; Olivenhain. Dec.-May. *n*=5.

PAPAVERACEAE. Poppy Family

KEY TO GENERA

Leaves mainly opposite or whorled
  Stamens many; carpels 6-25, torulose and separating when mature; petals rather persistent . . . . *Platystemon*
  Stamens 6-12; carpels mostly 3, united; petals falling early . . . . *Meconella*
Leaves mainly alternate or basal
  Plants shrubby, at least at base, mostly 1m or more tall
    Flowers white; leaves lobed . . . . *Romneya*
    Flowers yellow; leave entire . . . . *Dendromecon*
  Plants herbaceous, usually less than 1m tall
    Plant spiny or conspicuously long-hairy . . . . *Argemone*
    Plant not spiny or conspicuously long-hairy
      Leaves multifid into linear segments . . . . *Eschscholzia*
      Leaves not multifid into linear segments
        Style slender, short; stigma capitate . . . . *Stylomecon*
        Style lacking; stigma discoid . . . . *Papaver*

Argemone L. Prickly Poppy

Stems with about 100-300 prickles per square cm at 5cm below oldest capsule . . . . *A. munita* ssp. *argentea*
Stems with about 10-20 prickles per square cm at 5cm below oldest capsule . . . . *A. munita* ssp. *m.*

A. munita Dur. & Hilg. ssp. **argentea** G. Ownbey. Pale-stem Prickly Poppy. Scarce, in rocky areas; desert, 500-1000m: Split Mountain. Mar.-May. *n*=14.

A. munita ssp. **munita**. [*A. platyceras* auct. non Link & Otto] Chocolate Prickly Poppy. Occasional, in disturbed, dry areas; desert and foothill, above 1200m: Palomar Mountain; San Felipe; Rincon; Borrego Palm Canyon; Laguna Junction; Laguna Mountains; Cottonwood Valley; Jacumba; Potrero. Mar.-Apr. *n*=14.

Dendromecon Benth. Tree Poppy

D. rigida ssp. **rigida** Benth. [*D. saligna* Greene, the type from Encenitas (Encinitas).] California Bush Poppy. Frequent, on open slopes; cismontane: e.g. Encinitas; Del Mar; Sorrento Valley; La Jolla; Fallbrook; Agua Tibia Mountains; Montezuma Valley; Banner; Stonewall Peak; Monument Peak; Noble Canyon; McCain Valley; Dubber; Buckman Springs; Morena Dam; Jacumba; El Cajon Mountain; Otay Mountain. Apr.-Aug. *n*=28.

Eschscholzia Cham. in Nees California Poppy

Torus with 2 rims, the inner erect and hyaline, the outer spreading; perennials or annuals with 2-cleft cotyledons
  Plants perennial
    Plants compact, remarkably glaucous . . . . *E. californica* var. *c.*
    Plants open . . . . *E. californica* var. *douglasii*
  Plants annual . . . . *E. californica* var. *peninsularis*
Torus with only an erect hyaline rim, the outer rim absent or rudimentary; annuals with entire cotyledons
  Stems scapose, the leaves mostly crowded in a basal tuft . . . . *E. caespitosa*
  Stems leafy
    Petals 3-8mm long; deserts . . . . *E. minutiflora*
    Petals 10mm or more long
      Leaves mostly 3 times ternate into blunt lobes . . . . *E. caespitosa*
      Leaves mostly 4 times ternate . . . . *E. parishii*

E. caespitosa Benth. Tufted Gold-Poppy. Rare, known in the county from a single collection from Witch Creek, 800m. Apr. *n*=6.

***E. californica** Cham. var. **californica**. Coastal California Poppy. Native of the northern portion of California. Occasional, introduction on grassy slopes; cismontane: e.g. Hot Springs Mountain; Mount Woodson; Cardiff; Lake Hodges; Moosa Canyon; Sweetwater Valley; Silver Strand. Feb.-Sep. *n*=6.

E. californica var. **douglasii** (Benth.)Gray. [*E. c.* var. *crocea* (Benth.)Jeps.; *E. arvensis* var. *dilatata* Greene] Common California Poppy. Rare, known in the county only from a collection from Witch Creek, 800m. Apr.

E. californica var. **peninsularis** (Greene)Munz. [*E. clevelandii* Greene, the type collected at or near San Diego; *E. floribunda* Greene, the type from near Santa Ysabel; *E. physodes* Greene, the type from Witch Creek; *E. sanctum* Greene, the type from Grand Mesa (Mesa Grande).] Annual California Poppy. Frequent, on disturbed slopes and burns; cismontane, below 1450m: e.g. upper San Onofre Canyon; Warner Ranch; Montezuma View Point; Pala; De Luz; Banner; Rattlesnake Valley; Eckener Pass; Boulevard; Campo; Cottonwood; Foster; Descanso; Rolando; Point Loma; Boundary Monument No. 258. Feb.-Sep.

E. minutiflora Wats. Pygmy Gold-Poppy. Occasional, in sandy areas; desert, below 800m: Coyote Creek; Borrego Palm Canyon; Hellhole Canyon; northeast corner of county; The Narrows; Scissors Crossing; Vallecito Stage Station; Box Canyon; Mountain Springs Grade. Mar.-Apr. 2*n*=36.

E. parishii Greene. Parish's Gold-Poppy. Infrequent, in rocky places; desert, below 1000m: Borrego Valley; northeast corner of county; Yaqui Pass; The Narrows; Carrizo Stage Station; Lava Flow Wash; Plum Canyon; Alta Bisnaga Wash; south Coyote Mountain; Mortero Falls. Mar.-Apr. 2*n*=36.

Meconella Nutt. in T. & G.

M. denticulata Greene. [*M. kakoethes* Fedde, the type from San Diego.] Small-flowered Meconella. Infrequent, on grassy slopes; foothill, 300-1100m: Ramona; San Felipe Valley; Descanso; Santa Ysabel; Potrero Grade. Apr.-July. Type from Temecula Canyon, north of Mission San Luis Rey. *n*=8.

Papaver L. Poppy

P. californicum Gray. Fire Poppy. Occasional, in disturbed areas and burns; coastal and foothill, below 1200m: San Onofre; Ramona; Pala; Lilac; San Marcos Mountain; Lake Hodges; Banner; Harbison Canyon; Tecate Peak; Kimball Grade; Otay Mountain; Foster; San Diego; Kearny Mesa; Cowles Mountain. Apr.-May. 2*n*=28.

Platystemon Benth. Cream Cups

Mature fruit nodding . . . . *P. californicus* var. *nutans*
Mature fruit erect
  Petals cream; plant moderately hairy . . . . *P. californicus* var. *c.*
  Petals yellow; plant excessively long-hairy . . . . *P. californicus* var. *crinitus*

P. californicus Benth. var. **californicus**. [*P. obtectus* Greene, the type from Witch Creek; *P. verecundus* var. *glabrifructifer* Fedde, the type from San Diego.] California Cream Cups. Frequent, on grassy slopes; cismontane: Lake Hodges; Culp Valley; Laguna Lakes; Viejas Valley; Witch Creek; Banner; Foster; Pacific Beach. Feb.-July. *n*=6.

P. californicus var. **crinitus** (Greene)Greene. Yellow Cream Cups. Occasional, on open slopes; foothill and montane, above 600m: Palomar Mountain; Mesa Grande; Banner; Warner Ranch; Montezuma View Point; Cuyamaca Lake; Laguna Mountains; Descanso; Dubber; Mountain Springs. Apr.-June.

P. californicus var. **nutans** Bdg. San Diego Cream Cups. Occasional, on grassy slopes; coastal and foothill, below 400m: Fallbrook; Rainbow; Bonsall; Escondido; Bernardo; San Pasqual; Ramona; El Cajon Valley; Tecate Peak; Dehesa; Otay Mountain; Point Loma; Kearny Mesa; Silver Strand; Boundary Monument No. 258. Type from about San Diego and on many of the islands off the coast.

Romneya Harv. Matilija Poppy

Sepals and summit of peduncle glabrous; peduncles not conspicuously leafy near summit; lobes of leaves mostly more than 12mm wide . . . . *R. coulteri* var. *c.*
Sepals and summit of peduncles setose; peduncles leafy to top; lobes of leaves mostly less than 10mm wide . . . . *R. coulteri* var. *trichocalyx*

R. coulteri Harv. var. **coulteri**. Canyon Matilija Poppy. Infrequent, along creek beds and canyons; coastal and foothill, below 300m: San Diego; Sweetwater Valley; El Nido. May-July. 2*n*=38.

R. coulteri var. **trichocalyx** (Eastw.)Jeps. [*R. t.* Eastw.] Coastal Matilija Poppy. Occasional, on slopes and along creek beds; south coastal and foothill, below 1050m: Barrett Junction; Pine Creek; Espinosa Trail; Dulzura; Tecate Mountain; Potrero Grade; Jamul Mountain; Jamul Valley. May-July. *n*=19.

Stylomecon G. Taylor

S. heterophylla (Benth.)G. Taylor. Wind Poppy. Infrequent, on grassy slopes and burns; coastal and foothill, below 300m: Ysidora; La Costa; San Elijo Lagoon; Ocean Beach; San Diego; La Jolla; Mission Valley; Jamacha; Harbison Canyon; Imperial Beach. Feb.-Apr. 2*n*=56.

PHYTOLACCACEAE. Pokeweed Family

Phytolacca L. Pokeweed

***P. americana** L. Native of eastern United States. Infrequent, weed in cultivated areas; coastal, below 150m: Fallbrook; Lakeside; San Diego; National City; Chula Vista. Aug.-Oct. 2*n*=36.

PITTOSPORACEAE. Pittosporum Family

Pittosporum Banks

*__P. undulatum__ Vent. Victoria-Box. Native of Australia. Uncommon, escaping from cultivation; coastal, below 200m: Balboa Park; San Diego State University. Jan.-Mar.

PLANTAGINACEAE. Plantain Family

Plantago L. Plantain

Corolla-lobes spreading or reflexed; stamens 4
  Corolla-tube pubescent externally; strictly coastal . . . . . . *P. coronopus*
  Corolla-tube glabrous externally; generally distributed
    Anterior sepals connate; stamens conspicuously exserted . . . . . . *P. lanceolata*
    Anterior sepals separate; stamens not conspicuous
      Plant heavy-rooted perennials; bracts not longer than calyx; leaves broadly elliptic to cordate-ovate . . *P. major*
      Plants annual, or if perennial, with lower bracts much exceeding calyx
        Lowest bracts lanceolate to subulate, scarcely membranous on sides, 1-3 times as long as sepals; seeds brown, dull . . . . . . *P. purshii*
        Lowest bracts ovate with broad membranous margins and not exceeding calyx
          Seeds brown, dull; bracts not like sepals and about half their length
            Plants villous . . . . . . *P. erecta* ssp. *e.*
            Plants with more rigid, spreading hairs . . . . . . *P. erecta* ssp. *rigidior*
          Seeds reddish-yellow, shining; bracts almost exactly like sepals . . . . . . *P. ovata*
Corolla-lobes often erect in fertile flowers and forming a beak over the capsule
  Leaves linear to subfiliform; stamens 2
    Corolla-lobes mostly erect in age, forming a beak; seeds 4, about 0.8-1.8mm long . . . . . . *P. pusilla*
    Corolla-lobes spreading or reflexed in age, not forming a beak; seeds 4-9, 1.5-2mm long
      Seeds 4-5 per capsule . . . . . . *P. bigelovii* ssp. *b.*
      Seeds 6-9 per capsule . . . . . . *P. bigelovii* ssp. *californica*
  Leaves linear-lanceolate to ovate; stamens 4
    Plants annual; rarely biennial; spikes less than 1dm long; introduced weeds
      Sepals rounded at apex; seeds pale brown . . . . . . *P. virginica*
      Sepals long-attenuate; seeds red . . . . . . *P. rhodosperma*
    Plants perennial; spikes 1-2.5dm long; native . . . . . . *P. hirtella*

P. bigelovii Gray ssp. **bigelovii**. Uncommon, in vernal pools, cismontane, below 250m: Fletcher Hills; Grossmont College Vernal Pools. Mar.-June.

P. bigelovii Gray ssp. **californica** (Greene)Bassett. Infrequent, along shorelines of lakes and vernal pools; coastal and foothill, below 500m: Ramona; Kearny Mesa. Mar.-June.

*__P. coronopus__ L. Cut-leaf Plantain. Native of Europe. Introduced at Penasquitos. Several locations in alkaline areas along road down Penasquitos Canyon up to 5 miles from Black Mountain Road. Apr.-May.

P. erecta Morris ssp. **erecta**. [*P. hookeriana* var. *californica* (Greene)Poe; *P. obversa* Morris, the type from Del Mar.] Dot-seed Plantain. Frequent, in open areas; coastal and foothill, below 300m: San Onofre Canyon; Fallbrook; Pauma Valley; Rancho Santa Fe; Lakeside; Jamacha; Dulzura; Otay Mesa. Apr.-June. $2n=20$.

P. erecta ssp. **rigidior** Pilger. Vernal Pool Plantain. Uncommon, around vernal pools; coastal, below 200m: Kearny Mesa; San Diego; Otay Mesa. Mar.-May.

P. hirtella H.B.K. ssp. **galeottiana** (Dcne. in A.DC.)Thorne. Mexican Plantain. Infrequent, in moist, disturbed places; coastal, below 150m: Rancho Santa Fe; Mission Gorge; Sweetwater Valley; Balboa Park. Apr.-June.

*__P. lanceolata__ L. Rib-Grass. Native of Europe. Common, in disturbed and moist places; cismontane: e.g. Palomar Mountain; Santa Ysabel; Pine Hills; Valley Center; Ramona; Descanso; Cardiff; San Diego; Sweetwater Valley; National City. Apr.-Aug. $2n=12, 24, 96$.

*__P. major__ L. Common Plantain. Native of Europe. Common, in disturbed and moist places; cismontane: e.g. Palomar Mountain; Lake Henshaw; Ballena; Hot Springs Mountain; San Felipe; Laguna and Cuyamaca Mountains; Buckman Spring; Lakeside; Campo. May-Aug. $2n=12$.

*__P. ovata__ Forsk. [*P. insularis* Eastw.; *P. i.* var. *fastigiata* (Morris)Jeps.] Woolly Plantain. Apparently native to the Mediterranean. Frequent, on flats and slopes; coastal and desert, below 500m: e.g. Borrego Valley; Benson Dry Lake; Vallecito and Carrizo Stage Stations; San Diego; Coronado; Chula Vista. Mar.-May. $2n=4$ & 8 pairs.

P. purshii R. & S. var. **oblonga** (Morris)Shinners. Desert Plantain. Frequent, in open areas of Chaparral and Montane Coniferous Forest; foothill and montane, above 800m: e.g. Warner Hot Springs; Lake Henshaw; Eagles Nest; Anahuac School; Julian; Oriflamme Canyon. Apr.-June.

*__P. pusilla__ Nutt. [*P. heterophylla* auct. non Nutt.] Slender Plantain. Native of eastern United States. Scarce, adventive on drying mud flats; foothill and coastal, below 400m: Ramona; Soledad. Apr.-Aug. $2n=12$.

*__P. rhodosperma__ Dcne. Red-seed Plantain. Native of central United States and Mexico. Infrequent, naturalized in disturbed places; coastal, below 150m: San Diego; Sweetwater Valley. May.

*__P. virginica__ L. Dwarf Plantain. Rare, known in the county only by a collection at San Diego. May-July.

PLATANACEAE. Sycamore Family

Platanus L. Sycamore

P. racemosa Nutt. California Sycamore. Common, along streams and dry water courses; coastal and foothill, below 1200m: e.g. De Luz; San Mateo; Pala; Lilac; Palomar Mountain; Oak Grove; upper Borrego Palm Canyon; San Felipe; Banner; upper Agua Hedionda Lagoon; Buckman Spring; Barrett; Dulzura; Woodwardia Canyon; Lyons Peak; Poway; San Clemente Canyon. Feb.-Apr. 2*n*=21 pairs.

PLUMBAGINACEAE. Leadwort Family.

Limonium Mill. Sea-Lavender

Leaves not pinnatifid; branches of inflorescence not winged
  Flowers about 5-6mm long, and 2mm wide
    Calyx ribs pubescent . . . . *L. californicum* var. *c.*
    Calyx glabrous . . . . *L. californicum* var. *mexicanum*
  Flowers about 9mm long and 4-5mm wide . . . . *L. perezii*
Leaves pinnatifid; branches of inflorescence winged . . . . *L. sinuatum*

L. californicum (Boiss.)Heller var. **californicum**. California Marsh-Rosemary. Rare, in sandy areas; coastal, below 50m: San Diego Bay. July-Dec. 2*n*=18.

L. californicum var. **mexicanum** (Blake)Munz. [*L. m.* Blake] San Diego Rosemary. Frequent, in Coastal Strand and Coastal Salt Marsh; coastal, below 20m: e.g. Carlsbad; Moonlight State Beach; Santa Margarita River estuary; Cardiff; Solana Beach; Pacific Beach; Mission Bay; Silver Strand; Coronado; Imperial Beach; Salt Works. July-Nov. Type from San Diego.

***L. perezii** (Stapf.)F.T. Hubb. Perez Rosemary. Native of the Canary Islands. Infrequent, escaping from cultivation and naturalized in sandy areas; coastal, below 100m: Carlsbad; Bonita; Imperial Beach. Mar.-Sep.

***L. sinuatum** (L.)Mill. Notchleaf Marsh Rosemary. Native of Europe. Occasional, naturalized in disturbed, sandy areas; coastal, below 20m: San Elijo Lagoon; Moonlight State Beach; Del Mar; Imperial Beach. Also Lake Murray. June-Oct.

POLEMONIACEAE. Phlox Family

KEY TO GENERA

Calyx growing with the capsule and not ruptured by it, becoming chartaceous in age . . . . *Collomia*
Calyx at length ruptured by the maturing capsule, commonly with a membranous pseudotube formed by the coalescence of the membranes bordering the sepals
  Leaves usually opposite, at least near base of plant, the upper sometimes alternate (if the lower are alternate, plant usually woody and leaves prickly)
    Stamens unequally inserted; leaves entire
      Plants perennial; flowers showy . . . . *Phlox*
      Plants annual; flowers inconspicuous . . . . *Microsteris*
    Stamens equally inserted; leaves mostly palmately cleft
      Plants annual or perennial herbs, usually not very woody below; leaves not spinose . . . . *Linanthus*
      Plants subshrubby; leaf-lobes mostly ending in spinose tips . . . . *Leptodactylon*
  Leaves mostly alternate, entire to pinnately dissected
    Calyx-lobes subequal; flowers solitary, cymose or in heads
      Corolla-lobes essentially alike; leaf-lobes mostly not setose- or spine tipped
        Upper cauline leaves usually much reduced; tips of leaf-lobes not conspicuously mucronate; bracts mostly subtending groups of flowers; seeds small, rounded, brown . . . . *Gilia*
        Upper cauline leaves quite well developed
          Tips of leaf-lobes not conspicuously mucronate; bracts mostly subtending groups of 2-8 flowers; upper leaves digitately lobed; seeds round-ovoid, mostly dark brown to black . . . . *Allophyllum*
          Tips of leaf-lobes bearing horny mucros; bracts subtending individual flowers; upper leaves pinnatifid to entire; seeds long and slender to shorter and angled, curved, pale brown to straw-colored . . *Ipomopsis*
      Corolla-lobes unlike; leaf-lobes setose- or spine-tipped . . . . *Langloisia*
    Calyx-lobes unequal; flowers in dense bracteate heads
      Plants cobwebby-pubescent, at least the inflorescence with a feltlike mass of interlaced hairs; capsule dehiscent from top; leaves and bracts rarely with rigid spinose lobes . . . . *Eriastrum*
      Plants not having a feltlike mat of interlaced hairs; capsule dehiscent mostly from below or indehiscent; leaves and bracts usually with rigid spinose lobes . . . . *Navarretia*

Allophyllum (Nutt.)A. & V. Grant

Segments of cauline leaves lanceolate to oblong, 3-15mm wide, the terminal segments 3-8 times longer than wide; flowers pale to medium blue-violet, white or red-violet to purple with pink lobes . . . . *A. glutinosum*
Segments of leaves linear to linear-lanceolate, not over 3mm wide, the terminal segments of upper leaves 8-14 times longer than broad; flowers dark blue-violet
  Inflorescence glomerate with 4-8 flowers subtended by a leaf; many leaves pinnately lobed; inflorescence-branches usually arising above the middle of a plant . . . . *A. gilioides*
  Inflorescence loose with 2-3 flowers subtended by a leaf; few leaves pinnately lobed; inflorescence-branches arising throughout the plant . . . . *A. violaceum*

A. gilioides (Benth.)A.& V. Grant. [*Gilia g.* (Benth.)Greene] Straggling False-Gilia. Occasional, on dry slopes; cismontane, above 1000m: El Cajon Mountain; Mission Valley; Green Valley; Eagles Nest; Laguna Lakes; Garnet Peak. Apr.-June. *n*=9.

A. glutinosum (Benth.)A.& V. Grant. [*Gilia gilioides* var. *g.* (Benth.)Jeps.] Blue False-Gilia. Frequent, on shaded slopes; cismontane, below 1500m: e.g. Fallbrook; Hormiguero Ranch; Doane Valley; Oak Grove;

Chihuahua Valley; Hot Springs Mountain; Middle Peak; Laguna Mountains; Pine Valley; Campo; Dulzura; Otay Mountain; Harbison Canyon; Mount Woodson; Mission Gorge. Mar.-June. *n*=9.

A. violaceum (Heller)A.& V. Grant. Violet False-Gilia. Uncommon, in sandy and gravelly places; foothill and montane, above 400m: Barona Valley; Cuyamaca Mountains. May-July. *n*=9.

Collomia Nutt.

C. grandiflora Dougl. ex Lindl. Grand Collomia. Frequent, in Montane Coniferous Forest; foothill and montane, above 1000m: e.g. Palomar, Cuyamaca, and Laguna Mountains; Julian; Pine Hills; Pine Valley; Kitchen Creek; Descanso. May-July. *n*=8.

Eriastrum Woot. & Standl.

Plants perennial, with woody base; corolla 15-30mm long, the tube 2-3 times as long as calyx
- Bracts mostly 1-5 lobed; many flower-clusters axillary, mostly not more than 10 (15)-flowered; leaves lobed to entire . . . . . *E. densifolium* ssp. *elongatum*
- Bracts commonly 5-9 lobed; flower-clusters mostly terminal, commonly 15-20-flowered; leaves 22-15-lobed . . . . . *E. densifolium* ssp. *austromontanum*

Plants annual, not at all woody; corolla 7-18mm long, the tube rarely much longer than the calyx
- Corolla 15-16mm long, irregular, the lobes about 6mm long; stamens very unequal; bracts 5-9-lobed; desert . . . . . *E. eremicum*
- Corolla smaller, or if as long, regular and with stamens subequal and bracts 3-lobed
  - Lobes of corolla about as long as the tube
    - Flower heads 1-3 (-4)-flowered; corolla usually less than 12mm long, the lobes as long as the tube
      - Calyx rarely at all lanate, frequently glandular; bracts 1-3-lobed; corolla usually 10-12mm long
        - Flowers sessile, mostly more than one . . . . . *E. sapphirinum* ssp. *s.*
        - Flowers pedicelled, solitary . . . . . *E. sapphirinum* ssp. *gymnocephalum*
      - Calyx always lanate; bracts 2-5-lobed; corolla usually less than 10mm; desert *E. sapphirinum* ssp. *ambiguum*
    - Flower heads 3-10-flowered; corolla about 14mm long . . . . . *E. sapphirrinum* ssp. *dasyanthum*
  - Lobes of corolla conspicuously shorter than tube
    - Filaments of stamens long-exserted . . . . . *E. filifolium*
    - Filaments of stamens not long-exserted, the anthers sometimes showing . . . . . *E. diffusum*

E. densifolium (Benth.)Mason ssp. **austromontanum** (Craig)Mason. [*Gilia d.* var. *a.* Craig] Montane Woolly-Star. Occasional, in Montane Coniferous Forest; montane, above 1200m: Nellie; Henshaw Grade; Laguna Mountains. Also Morro Hill; Campo. Apr.-Oct. Type from dry slope near Nellie, Palomar Mountains. *n*=7.

E. densifolium ssp. **elongatum** (Benth.)Mason. [*Gilia d.* var. *e.* (Benth.)Gray] Chaparral Woolly-Star. Occasional, on dry slopes in Chaparral; foothill, 800-1400m: Palomar Mountains; Oak Grove; Boulevard; Jacumba; McCain Valley. June-Oct.

E. diffusum (Gray)Mason. [*Gilia filifolia* var. *d.* Gray] Miniature Woolly-Star. Infrequent, in sandy areas; desert, below 400m: Borrego Palm Canyon; San Felipe; Vallecito and Carrizo Stage Stations; Potrero; Inner Pasture. Mar.-Apr.

E. eremicum (Jeps.)Mason. [*Gilia e.* (Jeps.)Craig; *Navarretia densifolia* var. *jacumbana* Brand, the type from desert slopes, Jacumba.] Desert Woolly-Star. Frequent, in sandy areas; desert, below 1200m: e.g. Montezuma Valley; Borrego Palm Canyon; San Felipe Wash; The Narrows; Sentenac Canyon; Box Canyon; Vallecito Stage Station; Jacumba; southeast corner of county. Apr.-June.

E. filifolium (Nutt.)Woot. & Standl. [*Gilia f.* Nutt.; *Huegelia f.* (Nutt.) Jeps.] Thread-leaf Woolly-Star. Occasional, on dry slopes; coastal and foothill, below 700m: Bonsall; Encinitas; Mount Woodson; Miramar Naval Air Station; Lakeside; La Jolla; Jamacha; Otay Mesa; Potrero. Also Cuyamaca Lake. Apr.-June.

E. sapphirinum (Eastw.)Mason ssp. **ambiguum** (Jones)Mason. [*Gilia virgata* var. *a.* (Jones)Craig] Mojave Woolly-Star. Infrequent, in Chaparral; foothill and montane, 600-1400m: Oak Grove; Witch Creek; Cuyamaca Mountains; Pine Valley; Campo; Buckman Spring. May-Sep.

E. sapphirinum ssp. **dasyanthum** (Brand)Mason. [*Gilia virgata* var. *d.* (Brand)Craig; *Huegelia v.* var. *d.* (Brand)Jeps.] Frequent, on open slopes; coastal and foothill, below 1200m: e.g. Pala; Moosa Canyon; Palomar Mountain; Wyss Ranch; San Pasqual; Santa Ysabel; Julian; Jacumba; San Felipe. May-Aug.

E. sapphirinum ssp. **gymnocephalum** (Brand)Mason. San Diego Woolly-Star. Uncommon, in Montane Coniferous Forest and Chaparral; foothill, 800-1400m: Palomar Mountains; Oak Grove. Type from "Sud und Nieder Californien", the former involving San Diego County as the range of the taxon. June-July.

E. sapphirinum ssp. **sapphirinum**. [*Gilia virgata* var. *s.* (Eastw.)Macbr.] Sapphire Woolly-Star. Frequent, in dry places in Chaparral and Montane Coniferous Forest; foothill and montane, above 600m: e.g. Doane Valley; Palomar Mountains; Warner Hot Springs; Cuyamaca Peak; Laguna Recreation Area; Pine Valley; Ramona; Descanso; Campo; Buckman Spring. June-Nov. *n*=7.

Gilia Ruiz & Pavon

Pollen cream or yellow; calyx-lobes broadest at base, narrowing to a point, joined by sinus-membrane about to middle; corolla campanulate, rotate or funnelform; desert
- Corolla funnelform, pink . . . . . *G. latifolia*
- Corolla campanulate to turbinate, golden yellow . . . . . *G. filiformis*

Pollen blue (sometimes white); calyx-lobes narrow or broad, not wider at base than middle, joined by sinus-membrane to above middle; corolla salverform or funnelform
- Plants arachnoid-woolly on lower stems and leaves
  - Corolla-tube and -throat combined 1 1/2-2 times as long as calyx; corolla 8.2-12mm long . . . . . *G. diegensis*

Corolla-throat broadly expanded and with 2 bands of color (yellow at base, blue-violet above); pedicels and calyces glabrous or pedicels glandular immediately beneath the flowers and then the calyces cobwebby-pubescent . . . . . *G. exilis*

Plants not arachnoid-woolly on lower stems or leaves

Inflorescence with loose to headlike glomerules mostly of 3-many flowers subtended by a single leaf; upper cauline leaves well developed and pinnately dissected; cismontane

Flowers in dense spherical heads; calyx composed of relatively narrow herbaceous bands and narrow to broad hyaline sinuses; stamens as long as or longer than corolla-lobes. . . . . . *C. capitata*

Flowers solitary, in loose cymes, or glomerate to loosely capitate with 12-25 fowers in a head; calyx composed of broad herbaceous bands and narrow hyaline sinuses; stamens shorter than to subequal to corolla-lobes

Corolla campanulate, 5-12mm wide when pressed; style exserted, the stigmas beyond the anthers; corolla deciduous after anthesis. . . . . *G. angelensis*

Corolla funnelform, 3-5mm wide when pressed; style included, the stigmas scarcely beyond the anthers . . . . . *G. clivorum*

Inflorescence loose, consisting of 1-8 flowers subtended by a single leaf; upper cauline leaves much reduced, or, if well developed, then linear to linear-oblong

Plant body leafy throughout, the basal rosette absent or very little developed; leaves linear to linear-oblong, usually entire. . . . . *G. capillaris*

Plant scapose with reduced leaves above and basal rosette well developed; leaves sinuate to much dissected or the cauline linear and entire

Capsule oblong-ovoid; leaves finely divided; basal leaves pubescent with multicellular hairs, uppermost leaves usually entire

Stamens inserted in corolla-sinuses, shorter than corolla-lobes; flowers pink to whitish or pale violet . . . . . *G. australis*

Stamens inserted in middle of corolla-throat, exserted beyond corolla-lobes; flowers blue-violet . . . . . *G. caruifolia*

Capsule broadly ovoid; leaves finely to coarsely divided; basal leaves pubescent with multicellular or geniculate hairs; uppermost leaves toothed or lobed . . . . . *G. stellata*

G. angelensis V. Grant. Grassland Gilia. Occasional, in sandy and gravelly places; coastal and foothill, below 1200m: e.g. upper San Onofre Canyon; Fallbrook; Moro Hill; Lilac; Miramar; San Diego; Ramona; San Pasqual; Santa Ysabel; Warner Hot Springs; Echo Dell; Pine Valley; El Cajon Mountain; Potrero Grade; Lyons Valley; Point Loma. Mar.-May. n=9.

G. australis (Mason & A. Grant)V. Grant. Southern Gilia. Infrequent, in sandy places; cismontane: Oak Grove; Chihuahua Valley; Hot Springs Mountain; San Felipe Wash; Cuyamaca Peak; Hauser Creek; Foster; Campo; Tecate; Dubber; Balboa Park. Mar.-May. n=9.

G. capillaris Kell. Miniature Gilia. Rare, in open areas of Montane Coniferous Forest; montane, above 1500m: Doane Valley, Palomar Mountain. June-July. n=9.

G. capitata Sims ssp. **abrotanifolia** (Nutt. ex Greene)V. Grant. [*G. achilleaefolia* auct. non Benth.; *G. multicaulis* auct. non Benth.; *G. peduncularis* Eastw. ex Milkn.] Ball Gilia. Frequent, on open slopes and burns; cismontane: e.g. De Luz; Fallbrook; Bonsall; Rainbow; Mesa Grande; Nellie; Fry Creek; Springhill; Eagles Nest; Warner Ranch; upper Borrego Palm Canyon; San Felipe Creek; Julian; Cuyamaca Peak; Garnet Peak; Descanso; Buckman Spring; Jacumba; Hipass; Boulevard; Campo; Harbison Canyon. Apr.-May. n=9.

G. caruifolia Abrams. [*G. tenuifolia* var. *caruifolia* (Abrams)Munz] Caraway-leaf Gilia. Occasional, in openings of Montane Coniferous Forest and Chaparral; foothill and montane; above 900m: e.g. Pala; Doane Valley; Deer Flat; Palomar Mountain; Roderick Mountain; Oak Grove Trail; Hot Springs Mountain; Julian; Cuyamaca Lake; Banner; Oasis Spring; Pinyon Point; Barona Valley; Boulder Oaks; Campo; Jacumba; Otay Mountain. May-Aug. Type from dry hills between Cuyamaca Lake and Oriflamme Canyon. n=9.

G. clivorum (Jeps.)V. Grant. Rare, in grassy areas; cismontane, below 1000m: Fallbrook; Mesa Grande; Anahuac School. Mar.-May.

G. diegensis (Munz)A. & V. Grant. [*G. inconspicua* var. *d.* Munz] San Diego Gilia. Occasional, in sandy openings of Chaparral, Southern Oak Woodland, and Montane Coniferous Forest; foothill and montane, above 600m: e.g. Fallbrook; Doane Valley; Oak Grove Trail; Eagles Nest; San Felipe Wash; Banner; Garnet Peak; Laguna Lakes; Pine Valley; Buckman Spring; Jacumba; Boulevard; Mountain Springs. Apr.-June. Type from under pines, 5000 feet, Laguna Mountains.

G. exilis Gray. [*G. ochroleuca* Jones ssp. *e.*(Gray)A.& V. Grant; *G. o.* var. *abramsii* Brand, the type from Tijuana River, San Diego County.] Infrequent, in open areas; coastal and foothill, below 900m: Oak Grove; Warner Hot Springs; Rincon; Corte Madera Ranch; Campo; Miramar; San Diego; Tijuana River Valley. Mar.-July. n=9.

G. filiformis Parry ex Gray. Thin Gilia. Scarce, on volcanic gravel slopes; desert: Sweeney Pass; southwest of Lava Flow Wash. Mar.-May. n=9.

G. latifolia Wats. Broadleaf Gilia. Infrequent, in sandy areas; desert, below 300m: Borrego Palm Canyon; northeast corner of county; Calcite Mine; Split Mountain; The Narrows; Seventeen Palms; Vallecito and Carrizo Stage Stations; Ocotillo Wells; June Wash; Inner Pasture Outlet Canyon; south of Coyote Mountain. Mar.-May. n=18.

G. stellata Heller. Uncommon, in sandy areas; desert, below 900m: Coyote Creek; Borrego Palm Canyon; Hellhole Canyon; Butler Canyon; San Felipe Creek; Tamarisk Grove; Vallecito Stage Station; southeast corner of county; Mountain Springs. Mar.-May. n=9.

Ipomopsis Michx.

I. tenuifolia (Gray)V. Grant [*Gilia t.* (Gray)Gray; *Loeselia t.* Gray; *G. truncata* A. Davids., the type from near Jacumba] Local, in rocky areas; south desert, 700-1000m: Carrizo Gorge; Dubber; Jacumba; Mountain Springs. Mar.-May. *n*=7.

Langloisia Greene

Corolla distinctly 2-lipped; leaves pinnatifid with ligulate or spatulate rachis and single marginal bristles
  Calyx 5-6mm long; corolla-lobes rounded or truncate at apex, almost as long as tube . . . . *L. matthewsii*
  Calyx 3-4mm long; corolla-lobes pointed at apex, 1/3 to 1/2 as long as tube . . . . *L. schottii*
Corolla almost regular; leaves abruptly dilated toward apex, the marginal bristles mostly in pairs . . . . . . *L. setosissima*

L. matthewsii (Gray)Greene. [*Loeseliastrum m.* (Gray)Timbrook] Infrequent, in sandy areas; desert, below 300m: Borrego Palm Canyon; Kane Springs Road; Sentenac Canyon; Vallecito and Carrizo Stage Stations. Apr.-June.

L. schottii (Torr.)Greene. [*Gilia s.* (Torr.)Wats.; *Loeseliastrum s.* (Torr.)Timbrook] Occasional, in sandy areas; desert, below 900m: Borrego Palm Canyon; Borrego Valley; The Narrows; Sentenac Canyon; San Felipe Valley; Vallecito and Carrizo Stage Stations; Mountain Palm Springs; Dos Cabezas; Jacumba; southeast corner of county. Also Palomar Mountain. Jan.-Apr. *n*=7.

L. setosissima (T.& G.)Greene ssp. **setosissima**. [*Gilia s.* (T. & G.)Gray] Uncommon, in sandy areas; desert, below 300m: Ocotillo Wells; The Narrows; Seventeen Palms; Silent Canyon; Carrizo Stage Station. Apr.-June. *n*=7.

Leptodactylon H. & A.

Plants mostly 1-3dm high; corolla tubular-funnelform, mostly pale or whitish; corolla-lobes 7-10mm long . . *L. pungens*
Plants usually 3-10dm high; corolla salverform, rose to lilac; corolla-lobes 10-15mm long. . . . . *L. californicum*

L. californicum H.& A. ssp. **glandulosum** (Eastw.)Mason. [*Gilia c.* (H.& A.)Benth.; *L. c.* ssp. *leptotrichomum* P.Gordon, nom.nud.] Prickly-Phlox. Rare, in Chaparral; foothill, 100-300m: Richman Ranch near De Luz. Apr.

L. pungens (Torr.) Rydb. ssp. **hallii** (Parish)Mason. [*Gilia p.* var. *h.* (Parish)Milliken] Hall's Prickly-Phlox. Infrequent, on dry, rocky slopes; foothill and desert, above 700m: Oriflamme Canyon; Inspiration Point; Canebrake Canyon; Split Mountain; Dos Cabezas; InKoPah County Park; Jacumba; Mountain Springs. Mar.-Apr.

Linanthus Benth.

Plants annual
  Calyx with a conspicuous hyaline membrane in the sinuses, either forming a pseudo-tube or present on the margins of the lobes
    Flowers on pedicels at least 5mm long
      Corolla barely exceeding calyx, glabrous within; filaments glabrous; stems 2-10cm long; leaves with segments 2-6mm long. . . . . *L. pygmaeus*
      Corolla mostly 2-5 times as long as calyx, with a hairy ring within or filaments hairy at base
        Filaments pubescent at base, inserted on lower part of corolla-throat; corolla 1-3cm long; stems 1-5dm long . . . . *L. liniflorus*
        Filaments glabrous, inserted on upper half of throat; corolla 0.6-1.3cm long; stems 0.5-1.6dm long
          Corolla white. . . . . *L. aureus* var. *decorus*
          Corolla deep to pale yellow . . . . *L. aureus* var. *a.*
    Flowers sessile or on pedicels less than 5mm long
      Flowers yellow or cream
        Calyx 4-5mm long; corolla 5-8mm long. . . . . *L. lemmonii*
        Calyx 6-8mm long
          Leaves entire; pedicels and calyces with stalked glands; deserts . . . . *L. jonesii*
          Leaves 5-9-parted; pedicels and calyces without stalked glands; mostly cismontane *L. androsaceus* et sspp.
      Flowers white to pink or lilac
        Corolla-tube proper (below the throat) included in calyx
          Calyx 3-4mm long; leaves 3-cleft, 2-4mm long . . . . *L. bellus*
          Calyx 6-14mm long
            Flowers white to cream or with some purplish tinge; mostly vespertine, the corolla-lobes entire
              Calyx glabrous
                Leaves divided, 3-7-parted, each stamen with a basal hairy pad; corolla 1.5-3cm long *L. dichotomus*
                Leaves mostly entire, rarely 2-3-cleft; stamens without basal pads; corolla 1-1.5cm long . *L. bigelovii*
              Calyx with stalked glands; corolla 9-10mm long . . . . *L. jonesii*
            Flowers pink to lilac, rarely white, diurnal, the corolla-lobes dentate
              Leaves entire, 10-20mm long; calyx 10-16mm long
                Calyx-lobes linear . . . . *L. dianthiflorus* ssp. *d.*
                Calyx-lobed spatulate, farinose-pubescent . . . . *L. dianthiflorus* ssp. *farinosus*
              Leaves mostly 3-lobed, 2-4mm long; calyx 10-16mm long . . . . *L. bellus*
        Corolla-tube proper well exserted from the calyx
          Corolla funnelform, the tube stout, 1-1 1/2 times as long as calyx . . . . *L. orcuttii*
          Corolla salverform, the tube slender, subfiliform, 2-4 times as long as calyx
            Bracts of inflorescence pilose to short-hispid, mostly not conspicuously coarsely ciliate
              Styles conspicuously long-exserted . . . . *L. androsaceus* ssp. *micranthus*
              Styles scarcely, if at all, exserted . . . . *L. androsaceus* ssp. *luteus*
            Bracts of inflorescence conspicuously and coarsely ciliate; calyx 7-10mm long; corolla 12-25mm long . . . . *L. ciliatus*
  Calyx not membranous in sinuses or on margins of lobes, or very inconspicuously so . . . . . *L. androsaceus* et sspp.

Plants suffrutescent perennials, 1-4dm high; leaves 3-9-cleft; calyx not membranous below sinuses
Leaves mostly entire . . . . *L. floribundus* ssp. *hallii*
Leaves 3-lobed
Leaves and calyces subglabrous to moderately pubescent . . . . *L. floribundus* ssp. *f.*
Leaves and calyces glabrous . . . . *L. floribundus* ssp. *glabrus*

L. androsaceus (Benth.)Greene ssp. **luteolus** (Greene)Mason. Spreading Baby-Stars. Occasional, on open slopes; montane, above 1200m: Doane Valley; Volcan Mountain; Julian; Cuyamaca Lake; Garnet Peak; Harper Ranch; Laguna Camp; Pine Valley; Corte Madera. Mar.-June. Type from Cuyamaca Mountains. $2n$=18.

L. androsaceus ssp. **micranthus** (Steud.)Mason. [*Gilia lutea* var. *longistylis* Munz, the type from San Diego.] Coast Baby-Star. Frequent, on open slopes; cismontane: e.g. Fallbrook; Herron Ranch; Moro Hill; Palomar Mountain; Sunshine Summit; Henshaw Grade; Julian; Mesa Grande; Anahuac School; Hauser Creek; Pine Valley; Wildwood Glen; Echo Valley; El Cajon Mountain; Jamul; Dulzura; Woodwardia Canyon; Imperial Beach; Point Loma; La Jolla. Apr.-May n=9.

L. aureus (Nutt.) Greene var. **aureus**. [*Gilia a.* Nutt.] Desert Gold. Scare, in sandy areas; desert, below 700m: Mason Valley; Vallecito Stage Station; Pinyon Mountain; Mountain Springs. Mar.-June. n=9.

L. aureus var. **decorus** (Gray)Jeps. Uncommon, local in sandy desert areas: Dos Cabezas; Volcanic Hills. $2n$=18.

L. bellus (Gray)Greene. [*Gilia b.* Gray] Desert Beauty. Scarce, in open areas of Chaparral; south foothill and transmontane, 700-1200m: Tierra del Sol; Hipass; Boulevard; McCain Valley; Jacumba. Apr.-May.

L. bigelovii (Gray)Greene. [*Gilia b.* Gray] Bigelow's Linanthus. Uncommon, in dry areas; desert, below 1200m: Montezuma View Point; Blair Valley; Pinyon Mountain; Mountain Springs; Table Mountain; Jacumba. Apr.-May. $2n$=18

L. ciliatus (Benth.)Greene. [*Gilia c.* Benth.] Wisker-Brush. Uncommon, in open areas of Montane Coniferous Forest; montane, above 1400m: Doane Valley; Harrison Grade; French Valley; Morgan Hill; Harrison Park; Cuyamaca Mountains; Laguna Mountain Rim Trail; Airplane Ridge. Mar.-July. $2n$=18.

L. dianthiflorus (Benth.)Greene ssp. **dianthiflorus**. [*Gilia dianthoides* Endl.; *Fenzlia concinna* Nutt., the type from "near Santa Diego".] Ground Pink. Common, on open slopes; cismontane, below 1200m: e.g. San Onofre; Lilac; Boucher Hill; Oak Grove; Warner Hot Springs; Warner Pass; Julian; Montezuma View Point; Banner; Santa Ysabel; Ramona; Poway Grade; Harbison Canyon; Lee Valley; Bostonia; Otay Mountain; Lyons Valley; Foster; San Marcos; Oceanside; Pacific Beach; Cowles Mountain; Mission Hills; Imperial Beach. Feb.-May. n=9.

L. dianthiflorus ssp. **farinosus** (Brand)Mason. [*Gilia dianthoides* var. *f.* Brand] Farinose Ground Pink. Infrequent, on open slopes; coastal and foothill, below 900m: Oak Grove; Banner; Lilac; La Jolla. Feb.-May.

L. dichotomus Benth. ssp. **dichotomus**. Evening Snow. Rare, known in the county only from a collection from Jacumba, 800m. Apr.-May. $2n$=18

L. floribundus (Gray)Greene ex Milkn. ssp. **floribundus**. [*L. nuttallii* Gray; *L. nuttallii* (Gray)Greene ssp. *f.* (Gray)Munz; *Gilia n.* var. *f.* (Gray)Munz] Many-flower Linanthus. Infrequent, in Chaparral, Southern Oak Woodland and Montane Coniferous Forest; cismontane, above 600m: Herron Ranch; De Luz; Laguna Mountains. Apr.-July. Type taken by Coulter probably near Warner's Pass.

L. floribundus ssp. **glabrus** R. Patterson. Summer Snow. Common, in Chaparral, Southern Oak Woodland, and Montane Coniferous Forest; foothill and montane, above 800m: e.g. Palomar Mountain; Elder Canyon of Coyote Creek; Borrego Palm Canyon; Santa Rosa Mountains; Harper Ranch; Oakzanita; Cuyamaca Lake; Pine Valley; Laguna Junction; Descanso; Buckman Spring; Kitchen Creek; Corte Madera Lake; Jacumba; Campo. Apr.-July.

L. floribundus ssp. **hallii** (Jeps.)Mason in Abrams. Uncommon, dry washes; desert, below 1000m: Rockhouse Canyon, north of Clark Dry Lake; Borrego Palm Canyon; canyon east of Clark Dry Lake. Mar.-May. The type locality is given as, "Coyote Canyon, San Diego County, in the lower Sonoran Zone, 600 ft., Apr. 1902, Hall 2767a,". Several of Dr. Hall's collections cited from Coyote Canyon appear not to have been otherwise found in San Diego County. There is another Coyote Canyon in the Santa Rosa Mountains in Riverside which may be the site of this type collection. To further confuse the issue, until 1902, San Diego County included all of what is now Riverside County.

L. jonesii (Gray)Greene. [*Gilia bigelovii* var. *j.* (Gray)Brand] Uncommon, in sandy areas; desert, below 700m: northeast corner of county; Lone Palm; Mason Valley; southwest of Lava Flow Wash; north Mortero Wash; south Mortero Wash; Highway S-2 by Mortero Wash Road Turnoff. $2n$=18

L. lemmonii (Gray)Greene. [*Gilia l.* Gray] Lemmon Linanthus. Occasional, in sandy areas; desert and foothill, below 1500m: Collins Valley; San Felipe Valley; Oak Grove Trail; Sunshine Summit; Oak Grove; Montezuma View Point; Sentenac Canyon; Banner; Box Canyon; Pinyon Mountain; Whale Peak; Mason Valley; Vallecito Stage Station; Jacumba; Campo; Boulevard; Hipass. Mar.-July. $2n$=18

L. liniflorus (Benth.)Greene ssp. **pharnaceoides** (Benth.)Mason. Great Basin Linanthus. Occasional, on grassy slopes and mesas; cismontane, below 1500m: San Onofre Canyon; Warner Springs; Lake Henshaw; Julian; Ballena; Descanso; Laguna Mountains; Spring Valley; Mission Valley; El Cajon Mountain; Sweetwater Dam; Otay Mesa; San Diego. Apr.-Sep. $2n$=18

L. orcuttii (Parry & Gray)Jeps. [*Gilia o.* Parry & Gray; *L. o.* ssp. *pacificus* (Mlkn.)Mason in Abrams , the type from Cootca, Palomar Mountains.] Laguna Linanthus. Scarce, on floor of Montane Coniferous

Forest; montane, above 1400m: Cutca Valley; Doane Valley; Nate Harrison Grade; Hot Spring Mountain; Cedar Grove; Morgan Hill; Monument Peak; Cuyamaca Mountain; Oakzanita Peak; Arroyo Seco Trail. May-June.

L. pygmaeus (Brand)J.T. Howell ssp. **continentalis** Raven. [*Gilia pusilla* auct. non Benth.] Pygmy Linanthus. Infrequent, on dry slopes; foothill and montane, above 400m: Lilac; Banner; Inspiration Point; Julian; Laguna Lakes; Potrero Grade; Dulzura; Otay Mesa; Morena Dam. Apr.-May. $2n=18$.

Microsteris Greene

Corolla 5-8mm long, barely exserted from calyx . . . . *M. gracilis* ssp. *humilis*
Corolla 8-12mm long, well exserted . . . . *M. gracilis*. ssp. . *g.*

M. gracilis (Dougl. ex Hook.)Greene ssp. **gracilis**. [*Gilia g.* Hook.] Slender Phlox. Occasional, in grassy areas; montane, above 1100m: Palomar Mountain State Park; Cedar Creek; Montezuma Valley; Springhill; Cuyamaca Lake; Cuyamaca Rancho State Park headquarters; Shrine Camp; Crouch Meadows. Mar.-Aug. $n=7$.

M. gracilis ssp. **humilis** (Greene)V. Grant. Diffuse Phlox. Infrequent, in meadows; montane, above 1300m: Nellie; Laguna Lakes. May.

Navarretia Ruiz & Pavon

Main stems finely retrose-pubescent with crisped white hairs; mature capsule thin-membranous, the walls not splitting into distinct valves
  Stigma minutely 2-lobed; lobes of the foliaceous bracts soft-herbaceous when fresh
    Flower head sessile in a broad basal rosette; leaves 2-8 (-13)cm long . . . . *N. prostrata*
    Flower head more caulescent; leaves less than 3cm long . . . . *N. fossalis*
  Stigma deeply 2-3-lobed; lobes of the foliaceous bracts at base of the head rigidly acerose, the rachis expanded at base
    Stigma 2-cleft . . . . *N. intertexta*
    Stigma 3-cleft, each corolla-tube with 3 main veins . . . . *N. tagetina*
Main stems with spreading hairs to almost glabrous, or if the hairs introrse, not crisped and appressed against the stems; mature capsule with thicker wall, with definite scheme of dehiscence
  Leaves with filiform rachis, the lateral linear lobes much exceeded by the elongate terminal segment; plants not heavily glandular . . . . *N. peninsularis*
  Leaves with broader rachis, the terminal lobe not greatly elongate; plants heavily glandular
    Terminal segment of bract with 3 diverging acerose teeth; corolla-throat narrow; stamens included, unequally inserted . . . . *N. atractyloides*
    Terminal segment of bract as above; stamens exserted, equally inserted
      Corolla-tube long-exserted . . . . *N. hamata* ssp. *leptantha*
      Corolla-tube not so
        Flowers whitish . . . . *N. hamata* ssp. *foliacea*
        Flowers purplish . . . . *N. hamata* ssp. *h.*

N. atractyloides (Benth.)H. & A. [*Gilia a.* Benth.)Steud.] Holly-leaf Skunkweed. Frequent, in open areas of Chaparral and Coastal Sage Scrub; foothill, 200-800m: e.g. Pala; Palomar Mountain; Ramona; Descanso; Witch Creek; Dulzura; falls of the San Diego River; Lakeside. May-July. $n=9$.

N. fossalis Moran. [*Gilia prostrata* auct. non Greene] Rare, in vernal pools and roadside ditches, coastal, below 900m: San Marcos; Ramona; National City; Otay Mesa. $n=9$.

N. hamata Greene ssp. **foliacea** (Greene)Mason. [*Gilia atractyloides* var. *f.* (Greene)Munz] Soft Skunkweed. Occasional, in Chaparral and Coastal Sage Scrub; coastal and foothill, below 500m: San Diego; Potrero; Boulevard. Apr.-June. Type from San Diego.

N. hamata ssp. **hamata**. [*N. macrantha* Brand, the type from hills between Foster and Ramona; *Gilia h.* (Greene)Munz] Hooked Skunkweed. Common, in open areas of Chaparral and Coastal Sage Scrub; coastal and foothill, below 500m: e.g. Pala; Ramona; Encinitas; Point Loma; Del Mar; National City; El Cajon; Foster; Barona Valley; Alpine; Potrero. Apr.-June.

N. hamata ssp. **leptantha** (Greene)Mason. Frequent, in open areas of Chaparral and Coastal Sage Scrub; coastal and foothill, below 800m: e.g. Miramar Naval Air Station; De Luz; Chollas Valley; Ocean Beach; Otay Mountain; Oak Grove; San Felipe Canyon; Dehesa; Boulevard. May-July.

N. intertexta (Benth.)Hook. [*Gilia i.* (Benth.)Steud.] Uncommon, wet meadows, and muddy shorelines; montane, 1200m: Cuyamaca Lake; Cuyamaca Peak; Stonewall Mine; Laguna Lakes; King Creek; Corte Madera Ranch; Otay Mesa; National City. May-July.

N. peninsularis Greene. Peninsular Navarretia. Rare, known in the county only from a collection from the Cuyamaca Mountains, above 1500m. July.

N. prostrata (Gray)Greene. [*Gilia p.* Gray] Flat Navarretia. Rare, known in the county only from a collection from Kearny Mesa. Apr.-May.

N. tagetina Greene. Marigold Navarretia. Rare, known in the county only from a collection from near Cuyamaca Lake, 1380m. May.

Phlox L.

P. austromontana Cov. [*P. douglasii* var. *a.* (Cov.)Jeps.] Desert Mountain Phlox. Occasional, in rocky places; montane, above 1400m: Julian; Talley's Ranch; Inspiration Point; Cuyamaca Mountains; Rattlesnake Valley; Garnet Peak. Apr.-June.

POLYGALACEAE. Milkwort Family

Polygala L. Milkwort

P. cornuta Kell. ssp. **fishiae** (Parry)Munz. [*P. f.* Parry] Fish's Milkwort. Scarce, in brushy canyons; foothill, 200-900m: Santa Margarita River gorge; Castro Canyon; Pamo Valley; Barona Valley; Poser Mountain; McGinty Mountain; Sequan Peak; Alpine; Otay Mountain. June-Aug. *n*=9.

POLYGONACEAE. Buckwheat Family

KEY TO GENERA

Leaves without stipules
  Flowers subtended by bracts or not, but not enclosed in an involucre
    Bracts woolly; flowers in headlike clusters; calyx glabrous . . . . *Nemacaulis*
    Bracts not woolly; flowers not in heads
      Bracts solitary, 2-lobed, enlarged, reticulate and saccate in fruit; leaves rounded . . . . *Pterostegia*
      Bracts whorled, hooked at tip; leaves linear . . . . *Chorizanthe*
  Flowers enclosed in a tubular to campanulate involucre
    Involucre with spine- or bristle-tipped teeth
      Involucre commonly 1-flowered and mostly 4-5-toothed, with the tube generally cylindric or prismatic and the teeth often hooked at their tips . . . . *Chorizanthe*
      Involucre commonly 2-several-flowered and mostly 4-5-lobed, with the tube turbinate and the lobes ending in straight spines or bristles . . . . *Oxytheca*
    Involucre with 3-8 teeth or lobes, these not bristle- or spine-tipped . . . . *Eriogonum*
Leaves with evident stipular sheaths
  Calyx 4- or 6-parted or urn-shaped and 6-lobed
    Calyx urn-shaped, becoming hard and burlike in fruit; outer lobes spine-tipped . . . . *Emex*
    Calyx 4- or 6-parted, the outer lobes not spine-tipped . . . . *Rumex*
  Calyx 5-parted; akenes often enclosed by the somewhat enlarged fruiting calyx
    Outer 3 calyx-lobes keeled or winged, especially in fruit . . . . *Fallopia*
    Calyx wings and keeles none
      Stipules red or brown, cylindric or funnel-like, margin sometines bristly; leaves generally over 5cm long . . . . *Persicaria*
      Stiples mostly colorless, irregular in shape, eventually fringed; leaves usually less than 5cm long . . . *Polygonum*

Chorizanthe R. Br. ex Benth.

Involucres lacking; leaves linear, the cauline in whorls of 4-5 . . . . *C. coriacea*
Involucres present
  Bracts entire
    Involucres 6-toothed and -ribbed (see also *C. polygonoides*).
      Calyx-lobes entire or nearly so
        Outer and inner calyx-lobes alike or nearly so
          Plants erect; stems easily breaking up into joints when mature; deserts . . . . *C. brevicornu*
          Plants prostrate to ascending; stems not very fragile; cismontane
            Calyx yellow . . . . *C. procumbens* var. *p.*
            Calyx white. . . . . *C. procumbens* var. *albiflora*
        Outer and inner calyx-lobes definitely unlike, the inner much smaller
          Involucral teeth straight, not hooked at tip; calyx-lobes erose; plants decumbent . . . . *C. parryi*
          Involucral teeth uncinate
            Leaf-blades 1-3cm long; involucres mostly in subcapitate clusters; calyx not half-exserted . *C. staticoides*
            Leaf-blades 0.5-1.5cm long; involucres scattered, not crowded; calyx almost half-exserted *C. leptotheca*
      Calyx-lobes, at least the inner 3, fimbriate
        Calyx 7-9mm long, the segments finely laciniate so that the terminal lobe is scarcely broader than the others . . . . *C. fimbriata* var. *laciniata*
        Calyx 6-7mm long, the segments with fimbriate margins and linear entire terminal portion *C. fimbriata* var. *f.*
    Involucres not both 6-ribbed and 6-toothed
      Involucral tube angled or ribbed, not transversely corrugated
        Involucral teeth 6, with alternate shorter, cismontane. . . . . *C. polygonoides*
        Involucral teeth 3
          Plants prostrate; involucral teeth uncinate, equal . . . . *C. orcuttiana*
          Plants erect; involucral teeth straight, unequal; deserts . . . . *C. rigida*
      Involucral tube cylindric, corrugated but not ribbed or angled . . . . *C. corrugata*
  Bracts 3-lobed
    Involucres with 3-6 divaricate spurs at base; bracts small. . . . . *C. thurberi*
    Involucres not spurred at base. . . . . *C. californica*

C. brevicornu Torr. ssp. **brevicornu**. Brittle Spine-Flower. Frequent, in sandy areas; desert, below 800m: e.g. Borrego Palm Canyon; Lone Palm; northeast corner of county; Borrego Springs; Benson Dunes; Kane Springs Road; San Felipe Wash; Scissors Crossing; Vallecito Stage Station; Mountain Springs; southeast corner of county. Mar.-June.

C. californica (Benth.)Gray. California Spine-Flower. Infrequent, in dry, sandy places; coastal, below 200m: San Luis Rey; Torrey Pines; Pacific Beach; Point Loma; San Diego. Apr.-July.

C. coriacea Goodm. [*Lastarriaea chilensis* auct. non Remy in Gay] Lastarriaea. Occasional, in open and disturbed areas; coastal, below 200m: Fallbrook; San Luis Rey; Rainbow; Miramar; Silver Strand; Point Loma; Initial Boundary Monument; Dehesa; Morena; Harbison Canyon. Also Vallecito Stage Station. Apr.-June.

C. corrugata (Torr.)T. & G. Corrugate Spine-Flower. Scarce, in sandy areas; desert, below 500m: Yaqui Well; Kane Springs Road; Ocotillo Wells; Carrizo Stage Station; Mortero Wash; East Dolomite Mine Road. Feb.-May.

C. fimbriata Nutt. var. **fimbriata**. Fringed Spine-Flower. Occasional, on dry slopes and burns; coastal, below 250m: San Onofre Canyon; Pala; Fallbrook; Escondido; Witch Creek; Santa Maria Valley; Monserate Grade; Carlsbad; Poway Grade; Lakeside; Flinn Springs; Dulzura; Spring Valley; Point Loma; San Diego; Mission Hills; Miramar Naval Air Station; Lake Hodges. Apr.-June. Type from "St. Diego".

C. fimbriata var. **laciniata** (Torr.)Jeps. Laciniate Spine-Flower. Common, on open slopes and burns; foothill and montane, above 200m: e.g. Herron Ranch; Lilac; Palomar Mountain; Ballena; Eagles Nest; Lake Henshaw; Chihuahua Valley; Witch Creek; Coyote Canyon; Box Canyon; San Diego River falls; Pine Valley; Dubber; southeast corner of county; Garnet Peak; Guatay Mountain; Descanso; Wildwood Glen; Hauser Creek; El Cajon Mountain; Otay Mountain; Lee Valley; Alpine. May-June. Type from San Felipe.

C. leptotheca Goodm. Ramona Spine-Flower. Occasional, on open slopes and burns; foothill, 700-1200m: Doane Valley; Santa Ysabel; Mesa Grande; Pala; Ramona; Mount Woodson; Campo; Otay Mountain. May-June. Type from dry hills between Ramona and Ballena.

C. orcuttiana Parry. Orcutt's Spine-Flower. Endemic to San Diego County. Scarce, on mesas; coastal, below 150m: Encinitas; Del Mar; Point Loma; Kearny Mesa. Mar.-Apr. Type from exposed sandy soil on Point Loma.

C. parryi Wats. var. **fernandina** (Wats.)Jeps. San Fernando Spine-Flower. Rare, on mesas; coastal, below 150m: Del Mar Heights; Soledad; Sorrento Valley. Feb.-Mar.

C. polygonoides T. & G. ssp. **longispina** (Goodm.)Munz. [*C. p.* T. & G.] Knotweed Spine-Flower. Uncommon, on mesas; coastal, below 150m: Kearny Mesa; Miramar Naval Air Station; San Diego; Point Loma. Also Cuyamaca Lake. Apr.-July.

C. procumbens Nutt. var. **albiflora** Goodm. Pala Spine-Flower. Scarce, in open areas of Chaparral; north foothill, 50-200m: Fallbrook; Red Mountain Grade; Lilac; Dulin Ranch; Pala; Pauma. Apr.-June. Type from dry slopes 2 miles east of Pala.

C. procumbens var. **procumbens**. [*C. uncinata* Nutt., the type from "St. Diego".] Prostrate Spine-Flower. Occasional, in sandy areas; coastal and foothill, below 800m: Carlsbad; Cardiff; Encinitas; Del Mar; Soledad; Point Loma; Balboa Park; Silver Strand; Murphy Canyon; Initial Boundary Monument; Harbison Canyon; Witch Creek; Morena; Hidden Glen; Alpine; Potrero. Also Mason Valley. Apr.-June. Type from "St. Diego".

C. rigida (Torr.)T. & G. Rigid Spine-Flower. Infrequent, on sandy flats; desert, below 500m: northeast corner of county; Oil Well; Kane Springs Road; Borrego Springs; Vallecito and Carrizo Stage Stations. Mar.-May.

C. staticoides Benth. ssp. **staticoides**. [*C. discolor* Nutt., the type from "St. Diego".] Turkish Rugging. Common, on otherwise-bare slopes, mesas and burns; foothill and coastal, below 800m: e.g. Olivenhain; Carlsbad; Encinitas; Torrey Pines; Otay Mountain; Descanso; Hauser Creek; Witch Creek. Apr.-July.

C. thurberi (Gray)Wats. [*Centrostegia t.* Gray ex Benth.] Thurber's Spine-Flower. Uncommon, on open slopes; desert, 300-900m: San Felipe; Banner; Vallecito Stage Station; Bow Willow Palms; Walker Ranch; Jacumba; north Pinyon Mountain; Pinyon Mountain Valley. Apr.-June. Type from sandy areas near San Felipe.

Emex Neck.

***E. spinosa** (L.)Campd. Spiny Emex. Native of Europe. Scarce, naturalized in cultivated areas; foothill and coastal, below 1500m: Crescent Valley near Escondido; Lemon Grove; Sweetwater Valley; Monument Road. July-Nov. *n*=10.

Eriogonum Michx. Wild Buckwheat

Involucres campanulate to turbinate, not angled or ribbed, 4- or 5-toothed or -lobed, rarely obscurely nerved at the base, mostly peduncled
- Leaves basal and also on the lower nodes, tomentose or floccose
  - Involucres somewhat glandular-puberulent to pubescent externally
    - Outer segments obovate to elliptic, not obviously inflated, or if so, only near the base, the inner segments spatulate; stamens conspicuously exserted . . . . . *E. angulosum*
    - Outer segments elliptic to roundish or obovate, obviously inflated at maturity, the inner segments narrowly lanceolate; stamens included . . . . . *E. maculatum*
  - Involucres glabrous externally; calyx hispid; plants to 7dm tall . . . . . *E. ordii*
- Leaves strictly basal, pilose or tomentose at least beneath
  - Involucres 4-lobed or -toothed
    - Calyx white, 1.5-2mm long, apex notched or apiculate; leaves spatulate, ciliate or pilose . . . . . *E. apiculatum*
    - Calyx pink or yellow, 0.5-1mm long, the segments not apiculate
      - Calyx pink, 0.5-0.7mm long; leaves spatulate, hirsute; involucres 0.5-0.7mm long; montane . . . . . *E. parishii*
      - Calyx yellow, 1-2mm long; involucres 0.7-2mm long; deserts
        - Leaves short-hirsute, suborbicular, 1-2.5cm long; involucres 0.7-1mm long . . . . . *E. trichopes*
        - Leaves floccose or glabrous, obovate to oblanceolate, 2-8cm long; involucres 1-2mm long . . . . . *E. ordii*
  - Involucres 5-lobed or -toothed
    - Calyx pubescent or puberulent externally
      - Involucres glabrous externally
        - Outer calyx-segments not sacate-dilated
          - Leaf-blades short-hirsute; involucres turbinate
            - Plants strictly annual; branchlets numerous and whorled at each node; involucres often 4-lobed, occasionally 5-lobed . . . . . *E. trichopes*

Plants perennial but flowering the first year; branchlets few at each node, not in whorls . . *E. inflatum*
Leaf-blades woolly; calyx 1mm long; involucres broadly campanulate . . . . *E. reniforme*
Outer calyx-segments saccate-dilated at each side of the cordate base at maturity, sparsely puberulent at the base of the perianth-tube, yellow maturing reddish or reddish with white lobes . . . . *E. thomasii*
Involucres glandular-puberulent externally . . . . *E. thurberi*
Calyx glabrous externally
Peduncles deflexed . . . . *E. deflexum*
Involucres erect on peduncles, sessile or peduncles less than 5mm long . . . . *E. insigne*
Involucres cylindric to cylindric-turbinate to turbinate or prismatic, often 5-6-nerved, angled or ribbed, mostly sessile, solitary or congested into heads, the teeth usually short
Plants annual; leaves mostly in basal rosettes
Involucres 2-5mm long
Leaves oblong-obovate to oblanceolate; stems tomentose
Involucres 4-5mm long, cylindric, with minute teeth; outer calyx-segments narrowly obovate; akenes 2mm long . . . . *E. roseum*
Involucres 2-3mm long, turbinate, with prominent teeth; outer calyx segments broadly obovate; akenes about 1mm long . . . . *E. gracile*
Leaves rounded or nearly so; stems usually glabrous or only sparsely floccose; outer calyx-segments more than twice as long as wide . . . . *E. davidsonii*
Involucres 1-1.5mm long . . . . *E. foliosum*
Plants perennial; leaves often cauline as well as basal
Involucres solitary at the nodes, the lateral ones appressed to the branches
Calyx stipitate with long, winged attenuate bases, 5-7mm long . . . . *E. saxatile*
Calyx astipitate, 2-5mm long
Involucres 6-7mm long; loosely branched white tomentulose herbs 8-18dm high; cismontane. . *E. elongatum*
Involucres 2-6mm long
Inflorescence with involucres in cymes or panicles
Leaf-blades more than 2cm long . . . . *E. nudum*
Leaf-blades less than 2cm long
Calyx glabrous . . . . *E. plumatella*
Calyx silky-villous, yellowish; plants shrubby, 6-15dm high . . . . *E. deserticola*
Inflorescence with involucres placed along the branches; involucres tomentose
Petiole bases as brownish sheaths clasping the stem completely . . . . . . *E. wrightii* ssp. *membranaceum*
Petiole bases not completely clasping
Leaves 8-12mm long . . . . *E. wrightii* ssp. *nodosum*
Leaves 1.5-3cm long . . . . *E. wrightii* ssp. *trachygonum*
Involucres mostly clustered or in heads
Plants essentially herbaceous, only the base woody . . . . *E. nudum*
Plants definitely shrubby
Leaves narrowly linear or nearly so, the blades less than 2cm long, strongly fascicled; shrubs with terminal cymose or subumbellate inflorescence
Flower clusters solitary or in reduced cymes . . . . *E. fasciculatum* ssp. *polifolium*
Flower clusters many
Upper surface of foliage pubescent . . . . *E. fasciculatum* ssp. *foliolosum*
Foliage pubescent or subglabrous only on lower surfaces . . . . *E. fasciculatum* ssp. *f.*
Leaves linear-oblong to orbicular, or if linear, more that 2cm long
Heads in dense compound cymes . . . . *E. giganteum*
Heads terminal on 2-forked peduncles or scattered along the stems . . . . *E. parvifolium*

E. angulosum Benth. Anglestem Buckwheat. Uncommon, in dry, open places; coastal and desert, below 400m: Borrego Palm Canyon; San Felipe Wash; Mountain Springs; Jacumba; San Diego; Otay Mesa. May-Nov.

E. apiculatum Wats. San Jacinto Buckwheat. Occasional, in clearings of Montane Coniferous Forest; montane, above 1400m: Palomar Mountain; Deer Flat; Cuyamaca Mountains; Garnet Peak; Pinyon Mountain Valley. July-Aug.

E. davidsonii Greene. [*E. molestum* var. *d.* (Greene)Jeps.; *E. vimineum* ssp. *molestum* var. *glabrum* Stokes, the type from Boulevard to Buckman Springs.] Davidson's Buckwheat. Frequent, on dry slopes; foothill and montane, above 600m: e.g. Cole Canyon; Twin Oaks Valley; Doane Valley; Palomar Mountains; Eagles Nest; Warner Ranch; Montezuma Valley; Banner; Scissors Crossing; Cuyamaca Peak; Harper Ranch; Cold Spring; Laguna Junction; Pine Valley; Vallecito Stage Station; Buckman Spring; Jacumba; Boulevard; Campo; Tecate; Ballena. June-Sep.

E. deflexum Torr. in Ives. ssp. **deflexum**. Desert Skeleton Weed. Uncommon, in sandy areas; desert, below 300m: Arroyo Salado; Fish Creek; Jojoba Wash. May-Oct. n=20.

E. deserticola Wats. Imperial Buckwheat. Rare, in sandy areas, desert, below 300m: Palo Verde; Ocotillo Wells. July-Sep.

E. elongatum Benth. Tall Buckwheat. Common, on rocky slopes; montane and foothill, 100-1400m: e.g. De Luz; Rincon; Palomar Mountains; Chihuahua Valley; Oak Grove; Warner Hot Springs; San Felipe Valley; Witch Creek; Julian; Laguna Mountains; Anahuac School; Ballena; Richland; Mussey Grade; Bostonia; Padre Dam; Encanto; Otay Mountain; Potrero; Pine Valley; Buckman Spring. Aug.-Nov. 2n-34.

E. fasciculatum Benth. ssp. **fasciculatum**. [*E. f.* var. *maritimum* Parish, the type from Oceanside; *E. f.* var. *oleifolium* Gandoger, the type from San Diego.] Flat-Top Buckwheat. Prevalent, as an indicator of Coastal Sage Scrub; coastal and foothill, below 900m: e.g. San Mateo Creek; Bonsall; Carlsbad; Poway; Lake Hodges; Hatfield Creek; Ramona; Alvarado Canyon; Dulzura; Lee Valley; Campo; Otay Mountain; Initial Boundary Monument; Silver Strand; Mission Valley; Point Loma; Del Mar. Mar.-Oct. n=20.

E. fasciculatum ssp. **foliolosum** (Nutt.)S.Stokes. Interior Flat-top Buckwheat. Prevalent, in open areas of Chaparral and Southern Oak Woodland; foothill, 600-1400m: e.g. La Jolla Indian Reservation; Eagles Nest; Grapevine Canyon; Springhill; Ramona; Inaja Memorial; Spencer Valley; Barona Valley; Cuyamaca Lake; Garnet Peak; Pine Valley; Oakzanita; McCain Ranch; Campo; Jacumba; Guatay; Flinn Springs; Otay Mountain. Mar.-Oct. 2*n*=80.

E. fasciculatum ssp. **polifolium** (Benth.)S.Stokes. Rosemary Flat-top Buckwheat. Common, on rocky slopes, above 1000m: e.g. Chihuahua Valley; Hot Springs Mountain; Oak Grove; Borrego Valley; Grapevine Canyon; Cuyamaca Mountains; Banner; Vallecito Canyon; Shrine Camp; upper Blair Valley; Boulevard; Jacumba; Dubber; southeast corner of county. Apr.-Nov. 2*n*=40.

E. foliosum Wats. Rare, known in the county only from a collection from Pine Valley, 1060m. Sep.

***E. giganteum** Wats. ssp. **giganteum**. St. Catherine's Lace. Native of Santa Catalina Island. Local, introduced in disturbed areas; coastal, below 200m: La Jolla Shores; Point Loma. May-Aug. *n*=20.

E. gracile Benth. [*E. verticillatum* Nutt., the type from "near St. Diego".] Slender Buckwheat. Frequent, in dry washes and mesas; coastal and foothill, below 1100m: e.g. O'Neill Lake; De Luz; Fallbrook; Palomar Mountain; Chihuahua Valley; Warner Hot Springs; Mesa Grande; Ballena; Santa Ysabel; Alpine; Julian; Pine Valley; Cold Spring; Boulevard; Buckman Spring; Campo; Potrero; Dulzura; Jamacha; Point Loma; Chollas Valley. July-Oct. 2*n*=22.

E. inflatum Torr. & Frem. [*E. i.* var. *deflatum* Jtn.] Desert Trumpet. Occasional, in sandy areas; desert, below 900m: Plum Canyon; Borrego Valley; Fish Creek; Kane Springs Road; the Narrows; Sentenac Canyon; Vallecito Stage Station; Dos Cabezas; Mountain Springs. Sep.-Oct. *n*=16.

E. insigne Wats. Blue Buckwheat. Infrequent, in sandy areas; desert, below 800m: Fish Creek; Box Canyon; northwest of Carrizo Stage Station. May-Nov. *n*=20.

E. maculatum Heller. Spotted Buckwheat. Occasional, in sandy areas; desert, below 800m: Borrego Palm Canyon; San Felipe; Scissors Crossing; North Pinyon Mountain; La Puerta; Potrero; Vallecito Stage Station; Canyon from Inner Pasture. Mar.-Nov. *n*=20.

E. nudum Dougl. ex Benth. ssp. **pauciflorum** (Wats.)Munz. [*E. molestum* Wats.] Pine Buckwheat. Infrequent, on dry slopes of Montane Coniferous Forest; montane, above 1300m: Palomar Mountain; Eagles Nest; Cuyamaca Peak; Mount Laguna. Aug.-Oct.

E. ordii Wats. Ord Buckwheat. Rare, known in the county only from a collection at Split Mountain, 400m. Mar.-June.

E. parishii Wats. Parish Buckwheat. Occasional, in Chaparral; foothill, 200-1100m: Escondido; Pine Valley; Buckman Spring; Campo; Oak Knoll; Mountain Springs. July-Sep. *n*=20.

E. parvifolium Sm. in Rees. Bluff Buckwheat. Frequent, on sea bluffs; coastal, below 50m: San Mateo Creek; San Onofre Bluffs; Oceanside; Carlsbad; Solana Beach; Soledad Beach; Ocean Beach. Apr.-Nov. 2*n*=40.

E. plumatella Dur. & Hilg. [*E. palmeri* Wats., the type from Julian.] Yucca Buckwheat. Infrequent, on rocky slopes; desert and foothill, 600-1500m: San Felipe; Sentenac Canyon; Wagon Wash; Laguna Mountains; La Puerta Valley; Campo. Aug.-Oct.

E. reniforme Torr. & Frem. Kidney-leaf Buckwheat. Frequent, in sandy areas; desert, below 700m: Borrego Palm Canyon; Buttes Pass; Borrego Valley; Vallecito Stage Station. Mar.-June. *n*=16.

E. roseum Dur. & Hilg. Virgate Buckwheat. Uncommon, in Chaparral and Montane Coniferous Forest; foothill and montane, above 1000m: Warner Hot Springs; San Felipe; Laguna Meadows. June-Aug. 2*n*=18.

E. saxatile Wats. Rock Buckwheat. Rare, known in the county only from Oasis Spring, desert slope of the Laguna Mountains, above 1600m. Oct. *n*=20.

E. thomasii Torr. [*E. minutiflorum* Wats., the type from "Colorado Desert"] Thomas' Buckwheat. Frequent, in sandy areas; desert, below 600m: e.g. Lone Palm; northeast corner of county; Borrego Palm Canyon; Hellhole Canyon; Kane Springs Road; The Narrows; Vallecito and Carrizo Stage Stations; Palm Spring. Mar.-June. *n*=20.

E. thurberi Torr. [*E. cernuum* ssp. *viscosum* S.Stokes, the type from near road, Campo to Buckman Springs.] Thurber's Buckwheat. Common, in sandy places; desert and foothill, below 1200m: e.g. Pala; Borrego Valley; San Felipe Wash; Scissors Crossing; Mason Valley; Carrizo Stage Station; Jacumba; southeast corner of county; Boulevard; Campo; Descanso; Cottonwood; Buckman Spring; Sweetwater Dam; San Pasqual; Lakeside. Apr.-July. Type from sandy ravines, San Pasqual.

E. trichopes Torr. [*E. t.* var. *cordatum* (Torr.)S.Stokes, the type from around San Felipe.] Little Trumpet. Occasional, on dry slopes; foothill and desert, below 900m: Warner Ranch; San Felipe; Borrego Springs; La Puerta; Canebrake Canyon; Mountain Springs; Jacumba. Apr.-Aug. Type from eastern slopes of the Cordilleras of California. *n*=16.

E. wrightii Torr. ex Benth. ssp. **membranaceum** S.Stokes. Foothill Buckwheat. Prevalent, in Chaparral and Montane Coniferous Forest; foothill and montane, above 900m: e.g. Deer Flat; Wyss Ranch; Chihuahua Valley; Warner Hot Springs; Eagles Nest; San Felipe Pass; Julian; Cuyamaca Lake; Cuyamaca Peak; Pinyon Point; Descanso; Pine Valley; Boulevard; McCain Ranch; Campo; Jacumba. Aug.-Oct.

E. wrightii ssp. **nodosum** (Small)Munz. [*E. n.* Small] Knotty Buckwheat. Occasional, in rocky places; foothill and desert, below 1100m: Campo; InKoPah County Park; Canebrake Canyon; southwest of Lava Flow Wash; Box Canyon; Sentenac Canyon; North Pinyon Mountain; San Felipe. Aug.-Nov. Type from Dos Cabezas.

E. wrightii ssp. **trachygonum** (Torr. ex Benth.)S.Stokes. Uncommon, in Montane Coniferous Forest; montane, aove 1200m: Julian; Cuyamaca Mountains; Laguna Meadows. Aug.-Oct.

Fallopia Adans.

***F. convolvulus** (L.)A.Love. [*Polygonum c.* L.] Black Bindweed. Native of Europe. Uncommon, naturalized at Chula Vista, 10m. May-Sep. 2*n*=20, 40.

Nemacaulis Nutt.

Wool of inflorescence very long and copious, almost or quite concealing the flowers ...... *N. denudata* var. *gracilis*
Wool of inflorescence not concealing flowers .... *N. denudata* var. *d.*

N. denudata Nutt. var. **denudata**. [*N. foliosa* Nutt., the type from "sandy places near to the sea shore, St. Diego, Upper California"; *Eriogonum nemacaulis* S.Stokes] Coast Wooly-heads. Occasional, in sandy areas; coastal, below 20m: e.g. Oceanside; Del Mar; Mission Beach; Point Loma; Old Town; Silver Strand; National City; Initial Boundary Monument. Apr.-Sep. Type from "St. Diego".

N. denudata var. **gracilis** Goodm. & L.Benson in Munz. Uncommon, in sandy areas; desert, below 100m: Carrizo Creek. Mar.-June. Also attributed to coastal sites by Reveal and Ertter.

Oxytheca Nutt.

O. trilobata Gray. Spinebract. Common, on open slopes and burns; foothill and montane, above 600m: e.g. Palomar Mountain; High Point; Hot Springs Mountain; Montezuma View Point; Inaja Memorial; Echo Valley; Descanso; Cuyamaca Lake; Laguna Junction; Pine Valley; Green Valley Falls; Garnet Peak; Boulevard; Jacumba; Boulder Oaks; Clover Flats; Potrero. July-Sep. *n*=20.

Persicaria Mill. Smartweed

Ocreae eciliate or nearly so, at least in maturity
  Perennial plants with creeping rhizomes and stolons; in and about water
    Flowers in spikes 3-4cm long .... *P. amphibia* var. *stipulacea*
    Flowers in spikes 4-10cm long .... *P. amphibia* var. *emersa*
  Annuals; not aquatic
    Peduncles with stalked glands; spikes erect, oblong-cylindric; akenes suborbicular, 2.2-2.5mm in diameter .... *P. pensylvanica*
    Peduncles with or without sessile glands; spikes often arching, slender-cylindric; akenes subovate, 1.8-2.2mm long .... *P. lapathifolia*
Ocreae fringed with bristly cilia.. *P. punctata*

P. amphibia (L.)Gray var. **emersa** (Michx)Hickman. [*Polygonum coccineum* Muhl.; *Polygonum a.* var. *e.* Michx.] Marsh Knotweed. Common, in lakes, ponds, and quiet streams; cismontane: Lake Henshaw; Cuyamaca Lake; Lake Wohlford; Lake Murray; Sweetwater Lake; Potrero. June-Oct.

P. amphibia var. **stipulacea** (Colem.)Hara. [*Polygonum natans* (Michx.)A. Eaton; *Polygonum a.* L. var. *s.* Colem.] Floating Smartweed. Uncommon, in lakes and ponds; montane, above 1300m: Cuyamaca Lake; Milk Ranch; Laguna Lakes. July-Sep.

P. lapathifolia (L.)Gray. [*Polygonum l.* L.] Willow Smartweed. Frequent, in moist or marshy areas; foothill, 200-1400m: e.g. Witch Creek; Cuyamaca Lake; Hatfield Creek; Descanso Junction; Boulder Creek; Campo; White Star; Pine Valley; Lower Otay Reservoir; Jamacha Junction; Lindo Lake; Lake Hodges; San Pasqual. June-Oct. 2*n*=22.

***P. pensylvanica** (L.)Gomez. [*Polygonum p.* L.] Pinkweed. Native of eastern United States. Uncommon, naturalized in lakes and ponds; cismontane: Lake Hodges; Lindo Lake. July-Sep.

P. punctata (Ell.)Small. [*Polygonum p.* Ell.] Perennial Smartweed. Occasional, in moist places; foothill and montane, above 800m: Deadman Hole near Lake Henshaw; Witch Creek; Hatfield Creek; Ramona. Also Alvarado Canyon. July-Oct.

Polygonum L. Knotweed

Leaves jointed with the ocreae and 1-nerved; flowers mostly 2 or more in the axils, short-pedicelled
  Stems terete or nearly so, not sharply angled, often spreading to prostrate
    Akenes dull or scarcely shining; flowers borne in axils of reduced leaves; calyx with pinkish to purplish margins .... *P. arenastrum*
    Akenes smooth and shining
      Upper leaves bracteate; calyx with pinkish margins .... *P. argyrocoleon*
      Upper leaves not reduced to bracts; calyx with yellowish margins........................ *P. ramosissimum*
  Stems and branches strongly angled, usually suberect .... *P. douglasii*
    Pedicels erect, ca. 1mm long.. *P. douglasii* var. *johnstonii*
    Pedicels curved, 2-3mm long .... *P. douglasii* var. *d.*
Leaves not jointed with the ocreae, 3-nerved; flowers solitary in the axils, sessile, the calyx 1.5mm long..... *P. parryi*

***P. arenastrum** Bor. [*P. aviculare* auct. non L.] Yard Knotweed. Native of Europe. Frequent, in disturbed areas; cismontane: e.g. Fallbrook; Doane Valley; Ballena; Pine Valley; Ramona; Boulevard; Santee; East San Diego. May-Nov. 2*n*=40, 60.

***P. argyrocoleon** Steud. ex Kunze. Persian Wireweed. Native of Europe. Uncommon, weed in disturbed places; coastal and foothill, below 1000m: Moosa Canyon; Hidden Meadows; San Marcos. June-Oct. 2*n*=40.

P. douglasii Greene var. **douglasii**. Douglas' Knotweed. Infrequent, at meadow edges of Montane Coniferous Forest; montane, above 1300m: Doane Valley; Cuyamaca Peak; Laguna Public Camp. June-Sep. $2n=40$.

P. douglasii var. **johnstonii** Munz. Sawatch Knotweed. Rare, in meadows adjacent to Montane Coniferous Forest; montane, above 1600m: Laguna Mountains. July-Sep.

P. parryi Greene. Parry's Knotweed. Rare, known in the county only from a collection from Cuyamaca Peak, above 1500m. June.

***P. ramosissimum** Michx. Bushy Knotweed. Native of eastern North America. Occasional, in disturbed areas; coastal, below 150m: San Dieguito Valley; Balboa Park; Spring Valley; Otay. Also Banner. July-Sep.

Pterostegia F. & M.

P. drymarioides F.& M. Granny's Hairnet. Abundant, on dry slopes, burns, and shaded canyons; coastal, foothill, and desert, below 700m: e.g. Fallbrook; Red Mountain Grade; Lilac; San Luis Rey; Harrison Grade; Doane Valley; Eagles Nest; Montezuma View Point; Banner; Barona; Encinitas; San Pasqual; Kearny Mesa; Mission Hills; Silver Strand; Otay Mountain; Japatul Valley; Hipass; southeast corner of county; Borrego Palm Canyon; Lone Palm; Vallecito Stage Station. Mar.-July. $2n=28$.

Rumex L. Dock

Plant dioecious, low, slender, acid to the taste; valves small, without callosities . . . . *R. acetosella*
Plant monoecious, usually tall and weedy, not acid to the taste
  Stems ascending to decumbent, or axillary from a creeping rhizome, with axillary shoots or leaf-tufts; without basal leaves even when young
    Valves without callosities; mostly montane. . . . . *R. salicifolius* var. *denticulatus*
    Valves, at least one or two often callosity-bearing . . . . *R. salicifolius* var. *s.*
  Stems erect, without axillary shoots or tufts; with basal leaves when young
    Valves 8-14mm long, without callosities; nutlets 3-6mm long . . . . *R. hymenosepalus*
    Valves smaller, at least one valve with a distinct callosity
      Valves entire
        Leaves small, flat, truncate; valves very small, scarcely broader than the thick callosities; whorls remote, nearly all foliate . . . . *R. conglomeratus*
        Leaves large, somewhat crisped or undulate, often narrowed at base, seldom truncate; valves large, much broader than the callosities; only the lower whorls foliate and occasionally remote. . . . . *R. crispus*
      Valves denticulate
        Plants perennial . . . . *R. pulcher*
        Plants annual . . . . *R. maritimus*

***R. acetosella** L. Sheep-Sorrel. Native of Europe. Occasional, naturalized in moist, disturbed places; foothill and montane, above 900m: Wyss Ranch; Bailey Lodge; Henshaw Grade; Pine Hills; Cuyamaca Lake; Corte Madera; Glen Lonely. Mar.-Aug. $2n=42$.

***R. conglomeratus** Murr. Whorled Dock. Native of Europe. Occasional, naturalized in moist places; cismontane: De Luz; Witch Creek; Lake Henshaw; Mussey Grade; Balboa Park; Jamacha; Otay Lake. Apr.-Oct. $2n=18$.

***R. crispus** L. Curly Dock. Native of Eurasia. Common, naturalized in moist places; cismontane: e.g. San Onofre Bluffs; Moosa Canyon; Palomar Mountain; Eagles Nest; Pine Hills; Poway Grade; Kearny Mesa; Point Loma; San Ysidro; Otay Mountain; Alvarado Canyon; Cuyamaca Lake; White Star; Jacumba. Jan.-Dec. $2n=60$.

R. hymenosepalus Torr. Desert Rhubarb. Uncommon, in sandy areas; coastal, below 300m: Bonsall; Rincon; San Diego; Tijuana River Valley. Jan.-May. $2n=40$, 100.

R. maritimus L. var **fueginus** (Phil.)Dusen. [*R. fueginus* Phil.] Golden Dock. Infrequent, in moist areas; cismontane: Carlsbad; Del Dios; Lakeside; Cuyamaca Lake. May-Sep. $2n=40$.

***R. pulcher** L. Fiddle Dock. Native of Europe. Rare, adventive in Alvarado Canyon, near Adobe Falls, 76m. May.

R. salicifolius Weinm. var. **denticularus** Torr. [*R. californicus* Rech. f.] California Dock. Infrequent, in moist places; cismontane: Pala; Lake Hodges; Middle Peak; Potrero Creek. May-Sep. $2n=20$.

R. salicifolius var. **salicifolius**. Willow Dock. Common, in moist areas; coastal and foothill, below 1400m: e.g. San Mateo Creek; Oceanside; Rainbow; Doane Valley; Witch Creek; Cuyamaca Lake; Pine Valley; Japatul Valley; Harbison Canyon; El Capitan Mountain; Barona Valley; Clover Flats; Poway Junction; Encinitas; Mahogany Canyon; Dehesa. May-Sep. $2n=20$.

PORTULACACEAE. Purslane Family

KEY TO GENERA

Calyx adnate to lower part of ovary, its lobes coming off the summit of the circumscissile capsule . . . . *Portulaca*
Calyx and ovary free
  Capsule circumscissile; sepals 2-8; stamens 5 to many . . . . *Lewisia*
  Capsule opening by 2-3 valves; sepals 2; stamens 1-5, except in *Calandrinia*
    Style 1, with 2 stigmas; capsule 2-valved; inflorescence secund . . . . *Calyptridium*
    Style-branches 3; capsule 3-valved; inflorescence mostly not markedly secund
      Flowers in leafy racemes or panicles; petals mostly red; stamens 5 to many; seeds 18-25 . . . . *Calandrinia*
      Flowers in naked or bracteate racemes; petals white to pink; stamens 1-5; seeds 1-6
        Perennial or annual, with a basal rosette of leaves surmounting taproot, rhizome or deep-seated bulbous root; pollen tricolpate . . . . *Claytonia*

Perennial or annual branched above the base and lacking a basal rosette except in 2 spp. where it is early deciduous; pollen often dodecacolpate, sometimes polyrugate with each section of the wall with a hemispheric spine-covered saccus
Annuals
Cauline leaves opposite . . . . *Montia*
Cauline leaves alternate . . . . *Montiastrum*
Perennial with vegetative propagules at the ends of runners having opposite scalelike leaves . . *Crunocallis*

Calandrinia H.B.K.

Flowers in panicles of umbels; seeds naked at hilum; petals white; desert . . . . *C. ambigua*
Flowers somewhat racemose; seeds with a strophiole; petals usually red
Plants green; seeds black and shining
Capsule about twice as long as the calyx; leaves mostly basal; pedicels often reflexed in fruit . . . . *C. breweri*
Capsule scarcely exceeding calyx; leaves well distributed; pedicels ascending to erect . . . . *C. ciliata*
Plants glaucous; seeds dull and gray . . . . *C. maritima*

C. ambigua (Wats.)Howell. Desert Pot Herb. Occasional, in washes and on gypsum soil slopes; desert, below 200m: Borrego Springs; Benson Dunes; Oil Well; Kane Springs Road; Ocotillo Wells; Palm Spring; Carrizo Stage Station; McCain Valley; East Dolomite Mine Road; Mortero Wash turnoff from S-2. Jan.-Apr.

C. breweri Wats. Brewer's Calandrinia. Uncommon, on open slopes and burns; coastal and foothill, below 700m: Poway Grade; Point Loma; Mount San Miguel; Sweetwater Valley; Kimball Grade; Tecate Peak. Apr.-May.

C. ciliata (R. & P.)DC. var. **menziesii** (Hook.)Macbr. [*C. heterophylla* Rydb., the type from Old Mission Dam, San Diego County; *C. muricata* Rydb., the type from San Diego.] Red Maids. Common, on open slopes and burns; cismontane, below 1400m: e.g. San Onofre; Fallbrook; Bonsall; Pauma; Doane Valley; Warner Ranch; Witch Creek; Banner; El Cajon Valley; Springhill; Pine Hills; Julian; Cuyamaca Lake; Laguna Mountains; Volcan Indian Reservation; Tecate Peak; Bostonia; Mount Soledad; Kearny Mesa; Vista; San Marcos Creek. Feb.-Apr. $2n=24$.

C. maritima Nutt. Sea Kisses. Uncommon, in sandy places and sea bluffs; coastal, below 100m: Leucadia; Torrey Pines State Reserve; Mount Soledad; Point Loma; Balboa Park; Rice Canyon; Otay Mesa; Tijuana Hills. Mar.-May. Type from "St. Diego, California, on the sea coast".

Calyptridium Nutt. in T. & G.

C. monandrum Nutt. in T. & G. Common Calyptridium. Frequent, in sandy places and burns; cismontane: e.g. Fallbrook; Bonsall; Lilac; Pala; Harrison Grade; Witch Creek; Chihuahua Valley; Montezuma View Point; Eagles Nest; Lone Palm; Borrego Palm Canyon; Julian; Encinitas; Murphy Canyon; Jamacha; San Pasqual; Mount Woodson; Viejas Valley; Pine Valley; Garnet Peak; Campo; Hipass; Dubber; Jamul; Tecate Junction. Mar.-June. Type from "St. Diego".

Claytonia L.

Stem-leaves usually united on both sides, forming a flattish rounded somewhat angled disk
Basal leaves largely ovate to deltoid
Stems suberect, 1-3dm high; racemes elongate, often peduncled; sepals 3-5mm long . . . . *C. perfoliata* var. *p.*
Stems spreading, mostly less that 1dm long; racemes short, compact, sessile; sepals ca. 2mm long . . . . *C. perfoliata* var. *depressa*
Basal leaves largely linear to oblanceolate or spatulate
Inflorescence mostly elongate, open; sepals ca. 2mm long . . . . *C. perfoliata* var. *parviflora*
Inflorescence mostly short, compact; sepals ca. 3mm long . . . . *C. perfoliata* var. *angustifolia*
Stem-leaves united on part of one side, not forming a flattish disk
Stem-leaves 1-2cm long, not exceeding the inflorescence . . . . *C. spathulata* var. *exigua*
Stem-leaves 2-6cm long, surpassing the inflorescence
Plants glaucous; basal leaves linear or nearly so . . . . *C. spathulata* var. *tenuifolia*
Plants green; basal leaves oblanceolate . . . . *C. perfoliata* ssp. *viridis*

C. perfoliata Donn var. **angustifolia** Greene. Rare, known in the county only from a collection from Julian, 1230m. Apr.

C. perfoliata var. **depressa** (Gray)von Poellnitz. Reddish Miner's-Lettuce. Infrequent, in shaded places; foothill and montane, above 1000m: Palomar Mountain; Banner; Tecate Peak. Apr.-June.

C. perfoliata var. **parviflora** (Dougl. ex Hook.)Torr. Narrow-leaf Miner's-Lettuce. Occasional, in moist places; foothill, 300-1300m: Santa Margarita Ranch; Doane Valley; Chihuahua Valley; Mesa Grande; Eagles Nest; Springhill; Harper Ranch; San Felipe; Mason Valley; Cedar Canyon bridge; Pala; Poway Grade; El Capitan Mountain; Dulzura. Mar.-May.

C. perfoliata var. **perfoliata**. [*Montia p.* (Donn)Howell] Common Miner's-Lettuce. Common, in moist places; coastal and foothill, below 1200m: e.g. San Mateo; De Luz; Oceanside; Rainbow; Witch Creek; Mesa Grande; Pine Hills; Pine Valley; Harbison Canyon; Otay Mountain; Dulzura; Initial Boundary Monument; Otay Ranch; Lake Hodges; La Jolla; Vista. Mar.-May. $n=6, 12, 18$.

C. perfoliata ssp. **viridis** (A. Davids.)Fellows. [*Montia s.* var. *v.* A. Davids.] Occasional, in moist areas; montane, above 1200m: Oak Grove Trail; Julian; Cuyamaca Lake; Pine Valley; Laguna Camp; San Diego River falls. Apr.-June. $n=12$.

C. spathulata Dougl. ex Hook. var. **exigua** (T.& G.)Piper. Serpentine Montia. Infrequent, in moist areas; montane, above 1400m: Morgan Hill; Middle Peak. May-June. $n=24$.

C. spathulata var. **tenuifolia** (T.& G.)Gray. [*Montia s.* var. *t.* (T. & G.)Munz] Glaucous Montia. Occasional, in moist areas; foothill and montane, above 900m: Palomar Mountain; Hot Springs Mountain; Stonewall Mine; Banner; Laguna Camp; Descanso Creek; Pine Valley; Hipass. Apr.-June. n=24.

Crunocallis Rydb.

C. chamissoi (Ledeb.)Rydb. [*Montia c.* (Ledeb.)Dur. & Jackson] Toad Lily. Rare, moist montane habitats above 1000m: Doane Valley, Palomar Mountain; Paso Picacho; Cuyamaca Mountains; Scove Canyon, Laguna Mountains. June.

Lewisia Pursh

Bracts remote from sepals and not resembling them . . . . *L. nevadensis*
Bracts just below the sepals and resembling them . . . . *L. brachycalyx*

L. brachycalyx Engelm. ex Gray. Southwestern Bitterroot. Local, on grassy slopes about Cuyamaca Lake, 1400m; meadow on Azalea Creek and North Peak, Cuyamaca Mountains. May-June.

L. nevadensis (Gray)Rob. in Gray. Nevada Bitter-root. Rare, known in the county only at Azalea Spring, Cuyamaca Mountains.

Montia L.

M. fontana L. ssp. **amporitana** Sennen. [*M. hallii* (Gray)Greene]Water Montia. Infrequent, on shaded slopes, pool margins, and other moist places; montane, above 1200m: Witch Creek; Cuyamaca Lake; Harper Ranch; Laguna Mountains. Also Sloane Ranch. Mar.-July. 2n=18.

Montiastrum (Gray)Rydb.

M. lineare (Dougl.)Rydb. [*Montia l.* (Dougl.)Greene] Linear-leaf Montia. Uncommon, in moist spots of Montane Coniferous Forest; montane, above 1400m: Cuyamaca Lake; Shrine Camp. Apr.-June.

Portulaca L.

***P. oleracea** L. Common Purslane. Native of Europe. Frequent, weed in cultivated and other disturbed places; cismontane: e.g. Oak Grove; Angelina Springs; East San Diego; National City; Lakeside; Pine Valley; Jacumba. May-Sep. 2n=54.

PRIMULACEAE. Primrose Family

KEY TO GENERA

Ovary partly inferior, the lower portion included in and adnate to the calyx-tube; plant with leafy stems; flowers small, racemose . . . . *Samolus*
Ovary superior and free
  Plants scapose, the leaves all basal; flowers in bracteate umbels
    Corolla-lobes spreading or erect, 1-2mm long . . . . *Androsace*
    Corolla-lobes reflexed, much larger . . . . *Dodecatheon*
  Plants with leafy stems; flowers not in umbels; the leaves alternate or opposite . . . . *Anagallis*

Anagallis L.

Leaves 5-20mm long; pedicels 10-30mm long; petals to about 6mm broad
  Flowers blue . . . . *A. arvensis* var. *caerulea*
  Flowers salmon . . . . *A. arvensis* var. *a.*
Leaves 2-5mm long; pedicels shorter; petals smaller . . . . *A. minima*

***A. arvensis** L. var. **arvensis**. Scarlet Pimpernel. Native of Europe. Common, weed in disturbed areas; coastal and foothill, below 1100m: e.g. San Onofre; De Luz; Hormiguero Ranch; Rainbow; Lilac; Fernbrook; Pine Valley; Inaja Indian Reservation; Ramona; Dulzura; Lyons Valley; Jamul; Silver Strand; Otay Mountain; Sloane Ranch; Alpine; National City; La Mesa; Escondido. Mar.-July. 2n=40.

***A. arvensis** var. **caerulea** (Schreb.)Ledeb. Blue Pimpernel. Native of Europe. Infrequent, weed in disturbed areas; coastal and foothill, below 400m: San Onofre; Fallbrook; Ramona; Escondido; Ocean Beach. Mar.-July.

A. minima (L.)Krause. [*Centunculus m.* L.] Common Chaffweed. Frequent, in vernal pools and other moist spots; cismontane, below 1500m: e.g. Doane Valley; Wyss Ranch; Laguna Public Camp; Japatul Valley; Ramona; San Marcos; Kearny Mesa; Torrey Pines; La Jolla. Apr.-July. 2n=22.

Androsace L.

A. elongata L. ssp. **acuta** (Greene)Robbins. [*A. a.* Greene] California Androsace. Uncommon, on grassy slopes; foothill, 700-1100m: Montezuma Valley; Warner Hot Springs. Mar.-May.

Dodecatheon L. Shooting Star

D. clevelandii Greene ssp. **clevelandii**. Padre's Shooting Star. Common, on grassy slopes; coastal and foothill, below 700m: e.g. Case Spring; Fallbrook; Pala; Oceanside; San Marcos; Rancho Santa Fe; Springhill; Descanso; Bostonia; Harbison Canyon; Fletcher Hills; Lyons Valley: Woodwardia Canyon; Paradise Hills; Imperial Beach; Point Loma; Mission Valley; Rolando; Kearny Mesa. Type from dry hills and mesas in the southern part of California, about San Diego and San Bernardino. Jan.-Apr. 2n=22.

Samolus L. Water-Pimpernel

S. parviflorus Raf. [*S. floribundus* H.B.K.] Water-Pimpernel. Occasional, in moist places; coastal and foothill, below 1000m: San Onofre Canyon; Lone Palm Spring; Pala; Ramona; Descanso; San Diego River falls; Potrero; Jamul; Lakeside; Fletcher Hills; Kearny Mesa; Dulzura; Alvarado Canyon; Tecate Peak. June-Aug.

RAFFLESIACEAE. Rafflesia Family

Pilostyles Guill.

P. thurberi Gray. Thurber's Pilostyle. Rare, on *Psorothamnus emoryi*, desert slopes below 800m: Coyote Creek, below Borrego Valley; Font's Point Wash; below the Carrizo Overlook; ca. 4.5 miles west of the Imperial County line on Hwy. S-2. Jan.-May.

RANUNCULACEAE. Crowfoot Family

KEY TO GENERA

Pistils few- to several-ovuled, becoming follicles or berries when ripe
  Flowers irregular; upper sepal spurred; petals 4, the upper pair spurred and included in the spurred sepal *Delphinium*
  Flowers regular . . . . *Aquilegia*
Pistils 1-ovuled, becoming akenes
  Stem-leaves opposite or whorled; petals none
    Stem-leaves forming a single whorl; herbs . . . . *Anemone*
    Stem-leaves many, opposite; woody vines . . . . *Clematis*
  Stem-leaves alternate or none
    Petals present, rarely none in *Ranunculus*
      Sepals spurred; small annuals with basal linear leaves. . . . . *Myosurus*
      Sepals not spurred; leaves usually cauline as well as basal . . . . *Ranunculus*
    Petals absent; sepals greenish or petaloid; leaves decompound; flowers usually unisexual. . . . . *Thalictrum*

Anemone L.

A. tuberosa Rydb. Desert Anemone. Scarce, on rocky slopes; desert, above 700m: desert slope of the Laguna Mountains; Scissors Crossing; Table Mountain; Mountain Springs; Dubber; "Jade" Hill east of Jacumba. Apr.-May. $2n=16$.

Aquilegia L. Columbine

Stems viscid-pubescent . . . . *A. formosa* var. *hypolasia*
Stems glabrous to sparingly pubescent . . . . *A. formosa* var. *truncata*

A. formosa Fisch. in DC. var. **hypolasia** (Greene)Munz. Sticky Columbine. Frequent, in moist, shaded areas of Montane Coniferous Forest; montane, above 1200m: Palomar and Cuyamaca Mountains. June-Aug. Type from between Campbell and Cameron Ranches.

A. formosa var. **truncata** (F.& M.)Baker. California Columbine. Frequent, in moist, shaded areas of Chaparral and Montane Coniferous Forest; foothill and montane, above 1000m: Palomar Mountain; Hot Springs Mountain; Pine Hills; Cuyamaca Lake; Buckman Spring. May-July. $2n=14$.

Clematis L. Virgin's-Bower

Flowers many in cymose panicles; leaflets 5-7 . . . . *C. ligusticifolia*
Flowers 1-3 on a peduncle; leaflets 3-5
  Sepals 1.5-2.5cm long, pubescent above; akenes pubescent . . . . *C. lasiantha*
  Sepals mostly 0.8-1.2cm long, glabrous above; akenes glabrous . . . . *C. pauciflora*

C. lasiantha Nutt. in T.& G. Pipestem Virgin's-Bower. Occasional, in Chaparral; foothill, 600-1200m: Harrison Grade; Black Canyon; Banner Grade; west slope of Cuyamaca Mountains; Guatay Mountain; San Miguel Mountain; Otay Mountain. Apr.-July. Type from "near the sea-coast of St. Diego". $n=8$.

C. ligusticifolia Nutt. in T.& G. Yerba de Chivato. Occasional, along streams and in other moist areas; cismontane: Pala; Doane Valley; Pine Hills; Oasis Spring; Descanso Junction; La Posta; Lakeside; Jamacha; Buckman Spring. July-Sep. $n=8$.

C. pauciflora Nutt. in T.& G. Small-leaf Virgin's-Bower. Frequent, in Chaparral; coastal and foothill, below 1000m: e.g. Rainbow; Lilac; San Pasqual; Lakeside; Poway Grade; Mission Bay; Witch Creek; Springhill; Chollas Valley; Mission Valley; Lyons Peak; Dulzura; Dehesa; Otay Mountain; Canyon City. Also Banner Grade; San Felipe; Dubber. Jan.-June. Type from "near the sea-coast of St. Diego".

Delphinium L. Larkspur

Plants annual . . . . *D. ajacis*
Plants perennial
  Flowers usually scarlet, sometimes yellowish . . . . *D. cardinale*
  Flowers blue to purple or bluish-white
    Stems attenuate at base and very easily separable from the tuberous roots, glabrous. . . . . *D. patens*
    Stems not attenuate at base and rather firmly attached to woody or fibrous roots
      Flowers deep blue; sepals 8-24mm long . . . . *D. parishii*
      Flowers violet or pale blue to whitish
        Sepals deep blue to purple, 10-15mm long
          Pedicels 2.5-5cm long . . . . *D. parryi* var. *maritimum*
          Pedicels less than 2.5cm long . . . . *D. parryi* var. *p.*
        Sepals paler . . . . *D. hesperium*

***D. ajacis** L. [*Consolida ambigua* (L.)Ball & Heywood] Rocket Delphinium. Native of Europe. Infrequently escaping from cultivation: San Diego. May-July. n=8.

D. cardinale Hook. [*D. coccineum* Torr., the type from mountains east of San Diego.] Scarlet Larkspur. Frequent, in Chaparral; cismontane: e.g. Las Flores; Pala; Palomar Mountain; Mount Woodson; Monument Peak; Banner Grade; San Felipe; Oriflamme Canyon; Pine Hills; Cuyamaca Mountains; Descanso; Pine Valley; Potrero; Del Mar; Otay Mountain. June-Aug. $2n$=16.

D. hesperium Gray ssp. **cuyamacae** (Abrams)Lewis & Epling. [*D. c.* Abrams] Cuyamaca Larkspur. Uncommon, in meadows; montane, above 1200m: around Cuyamaca Lake; Pine Hills; French Valley. June-July. Type from grassy slopes bordering Cuyamaca Lake, 1550 meters.

D. parishii Gray ssp. **subglobosum** (Wiggins)Lewis & Epling. [*D. collinum* Ewan, the type from a ridge between Campbells Ranch and Mason Valley; *D. s.* Wiggins] Colorado Desert Larkspur. Occasional, on rocky slopes; desert, above 300m: Montezuma Valley; Borrego Valley; Borrego Palm Canyon; Hellhole Canyon; San Felipe Valley; Yaqui Well; Banner; Sentenac Canyon; Box Canyon; Pinyon Mountain; Garnet Peak; The Potrero; Agua Caliente Hot Springs; Jacumba; Mountain Springs; McCain Valley; Mortero Wash; north Mortero Wash; Mortero Palms. Mar.-May. Type from open grassy slope near Banner.

D. parryi Gray var. **maritimum** A. Davids. Maritime Larkspur. Infrequent, on sea bluffs and open slopes; coastal and foothill, below 800m: Pacific Beach; Silver Strand; National City; south fork of Featherstone Creek. Mar.-June.

D. parryi var. **parryi**. Parry's Larkspur. Common, on grassy slopes; cismontane: e.g. upper San Onofre Canyon; Doane Valley; Eagles Nest; Lilac; Pine Hills; Ramona; Mussey Grade; Santa Ysabel; Monument Peak; Laguna Lake; Descanso; Lakeside; Point Loma; Harbison Canyon; Hipass; Initial Boundary Monument. Apr.-July. $2n$=16.

D. patens Benth. ssp. **hepaticoideum** Ewan. [*D. decorum* auct. non F. & M.] Liver-leaf Larkspur. Occasional, in shaded areas; foothill and montane, above 1000m: Palomar Mountain; Warner Hot Springs; Cuyamaca Lake; Spencer Valley; Inspiration Point; Garnet Peak; Laguna Mountain by Al Bahr Shrine Camp; Morena Dam. Mar.-May.

Myosurus L. Mouse-Tail

Fruiting spikes extending beyond the leaves . . . . . . *M. minimus* var. *filiformis*
Fruiting spikes not extending beyond the leaves . . . . . . *M. minimus* var. *apus*

M. minimus L. var. **apus** Greene. Infrequent, in vernal pools; coastal, below 150m: National City; Proctor Valley; Otay Mesa. Feb.-Apr. Type from table-lands back of San Diego.

M. minimus var. **filiformis** Greene. Infrequent, in vernal pools; coastal, montane: East San Diego; Kearny Mesa; Ramona; granodiorite outcrops at Corte Madera. Apr.

Ranunculus L. Buttercup

Petals white; aquatics with submersed leaves dissected; akenes transversely ridged
  Submersed leaves once filiform . . . . . . *R. aquatilis* var. *capillaceus*
  Submersed leaves divided . . . . . . *R. aquatilis* var. *hispidulus*
Petals yellow or red, rarely white in age; plants mostly terrestrial; akenes not transversely ridged
  Plants annual; petals 1-3mm long . . . . . . *R. hebecarpus*
  Plants perennial
    Plants, at least the stems, hirsute or evidently pubescent
      Receptacle pubescent or hispid; sepals reflexed from base; petals 9-16
        Petals 6-8mm long . . . . . . *R. californicus* var. *austromontanus*
        Petals 8-15mm long . . . . . . *R. californicus* var. *c.*
      Receptacle glabrous
        Body of akene mostly 4-5mm long, the beak with broad thin base 0.6-1mm wide; petals 5-10 . . . . . . *R. canus*
        Body of akene mostly 2-3mm long, the beak with base less than 0.5mm wide . . . . . . *R. californicus* et vars.
    Plants glabrous or nearly so
      Lower leaves 3-7-parted or -foliate or even dissected . . . . . . *R. californicus* et vars.
      Lower leaves entire or shallowly lobed . . . . . . *R. cymbalaria*

R. aquatilis L. var. **capillaceus** (Thuill.)DC. Hairleaf Water Buttercup. Frequent, in persistent ponds, lakes and streams; cismontane: Morettis; Cuyamaca Lake; Laguna Lake; Descanso; Mission Valley. Mar.-July.

R. aquatilis var. **hispidulus** E.Drew. Floating Water Buttercup. Uncommon, in persistent ponds, lakes and streams; foothill and montane, above 1000m: Cuyamaca Lake; Pine Hills; below Corte Madera Lake. Apr.-July.

R. californicus Benth. var. **austromontanus** L.Benson. [*R. c.* var. *ludovicanus* auct. non Greene] Southern Buttercup. Common, in meadows; foothill and montane, above 1000m: e.g. Palomar Mountain; Lake Henshaw; Santa Ysabel; Julian; Rattlesnake Valley; Mount Laguna. May-July.

R. californicus var. **californicus**. California Buttercup. Occasional, in moist meadows; foothill and coastal, below 500m: San Onofre; San Marcos; Ramona; Poway; Mount Soledad; Paradise Valley; Otay River. Feb.-Apr. $2n$=28.

R. canus Benth. var. **ludovicianus** (Greene)L.Benson. Rare, known in the county only at Lake Henshaw, 900m. May.

R. cymbalaria Pursh ssp. **saximontanus** (Fern.)Thorne. Desert Buttercup. Occasional, in moist areas; below 1200m: Oceanside; San Pasqual Valley; Pala; Eagles Nest; Borrego Palm Canyon; upper Box Canyon; Buckman Spring; Lost Valley in McCain Valley Recreation Area. May-Aug.

R. hebecarpus H.& A. Pubescent-fruit Buttercup. Occasional, on shaded slopes; coastal and foothill, below 1000m: De Luz; Clevenger Canyon; Mission Valley; Whispering Oaks; Otay. Mar.-May.

Thalictrum L. Meadow-Rue

Carpels and under surface of upper leaves glandular-puberulent, the herbage green; akenes ovate-lanceolate, strongly 3-5-nerved, not obviously reflexed or inflated . . . . . *T. fendleri*
Carpels and leaves entirely glabrous, the herbage glaucous; akenes obovate, 1-nerved, or the several veins somewhat reticulate, somewhat reflexed and when fresh turgid. Below pine belt . . . . . *T. polycarpum*

T. fendleri Engelm. ex Gray. Fendler's Meadow-Rue. Frequent, in moist places of Montane Coniferous Forest; montane, above 1200m: Boucher Hill; Eagles Nest; Pine Hills; Cuyamaca and Laguna Mountains. May-Aug. n=14, 28, 35.

T. polycarpum (Torr.)Wats. [*T. coreospermum* Greene, the type from the Cuyamaca Mountains; *T. magarum* Greene, the type from Witch Creek.] Common Meadow-Rue. Frequent, on shaded slopes; coastal and foothill, below 1100m: e.g. upper San Onofre Canyon; Pala; Mesa Grande; Inspiration Point; Poway Grade; Morena Lake; Hipass; San Miguel Mountain; Jamul. Mar.-June. $2n$=14 pairs.

RESEDACEAE. Mignonette Family

KEY TO GENERA

Petals 2; disk lacking; leaves entire . . . . . *Oligomeris*
Petals 4-7; disk cup-shaped; leaves often pinnatifid . . . . . *Reseda*

Oligomeris Camb.

O. linifolia (Vahl)Macbr. Narrowleaf Oligomeris. Occasional, in open, alkaline areas; desert and south coastal, below 200m: Benson Dry Lake; Borrego Palm Canyon; Borrego Mud Hills; Vallecito Stage Station; Carrizo Stage Station; Otay Mesa; San Ysidro; Silver Strand; Paradise Valley; Point Loma; East San Diego. Feb.-July.

Reseda L. Mignonette

All leaves entire or with a few minute, hyaline teeth near the base
  Pedicels not more than 2.5mm long . . . . . *R. luteola*
  Pedicels 5mm or more . . . . . *R. odorata*
Some leaves ternate or pinnatifid
  Flowers greenish-white; stamens 12-15; all petals cleft . . . . . *R. alba*
  Flowers pale yellow; stamens 15-20; lowest petal not cleft . . . . . *R. lutea*

***R. alba** L. White Mignonette. Native of Europe. Rare, escaping from cultivation at Pala, 90m. Apr. $2n$=20.

***R. lutea** L. Yellow Mignonette. Native of Europe. Rare, escaping from cultivation at San Diego, 30m. May-Sep. $2n$=48.

***R. luteola** L. Dyer's Rocket. Native of Europe. Rare, waif along roads: Poway; Beeler Canyon; Miramar, 100-200m. May-June. $2n$=24, 26, 28.

***R. odorata** L. Sweet Mignonette. Native of north Africa. Rare, escaping from cultivation; coastal, below 150m: Solana Beach; San Marcos; Point Loma; San Diego. Mar.-June. $2n$=12

RHAMNACEAE. Buckthorn Family

KEY TO GENERA

Fruit fleshy, drupelike
  Drupe with 1 nutlet; petals clawed or none; spinose desert shrubs . . . . . *Ziziphus*
  Drupe with 2-3 nutlets; petals clawless or none; plants not usually spinose . . . . . *Rhamnus*
Fruit dry, capsular
  Calyx-tube free from lower part of ovary, the sepals persistent; style jointed. . . . . *Adolphia*
  Calyx-tube joined to base of ovary, the sepals deciduous in fruit; style not jointed . . . . . *Ceanothus*

Adolphia Meissn.

A. californica Wats. California Spinebush. Local, on dry slopes; foothill and coastal, below 300m: Moro Hill; Cerro de la Calavera; Agua Hedionda; Rancho Santa Fe; Mount Soledad; Bernardo; Chollas Valley; Barrett Junction; Proctor Valley; Otay. Dec.-May. Type from "Solidad (Soledad) and Chollas Valley, near San Diego".

Ceanothus L. California-Lilac

Stipules thin and early deciduous; leaves alternate, with stomata not sunken; capsule without horns, but often with ridges or crests; flowers in simple or compound panicles
  Ultimate branchlets flexible, not rigidly divaricate and spinose
    Leaves with 3 distinct veins from the base, the lateral sometimes obscure
      Leaves entire, sometimes deciduous, mostly more than 2cm long; flowers mostly white . . . . . *C. integerrimus*
      Leaves serrate, serrulate or glandular-denticulate; flowers blue or purple
        Branchlets angled and with scattered brownish glands . . . . . *C. cyaneus*
        Branchlets terete, not angled or striate
          Leaves tomentose beneath . . . . . *C. tomentosus*

Leaves somewhat pubescent beneath, chiefly on the veins . . . . *C. oliganthus*
Leaves with 1 main vein from the base, even though the basal lateral pair may be longer than the other pairs
Leaves entire, mostly rather glabrous
Shrubs evergreen; leaves rather firm and thick
Flowers white; capsule 5-7mm broad, conspicuously crested . . . . *C. palmeri*
Flowers blue to almost white; capsule 4-5mm broad, scarcely crested . . . . *C. spinosus*
Shrubs deciduous; leaves thinnish; flowers mostly white . . . . *C. integerrimus*
Leaves dentate to serrulate, usually pubescent at least beneath . . . . *C. foliosus*
Ultimate branchlets rigidly divaricate and spinose; leaves persistent, plane
Leaves with 1 main vein from base, mostly entire and glossy on both surfaces . . . . *C. spinosus*
Leaves with 3 main veins from the base, the 2 lateral sometimes obscure, gray-glaucous on both surfaces . . . . *C. leucodermis*
Stipules with thick corky persistent bases; leaves alternate or opposite, the stomata in sunken pits; capsule usually with 3 horns; flowers in lateral umbels; leaves firm-coriaceous, persistent
Leaves alternate at least in part
Leaves all alternate, broadly obovate to almost round; capsule 4-6mm in diameter . . . . *C. verrucosus*
Leaves both alternate and opposite, oblanceolate to broadly elliptic; capsule 8-12mm in diameter *C. megacarpus*
Leaves all opposite
Leaves densely white-tomentose beneath, mostly revolute . . . . *C. crassifolius*
Leaves not densely white-tomentose beneath, mostly plane and minutely canescent
Leaves very revolute . . . . *C. otayensis*
Leaves mostly plane
Leaves mostly not concave on upper surface; horns of capsule apical or subapical, conspicuous. . *C. cuneatus*
Leaves mostly concave on upper surface; horns of capsule lateral and spreading, usually small . . . . *C. greggii*

C. crassifolius Torr. Thick-leaf Wild-Lilac. Local in dense stands, in Chaparral; foothill, 200-1100m: e.g. De Luz; Pala; Warner Hot Springs; Ramona; Barona Valley; Mussey Grade. Jan.-May.

C. cuneatus (Hook.)Nutt. ex T.& G. Buckbrush. Local, in Chaparral; foothill, 700-1250m: e.g. Chihuahua Valley; Warner Hot Springs; Buckman Spring; Boulevard; Tierra del Sol; Campo; Jacumba. Feb.-June. *n*=12.

C. cyaneus Eastw. Lakeside Lilac. Local, in Chaparral; foothill, 150-1000m: Barona Valley; El Cajon Mountain; Mussey Grade; Crest; Harbison Canyon. Apr.-June. Type from Philbrook Ranch, Lakeside. *n*=12.

C. foliosus Parry. [*C. austromontanus* Abrams, the type from coniferous forests, between Julian and Cuiamaca (Cuyamaca).] Southern Blue Lilac. Local, in Chaparral and Montane Coniferous Forest; montane, 1200-1500m: Cuyamaca Mountains; Pine Hills; Los Pinos Mountain; Guatay Mountain; Inspiration Point. May-Aug. *n*=12.

C. greggii Gray ssp. **perplexans** (Trel.)Beauchamp. [*C. p.* Trel.] Cupleaf Lilac. Prevalent, in Chaparral; foothill and montane, 300-1500m: e.g. Oak Grove; Palomar Mountains; Warner Hot Springs; Pine Hills; Laguna Mountains; Banner Grade; Alpine; Potrero; Jacumba. May-July.

C. integerrimus H.& A. var. **puberulus** (Greene)Abrams. Southern Deerbrush. Infrequent, in Chaparral; foothill, 200-1400m: Spencer Valley; Alpine; Poway Junction; Cuyamaca Mountains. May-June.

C. leucodermis Greene. [*C. divaricatus* var. *laetiflorus* Jeps., the type from Palomar Mountain (Pala Mission).] Whitebark Lilac. Common, in Chaparral; montane and foothill, 400-1500m: e.g. Pala; Hot Springs Mountain; Doane Valley; El Monte; Barrett Reservoir; Flinn Springs; Jamul; Laguna Mountain; Campo; Dubber. Mar.-June. *n*=12.

C. megacarpus Nutt. ssp. **megacarpus**. Big-fruit Lilac. Rare, in canyons; north coastal, below 100m: San Onofre Mountains. Jan.-Apr.

C. oliganthus Nutt. ex T. & G. [*C. oliganthus* var. *orcuttii* (Parry)Jeps.; *C. orcuttii* Parry] Frequent, in Chaparral and Montane Coniferous Forest; foothill and montane, 400-1500m: e.g. Palomar Mountains; Julian; Cuyamaca Mountains; Ramona; Lyons Peak; Tecate Peak; Otay Mountain. Mar.-May. Type from high mountains east of San Diego.

C. otayensis McMinn. Otay Lilac. Occasional, in Chaparral; south foothill, 600-1100m: San Miguel and Otay Mountains. Jan.-Mar. Type from Otay Mountain.

C. palmeri Trel. Palmer's Lilac. Common, in Chaparral and Montane Coniferous Forest; montane, above 1100m: Palomar, Hot Springs; Volcan, Cuyamaca, and Laguna Mountains. May-Aug. Type from mountains of southern California.

C. spinosus Nutt. in T.& G. Greenbark Lilac. Uncommon, in Chaparral; cismontane: San Onofre Mountains; San Marcos; Ramona; Cold Spring. Feb.-May. *n*=12.

C. tomentosus Parry ssp. **olivaceus** (Jeps.)Munz. Ramona Lilac. Frequent, in Chaparral; coastal and foothill, below 1100m: e.g. Fallbrook; San Onofre; San Pasqual; Encinitas; Miramar; Potrero Grade; Otay Mountain. Also Banner Grade. Feb.-May. Type from Clevenger Canyon, near Ramona.

C. verrucosus Nutt. in T.& G. Coast White-Lilac. Frequent, in Chaparral; coastal, below 300m: Agua Hedionda; Encinitas; Leucadia; Torrey Pines State Reserve; Kearny Mesa; Lake Hodges; Point Loma. Dec.-June. Type from "low hills near the coast, St. Diego". *n*=12.

Rhamnus L. Cascara

Petals present; winter-buds without bud-scales
Underside of leaves glabrous or with a few hairs . . . . *R. californica* ssp. *c.*
Underside of leaves tomentose . . . . *R. californica* var. *tomentella*
Petals lacking; winter-buds with bud-scales
Branchlets somewhat divaricate, ending in weak spines; leaf-blades 10-17mm long . . . . *R. crocea*

Branchlets not divaricate, seldom ending in weak spines; leaf-blades 18-60mm long
  Leaves glabrous to pubescent, rarely slightly revolute . . . . . *R. ilicifolia*
  Leaves pilose on both surfaces, revolute . . . . . *R. pilosa*

R. californica Esch. ssp. **californica**. California Coffeeberry. Infrequent, on open slopes; coastal, below 200m: De Luz; Encinitas. May-July. $2n=24$.

R. californica ssp. **tomentella** (Benth.)C.B. Wolf. Chaparral Coffeeberry. Frequent, in Chaparral and Montane Coniferous Forest; cismontane, below 1500m: e.g. Palomar Mountain; Hot Springs Mountain; Santa Ysabel; Cuyamaca Mountains; Buckman Spring; Laguna Mountains; Campo. May-July.

R. crocea Nutt. in T.& G. Spiny Redberry. Common, in Coastal Sage Scrub and Chaparral; coastal and foothill, below 1000m: e.g. Encinitas; Torrey Pines; Moosa Canyon; Poway; Kearny Mesa; Jamul Mountain; Potrero Grade; Viejas Grade; Spring Valley; Echo Valley. Mar.-June.

R. ilicifolia Kell. [*R. crocea* var. *i.* (Kell.)Greene] Holly-leaf Redberry. Frequent, in Chaparral; foothill and montane, above 300m: e.g. San Onofre Canyon; Pala; Palomar Mountains; Santa Ysabel; Bear Valley; Laguna Mountains; Morena Reservoir; Jacumba; Campo; Otay Mountain. Feb.-June. $2n=12$ pairs.

R. pilosa (Trel.)Abrams. [*R. crocea* var. *p.* Trel.] Hairy-leaf Redberry. Infrequent, in Chaparral; foothill, 300-700m: Lilac; Ramona; Lakeside; Barona Valley; Viejas Grade; Alpine; Harbison Canyon; Otay Mountain; Poway Lake. Mar.-June. Type from Nuevo (Ramona).

Ziziphus Mill.

Z. parryi Torr. [*Condaliopsis p.* (Torr.)Suesseng. in Engler & Prantl; *Condalia p.* (Torr.)Weberb.] California Lote Bush. Infrequent, on rocky slopes; desert, 600-1300m: Montezuma Grade; San Felipe Valley; Oriflamme Canyon; Mason Valley; Pinyon Mountain; Vallecito Canyon; Jacumba; Mountain Springs. Mar.-May. Type from gravelly ravines near San Felipe.

ROSACEAE. Rose Family

KEY TO GENERA

Ovary or ovaries superior; fruit not a pome
  Fruit of dry akenes, which may be embedded in the surface of a fleshy receptacle (strawberry) or enclosed in a fleshy cup-shaped structure (rose)
    Leaves simple, sometimes 3-lobed; plants woody
      Petals present
        Flowers solitary; leaves cuneate, 3-lobed above . . . . . *Purshia*
        Flowers paniculate; leaves not lobed . . . . . *Adenostoma*
      Petals absent
        Floral-tube salver-shaped, the limb deciduous, the tubular base persistent; akene with long feather tail . . . . . *Cercocarpus*
        Floral-tube campanulate; akene not long-tailed . . . . . *Coleogyne*
    Leaves compound or pinnately dissected
      Plants shrubby
        Pistil usually 1
          Leaves 3-lobed; flowers solitary . . . . . *Purshia*
          Leaves pinnately dissected into minute segments; flowers cymose-paniculate . . . . . *Chamaebatia*
        Pistils 4-many
          Carpels not enclosed in a fleshy floral-tube; plants not prickly . . . . . *Potentilla*
          Carpels enclosed in the floral-tube which is fleshy in fruit; stems mostly prickly . . . . . *Rosa*
      Plants herbaceous
        Styles jointed with the ovary and deciduous
          Filaments dilated, 10 in number; floral-tube usually deep . . . . . *Horkelia*
          Filaments filiform or narrow, mostly 20 or more; floral-tube shallow or none
            Receptacle not enlarged in fruit . . . . . *Potentilla*
            Receptacle enlarged in fruit, fleshy . . . . . *Fragaria*
        Styles not jointed with the ovary, persistent
          Floral-tube not prickly; petals none
            Leaves palmately lobed; plants not over 1dm high . . . . . *Alchemilla*
            Leaves pinnate; plants 1-6dm high . . . . . *Sanguisorba*
          Floral-tube armed with prickles, these hooked; petals yellow; sepals 5 . . . . . *Agrimonia*
  Fruit of drupelets or drupes
    Carpels becoming drupelets, which are somewhat coherent and form a fleshy "berry"; plants somewhat woody, often prickly; leaves mostly pinnate . . . . . *Rubus*
    Carpels becoming larger solitary drupes; trees or shrubs with simple leaves; pistil 1 . . . . . *Prunus*
Ovary inferior; fruit a pome
  Plant with deciduous leaves . . . . . *Amelanchier*
  Plants with persistent evergreen leaves . . . . . *Heteromeles*

Adenostoma H. & A.

Leaves fascicled; bracts lance-linear, not scarious; flowers sessile
  Stems pubescent; leaves obtuse . . . . . *A. fasciculatum* var. *obtusifolium*
  Stems glabrous; leaves linear to narrow-clavate . . . . . *A. fasciculatum* var. *f.*
Leaves scattered; bracts broadly lanceolate, scarious-margined; flowers pedicelled . . . . . *A. sparsifolium*

A. fasciculatum H.& A. var. **fasciculatum**. Common Chamise. Prevalent, in Chaparral as an indicator species, often forming pure stands; cismontane: e.g. Camp Pendleton; Palomar Mountain; Montezuma View Point; Escondido; Cuyamaca Mountains; Monument Peak; Laguna Mountains; Buckman Spring; El Cajon Mountain; Del Mar; Ocean Beach; Alpine; Jacumba. Apr.-July. $2n=9$ pairs.

A. fasciculatum var. **obtusifolium** Wats. [*A. brevifolia* Nutt. ex T.& G.; *A. f.* var. *hirsutum* C.K. Schn.; *A. laxum* Gandoger, the type from San Diego.] San Diego Chamise. Common, in Chaparral; coastal and foothill, below 300m: e.g. upper San Onofre Canyon; Point Loma; Proctor Valley; Campo. Mar.-July. Type from near San Diego.

A. sparsifolium Torr. in Emory. Red Shank. Abundant, on dry slopes and mesas; foothill and montane, above 600m: e.g. Palomar Mountain; Oak Grove; Hot Springs Mountain; upper Borrego Palm Canyon; Laguna and Cuyamaca Mountains; Buckman Spring; Pine Valley; Campo; Hipass; Jacumba; Mountain Springs Grade. July-Oct. Type from Cordilleras of California. $2n=9$ pairs.

Agrimonia L.

A. gryposepala Wallr. Tall Hairy Agrimony. Occasional, in moist places; montane, above 1200m: Palomar Mountain; Dehr Creek; Cuyamaca and Laguna Mountains. July-Aug.

Alchemilla L.

A. occidentalis Nutt. ex T.& G. [*A. cuneifolia* Nutt.] Western Lady's-mantle. Occasional, on shaded slopes; coastal and foothill, below 1100m: Moro Hill; Moosa Canyon; San Marcos; Poway Grade; Penasquitos Canyon; Mahogany Canyon; Otay Mountain. Mar.-June.

Amelanchier Medic. Service-Berry

A. utahensis Koehne. [*A. pallida* Greene; *A. alnifolia* var. *cuyamacensis* Munz, the type from Cuyamaca Lake; *A. alnifolia* var. *venulosa* auct. non Jeps.; *A. alnifolia* var. *u.* (Koehne)Jones] Utah Service-Berry. Frequent, in Montane Coniferous Forest; montane, above 1050m: Doane Valley; French Valley; Julian; Banner Grade; Cuyamaca Lake; Oasis Spring; Garnet Peak Laguna Mountains; Vallecito Canyon; Guatay Mountain; Buckman Spring. Apr.-June.

Cercocarpus H.B.K. Mountain-Mahogany

Flowers 6-8mm broad; leaves whitish, somewhat hairy beneath . . . . *C. betuloides*
Flowers 2-4mm broad; leaves greenish-yellow and glabrous beneath . . . . *C. minutiflorus*

C. betuloides Nutt. ex T.& G. ssp. **betuloides**. Interior Mountain-Mahogany. Common, on dry slopes and canyons; foothill and montane, above 300m: Palomar Mountain; Warner Hot Springs; San Felipe Valley; Banner Grade; Rattlesnake Valley; Oriflamme Canyon; Laguna Mountain; Manzanita; Jacumba; Guatay Mountain; Barrett Dam; Miramar. Apr.-May. $2n=18$.

C. minutiflorus Abrams. San Diego Mountain-Mahogany. Occasional, on dry slopes and canyons; coastal and foothill, below 1100m: San Onofre Canyon; Fallbrook; Vista; Escondido; Encinitas; Clevenger Canyon; Mussey Grade; Ramona; Lyons Peak; Campo; Jamul; Otay Mountain. Mar.-May. Type from dry Chaparral-covered hills, near San Dieguito (Bernardo).

Chamaebatia Benth.

C. australis (Bdg.)Abrams. Southern Mountain Misery. Local, in Chaparral, foothill, 300-1050m: San Marcos Hills; Iron Mountain; San Miguel Mountain; McGinty Mountain; Jamul Mountains; Otay Mountain; Potrero Grade; Potrero Peak; Barber Mountain; Tecate Peak. Nov.-May. $n=9$.

Coleogyne Torr. Blackbush

C. ramosissima Torr. Blackbush. Uncommon, on rocky slopes; desert, 450-800m: Scissors Crossing; Pinyon Mountain; Yaqui Well; Smugglers Canyon; Mountain Springs. Apr.-June. $2n=16$.

Fragaria L. Strawberry

F. californica Cham. & Schlecht. [*F. vesca* ssp. *c.* (C. & S.)Staudt.] California Strawberry. Frequent, on floor of Montane Coniferous Forest; montane, above 1200m: Palomar, Hot Springs, Volcan, and Cuyamaca Mountains. Mar.-July. $n=7$.

Heteromeles M. Roem. Toyon

Fruits red . . . . *H. arbutifolia* ssp. *a.*
Fruits yellow . . . . *H. arbutifolia* forma *cerina*

H. arbutifolia (Ait.)M.Roem. ssp. **arbutifolia**. [*Photinia a.* (Ait.)Lindl.] Toyon, Hollywood. Common, on slopes; cismontane, below 1350m: e.g. Camp Pendleton; Palomar, Hot Springs, and Laguna Mountains; Banner Grade; San Felipe Valley; Lake Hodges; Echo Valley; Point Loma; Miramar. June-July.

H. arbutifolia forma **cerina** Jeps. Uncommon, on slopes; coastal, below 300m: San Onofre Mountains; Las Flores Creek. June-July.

Horkelia Cham. & Schlecht.

Bractlets ovate, sepaloid; pistils more than 50
  Leaflets 5-10 pairs; the uppermost confluent; petals oblanceolate to narrowly obovate
    Infloescence congested . . . . *H. cuneata* ssp. *c.*
    Flowers in open cymes . . . . *H. cuneata* ssp. *puberula*
  Leaflets 1-3 pairs; terminal leaflet petiolulate; petals orbicular . . . . *H. truncata*
Bractlets lanceolate to narrowly linear, much smaller than the sepals; pistils fewer . . . . *H. bolanderi*

H. bolanderi Gray ssp. **clevelandii** (Greene)Keck. [*Potentilla b.* var. *c.* (Greene)Munz & Jtn.] Cleveland's Horkelia. Infrequent, in moist areas of floor of Montane Coniferous Forest; montane, above

1350m: Palomar, Cuyamaca, and Laguna Mountains. June-Aug. Type from "Laguna, in the mountains back of San Diego".

H. cuneata Lndl. ssp. **cuneata**. [*Potentilla lindleyi* Greene] Wedgeleaf Horkelia. Rare, known in the county only at Carlsbad, 20m. Apr.-Sep.

H. cuneata ssp. **puberula** (Greene)Keck. Mesa Horkelia. Rare, known in the county only at Magee Ranch, Pala, 200m. Mar.-Apr.

H. truncata Rydb. [*Potentilla t.* (Rydb.)Munz & Jtn.] Infrequent, on dry slopes; foothill, 600-1200m: Roblar Grade; Black Mountain-Lusardi; Mesa Grande; El Cajon Mountain; Ramona; Barona Valley; McGinty Mountain; Lawson Peak; Lawson Valley. May-June. Type collected in Brandegee's garden at San Diego, cultivated plant brought from Ramona.

Potentilla L. Cinquefoil

Styles terminal or nearly so
  Styles fusiform and glandular; plants annual or biennial; leaflets not deeply cleft; petals shorter than sepals
    Basal leaves with 5 or more leaflets . . . . *P. rivalis*
    Basal leaves with 3 leaflets. . . . . *P. norvegica*
  Styles filiform, not glandular; plants perennial
    Basal leaves definitely pinnate into 5 or more leaflets . . . . *P. saxosa*
    Basal leaves palmate, 5-9-foliolate . . . . *P. gracilis*
Styles lateral or nearly basal, fusiform; leaves pinnate
  Petals broad, equaling or slightly exceeding petals. . . . . *P. glandulosa* ssp. *g.*
  Petals narrow, shorter than the sepals . . . . *P. glandulosa* ssp. *reflexa*

P. glandulosa Lindl. ssp. **glandulosa**. Sticky Cinquefoil. Occasional, on shaded slopes; foothill and montane, above 200m: e.g. Palomar, Hot Springs, Cuyamaca, Laguna, and El Cajon Mountains; Escondido; Barona Valley; Mussey Grade; Bratton Valley; San Miguel, Jamul, and Otay Mountains; Mission Gorge. May-July. n=7.

P. glandulosa ssp. **reflexa** (Greene)Keck. Greene's Cinquefoil. Occasional, on shaded slopes; montane, above 1200m: Doane Valley; Cuyamaca Lake; Banner Grade; Laguna Public Camp. May-June.

P. gracilis Dougl. ex Hook. ssp. **nuttallii** (Lehm.)Keck. [*P. parishii* Rydb., the type from Descanso.] Nuttall's Cinquefoil. Uncommon, in shaded areas of Montane Coniferous Forest; montane, above 1100m: Doane Valley; Cuyamaca Lake; Laguna Public Camp; Corte Madera. June-Aug. 2n=52-109.

P. norvegica L. ssp. **monspeliensis** (L.)Asch. & Graebn. [*P. n.* var. *hirsuta* (Michx.)Lehm.] Rough Cinquefoil. Local, on lake shores; cismontane: Cuyamaca Lake; Sweetwater Lake. June-Sep. n=28.

P. rivalis Nutt. var. **millegrana** (Engelm.)Wats. [*P. m.* Engelm.] Diffuse Cinquefoil. Rare, known in the county only at Mountain Springs, 700m. Apr.-Oct.

P. saxosa Lemmon ex Greene. Rock Cinquefoil. Uncommon, in rock crevices; south foothill, 600-1100m: Hipass; Boulevard; Bankhead Springs.

Prunus L. Stone-Fruits

Plants evergreen, with somewhat coriaceous or leathery leaves; flowers racemose . . . . *P. ilicifolia*
Plants with deciduous leaves
  Plants shrubby, many of branchlets spinose
    Leaves broadly ovate to rounded, the blades 1-2cm long; drupe puberulent. . . . . *P. fremontii*
    Leaves somewhat oblanceolate, obscurely if at all serrulate, mostly fascicled . . . . *P. fasciculata*
  Plants shrubby to arboreous, not having spinose branchlets
    Flowers 12 to many, in elongate racemes . . . . *P. virginiana*
    Flowers 1-10, not in well developed racemes . . . . *P. emarginata*

P. emarginata (Dougl.)Walp. Bitter Cherry. Occasional, in Montane Coniferous Forest; montane, above 1500m: Palomar and Cuyamaca Mountains. May-June.

P. fasciculata (Torr.)Gray. Desert Almond. Uncommon, in sandy areas; desert, 600-800m: San Felipe Valley; Earthquake Valley. Mar.-May.

P. fremontii Wats. [*P. f.* var. *pilulata* Jeps., the type from Wagon Wash near Sentenac Canyon.] Desert Apricot. Frequent, on rocky slopes; desert, below 1200m: e.g. Culp Valley; Borrego Palm Canyon; San Felipe Valley; Box Canyon; Mason Valley; Pinyon Mountain; Tierra Blanca Mountains; Jacumba; Mountain Springs Grade. Mar.-Apr. Type from Oriflamme Canyon.

P. ilicifolia (Nutt.)Walp. Holly-leaf Cherry. Common, on slopes and in canyons; foothill and montane, 200-1500m: e.g. De Luz; Oak Grove; Hot Springs Mountain; Mussey Grade; Oakzanita; Moosa Canyon; Poway Grade; Campo; Potrero; Jamul; Florida Canyon. Apr.-May.

P. virginiana L. var. **demissa** (Nutt.)Sarg. Western Choke Cherry. Frequent, in Montane Coniferous Forest; montane, above 1000m: e.g. Doane Valley; Mesa Grande; Hot Springs, Volcan, Cuyamaca, and Laguna Mountains. May-June.

Purshia DC. ex Poir. Antelope Bush

P. glandulosa Curran. Mojave Antelope Bush. Uncommon, on rocky slopes and alluvial areas; desert, 700-1100m: Dubber; Jacumba; Mountain Springs. Apr.-May.

Rosa L. Rose

Prickles mostly slender, straight
  Stems and petioles bristling with straight prickles and spines . . . . *R. minutifolia*
  Stems not so heavily armed . . . . *R. gymnocarpa*
Prickles stout, flattened, recurved . . . . *R. californica*

R. californica Cham. & Schlecht. [*R. aldersonii* Greene, the type from Witch Creek.] California Rose. Common, along streams and other moist places; cismontane: e.g. San Luis Rey Valley; Palomar, Volcan, Hot Springs, Cuyamaca, and Laguna Mountains; Buckman Spring; San Felipe Valley; Live Oak Springs; Barrett Junction; Sweetwater Valley. May-Aug. n=14.

R. gymnocarpa Nutt. ex T. & G. Baldhip Rose. Occasional, in shaded places; montane, above 1000m: Harrison Grade; Castro Canyon; Doane Valley. May-June. n=7.

R. minutifolia Engelm. in Parry. Desert Rose. Rare, in Sage Scrub; apparently at its northern distributional limit, western Otay Mesa, 170m.

Rubus L. Blackberry; Raspberry

Leaves simple, palmately lobed; stems unarmed, erect, with peeling or flaky bark; styles club-shaped . . *R. parviflorus*
Leaves mostly 3-5-foliolate; stems mostly prickly; styles filiform
  Drupelets adhering to the fleshy receptacle and falling with it or separately (Blackberries, etc.)
    Flowers borne in small clusters; prickles slender, somewhat setose . . . . *R. ursinus*
    Flowers borne in large terminal panicles; prickles stout, broad-based . . . . *R. laciniatus*
  Drupelets forming a hollow cone which separates from the dry receptacle as a single aggregate fruit (Raspberries)
    Sepals deflexed at anthesis; sepals and pedicels glandless or nearly so . . . . *R. leucodermis*
    Sepals not deflexed; pedicels and sepals with stipitate glands . . . . *R. glaucifolius*

R. glaucifolius Kell. ssp. **ganderi** (Bailey)Beauchamp. Gander's Raspberry. Endemic to San Diego County. Uncommon, in Montane Coniferous Forest; montane, above 1200m: Palomar Mountain; Nellie; around Cuyamaca Lake; Greene Valley; Cuyamaca Mountains; north slope of Middle Peak. June. Type from North Peak, Cuyamaca Mountains.

***R. laciniatus** Willd. Evergreen Blackberry. Native of Europe. Local, naturalized at Sunday School Flat, Palomar Mountain, and at Julian. May-July.

R. leucodermis Dougl. ex T.& G. ssp **bernardinus** (Greene)Thorne. [*R. l.* Dougl.] San Bernardino Raspberry. Infrequent, in shaded areas; montane, above 1200m: Harrison Grade; Palomar Mountain; North Peak. July-Aug.

R. parviflorus Nutt. [*R. nutkanus* var. *velutinus* (H.& A.)Brew.; *R. p.* forma *villosus* Fassett, the type from near head of Nate Harrison Grade, Palomar Mountain.] Thimbleberry. Occasional, along streams; montane, above 1350m: Doane Valley; around Cuyamaca Lake; Paso Picacho; Middle Peak. Mar.-Aug. n=7.

R. ursinus Cham. & Schlecht. California Blackberry. Frequent, in moist areas; montane and foothill, 200-1200m: e.g. De Luz; Pala; Rainbow; Palomar Mountain; Lake Henshaw; Eagles Nest; Banner Grade; Witch Creek; Descanso; Ramona; San Miguel Mountain. Mar.-July. n=21, 28, 42.

Sanguisorba L. Burnet

Plants annual or biennial; leaflets incisely pinnatifid . . . . *S. annua*
Plants perennial; leaflets toothed . . . . *S. minor*

S. annua (Nutt. ex Hook.)T.& G. [*S. occidentalis* Nutt. ex T.& G.] Western Great Burnet. Uncommon, in the Cuyamaca Mountains; above 1350m: Cuyamaca Lake; North Peak. May-July.

***S. minor** Scop. Garden Burnet. Native of Eurasia. Infrequent, naturalized in disturbed places; cismontane: San Marcos Valley; Barona Valley; Julian; Descanso; Spring Valley. May-July. 2n=28, 56.

RUBIACEAE. Madder Family

KEY TO GENERA

Leaves in whorls mostly of 4 or more; plants herbs or low shrubs
  Flowers solitary or in cymes, pedicelled; corolla rotate . . . . *Galium*
  Flowers in involucrate heads; corolla funnelform . . . . *Sherardia*
Leaves opposite; low perennial herb of montane pine belt . . . . *Kelloggia*

Galium L. Bedstraw; Cleavers

Leaves 5-8 in a whorl . . . . *G. aparine*
Leaves mostly 4 in a whorl, occasionally 2 or 5
  Mature fruit glabrous or slightly pubescent, the hairs if present much shorter than the body of the fruit
    Leaves linear-subulate; plants often forming low dense mats, the stems densely leafy . . . . *G. andrewsii*
    Leaves lance-linear or broader; plants taller, more open
      Leaves generally rounded or apiculate, the terminal hair ephemeral; plant usually hanging in tangled masses from supporting shrubs . . . . *G. porrigens*
      Leaves gradually narrowed to a generally acute, frequently pungent apex, the terminal hair usually persistent . . . . *G. nuttallii*
  Mature fruits densely hispid with hairs usually almost as long as the body of the fruit
    Corolla-lobes glabrous or minutely pubescent
      Corolla hispid . . . . *G. angustifolium* ssp. *borregoense*
      Corolla usually glabrous . . . . *G. angustifolium* ssp. *a.*
    Corolla-lobes mostly hairy to hispid . . . . *G. stellatum*

G. andrewsii Gray ssp. **andrewsii**. Moss Bedstraw. Common, in shaded areas of Chaparral and Montane Coniferous Forest; foothill and montane, above 700m: e.g. Palomar Mountains; Montezuma View Point;

Volcan Mountain; Banner Grade; Hauser Creek; Cuyamaca and Laguna Mountains; Pine Valley; Corte Madera; Campo; Descanso; Carrizo Gorge. Apr.-June. $2n=22$.

G. angustifolium Nutt. ex T. & G. ssp. **angustifolium**. [*G. a.* var. *bernardinum* Hilend & Howell; *G. a.* var. *siccatum* (Wight)Hilend & Howell, the type from Del Mar.] Narrow-leaf Bedstraw. Common, on brushy slopes; cismontane: e.g. San Onofre Canyon; Rainbow; Palomar Mountain; Eagles Nest; Santa Ysabel; Oasis Spring; Cuyamaca Mountains; Pine Valley; Buckman Spring; Campo; Boulevard; Otay Mountain; Mount Soledad; Torrey Pines; Del Mar; Encinitas. Apr.-July. Type from San Diego. $2n=22$, 44.

G. angustifolium ssp. **borregoense** Demp. & Steb. Borrego Bedstraw. Endemic to San Diego County. Rare, known from the type locality, "dry base of slope, Palm Canyon, Borrego Valley"; Hellhole Canyon. May.

***G. aparine** L. Common Bedstraw. Native of Europe. Common, in vernally moist places; cismontane: e.g. Fallbrook; De Luz; Doane Valley; Eagles Nest; Clevenger Canyon; Chariot Canyon; Descanso; Jacumba; Dulzura; San Diego. Mar.-July. $2n=44$, 66.

G. nuttallii Gray ssp. **nuttallii**. [*Galium suffruticosum* Nutt. non H. & A.] Nuttall's Bedstraw. Frequent, in Coastal Sage Scrub; coastal and foothill, below 500m: e.g. San Mateo Creek; Lilac; Encinitas; Mount Soledad; Point Loma; Chollas Valley; Cowles Mountain; Dehesa; Lakeside; Mount Helix. Mar.-July. Type from San Diego. $2n=22$.

G. porrigens Demp. var. **porrigens**. [*G. nuttallii* ssp. *ovatifolium* (Demp.)Demp. & Steb.] Oval-leaf Bedstraw. Occasional, on shrubby slopes; cismontane, below 1500m: San Onofre Bluffs; Julian; Echo Dell; Japatul Valley; Lake Hodges; Del Mar; La Jolla; San Ysidro. Feb.-June.

G. stellatum ssp. **eremicum** (Hilend & Howell)Ehrend. [*G. stellatum* Kell.] Desert Bedstraw. Frequent, on rocky slopes; desert, below 750m: Borrego Valley; Yaqui Well; Sentenac Canyon; San Felipe Hill; Mason Valley; Mountain Palm Springs; Jacumba; Mountain Springs. Mar.-Apr. $2n=22$.

Kelloggia Torr.

K. galioides Torr. Rare, known in the county only along Chimney Creek, Palomar Mountains, 1400m. June.

Sherardia L. Field Madder

***S. arvensis** L. Blue Field Madder. Native of Europe. Occasional, weed in lawns and other cultivated areas: San Diego. Jan.-July. $2n=22$.

RUTACEAE. Citrus Family

KEY TO GENERA

Leaves simple
  Leaves opposite; ovary 1-loculed, forming a round berry . . . . . *Cneoridium*
  Leaves alternate; ovary 2-loculed, forming a deeply 2-lobed capsule . . . . . *Thamnosma*
Leaves pinnate; fruit a 4-5-lobed capsule . . . . . *Ruta*

Cneoridium Hook. f. in Benth. & Hook.

C. dumosum (Nutt.)Hook. f. Coast Spice Bush. Frequent, in Coastal Sage Scrub and Chaparral; coastal and foothill, below 700m: e.g. Vista; Camp Pendleton; Barona Valley; Alpine; Potrero Grade; Point Loma. Nov.-Mar. Type from "St. Diego". $2n=18$ pairs.

Ruta L. Rue

***R. chalepensis** L. Fringe Rue. Native of Old World. Rare, adventive along highway south of Bonsall, 25m. Feb.-June. $2n=36$.

Thamnosma Torr. & Frem. in Frem.

T. montana Torr. & Frem. Turpentine-Broom. Occasional, on rocky slopes and sandy flats; desert, below 1000m: San Felipe Valley; Sentenac Canyon; Volcanic Hills; south of Lava Flow Wash; Jacumba. Also north of Escondido. Mar.-July.

SALICACEAE. Willow Family

KEY TO GENERA

Scales of the catkins laciniate; flowers surrounded by a cup-shaped often oblique disc; stamens many; buds with numerous scales . . . . . *Populus*
Scales of the catkins entire; disc a minute glandlike structure; stamens 1, 2, or more; buds with a single scale . . . . *Salix*

Populus L. Poplar

Petioles terete; leaves dark green above, rusty or silvery beneath . . . . . *P. trichocarpa*
Petioles laterally compressed; leaves yellowish-green, alike on both surfaces
  Leaves and stems pubescent . . . . . *P. fremontii* var. *pubescens*
  Leaves and stems glabrous . . . . . *P. fremontii* var. *f.*

P. fremontii Wats. var. **fremontii**. Western Cottonwood. Common, along streams and other moist places; cismontane and desert: e.g. De Luz; San Onofre; Palomar Mountain; Oak Grove; San Felipe; Mason Valley; Banner; Mussey Grade; San Dieguito River; Sloane Ranch; Marron Valley; Jamacha; Dehesa; Mission Valley; Sweetwater Valley. Mar.-Apr.

P. fremontii var. **pubescens** Sarg. Arizona Cottonwood. Scarce, known in the county only at Barrett. Mar.

P. trichocarpa T.& G. ssp. **trichocarpa**. Black Cottonwood. Occasional, along streams; north foothill, 200-1100m: De Luz; Sandia Creek; Bonsall; Pala; Cutca Valley; Vista; Bear Valley. Also Foster. Feb.-Apr. $2n=28$.

Salix L. Willow

Leaf-blades entire or nearly so
  Leaves glabrous beneath, except possibly when very young
    Catkin-scales yellow; stamens 3-9; trees . . . . . *S. gooddingii* var. *variabilis*
    Catkin-scales dark; stamens 2; tree or shrub
      Leaves subrevolute, oblanceolate . . . . . *S. lasiolepis* var. *l.*
      Leaves plane, linear to lanceolate . . . . . *S. lasiolepis* var. *bracelinae*
  Leaves definitely pubescent or hairy beneath, even when quite mature
    Leaves linear or lance-linear, less than 1cm wide
      Stamen-filaments pubescent near the base; catkin-scales yellow
        Stigmas about 1mm long, on styles about 0.5mm long; capsule hairy . . . . . *S. hindsiana* var. *h.*
        Stigmas about 0.5mm long, sessile; capsule glabrous or nearly so, at least in maturity. . . . . *S. exigua*
    Leaves oblanceolate or obovate, mostly more than 1cm wide . . . . . *S. lasiolepis* var. *l.*
Leaf-blades distinctly serrulate
  Leaf-blades glabrous beneath except sometimes when very young
    Leaf-blades 0.6-1.5cm wide; capsule 5.5-7mm long; stamens 4-5; leaves gray-green. . *S. gooddingii* var. *variabilis*
    Leaf-blades 1.5-4cm wide
      Petioles glandular above. . . . . *S. lasiandra*
      Petioles not glandular
        Stems glabrous . . . . . *S. laevigata* var. *l.*
        Stems and petioles pilose . . . . . *S. laevigata* var. *araquipa*
  Leaves rather permanently pubescent beneath
    Trees or large shrubs mostly 3-15m high
      Stamens 4-6; leaves 1.5-3cm wide, 5-12cm long; style and stigmas minute . . . . . *S. laevigata* var. *araquipa*
      Stamens 2; leaves 0.2-1.4cm wide, 4-8cm long; styles to about 0.5mm long; catkin-scales yellow
        Leaves silvery-gray, 2-3.5mm wide . . . . . *S. hindsiana* var. *parishiana*
        Leaves gray-tomentose, to 1cm wide . . . . . *S. hindsiana* var. *leucodendroides*
    Shrubs mostly not over 3m tall; catkin-scales yellowish; leaves linear . . . . . *S. exigua*

S. exigua Nutt. Slender Willow. Common, along streams; desert and south foothill, below 900m: e.g. Lone Palm; Borrego Palm Canyon; Thousand Palm Canyon; San Felipe; Sentenac Canyon; Santa Ysabel Creek; Jacumba; Canyon City; Jamacha. Also Santa Margarita River; San Diego River; Tijuana River. Mar.-May. $2n=38$.

S. gooddingii Ball var. **variabilis** Ball. Southwestern Willow. Occasional, along streams; coastal, below 300m: Moosa Canyon at Hwy. 395; San Pasqual; Lakeside; San Diego River; Tijuana River. Also Scissors Crossing. Mar.-Apr. $2n=19$ pairs.

S. hindsiana Benth. var. **hindsiana**. Valley Willow. Uncommon, along streams; desert and coastal, below 500m: Jamacha Junction; Borrego Palm Canyon. Feb.-Apr.

S. hindsiana var. **leucodendroides** (Rowlee)Ball. Sandbar Willow. Frequent, along streams; coastal, below 300m: San Luis Rey; Lilac; Moosa Canyon; El Capitan; Olivenhain; Lakeside; Old Town; Sweetwater Valley; Otay. Mar.-Apr.

S. hindsiana var. **parishiana** (Rowlee)Ball. Parish's Willow. Uncommon, along streams; coastal, below 200m: Sweetwater Valley. Mar.-May.

S. laevigata Bebb var. **araquipa** (Jeps.)Ball. Large-leaf Willow. Common, along streams; coastal and foothill, below 1100m: e.g. Rincon; Lake Henshaw; Cuyamaca Lake; Spencer Valley; Oriflamme Canyon; Ballena; Alpine; Dulzura; La Mesa; Sloane Ranch; Jamacha Junction; Canyon City; Campo; Boulevard. Feb.-May.

S. laevigata var. **laevigata**. Red Willow. Occasional, along streams; cismontane, below 1400m: Bonsall; Cuyamaca Lake; Mason Valley; Campo; Carrizo Gorge; Potrero Grade. Mar.-May.

S. lasiandra Benth. var. **lancifolia** (Anderss.)Bebb. Lance-leaf Willow. Uncommon, along streams; coastal and foothill, below 1200m: San Mateo Creek; Eagles Nest; Murphy Canyon; Sweetwater Valley. Mar.-May.

S. lasiolepis Benth. var. **bracelinae** Ball. Bracelin's Willow. Frequent, along streams; cismontane, below 1300m: e.g. San Mateo Creek; San Onofre Bluffs; Mission San Luis Rey; Lake Hodges; Solana Beach; Point Loma; Rolando; Alvarado Canyon; East San Diego; Silver Strand; Julian; Potrero Grade. Feb.-Apr.

S. lasiolepis var. **lasiolepis**. Arroyo Willow. Common, along streams; foothill, 200-1400m: e.g. Fallbrook; San Luis Rey; Lake Henshaw; Mesa Grande; Santa Ysabel; Julian; Spencer Valley; Pine Valley; Buckman Spring; Ramona; Poway; Kearny Mesa; Sweetwater Valley. Feb.-Apr. $2n=38$ pairs.

SAURURACEAE. Lizard-Tail Family

Anemopsis Hook.

A. californica (Nutt.)H.& A. Yerba Mansa. Common, in wet and brackish areas; cismontane and desert: e.g. San Onofre; O'Neill Lake; Del Mar; Sweetwater Valley; Silver Strand; Oak Grove; Cuyamaca Lake; Hot Springs Mountain; Cottonwood Valley; Sentenac Canyon; Vallecito Stage Station; Boulevard. Mar.-Sep. $2n=22$ pairs.

SAXIFRAGACEAE. Saxifrage Family

KEY TO GENERA

Flowers appearing before the leaves; petals spatulate; stems from cormlike root . . . . *Jepsonia*
Flowers normally appearing with or after the leaves
  Fertile stamens 10
    Styles 3; petals clawed, usually laciniate or toothed; rootstocks usually slender, tuberous . . . . *Lithophragma*
    Styles 2; petals clawed or sessile, entire; rootstocks various . . . . *Saxifraga*
  Fertile stamens 5 or 3
    Ovary 2-celled with placentae axile; seeds minutely punctate. . . . . *Boykinia*
    Ovary 1-celled with placentae parietal; seeds echinate with minute spines . . . . *Heuchera*

Boykinia Nutt.

B. rotundifolia Parry. Uncommon, cool moist sites, montane, above 1500m: Agua Tibia Wilderness; Volcan Mountain.

Heuchera L. Alum-Root

Stamens long exserted, surpassing the sepals and about equal to petals; free portion of the floral-tube less than 1mm long; petals very narrow, filament-like; styles long exserted . . . . *H. leptomeria*
Stamens mostly included, shorter than petals; free portion of floral-tube more than 1mm long on its shorter side; petals distinctly wider than filaments; styles scarcely exserted . . . . *H. brevistaminea*

H. brevistaminea Wiggins. Laguna Alum-Root. Endemic to San Diego County. Scarce, on rocky cliffs; montane, above 1300m: Potrero Canyon; Garnet Peak; Foster Point; Shrine Camp; Oasis Spring. May-July. Type from shallow soil on a rocky canyon slope a half-mile below Vallecitos View, Laguna Mountains.

H. leptomeria Greene var. **peninsularis** Rosend., Butt. and Lak. [*H. rubescens* var. *glandulosa* auct. non Kell.] Peninsular Alum-Root. Infrequent, on moist, shaded slopes; montane, above 1500m: Hot Springs and Cuyamaca Mountains. May-June.

Jepsonia Small

J. parryi (Torr.)Small. Coast Jepsonia. Common, on mesas and shaded slopes; coastal and foothill, below 1200m: e.g. De Luz; San Onofre Mountains; Moosa Canyon; San Diego; Boulder Creek; Descanso Junction; Morena Dam; Tecate Peak; San Ysidro. Oct.-Jan. Type from dry hills near San Diego and San Luis Rey. n=7.

Lithophragma Nutt. Woodland-Star

Basal leaves not lobed to near their bases
  Petals entire or shallowly toothed . . . . *L. cymbalaria*
  Petals deeply parted. . . . . *L. affine*
Basal leaves lobed almost to their bases
  Petals mostly 3-5mm long; floral-tube somewhat rounded at base . . . . *L. tenellum*
  Petals mostly 6-10mm long; floral-tube somewhat acute at base . . . . *L. parviflorum*

L. affine Gray ssp. **mixtum** R.L. Taylor. [*L. a.* Gray; *L. trifida* Eastw. ex Rydb., the type from near Sweetwater Dam.] Woodland-Star. Occasional, on shaded slopes; cismontane: e.g. Doane Valley; Oak Grove; Ysidora; La Costa; Julian; Mount Laguna; Barona Valley; Mount Soledad; Mission Gorge; Dulzura; Lower Otay Lake. Mar.-June. 2n=28.

L. cymbalaria T.& G. Mission Star. Rare, known in the county only along Oak Grove Trail, Palomar Mountain, above 1200m. Mar.-May. 2n=14 + 1.

L. parviflorum (Hook.)T.& G. [*L. tenella* auct. non Nutt.] Prairie Star. Uncommon, on shaded slopes about Cuyamaca Lake, 1400m. May. 2n=21.

L. tenellum Nutt. in T.& G. Rare, known in the county only from Corte Madera Ranch, 1200m. 2n=14,35.

Saxifraga L. Saxifrage

S. californica Greene. California Saxifrage. Occasional, on shaded slopes; coastal and foothill, below 1100m: e.g. Moosa Canyon; Del Dios; San Diego; Descanso; Potrero; Jamul; Otay Mountain. Mar.-May.

SCROPHULARIACEAE. Figwort Family

KEY TO GENERA

Fertile stamens 5; leaves alternate; corolla nearly regular, rotate; tall introduced weedy plants . . . . *Verbascum*
Fertile stamens 4 or 2; leaves opposite, corolla usually somewhat labiate
  Plants acaulescent; corolla nearly rotate, 1.5mm long; anther-cells wholly confluent . . . . *Limosella*
  Plants mostly caulescent; corolla usually not rotate, larger; anther-cells distinct
    Stigmas distinct, flattened or platelike
      Sepals nearly or quite distinct; flowers blue-violet . . . . *Stemodia*
      Calyx of united sepals, usually angled; corolla usually purple or red to yellow or buff
        Suffrutescent perennial plants or with woody bases . . . . *Diplacus*
        Herbaceous perennial or annual plants . . . . *Mimulus*
    Stigmas wholly united, capitate or punctiform
      Corolla spurred or saccate on lower side of base, usually also with prominent palate; capsule opening by pores or irregularly; leaves usually alternate
        Fertile stamens 2; corolla 15-30mm long; desert annuals . . . . *Mohavea*

Fertile stamens 4; corolla smaller if from desert, and plants annuals
Corolla gibbous or saccate at base . . . . *Antirrhinum*
Corolla with narrow spur at base
Flowers in terminal spikes or racemes; stems somewhat erect at least in upper part; leaves linear to lanceolate or ovate . . . . *Linaria*
Flowers in axils of broad leaves; stems trailing or twining; flowers white or yellowish-white . . . *Kickxia*
Corolla not spurred or saccate on lower side at base, 2-lipped or nearly regular, without prominent palate; capsule valvate
Corolla with upper lip flattened or widely arched, not forming a galea
Upper lip of corolla 2-lobed
Plants annual; corolla gibbous on upper side at base, the lower lip appearing 2-lobed, the middle lobe keel-shaped and concealed by the lateral lobes . . . . *Collinsia*
Plants perennial; corolla not gibbous on upper side
Corolla somewhat brownish or darker, mostly less than 1cm long, inflated, with 4 erect lobes and 1 reflexed; sterile stamens reduced to a scale or wanting . . . . *Scrophularia*
Corolla mostly blue, lavender, rose, white, rarely brownish or yellow, 10-40mm long, tubular or funnelform, somewhat bilabiate; sterile stamen usually about as long as the fertile ones
Nectary a hypogynous disc; plant a definite shrub; inflorescence cymose . . . . *Keckiella*
Nectary of glandular epistaminal hairs; plant mostly somewhat herbaceous; inflorescence racemose to cymose . . . . *Penstemon*
Upper lip of corolla appearing 1-lobed by fusion of 2 flowers; small, mostly bluish . . . . *Veronica*
Corolla with upper lip narrowly arched, forming a galea which encloses the anthers
Anther-cells equal in size and position; seed-coat not loose and reticulate; perennials . . . . *Pedicularis*
Anther-cells unequally placed, the upper attached by its middle, the lower normally attached or abortive; seed-coat mostly loose and reticulate
Calyx-tube surrounding the lower portion of the corolla, its 4 (or 2 by fusion) lobes placed laterally
Lower lip of corolla about as long as and larger than galea; plants annual
Plants pubescent to glandular . . . . *Orthocarpus*
Plants extremely viscid, often with ensnared insects . . . . *Parentucellia*
Lower lip of corolla definitely shorter and smaller than galea; plants mostly perennial . . . . . *Castilleja*
Calyx-tube surrounding only base of corolla, or mostly entirely to the dorsal side and consisting of a bractlike structure, entire or minutely bidentate at apex . . . . *Cordylanthus*

Antirrhinum L. Snapdragon

Capsule somewhat oblique, dehiscing by fairly definite terminal or subterminal pores; stems self-supporting or supporting themselves by tortile branchlets
Plant simple below, erect, glabrous below the glandular-villous minutely bracted spicate raceme; flowers whitish with the lower lip much enlarged . . . . *A. coulterianum*
Plant usually branched below and somewhat pubescent along the stem; inflorescence lax or fairly dense but not set off sharply by its pubescence and leaflessness from the upper stem . . . . *A. nuttallianum*
Capsule not oblique, dehiscing by irregular bursting; stems twining and supported by the long capillary pedicels
Flowers yellow, 11-13mm long; stems very slender; deserts . . . . *A. filipes*
Flowers blue, 13-15mm long; stems fairly stout at base; cismontane . . . . *A. kelloggii*

A. coulterianum Benth. in A. DC. [*A. c.* ssp. *orcuttianum* (Gray)Pennell, the type from near San Diego] Frequent, on grassy slopes and burns; cismontane and desert, below 1500m: e.g. Palomar Mountain; Chihuahua Valley; Bear Valley; Escondido; Eagles Nest; San Felipe Valley; upper Mason Valley; Laguna Mountains; Pine Valley; Ramona; Campo; Jacumba; Lake Hodges; Chollas Valley. Apr.-July.

A. filipes Gray. Desert Snapdragon. Occasional, on slopes and in washes; desert, below 700m: northeast corner of county; Borrego Palm Canyon; Agua Caliente; Canebrake Canyon; Rockhouse Canyon; Yaqui Well; Vallecito Stage Station; south of Mortero Wash; Plum Canyon; Mountain Springs. Jan.-Apr.

A. kelloggii Greene. [*A. hookerianum* Pennell ex Millsp.] Climbing Snapdragon. Occasional, on brushy slopes and burns; coastal and foothill, below 900m: San Onofre; Lilac; Pala; El Cajon Mountain; Lake Hodges; Kearny Mesa; Harbison Canyon; Alpine; Sweetwater; Dulzura; Tecate Peak. Mar.-May.

A. nuttallianum Benth. in DC. Nuttall's Snapdragon. Frequent, on grassy slopes and burns; coastal and foothill, below 1100m: e.g. San Onofre; Pala; Fallbrook; Oak Grove; San Pasqual; Spencer Valley; El Cajon Mountain; Santa Ysabel; Barrett Dam; Pine Valley; Moreno Valley; Dulzura; Carlsbad; Del Mar; Torrey Pines; Jamul Mountain; Campo; Jacumba. Feb.-Aug. Type from "S. Diego". 2*n*=16 pairs.

Castilleja Mutis ex L. f. Paint-Brush

Plants annual; leaves and bracts entire, linear-lanceolate . . . . *C. stenantha*
Plants perennial; leaves often and upper bracts usually divided
Calyx lobes united much further dorsally than ventrally and their tips usually curved upward; corolla usually curved forward through lower calyx sinus . . . . *C. jepsonii*
Calyx lobes not united much further dorsally than ventrally, the tips straight
Herbage with simple (not branched) hairs, green to grayish-green
Plant evidently glandular pubescent below the inflorescence . . . . *C. martinii*
Plant not evidently glandular-pubescent below the inflorescence
Main cauline leaves entire
Herbage glabrous or inconspicuously pubescent; leaves mostly 2-5cm long . . . . *C. miniata*
Herbage stiff-pubescent; leaves 3-9cm long . . . . *C. affinis*
Main cauline leaves mostly with 1 or 2 pairs of lobes . . . . *C. affinis*
Herbage with branched hairs . . . . *C. foliolosa*

C. affinis H.& A. ssp. **affinis**. [*C. douglasii* Benth.] Coast Paint-Brush. Common, on shaded slopes; cismontane: e.g. Santa Margarita River estuary; Red Mountain; Palomar Mountain; Black Mountain;

Cuyamaca Lake; Laguna Mountains; Torrey Pines; San Pasqual; San Diego; Mount Soledad; Telegraph Canyon; San Miguel Mountain; Jamul Mountain; Pine Valley; Boulevard. Jan.-July.

C. foliolosa H.& A. Felt Paint-Brush. Frequent, in dry, rocky places; foothill, montane and desert: e.g. upper San Onofre Canyon; Pala; Palomar Mountain; Montezuma View Point; Eagles Nest; Moosa Canyon; Oriflamme Canyon; Grapevine Canyon; Julian; Laguna Junction; Echo Dell; El Cajon Mountain; Lyons Peak; Potrero Grade; Tecate Peak; Mountain Springs; Table Mountain; Jacumba. Also Mission Gorge; Torrey Pines; Ocean Beach. Feb.-June. *n*=12.

C. jepsonii Bacig. & Heckard. Jepson's Paint-Brush. Uncommon, in Chaparral; foothill, 600-1300m: west of Potrero; Pine Valley; Cameron Corners. May-June. *n*=12.

C. martinii Abrams. Martin's Paint-Brush. Frequent, in shaded glens; montane, above 1000m: e.g. Palomar and Hot Springs Mountains; Castro Canyon; Mesa Grande; Pine Hills; Banner Grade; Cuyamaca and Laguna Mountains. Apr.-June. 2*n*=24, 48, 72.

C. miniata Dougl. ex Hook. [*C. oblongifolia* Gray, the type from southern borders of San Diego County; *C. montana* Congd.] Giant Red Paint-Brush. Occasional, in meadows; montane, above 1300m: Cuyamaca and Laguna Mountains. June-Oct. *n*=24.

C. stenantha Gray. California Thread-torch. Occasional, in moist places; foothill, montane, and desert, above 200m: Palomar Mountain; Lost Valley; Ballena; Escondido; Descanso; Julian; Cuyamaca Mountains; Pine Valley; Chariot Creek; Borrego Palm Canyon; Barona Valley; Japatul Valley; Tecate Peak; Campo; Jacumba; Bow Willow Palms. Mar.-Oct. *n*=12.

Collinsia Nutt.

Flowers congested in whorls, the pedicels mostly shorter than calyces in the lower whorls
  Upper pair of filaments with distinct basal appendages 1-2mm long; upper lip of corolla usually distinctly paler than lower
    Stems subglabrous, sometimes glandular . . . . *C. heterophylla* var. *h.*
    Stems pubescent . . . . *C. heterophylla* var. *austromontana*
  Upper pair of filaments without or with very rudimentary basal appendages . . . . *C. concolor*
Flowers pedicelled, solitary or in whorls, the lower pedicels as long as or longer than calyces
  Inflorescence not glandular-pubescent, or if so, the glands minute and scarcely thicker than the supporting hairs; calyx-lobes exceeding capsule; seeds usually 3 or more to a locule . . . . *C. parviflora*
  Inflorescence glandular-pubescent; calyx-lobes scarcely or not longer than capsule; seeds mostly 1-2 (-3) to a locule . . . . *C. childii*

C. childii Parry ex Gray. Child's Blue-eyed Mary. Occasional, in open grassy areas of Chaparral and Montane Coniferous Forest; montane, above 900m: Fry Creek; Doane Valley; Hot Springs Mountain; Cuyamaca Mountains; Descanso Creek. May-June. *n*=7.

C. concolor Greene. Southern Chinese Houses. Common, on grassy slopes and meadows; foothill and montane, above 700m: e.g. Palomar Mountain; Chihuahua Valley; Hot Springs Mountain; Culp Valley; Cuyamaca and Laguna Mountains; Witch Creek; Boulevard; Campo; Buckman Spring. Apr.-July. Type presumably from the southern part of San Diego County. *n*=7.

C. heterophylla Buist ex Grah. var. **austromontana** (Newsom)Munz. Downy-leaf Chinese Houses. Infrequent, on grassy slopes; foothill, 600-1500m: northeast of Escondido; Julian; Pine Hills; Mussey Grade. Apr.-June.

C. heterophylla var. **heterophylla**. [*C. bicolor* Benth. non Raf.] Purple Chinese Houses. Common, on shaded slopes; coastal and foothill, below 1500m: e.g. Rainbow Valley; Castro Canyon; Lilac; Valley Center; Volcan Mountain; Santa Ysabel; Green Valley; Escondido; Poway; La Jolla; Barrett Junction; Harbison Canyon; Otay Mountain; San Diego. Mar.-July. *n*=7.

C. parviflora Dougl. ex Lindl. Small-flower Blue-eyed Mary. Frequent, in moist, shaded places; foothill and montane, above 800m: Cuyamaca Lake; Santa Ysabel; Pine Valley; Cuyamaca and Laguna Mountains. Apr.-May. *n*=7.

Cordylanthus Nutt. ex Benth. in DC.

Leaves oblong to oblanceolate, at least the lower entire; flowers in an elongate spike; calyx spathelike, enclosing proximal part of corolla
  Corolla shorter than calyx . . . . *C. maritimus* ssp. *m.*
  Corolla longer than calyx . . . . *C. maritimus* ssp. *canescens*
Leaves or their segments filiform to linear; flowers in heads or scattered; calyx proper narrow, enclosing corolla at base only (the bract below the flower often confused with the calyx)
  Corolla less than to about twice as long as wide, inverted with ventral side up; flowering bracts setose . . . *C. nevinii*
  Corolla more than twice as long as wide, the dorsal side up; flowering bracts with hairs like those of other leaves of inflorescence
    Leaf-blades mostly 3-lobed, the lateral lobes nearly the same length and width as the central one; anthers 2-celled, all cells essentially alike . . . . *C. filifolius*
    Leaf-blades 5- to 11-lobed, the lobes about 1mm wide or less, the central rachis 3-3.5mm wide; lower anther cell functional, the upper markedly smaller or lacking . . . . *C. orcuttianus*

C. filifolius Nutt. ex Benth. in DC. [*Adenostegia f.* (Nutt.) Abrams] Dark-tip Bird's Beak. Common, in disturbed and open areas; cismontane: e.g. De Luz; Pala; Palomar Mountain; Oak Grove; Warner Hot Springs; Hot Springs Mountain; Cuyamaca and Laguna Mountains; Campo; Potrero; Otay Mountain; Poway; La Jolla; Encinitas. Apr.-Sep. Type from San Diego.

C. maritimus Nutt. ex Benth. in DC. ssp. **canescens** (Gray)Chuang & Heckard. Alkali Bird's Beak. Rare, known in the county only from a collection from Oak Grove, 840m. May-July. *n*=15.

C. maritimus ssp. **maritimus**. Salt Marsh Bird's Beak. Rare, in Coastal Salt Marsh; coastal, below 20m: San Diego; Chula Vista; Imperial Beach, in the Tijuana River estuary. Mar.-Oct. Type from San Diego. *n*=15.

C. nevinii Gray. [*Adenostegia n.* (Gray)Greene] Nevin's Bird's Beak. Rare, known in the county from a collection from Cuyamaca Peak and San Diego State Observatory Site on Mount Laguna, above 1400m. July-Sep.

C. orcuttianus Gray. [*Dicranostegia o.* (Gray)Pennell; *Adenostegia o.* (Gray)Greene] Orcutt's Bird's Beak. Rare, on sunny slopes, coastal, 10-30m: Rice Canyon; Tijuana Hills; Otay Mesa; Otay Valley at I-805; east of Greg Rogers Park. Mar.-Sep. *n*=16.

Diplacus Nutt.

Plants woody at base only; leaves finely glandular-pubescent; calyx usually swollen at base . . . . *D. clevelandii*
Plants shrubby
  Corollas orange, copper, yellow, buff, cream or almost white, never true red
    Calyx, lower leaf-surfaces, and upper branchlets glabrous or microscopically glandular-puberulent
      Leaves slightly to distinctly impressed-veiny above; corolla-tube not usually exserted from calyx *D. aurantiacus*
      Leaves not impressed-veiny above; corolla-tube exserted from calyx, the corolla light lemon-yellow . *D. aridus*
    Calyx, lower leaf-surfaces, and upper branchlets obviously pubescent to almost woolly
      Corolla light lemon-yellow . . . . *D. longiflorus* var. *calycinus*
      Corolla orange-yellow to deep orange, buff or white . . . . *D. longiflorus* var. *l.*
  Corollas red or reddish
    Calyx glandular-hairy; pedicels mostly less than 10mm long . . . . *D. longiflorus* var. *rutilus*
    Calyx glabrous or nearly so; pedicels mostly 10-25mm long . . . . *D. puniceus*

D. aridus Abrams. [*Mimulus a.* (Abrams)Grant] Jacumba Monkey Flower. Local, among large rocks in Chaparral; south foothill, 750-1100m: San Ysidro Mountains; Ranchita; Canebrake Canyon; Mountain Springs; Grapevine Mountain; Bankhead Springs; Jacumba; Dubber; Hipass; Mortero Palms; desert slopes of Laguna Mountains. Mar.-July. Type from dry rocky ridges at Jacumba, near the boundary monument. *n*=10.

D. aurantiacus (Curt.)Jeps. ssp. **australis** (McMinn)R.M.Beeks ex Thorne. [*D. australis* McMinn; *Mimulus aurantiacus* ssp. *a.* (McMinn)Munz; *M. longiflorus* var. *linearis* (Benth.)McMinn] San Diego Monkey Flower. Frequent, on dry slopes; cismontane: e.g. Ysidora; Castro Canyon; Palomar Mountain; Oak Grove; Julian; Ramona; San Pasqual Grade; Flinn Springs; Descanso; Dulzura Grade; Campo. Apr.-Sep.

D. clevelandii (Bdg.)Greene. [*Mimulus c.* Bdg.] Cleveland Monkey Flower. Infrequent, in dry, rocky places; foothill and montane, above 1000m: Harrison Grade; Cuyamaca and Laguna Mountains; Agua Tibia Mountains; Pine Valley; Guatay Peak; Viejas Mountain; Descanso Grade; Lawson Peak; Otay Mountain; Tecate Peak. May-July. Type from "south side of Cuyamaca Peak...not far from the signal station on its summit". *n*=10.

D. longiflorus Nutt. var. **calycinus** (Eastw.)Jeps. [*Mimulus l.* var. *c.* (Eastw.) Grant] Buff Monkey Flower. Infrequent, Chaparral; foothill, 1000-1400m: Palomar Mountain. July. *n*=10.

D. longiflorus var. **longiflorus**. [*Mimulus l.* (Nutt.)Grant] Salmon Monkey Flower. Infrequent, on dry rocky slopes; foothill and montane, above 300m: e.g. Palomar Mountain; below Lake Henshaw; Chariot Canyon; Hot Springs Mountain; Borrego Palm Canyon; Descanso Grade. May-June.

D. longiflorus var. **rutilus** (Grant)McMinn. [*Mimulus l.* var. *r.* Grant] Velvet-red Monkey Flower. Scarce, known in the county only on Palomar Mountain, above 1200m. July.

D. puniceus Nutt. ex Taylor. [*Mimulus p.* (Nutt.)Steud.] Coast Monkey Flower. Frequent, in Coastal Sage Scrub and Chaparral; coastal and foothill, below 750m: e.g. Fallbrook; Cuca Grade; Lake Hodges; Poway; Moosa Canyon; Ramona; Torrey Pines; Mission Hills; San Diego; Jamul Valley. Mar.-July. Type from sandy loam by the borders of small winter streams...near St. Diego". *n*=10.

Keckiella Straw

Corolla whitish, yellowish or fulvous, not distinctly tubular
  Calyx canescent . . . . *K. antirrhinoides* ssp. *microphylla*
  Calyx puberulent . . . . *K. antirrhinoides* ssp. *a.*
Corolla red, distinctly tubular
  Leaves opposite; stems not glaucous . . . . *K. cordifolia*
  Leaves ternate; stems glaucous . . . . *K. ternata*

K. antirrhinoides (Benth.)Straw ssp. **antirrhinoides**. [*Penstemon a.* Benth.] Yellow Bush Penstemon. Frequent, on dry slopes; foothill, 300-1300m: e.g. Rainbow; Pala; Oak Grove; San Felipe Valley; Witch Creek; Moosa Canyon; San Pasqual; Ramona; Lake Hodges; Dulzura; Jamul; Campo; Dubber. Mar.-Aug.

K. antirrhinoides ssp. **microphylla** (Gray)Straw. [*Penstemon a.* var. *m.* (Gray)Munz & Jtn.] Desert Bush Penstemon. Occasional, on rocky slopes; desert, below 1000m: Borrego Palm Canyon; Vallecitos Canyon; upper Box Canyon; Jacumba. Apr.-May.

K. cordifolia (Benth.)Straw. [*Penstemon c.* Benth.] Climbing Bush Penstemon. Frequent, on shaded slopes; coastal and foothill, below 1200m: e.g. De Luz; Palomar Mountain; Castro Canyon; Ramona; Rincon Grade; Santa Ysabel; San Marcos; Bear Valley; Poway Grade; Flinn Springs; El Capitan, Jamul, and Otay Mountains. May-July. *n*=8.

K. ternata (Torr. ex Gray)Straw. [*Penstemon t.* Torr. ex Gray] Summer Bush Penstemon. Frequent, in Chaparral and Southern Oak Woodland, above 400m: e.g. Palomar Mountain; Chihuahua Valley; Hot

Springs Mountain; Julian; Witch Creek; Vallecitos Canyon; Mount Laguna; Pine Valley; Descanso; Campo; Barrett Dam; Boulevard; Otay Mountain. May-Sep. Type from mountains east of San Diego. n=8.

Kickxia Dumort. Fluellin

***K. elatine** (L.)Dumort. Native of Europe. Uncommon weed in orchards, as at Canyon Creek, Rancho Santa Fe.

Limosella L. Mudwort

L. acaulis Ses. & Moc. [*L. aquatica* auct. non L.] Southern Mudwort. Infrequent, on muddy shorelines; montane, above 1200m: Pine Hills; Cuyamaca Lake; Laguna Lakes. May-Sep.

Linaria Mill. Toadflax

Flowers yellow
  Leaves linear; corolla (including the spur) 2-3cm long . . . . *L. vulgaris*
  Leaves ovate to lanceolate; corolla (including the spur) 3.5-4cm long . . . . *L. dalmatica*
Flowers not yellow
  Corolla violet with whitish ridges instead of well formed palate; native . . . . *L. canadensis*
  Corolla violet-purple with well-formed orange palate; occasional waif . . . . *L. bipartita*

***L. bipartita** Willd. Clovenlip Toadflax. Native of Europe. Rare, escaping from cultivation: San Diego. May-July.

L. canadensis (L.)Dum.-Cours. var. **texana** (Scheele)Penn. Large Blue Toadflax. Occasional, in disturbed areas and burns; coastal and foothill, below 1000m: San Luis Rey; Torrey Pines; Kearny Mesa; Silver Strand; San Diego; Potrero Grade; Japatul Valley; Barrett Dam; Tecate Peak. Mar.-May. $2n$=12.

***L. dalmatica** (L.)Mill. Dalmatia Toadflax. Native of Europe. Infrequent, waif along roads and in disturbed areas; coastal and foothill, below 1000m: Fallbrook; Vista. May-June. $2n$=12.

***L. vulgaris** Hill. Butter-and-Eggs. Native of Europe. Rare, adventive at Julian, 1200m. June-Aug. n=6.

Mimulus L. Monkey Flower

Plants perennial
  Corolla scarlet, the upper lobes arched; stamens exserted . . . . *M. cardinalis*
  Corolla yellow
    Mature calyx little or not much inflated, the lobes practically straight; corolla-throat open, the limb not conspicuously 2-lipped . . . . *M. moschatus*
    Mature calyx strongly inflated, the lower lobes curved against the others; corolla-throat nearly closed by the palate formed by the large ventral ridges, the limb strongly 2-lipped . . . . *M. guttatus*
Plants annual
  Pedicels conspicuous, normally longer than calyx; corolla mostly deciduous, dropping before shriveling and leaving the styles exposed
    Mature calyx strongly inflated, the lower lobes curving up; flowers yellow
      Uppermost calyx-lobe rarely more than twice the length of the others, the lower teeth in maturity only partly closing the orifice . . . . *M. guttatus*
      Uppermost calyx-lobe almost 3 times as long as others, the lower teeth in maturity folding up so as almost to close the orifice . . . . *M. nasutus*
    Mature calyx not strongly inflated or if somewhat so, cylindric, the lobes permanently straight or nearly so
      Flowers predominantly yellow, or sometimes partly white, sometimes with reddish spots
        Calyx not ridged or low-winged along the sepal midribs; capsule glandular-puberulent . . . . *M. pilosus*
        Calyx ridged or low-winged along the sepal midribs; capsule glabrous
          Calyx-lobes 1.5-3mm long; corolla 10-12mm long . . . . *M. latidens*
          Calyx-lobes 0.5-1mm long; corolla 4-15 (-20)mm long
            Calyx-lobes triangular-acute to -acuminate; stem and leaves strongly pubescent, the leaves rounded to cordate at base . . . . *M. floribundus*
            Calyx-lobes rounded, either bluntly so or mucronate . . . . *M. rubellus*
      Flowers predominantly purple, red or magenta
        Calyx-lobes mostly 1-2mm long, acute to subulate-tipped
          Plant minutely glandular-puberulent; leaves 1-2cm long; calyx inflated in fruit, cismontane . . . . *M. latidens*
          Plant densely glandular-villous; leaves 1.5-4cm long; calyx not much inflated in fruit . . . . *M. parishii*
        Calyx-lobes mostly about 0.5mm long, broader than long, obtuse to rounded
          Stigmas not fringed; basal tube or corolla scarcely or not exceeding calyx; corollas 6-10mm long *M. rubellus*
          Stigmas ciliate-fringed; basal tube or corolla exceeding calyx; corollas 10-15mm long . . . . *M. diffusus*
  Pedicels normally shorter than calyces; corolla mostly shriveling and persistent on the developing capsule
    Flowers yellow . . . . *M. brevipes*
    Flowers purple to rose or magenta
      Stigmas and longer stamens usually exserted from the corolla-throat . . . . *M. johnstonii*
      Stigmas and anthers included in corolla-throat
        Inner face of corolla-lobes glabrous; leaves somewhat lanceolate; corolla 2-2.5cm long; the lobes abruptly widely spreading . . . . *M. fremontii*
        Inner face of corolla-lobes pilose; leaves somewhat elliptic
          Lowest pair of calyx-lobes more than half as long as uppermost lobe; corolla 12-30mm long; bracts acuminate to cuspidate; deserts . . . . *M. bigelovii*
          Lowest pair of calyx-lobes not more than half as long as uppermost lobe; corolla 13-20mm long; bracts obtuse to acute . . . . *M. johnstonii*

M. bigelovii (Gray)Gray var. **bigelovii**. Bigelow's Monkey Flower. Frequent, on sandy flats; desert, below 1000m: e.g. Borrego Palm Canyon; Borrego Valley; Hellhole Canyon; San Felipe Canyon; Earthquake Valley; Vallecito Valley; Agua Caliente Hot Springs; Carrizo Creek; Dos Cabezas. Jan.-Apr.

M. brevipes Benth. Slope Semiphore. Occasional, on open slopes and burns; cismontane: e.g. San Onofre; Palomar Mountain; Hot Springs Mountain; San Felipe Valley; Hatfield Creek; Julian; Boulder Oaks; Jacumba; Descanso; Alvarado Canyon; Lee Valley; El Cajon Mountains; Pala; San Luis Rey; Lake Henshaw; Ramona; Pine Valley; San Diego; Point Loma; Tijuana Hills; Otay Mesa. Apr.-July. *n*=8.

M. cardinalis Dougl. ex Benth. Scarlet Monkey Flower. Occasional, along streams, seepages and other moist places; cismontane: e.g. San Onofre; Palomar Mountain; Hot Springs Mountain; San Felipe Valley; Julian; Hatfield Creek; Alpine; Boulder Oaks; Jacumba; Descanso; Jamul Valley; Lee Valley; Alvarado Canyon; Carrizo Wash near S-2. June-Sep. *n*=8.

M. diffusus Grant. [*M. grantianus* Eastw., the type from Campo...in sandy soil amid the brush.] Palomar Monkey Flower. Occasional, in Chaparral understory; cismontane: Fallbrook; Encinitas; Palomar Mountain; Morgan Hill; Cutca; Witch Creek; Banner; Santa Ysabel; Laguna Mountains; Pine Valley; Boulder Creek; Silverwood; Campo; Potrero; McCain Valley. Apr.-July. Type from Palomar.

M. floribundus Dougl. ex Lindl. var. **floribundus**. Showy Monkey Flower. Infrequent, on open slopes and burns; cismontane: Pala; Palomar Mountain; Dodge Valley; Laguna Mountains; Pine Valley; McCain Valley; Dulzura; Jamul Valley; San Onofre. Apr.-Aug.

M.fremontii (Benth.)Gray. Fremont's Monkey Flower. Occasional, on dry slopes; foothill and montane, above 600m: Palomar Mountain; Cuyamaca Mountains; Oak Grove; Hot Springs Mountain; Chihuahua Valley; San Felipe Valley; Pinyon Mountain; Ramona; El Cajon Mountain; Buckman Spring; Campo; Jacumba. Apr.-June.

M. guttatus Fisch. ex DC. ssp. **guttatus**. [*M. paniculatus* Greene, the type from Witch Creek.] Seep Monkey Flower. Common, in wet places; cismontane: e.g. Palomar Mountain; Camp Pendleton; Hot Springs Mountain; Banner Grade; Cuyamaca Lake; Laguna Mountains; Hatfield Creek; Jacumba; Otay Mountain; Carrizo Wash near S-2. June-Aug. *n*=14, 28, 31.

M. johnstonii Grant. Rare, known in the county from a collection at Descanso. May-July

M. latidens (Gray)Greene. Broad-tooth Monkey Flower. Rare, in vernal pool sites; coastal, below300m. Chollas Valley; Otay Mesa. Apr.-June. Type from Chollas Valley.

M. moschatus Dougl. ex Lindl. var. **longiflorus** Gray. Long-tube Musk-Flower. Scarce, in wet places on Palomar Mountain; montane, above 1300m: Doane Valley. June. *n*=16.

M. nasutus Greene. Snout Monkey Flower. Occasional, in moist and disturbed places; cismontane: e.g. San Mateo Creek; Pala; Warner Ranch; Chariot Canyon; Santa Ysabel; Cuyamaca Lake; Descanso Junction; Mussey Grade; Alpine; Dulzura; San Diego; La Jolla; McCain Valley. Apr.-Sep. *n*=14.

M. parishii Greene. Parish's Monkey Flower. Uncommon, in wet, sandy places; desert and foothill, 600-1200m: Dodge Valley; Oak Grove; Warner Ranch; Borrego Palm Canyon; Carrizo Wash near S-2; McCain Valley by Bow Willow Creek. June.

M. pilosus (Benth.)Wats. [*Mimetanthe p.* (Benth.)Greene] Downy Monkey Flower. Occasional, in sandy places; cismontane: Palomar Mountain; Lake Henshaw; Warner Ranch; San Felipe Valley; Hatfield Creek; Cuyamaca Peak; Mount Laguna; Buckman Spring; Jacumba; Descanso; Dulzura; Miramar; McCain Valley; Carrizo Wash near S-2. May-Sep.

M. rubellus Gray in Torr. Fringe-tooth Monkey Flower. Rare, in rocky areas; desert, 900-1200m: Vallecito Canyon. May.

Mohavea Gray

M. confertiflora (Benth. in A. DC.)Heller. Ghost Flower. Occasional, in sandy areas; desert, below 700m: Borrego Valley; Rockhouse Canyon; Borrego Palm Canyon; The Narrows; Yaqui Well; San Felipe; Vallecito Wash; Carrizo Creek; Dos Cabezas; Canyon Sin Nombre. Jan.-Apr. 2*n*=15 pairs.

Orthocarpus Nutt. Owl's-Clover

Anthers 1-celled; seed-coat tight-fitting or ridged; sacs of lower lip of corolla 3-4mm deep, yellow, the galea purple . . . . *O. erianthus*
Anthers 2-celled; lower lip of corolla rather shallow, the sacs mostly 1-2mm deep; seed-coat loose-fitting
  Bracts green throughout; galea equaling or barely exceeding lower lip
    Galea finely puberulent or pubescent; upper leaves 3-5-cleft . . . . *O. hispidus*
    Galea densely white-villous; upper leaves with 2 small lateral lobes . . . . *O. lasiorhynchus*
  Bracts tipped with purple or yellow; galea exceeding lower lip
    Corolla widened upward; spike usually broad and conspicuous
      Galea nearly straight, pubescent; lower leaves entire, the upper with a pair of lateral lobes . . . . . *O. densiflorus*
      Galea hooked at tip, densely bearded; leaves deeply pinnatifid into several filiform divisions
        Lower corolla lip mostly purple, white or yellowish only toward tip . . . . *O. purpurascens* var. *p.*
        Lower corolla lip white or pale yellow . . . . *O. purpurascens* var. *pallidus*
    Corolla linear, the lower lip shallow; spike narrow, pale and rather inconspicuous . . . . *O. attenuatus*

O. attenuatus Gray. Narrow-leaf Owl's-Clover. Local, in grassy areas; foothill and montane, above 1000m: Cuyamaca Lake; Rattlesnake Valley; Campo; Temecula Canyon; Descanso. May-June. *n*=12.

O. densiflorus Benth. var. **gracilis** (Benth.)Keck. [*O. parishii* Gray, the type from meadows near Stonewall Mine, in the Cuyamaca Mountains.] Parish's Owl's-Clover. Frequent, in grassy areas; cismontane: e.g. Palomar Mountain; Lake Henshaw; Banner Grade; Cuyamaca Lake; Campo; Otay Mountain; Torrey Pines State Reserve; Kearny Mesa; San Diego. Mar.-May. *n*=12.

O. erianthus Benth. Johnny-Tuck. Scarce, in grassy areas; cismontane: Foster; Boulder Creek. Mar.-May. *n*=11.

O. hispidus Benth. [*O. falcatus* Eastw., the type from Smith or Palomar Mountain, at an elevation of 1500-1800 meters.] Hairy Owl's-Clover. Rare, known in the county only on Palomar Mountain, above 1500m. June. *n*=12,24.

O. lasiorhynchus Gray. San Bernardino Owl's-Clover. Rare, wet meadow; montane: Cuyamaca Lake, 1400m. June-July. *n*=12.

O. purpurascens Benth. var. **pallidus** Keck. Pallid Owl's-Clover. Rare, on grassy slopes, cismontane, below 800m: Ysidora; Richman Ranch; De Luz; Pala. Mar. *n*=12.

O. purpurascens var. **purpurascens**. [*O. p.* var. *palmeri* Gray] Red Owl's-Clover. Common, in grassy areas; coastal and foothill, below 1400m: e.g. San Onofre Canyon; Pala; Lake Henshaw; Rincon; Cuyamaca Lake; Banner Grade; Potrero Grade; Dulzura; Otay Mountain; Kearny Mesa. Mar.-June. *n*=12.

Parentucellia Viviani

***P. viscosa** (L.)Caruel. Native of the Mediterranean region. Known from moist sand bank, Santa Ysabel Creek at Highway 78, San Pasqual Valley, 130m. May.

Pedicularis L. Lousewort

Corolla mostly purple-red, about 25mm long . . . . *P. densiflora*
Corolla pale yellow, 15-20mm long . . . . *P. semibarbata*

P. densiflora Benth. ex Hook. Indian Warrior. Occasional, in Chaparral understory and moist places; coastal and foothill, below 1200m: Camp Pendleton; Pala; Cutca Valley; Encinitas; Mount Soledad; Mount Woodson; El Cajon Mountain; Otay Mountain; Potrero Grade. Jan.-Apr.

P. semibarbata Gray. Dwarf Lousewort. Rare, known in the county only on North Peak, Cuyamaca Mountains, 1800m. July.

Penstemon Mitch. Beard-Tongue

Anther-sacs opening from free tips throughout or partially, almost always divaricate after dehiscence
  Corolla scarlet
    Herbage glaucous; anther-sacs dehiscent by a continuous slit extending across their contiguous parts; corolla subtubular; cismontane . . . . *P. centranthifolius*
    Herbage not glaucous; anther-sacs opening at their distal ends, the slits not confluent . . . . *P. labrosus*
  Corolla not scarlet, although it may be rose or carmine to blue, purple, etc.
    Leaves of lower and middle stems entire or obscurely denticulate
      Corolla 18-35mm long . . . . *P. × parishii*
      Corolla 7-18mm long, pink to lavender-rose . . . . *P. thurberi*
    Leaves of lower and middle stems definitely toothed, usually coarsely so
      Anther-sacs peltately explanate and glabrous; thyrsus virgate, secund; corolla pink to rose-purple . . . . *P. clevelandii* ssp. *c.*
      Anther-sacs not explanate, somewhat scabro-ciliate at suture
        Corolla pink to rose-purple, rotate . . . . *P. clevelandii* ssp. *connatus*
        Corolla purplish with blue limb, obviously 2-lipped . . . . *P. spectabilis*
Anther-sacs opening across their continuous apices, the free tips remaining saccate, parallel even after dehiscence, glabrous or somewhat hairy
  Corolla scarlet, tubular; limb relatively long, the upper lip erect, the lower sharply reflexed; montane . . . *P. bridgesii*
  Corolla bluish or purplish, usually ampliate; limb relatively short, with both lips somewhat spreading
    Plants puberulent throughout . . . . *P. heterophyllus* var. *australis*
    Plants glabrous or sometimes minutely puberulent . . . . *P. heterophyllus* var. *h.*

P. bridgesii Gray. Bridge's Penstemon. Infrequent, on open slopes; montane, above 1200m: Palomar Mountain; Little Laguna Lake; desert slopes of Laguna Mountains. July-Sep. *n*=8.

P. centranthifolius Benth. Scarlet Bugler. Frequent, in open, sandy places; cismontane: e.g. Palomar, Hot Springs, Cuyamaca, and Laguna Mountains; Ysidora; Pala; Olivenhain; Encinitas; Jacumba; Scissors Crossing; San Felipe Valley; Montezuma View Point; Lost Valley. Apr.-July. *n*=8.

P. clevelandii Gray ssp. **clevelandii**. Cleveland's Penstemon. Occasional, on rocky slopes; desert, below 1200m: Borrego Palm Canyon; Scissors Crossing; desert slopes of Laguna Mountains; McCain Valley; Sentenac Canyon; Rockhouse Canyon; Jacumba; Mortero Palms. Also Campo. Mar.-May. *n*=8.

P. clevelandii ssp. **connatus** (Munz & Jtn.)Keck. Blue-leaf Penstemon. Rare, in rocky areas; desert, 300-1000m: Granite Peak road cut; Campbell Ranch; Vallecito Canyon. Apr.-May.

P. heterophyllus var. **australis** Munz & Jtn. Chaparral Penstemon. Occasional, on open slopes; foothill and montane, above 1000m: Pala; Lusardi Truck Trail; Spencer Valley; Julian; Cuyamaca Lake; Pine Valley; Guatay. June-July.

P. heterophyllus var. **heterophyllus**. Foothill Penstemon. Occasional, on slopes in Montane Coniferous Forest; montane, above 1200m: Pine Hills; Cuyamaca Lake; Cuyamaca Peak. June-Sep. *n*=8, 16.

P. labrosus (Gray)Hook. f. San Gabriel Penstemon. Infrequent, in glens of Montane Coniferous Forest; montane, above 1300m: Palomar Mountain; Little Laguna Lake; Laguna Recreation Areas; Guymon Lodge. June-Sep. *n*=8.

P. × parishii Gray. (*P. centranthifolius* × *P. spectabilis*). Infrequent, in disturbed areas; foothill and montane, above 200m: Palomar Mountain; Laguna Rim Trail; Buckman Spring; Green Valley; Hipass; Lost Valley. Apr.-July. n=8.

P. spectabilis Thurb. ex Gray. Showy Penstemon. Frequent, in dry, sandy places; cismontane: e.g. De Luz; Pala; Palomar Mountain; Hot Springs Mountain; Rattlesnake Valley; Descanso; Campo; Hipass; Alpine; Poway; Jamul. Apr.-June. Type from San Pasqual. *n*=8.

P. thurberi Torr. ssp. **thurberi**. [*P. ambiguus* var. *t.* (Torr.)Gray] Thurber's Penstemon. Infrequent, in sandy areas; desert, below 800m: San Felipe Valley; Scissors Crossing; Palm Spring. May-June.

Scrophularia L. Figwort

S. californica Cham. & Schlecht. ssp. **floribunda** (Greene)Shaw. California Bee Plant. Common, in canyons and on shaded slopes; cismontane: e.g. Santa Margarita River estuary; De Luz; Pala Grade; Hot Springs Mountain; Cuyamaca Mountains; Borrego Palm Canyon; Mussey Grade; Witch Creek; Banner Grade; Laguna Junction; Dulzura; Alpine; Jamul; Mission Valley; Otay Mountain. Mar.-May. *n*=ca. 45, 46, 47, 48.

Stemodia L.

S. durantifolia (L.)Sw. Blue Steamwort. Infrequent, in reservoir and aquatic habitats; coastal, below 200m: Mission Gorge; San Diego; Sweetwater Lake; Otay Mountain; Cedar Canyon; Otay Lake. Jan.-Dec.

Verbascum L.

***V. thapsus** L. Common Mullein. Native of Eurasia. Occasional, naturalized in meadows; montane, above 900m: Palomar Mountain; Cuyamaca Mountains; Singing Hills; Batiquitos Lagoon; Dehesa; Descanso. June-Sep. 2*n*=24.

Veronica L. Speedwell

Main stem ending in a single racemelike inflorescence
  Leaves linear-oblong to spatulate; corolla white . . . . . *V. peregrina*
  Leaves rounded to oval; corolla bright blue . . . . . *V. arvensis*
Main stem with lateral racemes below the tip
  Pedicels in fruit ascending, 6-8mm long; calyx mostly exceeding the capsule . . . . . *V. anagallis-aquatica*
  Pedicels in fruit somewhat horizontal, 4-6mm long; calyx shorter than the capsule . . . . . *V. comosa*

***V. anagallis-aquatica** L. Native of Europe. Uncommon, in moist areas: De Luz; Palomar Mountain; Lake Hodges. May-Sept.

***V. arvensis** L. Corn Speedwell. Native of Europe. Scarce, weed in lawns and meadows: San Diego; Cuyamaca Lake. Apr.-July. 2*n*=24, 16.

***V. comosa** Richt. Broad-fruit Speedwell. Native of eastern United States. Rare, introduced at Lower Otay Reservoir, 120m. July-Sep.

V. peregrina L. ssp. **xalapensis** (H.B.K.)Penn. Mexican Speedwell. Frequent, in wet places; cismontane: e.g. Palomar, Cuyamaca, and Laguna Mountains; San Pasqual Valley; Ramona; San Diego; Kearny Mesa. Apr.-Aug.

SIMAROUBACEAE. Quassia Family

Ailanthus Desf. Tree of Heaven

***A. altissima** (Mill.)Swingle. [*A. glandulosa* Desf.] Tree of Heaven. Native of Asia. Infrequent, naturalized about old dwelling sites; cismontane: Rainbow Creek; Banner; Palomar Mountain; Jamacha Junction. June.

SIMMONDSIACEAE. Jojoba Family

Simmondsia Nutt. Goatnut; Jojoba

S. chinensis (Link)C.K. Schneid. [*S. californica* Nutt., the type from "sides of barren hills, in argillaceous soils, near the sea, in the vicinity of St. Diego".] Common, on dry slopes; coastal, foothill, and desert, below 120m: e.g. Sorrento Slough; Mahogany Canyon; Point Loma; Sunnyside; Telegraph Canyon; Otay; San Ysidro; Barrett Junction; Jacumba; Mountain Springs; Canebrake Canyon; upper Blair Valley; Oriflamme Canyon; Sentenac Canyon; Borrego Palm Canyon; The Narrows; Rockhouse Canyon. Feb.-May. 2*n*=26 pairs.

SOLANACEAE. Nightshade Family

KEY TO GENERA

Shrubs with spiny branches; fruit fleshy or dry and bony; corolla funnelform . . . . . *Lycium*
Shrubs without spines or more commonly herbs
  Corolla rotate or nearly so
    Anthers connivent, longer than the filaments; calyx remaining small
      Anthers dehiscing by terminal pores; plants prickly if with yellow flowers . . . . . *Solanum*
      Anthers dehiscing lengthwise; plants with yellow flowers, but not prickly . . . . . *Lycopersicon*
    Anthers not connivent, mostly shorter than the filaments . . . . . *Physalis*
  Corolla not rotate
    Fruit a berry; corolla urceolate; introduced woody perennial . . . . . *Salpichroa*
    Fruit a capsule; corolla tubular to funnelform
      Corolla 5-6mm long; plant a low annual, prostrate, with purplish-red flowers solitary in the axils . . . . . *Petunia*
      Corolla 15-200mm long; plants mostly larger, more erect
        Capsule prickly, 25-50mm long; flowers solitary, axillary . . . . . *Datura*
        Capsule not prickly, 5-13mm long; flowers in terminal racemes or panicles . . . . . *Nicotiana*

Datura L. Thorn-Apple; Jimsonweed

Corolla 15-20cm long; calyx tubular, not prismatic; capsule nodding . . . . *D. wrightii*
Corolla 6-12cm long; calyx prismatic
  Capsule nodding, pubescent as well as prickly; stems cinereous; desert native . . . . *D. discolor*
  Capsule erect, glabrous to puberulent; stems glabrous to puberulent, not cinereous; introduced weed *D. stramonium*

D. discolor Bernh. Occasional, in sandy places; desert, below 500m: Borrego Valley; Vallecito Wash; Mountain Palm Springs. Apr.-Oct. n=12.

***D. stramonium** L. Native of Neotropics. Infrequent, adventive in sandy areas; coastal, below 200m: Sweetwater Valley. June-Sep. n=12.

D. wrightii Regel. [*D. meteloides* A. DC.] Occasional, in sandy waste areas; cismontane and desert, below 1200m: San Onofre; Fallbrook; Borrego Palm Canyon; Vallecito Stage Station; Boulevard; Del Mar; Chollas Valley; San Diego. Apr.-Oct. n=12.

Lycium. Box-Thorn

Fruit 2-seeded; leaves fleshy-turgid, subterete in cross section; corolla-tube about as long as calyx; coastal . . . . *L. californicum*
Fruit several-seeded; leaves somewhat flattened; corolla-tube longer than calyx
  Calyx-lobes 2/3 as long as calyx-tube, or at least 2mm long
    Fruit hard, greenish or purplish . . . . *L. cooperi*
    Fruit soft, fleshy, red
      Calyx and rest of plant densely pubescent; flowers mostly borne singly . . . . *L. parishii*
      Calyx and rest of plant glabrous or sparsely pubescent; flowers mostly in 2's and 3's
        Calyx lobes unequal, linear . . . . *L. brevipes* var. *b.*
        Calyx lobes equal, somewhat spatulate . . . . *L. brevipes* var. *hassei*
  Calyx-lobes less than 2/3 as long as the tube, or less than 2mm long
    Scrambling shrubs, sometimes adventive near habitations of cismontane California as a plant introduction . . . . *L. barbarum*
    Stiff erect native shrubs of the deserts
      Plant densely pubescent
        Calyx-tube 3-6mm long, the lobes shorter; alkaline soil . . . . *L. fremontii*
        Calyx-tube not more than 3mm long, the lobes usually as long; not from alkaline soil . . . . *L. parishii*
      Plant not densely pubescent
        Corolla-lobes lanate-ciliate; some leaves usually over 5mm broad . . . . *L. torreyi*
        Corolla-lobes glabrous to ciliolate; no leaves over 5mm broad . . . . *L. andersonii*

L. andersonii Gray. [*L. a.*. var. *deserticola* (C.L. Hitchc.)Jeps.; *L. torreyi* var. *wrightii* Jeps. non Gray] Waterjacket. Occasional, on dry slopes; coastal and desert, below 1200m: e.g. Warner Hot Springs; San Felipe; Box Canyon; Vallecito Stage Station; Jacumba; Mountain Springs; Torrey Pines; Mission Hills; Mission Gorge; Rice Canyon. Mar.-May.

***L. barbarum** L. [*L. halimifolilum* Mill.] Matrimony Vine. Native of Eurasia. Local, naturalized in the Tijuana River Valley, at the Monument School site, 10m. July-Oct.

L. brevipes Benth. var. **brevipes**. [*L. richii* Gray] Common Desert Thorn. Uncommon, on dry slopes and in washes; coastal and desert, below 450m: Borrego Valley; Yaqui Well; Carrizo Stage Station; Bonita. Feb.-May.

L. brevipes var. **hassei** (Greene)C.L. Hitchc. Catalina Desert Thorn. Rare, from the T. S. Brandegee residence, San Diego, a specimen approaching this variety was once taken. June.

L. californicum Nutt. ex Gray. California Desert Thorn. Occasional, on bluffs and dry slopes; coastal, below 150m: San Onofre; Oceanside; Torrey Pines Glider Port; Mission San Diego; San Diego; Ocean Beach; Otay Mesa. Mar.-July. Type from near San Diego, on clayhill slopes.

L. cooperi Gray. Peach Desert Thorn. Occasional, on slopes and in washes; desert, below 1000m: San Felipe Valley; Yaqui Well; Scissors Crossing; Box Canyon; Mason Valley. Mar.-May.

L. fremontii Gray. Fremont's Desert Thorn. Infrequent, in alkaline areas; desert, below 450m: Borrego Valley; Split Mountain; Vallecito Stage Station; Canebrake Canyon. Mar.-Apr.

L. parishii Gray. Parish's Desert Thorn. Scarce, on dry slopes and sandy flats; desert, below 800m: Mountain Palm Springs; Vallecito Stage Station. Mar.-Apr.

L. torreyi Gray. Torrey's Desert-thorn. Rare, known in the county only at Carrizo Stage Station, 150m. Jan.

Lycopersicon Mill. Tomato

***L. esculentum** Mill. Cherry Tomato. Native of South America. Local, persisting in old fields; coastal, below 150m: Otay Mesa; Tijuana River Valley. June-Oct.

Nicotiana L. Tobacco

Shrub; flowers yellow . . . . *N. glauca*
Herbs; flowers whitish or cream, sometimes tinged violet
  Leaves auriculate-clasping at base; flowers open throughout day; desert perennial or biennial . . . . *N. trigonophylla*
  Leaves not auriculate-clasping; flowers closing in sun; annuals
    Cauline leaves petioled; calyx-teeth deltoid; corolla 2.5-3cm long . . . . *N. attenuata*
    Cauline leaves sessile or nearly so, except near base of stem
      Corolla 4-7cm long; calyx-teeth linear-lanceolate . . . . *N. bigelovii*
      Corolla 1.5-2cm long; calyx-teeth linear . . . . *N. clevelandii*

N. attenuata Torr. ex Wats. in King. Coyote Tobacco. Occasional, on open slopes and burns; montane and foothill, 200-1500m: Palomar Mountain; Descanso; Alpine; Laguna Mountains; Pine Valley; Campo; Jacumba. Mar.-Oct. *n*=12.

N. bigelovii (Torr.)Wats. var. **wallacei** Gray. Wallace's Tobacco. Occasional, in sandy areas; coastal and foothill, below 1050m: Palomar Mountain; Oak Grove; Fallbrook; Rancho Santa Fe; Chollas Valley; Pine Valley; Cottonwood Grade; San Diego. May-Oct.

N. clevelandii Gray. Cleveland's Tobacco. Occasional, in sandy places; coastal and desert, below 500m: Rancho Santa Fe; Point Loma; Chollas Valley; Foster; Vallecito and Carrizo Stage Stations; Mountain Palm Springs. Mar.-June. Type from dry bed of stream, Chollas Valley near San Diego. *n*=24.

***N. glauca** Grah. Tree Tobacco. Native of South America. Common, weed in disturbed areas; coastal and foothill, below 1000m: e.g. Camp Pendleton; Palomar Mountain; Ballena; Buckman Spring; San Diego; National City. Mar.-Sep. *n*=12.

N. trigonophylla Dunal in A. DC. Desert Tobacco. Frequent, in sandy flats and rocky slopes; desert, below 600m: e.g. Borrego Palm Canyon; Sentenac Canyon; Vallecito Wash; desert slope of Laguna Mountains; Fish Creek; Carrizo Gorge. Mar.-June. *n*=12.

Petunia Juss.

P. parviflora Juss. Wild Petunia. Infrequent, in moist places and drying mud flats; cismontane: San Luis Rey River; Rancho Santa Fe; Mount Soledad; Lake Hodges; Mission Bay; Mission Valley; Sweetwater Lake; Otay Mesa; Pine Valley; Cuyamaca Lake; Warner Ranch. Apr.-Aug. *n*=9.

Physalis L. Ground-Cherry

Leaves pubescent beneath with at least some stellate or forked hairs; plants perennial . . . . *P. hederaefolia*
Leaves lacking stellate or forked hairs beneath
  Plants perennial; anthers yellow, rarely purplish; desert plants
    Pedicels shorter than flowers; foliage somewhat canescent; calyx-lobes more than 2/3 the length of tube at anthesis . . . . *P. hederaefolia*
    Pedicels longer than flowers; foliage green; calyx-lobes not more than half as long as tube at anthesis *P. crassifolia*
  Plants annual; anthers purple, green or blue, rarely yellow (as in *P. greenei*)
    Plant distinctly pubescent
      Calyx-lobes deltoid, shorter than the tube at anthesis; pedicels longer, exceeding the fruiting calyx . . *P. greenei*
      Calyx-lobes somewhat lanceolate, at anthesis often as long as tube; pedicels short, not exceeding the fruiting calyx . . . . *P. pubescens*
    Plants subglabrous
      Pedicels exceeding fruiting calyx; corolla without dark center
        Corolla whitish, with yellow center and 10-20mm broad; leaves sinuate-dentate; fruiting calyx-lobes acuminate . . . . *P. acutifolia*
        Corolla yellow, 3-8mm broad; leaves shallowly sinuate; fruiting calyx-lobes acute . . . . *P. angulata*
      Pedicels shorter than fruiting calyx; corolla 10-15mm broad, yellow, with dark center . . . . *P. philadelphica*

P. acutifolia (Miers)Sandwith. [*P. wrightii* Gray] Wright's Ground-Cherry. Infrequent, as a native weed along roads and waste places; coastal and desert, below 1000m: San Diego; El Cajon Valley; Jacumba. July-Oct.

P. angulata L. var. **lanceifolia** (Nees)Waterfall. [*P. l.* Nees] Lance-leaf Ground-Cherry. Rare, native weed in disturbed areas; coastal and desert, below 200m: Fallbrook. July-Oct.

P. crassifolia Benth. var. **crassifolia**. Thick-leaf Ground-Cherry. Occasional, in sandy or rocky areas; desert, below 1200m: northeast corner of county; Fish Creek; Borrego Palm Canyon; Borrego Valley; San Felipe Valley; Agua Caliente Hot Springs; Mountain Palm Springs; Table Mountain. Mar.-May.

P. greenei Vasey & Rose. Greene's Ground-Cherry. Occasional, in moist areas; coastal, below 300m: Ysidora; Bonsall; Fallbrook; San Marcos; Santee; Mount Helix; Dehesa; near Barrett Lake; Potrero; San Diego. Mar.-June.

P. hederaefolia Gray var. **cordifolia** (Gray)Waterfall. [*P. h.* Gray] Scarce, in sandy or rocky areas; desert and foothill, 800-1200m: Jacumba; near Walker Ranch. June.

***P. philadelphica** Lam. [*P. ixocarpa* Brot. ex Hornem.] Tomatillo. Native of Mexico. Infrequent, escaping from cultivation; coastal, below 200m: San Luis Rey; San Pasqual Valley; San Diego. June-Sep. 2*n*=24.

***P. pubescens** L. var. **integrifolia** (Dunal)Waterfall. [*P. p.* L.] Native of eastern United States. Infrequent, escaping from cultivation; cismontane: Mission Valley; Lake Murray; Lakeside; Pine Valley. June-Dec. *n*=12.

Salpichroa Miers

***S. origanifolia** (Lam.)Thell. [*S. rhomboidea* (Gill. & Hook.)Meirs] Lily-of-the-Valley Vine. Local, weed in gardens and orchards: San Diego. July-Oct. *n*=12.

Solanum L. Nightshade

Plants not prickly
  Corolla deeply 5-cleft, whitish; peduncles mostly longer than pedicels
    Stems conspicuously and persistently villous or hirsute; ripe berries greenish, yellow or reddish . . *S. sarrachoides*
    Stems sparsely pubescent or glabrous, at least in maturity; ripe berries black
      Corolla 6-11mm long; anthers 2.6-4mm long; native half-shrub . . . . *S. douglasii*

Corolla 4-6mm long; anthers 1.2-2.6mm long; introduced weeds
Inflorescence umbelliform; calyx-lobes distinct, reflexed at maturity; surface of berry glossy; anthers 1-1.2mm long; frequent . . . . *S. americanum*
Inflorescence subracemose; calyx-lobes partly fused, not reflexed at maturity; surface of berry dull; anthers 1.8-2.4mm long; occasional . . . . *S. nigrum*
Corolla angulately shallowly lobed, often purple; peduncles mostly shorter than pedicels
Leaves mostly hastate with a pair of basal linear lobes . . . . *S. tenuilobatum*
Leaves seldom lobed
Pubescence of upper stems with at least some forked or branched hairs. . . . . *S. umbelliferum*
Pubescence on stems lacking or of simple hairs
Stems distinctly pubescent
Plants subshrubs. . . . . *S. xanti* var. *intermedium*
Plants strictly herbaceous, often prostrate mats. . . . . *S. xanti* var. *montanum*
Stems glabrous or nearly so
Leaves acute or tapering at base . . . . *S. parishii*
Leaves obtuse or subcordate at base . . . . *S. xanti* var. *hoffmannii*
Plants prickly
Plants annual; leaves pinnatifid or bipinnatifid; berry partly or wholly invested by spiny calyx; corolla yellow . . . . *S. rostratum*
Plants perennial or shurbby; leaves entire, sinuate or lobed, but not pinnatifid; berry not enclosed by the spiny or unarmed calyx
Leaves entire to repand-dentate, permanently gray-canescent on both surfaces, linear-lanceolate to narrowly oblong . . . . *S. elaeagnifolium*
Leaves lobed or deeply dentate, not permanently gray-canescent at least on upper surface . . . . . . *S. marginatum*

***S. americanum** Mill. [*S. nodiflorum* Jacq.] White Nightshade. Native of eastern North America. Infrequent, weed in waste places; cismontane: Del Mar; Casa de Oro, 180m. Apr.-Nov. *n*=12.

S. douglasii Dunal in DC. Douglas' Nightshade. Occasional, on shaded slopes; coastal and foothill, below 1100m: e.g. San Mateo Creek; Fallbrook; Silver Strand; National City; Potrero Grade; Vallecito Canyon; Borrego Palm Canyon. Jan.-Dec. *n*=12.

***S. elaeagnifolium** Cav. Silverleaf Horse-Nettle. Native of central United States. Infrequent, weed along roads and in disturbed places; cismontane: Oak Grove; San Marcos; Spring Valley; Chula Vista; Blair Valley. Mar.-Sep. *n*=12.

***S. marginatum** L. f. White-edged Horse-Nettle. Native of Africa. Local, weed at Pala, 150m. Apr. 2*n*=24.

***S. nigrum** L. Black Nightshade. Native of Europe. Occasional, weed in waste places; cismontane: De Luz; San Luis Rey; Encinitas; San Felipe; Dulzura. Mar.-Oct. *n*=36.

S. parishii Heller. [*S. xanti* var. *glabrescens* Parish, in part; *S. x.* var. *spencerae* Macbr., type from Torrey Pines.] Parish's Nightshade. Occasional, on dry slopes; coastal and foothill, below 1200m: Palomar Mountain; Oak Grove; Encinitas; Lake Hodges; Bernardo; Santa Ysabel; Tule Springs; Torrey Pines. Mar.-May.

***S. rostratum** Dunal. Buffalo-Bur. Native of central United States. Uncommon, weed in disturbed places: San Diego. July. *n*=12.

***S. sarrachoides** Sendt. ex Mart. [*S. villosum* auct. non L.] Hairy Nightshade. Native of Brazil. Uncommon, weed in moist places; cismontane, below 1300m: San Luis Rey River; Hormiguero Ranch; Point Loma; San Diego. May-Oct. *n*=12.

S. tenuilobatum Parish. Infrequent, in Chaparral; foothill, 200-1100m: Otay Mountain; Potrero Grade; Barrett Dam; Jamul; Lyons Valley; Silverwood; Campo. Mar.-Apr.

S. umbelliferum Eschs. var. **glabrescens** Torr. [*S. xanti* var. *g.* (Torr.)Parish] Blue Witch. Occasional, on shaded slopes; foothill and montane, above 700m: Laguna Mountains; Jacumba. May-June.

S. xanti Gray var. **hoffmannii** Munz. Hoffman's Nightshade. Rare, known in the county only at Warner Ranch, 950m. May.

S. xanti var. **intermedium** Parish. Chaparral Nightshade. Infrequent, on shades slopes; foothill, 300-1300m: Descanso; Campo. May-June.

S. xanti var. **montanum** Munz. Mountain Nightshade. Rare, in Chaparral, foothill and montane, 1200-1700m: Garnet Peak; Eckener Pass. July.

STERCULIACEAE. Cacao Family

Flowers brownish, 0.4-0.7mm across; shrub 1-3dm high; leaves not lobed. . . . . *Ayenia*
Flowers yellow, 3-8mm across; shrub 1-6dm tall; leaves usually lobed . . . . *Fremontodendron*

Ayenia Loefl.

A. compacta Rose. California Ayenia. Infrequent, on rocky slopes; desert, below 500m: Borrego Palm Canyon; Hellhole Canyon; Yaqui Well; Sentenac Canyon; Box Canyon; Potrero; Vallecito Canyon; Agua Caliente Hot Springs; Canebrake Canyon; Rockhouse Canyon; Alta Bisnaga Wash; Volcanic Hills; south of Lava Flow Wash. Mar.-Apr.

Fremontodendron Cov. Flannel Bush

Leaves 1-3 veined at base; glands at base of sepals hairy; flowers 3-6cm wide, borne on lateral branches; seeds dull, with large terminal caruncle . . . . *F. californicum*
Leaves palmately 5-7 veined from base; glands at base of sepals glabrous; flowers 6-9cm across, borne on main twigs; seeds shining, without caruncle . . . . *F. mexicanum*

F. californicum (Torr.)Cov. [*Fremontia c.* Torr.; *F. c.* var. *diegoensis* M. Harvey, the type from the bottom of Vallecito Canyon, Laguna Mountains.] California Fremontia. Infrequent, in shaded canyons; foothill and montane, 700-1600m: Agua Tibia Mountains; Oriflamme Canyon; Vallecito Canyon; Storm Canyon; Rockhouse Canyon; Viejas Grade. May-June. *n*=20.

F. mexicanum A. Davids. [*Fremontia m.* (A. Davids.)Macbr.] Southern Fremontia. Local, in shaded canyons; foothill, 300-900m: Jamul Mountains; Woodwardia Canyon; Otay Mountain. Mar.-June.

STYRACACEAE. Storax Family

Styrax L. Snowdrop Bush

S. officinalis L. ssp. **fulvescens** (Eastw.)Beauchamp ex Thorne. Snowdrop Bush. Infrequent, on shaded slopes; foothill and montane, 300-1500m: Dehesa; Santa Margarita Mountains; Agua Tibia Mountains; Palomar Mountains; Lusardi Canyon; Iron Mountain; El Cajon Mountains; Silverwood; Pala; Mesa Grande; Middle Peak. Apr.-July. 2*n*=8 pairs.

TAMARICACEAE. Tamarisk Family

Tamarix L. Tamarisk

Flowers 4-merous; petals oblong, scarcely 2mm long; bracts diaphanous . . . . *T. parviflora*
Flowers 5-merous
  Leaves vaginate; bracts not vaginate, but somewhat clasping . . . . *T. aphylla*
  Leaves sometimes auriculate, but not vaginate or amplexicaul
    Staminate flowers with filaments inserted below the disk
      Petals obovate, widened distally; bracts ovate, subobtuse; saline places. *T. ramosissima*
      Petals oblong-ovate, narrowed distally; non-saline habitats . . . . *T. chinensis*
    Staminate flowers with filaments inserted around the disk . . . . *T. gallica*

***T. aphylla** (L.)Karsten. Athel. Native of Eurasia. Infrequent, escaping from cultivation: Borrego Valley; Pine Valley. May-July. 2*n*=24.

***T. chinensis** Loureiro. Native of Asia. Occasional, along washes; desert, below 500m: Coyote Creek; Carrizo Stage Station. Mar.-June.

***T. gallica** L. Native of Europe. Uncommon, along washes and escaping from cultivation: Carrizo Stage Station; La Jolla. June-Aug. *n*-12.

***T. parviflora** DC. Native of Europe. Common, escaping from cultivation in moist, alkaline areas: San Luis Rey; Escondido; La Jolla; Jamacha; Pine Valley. Mar.-Apr. *n*=12.

***T. ramosissima** Ledeb. Native of Mediterranean region. Rare, escaping from cultivation along water courses and irrigation ditches: Coyote Creek. Mar.-Aug.

TROPAEOLACEAE. Tropaeolum Family

Tropaeolum L. Nasturtium

***T. majus** L. Garden Nasturtium. Native of South America. Local, escaping from cultivation; coastal, below 200m: La Jolla; Paradise Hills; San Diego. Mar.-Aug. 2*n*=28.

ULMACEAE. Elm Family

Celtis L. Hackberry

C. reticulata Torr. [*C. douglasii* Planch.] Netleaf Hackberry. Rare, montane valleys, 1200-1500m: Thing Valley; Laguna Mountains; Wynola; East Mesa, Cuyamaca Rancho State Park; Volcan Mountain. Apr.-May.

URITACEAE. Nettle Family

KEY TO GENERA

Plants with stinging hairs and opposite leaves
  Pistillate calyx saccate, 2-4-toothed at apex . . . . *Hesperocnide*
  Pistillate calyx 4-parted, the segments almost distinct, the inner much the longer and inclosing the akene . . . *Urtica*
Plants without stinging hairs and with alternate leaves
  Delicate herb . . . . *Parietaria*
  Coarse-leaved perennialwith milky sap . . . . *Ficus*

Ficus L. Fig.

***F. carica** L. Native of western Asia. Occasional, persisting around abandoned dwelling or naturalized in moist areas; coastal, below 150m; Las Flores; Bonita. Apr.-June.

Hesperocnide Torr.

H. tenella Torr. Western Nettle. Occasional, on shaded slopes; coastal and foothill, below 1200m: San Onofre; Whispering Oaks; Barona Valley; Silver Strand; Corte Madera. Apr.-June.

Parietaria L.

P. hespera Hinton var. **californica** Hinton. [*H. floridana* Calif. auct. non Nutt.] Western Pellitory. Frequent, on shaded slopes; coastal and foothill, below 1000m: e.g. San Onofre; Borrego Palm Canyon; San Felipe; Mission Gorge; Tecate Peak; Live Oak Springs; Mountain Palm Springs. Feb.-June.

Urtica L. Nettle

Plant perennial, staminate and pistillate flowers in separate clusters . . . . . *U. dioica*
Plant annual; staminate and pistillate flowers mixed in the same cluster . . . . . *U. urens*

U. dioica L. ssp. **holosericea** (Nutt.)Thorne. [*U. h.* Nutt.; *U. gracilis* var. *h.* (Nutt.)Jeps.; *U. dioica* var. *h.* (Nutt.)Nutt.] Hoary Nettle. Common, in damp places; cismontane, below 1200m: e.g. San Onofre; San Felipe; Palomar Mountain; Buckman Spring; Descanso; Del Mar. June-Sep.

***U. urens** L. Dwarf Nettle. Native of Europe. Occasional, weed in disturbed, moist places; coastal, below 400m: Vista; Rancho Santa Fe; San Pasqual; Mission Valley; San Diego. Jan.-Apr. 2*n*=24, 26, 52.

VALERIANACEAE. Valerian Family

KEY TO GENERA

Perennial; corolla with a tubular spur; stamen 1; naturalized . . . . . *Centranthus*
Annual; stamens 3; native . . . . . *Plectitis*

Centranthus DC.

***C. ruber** (L.)DC. in Lam. & DC. Red Valerian. Native of Europe. Local, escaping from cultivation; coastal, below 100m: Vista; La Jolla; Balboa Park. Apr.-Aug. 2*n*=14.

Plectritis DC.

Corolla bilabiate, the spur obsolete to usually less than 1/3 the length of the corolla; keel of the fruit rarely grooved
  Corolla with a slender spur . . . . . *P. congesta* ssp. *insignis*
  Corolla often lacking a spur . . . . . *P. congesta* ssp. *brachystemon*
Corolla regular to bilabiate, the spur usually more than 1/3 the length of the corolla; keel of the fruit often with a dorsal groove . . . . . *P. ciliosa*

P. ciliosa (Greene)Jeps. ssp. **insignis** (Suksd.)Morey. [*P. californica* (Suksd.)Dyal, the type from Sweetwater; *P. ciliosa* var. *patelliformis* (Suksd.)Demp., the type from damp or wet places, Stonewall Mine, Cuyamaca Mountains, 4600 feet elevation.] Occasional, on grassy slopes; foothill and montane, 300-1800m: Dyche Valley; Julian; Inspiration Point; Lake Henshaw; Cuyamaca Peak; Stonewall Mine; Laguna Lakes; Shrine Camp; Pine Valley; Boulevard; Barrett; Whispering Oaks; Harbison Canyon; Descanso; Ramona; Witch Creek. Apr.-May.

P. congesta (Lindl.)DC. ssp. **brachystemon** (F.& M.)Morey. Rare, known in the county from collections from Cuyamaca Lake, 1400m; below Fern Flat Spring; below Japacha Spring. June.

P. congesta ssp. **congesta**. Rare, known in the county only from a collection from Palomar Mountains, above 1300m. July.

VERBENACEAE. Vervain Family

KEY TO GENERA

Nutlets 2; flowers in dense spikes or heads . . . . . *Lippia*
Nutlets 4; flowers in terminal spikes
  Styles short, mostly 0.5-1.5mm long . . . . . *Verbena*
  Styles elongate, mostly 6-12mm long . . . . . *Glandularia*

Glandularia J. F. Gmel.

G. gooddingii (Briq.)Solbrig. [*Verbena g.* Briq.] Rare, known in the county only from a vague locality in the Santa Rosa Mountains, 500-1300m. Apr.-May. *n*=10.

Lippia L.

Leaves broadest below the middle, the blades 2-4cm long, serrate from below the middle to the apex; calyx-lobes about as long as tube . . . . . *L. lanceolata*
Leaves broadest above the middle, the blades mostly 0.5-2cm long, serrate only above the middle; calyx-lobes shorter than tube
  Leaf-margins with spreading teeth; leaves apically obtuse, tapering gradually to the somewhat sessile base *L. incisa*
  Leaf-margins with teeth pointing forward; leaves mostly acute at apex, abruptly narrowed to a petioled base . . . . . *L. nodiflora*

L. incisa (Small)Tides. Narrow-leaf Frog Fruit. Rare, known in the county only at Encinitas, 50m. Sep.

L. lanceolata Michx. [*Phyla l.* (Michx.)Greene] Lance-leaf Frog Fruit. Rare, known in the county only at Lake Hodges, 96m. Sep. 2*n*=32.

***L. nodiflora** Michx. var. **rosea** (D. Don)Munz. [*Phyla n.* var. *r.* (D. Don)Moldk.] Garden Lippia. Native of South America. Infrequent, weed in moist places; coastal, below 300m: Sweetwater Lake; San Diego. May-Oct.

Verbena L.

Bracts longer than or equalling flowers, conspicuous . . . . . *V. bracteata*
Bracts shorter than flowers, inconspicuous
  Spikes panicled, much elongate, slender . . . . . *V. menthaefolia*
  Spikes in 3's at the ends of the branches, often congested at anthesis
    Spikes 1-3dm long in fruit; leaves canescent, not scabrous above; inner face of nutlets without whitish papillae . . . . . *V. lasiostachys*
    Spikes 0.3-1dm long in fruit; leaves bright green, scabrous above, inner face of nutlets with whitish papillae . . . . . *V. robusta*

V. bracteata Lag. & Rodr. Bract Vervain. Infrequent, in wet places; coastal, below 150m: Sweetwater Lake; Sweetwater Valley; Penasquitos Canyon. May-Oct. *n*=7.

V. lasiostachys Link. [*V. abramsii* Mold., the type from Hot Springs (Jacumba), in the southern part of San Diego County; *V. prostrata* R. Br. non Savi] Western Vervain. Occasional, in moist places; cismontane, below 1500m: San Mateo Creek; Palomar and Cuyamaca Mountains; Mesa Grande; Japatul Valley; La Jolla; Point Loma; Jacumba. May-Sep. 2*n*=7 pairs.

V. menthaefolia Benth. Mint-leaf Vervain. Infrequent, in canyons and on shaded slopes; coastal, below 300m: Mission Hills; Sweetwater Valley; Otay Creek; San Ysidro. Apr.-June.

V. robusta Greene. Robust Vervain. Uncommon, in moist places; cismontane, below 1300m: Santa Ysabel; Jamul; Lake Murray; Sweetwater Lake. May-Oct.

VIOLACEAE. Violet Family

Viola L. Violet

Flowers with some yellow, at least at bases of petals and on spur; access to the spur-cavity blocked by the much-expanded retrorsely bearded thickened head of the style; bearding of lateral petals short and clavate; spur about as broad as long; dry habitats
- Leaves not cut or lobed, but with margins entire or serrate-crenate
  - Plants strictly erect with cauline leaves and flowers crowded at ends of stems which are naked below; capsule acute . . . . . . *V. lobata* var. *integrifolia*
  - Plants erect to prostrate, but with leaves and flowers scattered along the stem; capsule obtuse
    - Crown of rootstock near the surface; flowers 2m or less across; cleistogamous flowers present in upper axils
      - Blades of basal leaves as long as or longer than wide
        - Petals 6-10mm long; leaves rather spreading, greenish to grayish, usually tinted purple; upper leaves subentire or irregularly or few (2-4)-toothed; Montane Coniferous Forest or higher
          - Foliage green; stems well developed above ground, petals 8-10mm long . . . . . . *V. purpurea* ssp. *p.*
          - Foliage grayish; stems mostly buried; petals 6-8mm long . . . . . . *V. purpurea* ssp. *xerophyta*
        - Petals 10-12mm long; leaves erect, grayish, without purple tinting; upper leaves rather regularly 5-6-dentate on each side; below Montane Coniferous Forest . . . . . . *V. quercetorum*
      - Blades of basal leaves commonly wider than long
        - Teeth of basal leaves shallow and very irregular . . . . . . *V. purpurea* et vars.
        - Teeth of basal leaves deep (2-3mm) and regular . . . . . . *V. aurea*
    - Rootstocks deep, short (1-2cm long), soft; flowers more than 2cm across; cleistogamous flowers lacking; below Montane Coniferous Forest . . . . . . *V. pedunculata*
- Leaves deeply cut or lobed
  - Leaf-blades palmately 3-7-lobed, these divisions broad and not dissected; stems naked below . . . . *V. lobata* var. *l.*
  - Leaf-blades dissected into narrow ultimate segments . . . . . . *V. douglasii*

Flowers white, blue or purple, without any yellow; access to spur cavity not blocked by the style (the stigma held above the floor of the lower petal); bearding of the lateral petals longer, not clavate; spur longer than broad; damp places
- Plants with evident leaf-bearing stems; petals violet, 8-12mm long; montane . . . . . . *V. adunca*
- Plants lacking leaf-bearing stems (although they may have stolons), the leaves and peduncles arising from the base or crown . . . . . . *V. nephrophylla*

V. adunca Sm. Western Dog-Violet. Infrequent, in meadows; montane, above 1200m: upper Doane Valley; Middle Peak. Mar.-July. *n*=10.

V. aurea Kell. ssp. **mohavensis** Baker & Clausen. Mojave Golden Violet. Rare, known in the county only at Doane Valley, 1400m. June. *n*=6.

V. douglasii Steud. Douglas' Violet. Occasional, in meadows; montane and foothill, above 750m: Henshaw Dam; Santa Ysabel; Wynola; Julian; Vallecito Canyon; Cuyamaca Lake; Cuyamaca and Laguna Mountains. Mar.-June.

V. lobata Benth. var. **integrifolia** Wats. Pine Yellow Violet. Infrequent, on floor of Montane Coniferous Forest; montane, above 1200m: North Peak. Apr.-July.

V. lobata var. **lobata**. Moose-Horn Yellow Violet. Infrequent, on floor of Montane Coniferous Forest; montane, above 1200m: Julian; Inspiration Point; Stonewall Mine; Cuyamaca Mountains. Apr.-July. *n*=6.

V. nephrophylla Greene. Kidney-Leaf Blue Violet. Occasional, in wet places; montane, above 1000m: Palomar, Hot Springs, and Cuyamaca Mountains; Lake Henshaw. May-June.

V. pedunculata T.& G. ssp.**pedunculata**. Yellow Johnny Jump-Ups. Frequent, on grassy slopes; coastal and foothill, below 900m: e.g. Bonsall; Castro Canyon; Springhill; near Ramona; Witch Creek; Ballena; Mount Helix; Torrey Pines; Mount Soledad; Point Loma; Jamul; Dulzura; La Mesa; Otay Mountain. Feb.-Apr. *n*=6.

V. purpurea Kell. ssp. **purpurea**. Mountain Yellow Violet. Occasional, in shaded areas; foothill and montane, above 700m: Palomar, Hot Springs, Cuyamaca, and Laguna Mountains; Lyons Valley; Morena Dam; Corte Madera; Descanso. Apr.-June. *n*=6.

V. purpurea ssp. **xerophyta** Baker & Clausen. [*V. p.* var. *pinetorum* (Greene)Greene] Grey-leaf Yellow Violet. Uncommon, in Montane Coniferous Forest; montane, above 1200m: Hot Springs Mountain; West Mesa of Cuyamaca Mountains; Laguna Mountains. May-July.

V. quercetorum Baker & Clausen. Oak Yellow Violet. Occasional, on dry slopes; foothill and montane, above 600m: Palomar Mountains; Lake Henshaw; Julian; Banner Grade; Cuyamaca Mountains; Laguna Public Camp; Lyons Valley; Pine Valley. Mar.-June. *n*=12.

VISCACEAE. Mistletoe Family

KEY TO GENERA

Anthers 1-celled; pistillate sepals 2; berry on a recurved pedicel, compressed. *Arceuthobium*
Anthers 2-celled; pistillate sepals 3; berry sessile, globose. . . . . *Phoradendron*

Arceuthobium Bieb.

Internodes about 1.5mm wide at or near the third internode above the base; mature fruit about 3.5mm high, about 2.0mm wide; parasite on *Pinus monophylla* . . . . *A. divaricatum*
Internodes about 2-4mm wide at or near the third internode above the base; mature fruit about 4.0mm or more high and 2.5mm or more wide; never parasitic on *Pinus monophylla*
Mature fruit about 4mm long and about 2.5mm wide; internodes not glaucous; anthesis primarily in July; parasite on *Pinus lambertiana* . . . . *A. californicum*
Mature fruit about 4.5-5.0mm long and about 3.0mm wide; some internodes at least usually slightly glaucous; anthesis primarily from late August to late November; parasite primarily on *Pinus ponderosa*, occasionally also on *P. coulteri* and *P. jeffreyi* . . . . *A. campylopodum*

A. californicum Hawksw. & Wiens. Sugar-Pine Dwarf Mistletoe. Scarce, in Montane Coniferous Forest; montane, above 1300m: Cuyamaca Peak. June-Aug. Known host: *Pinus lambertiana*. *n*=14.

A. campylopodum Engelm. in Gray. Western Dwarf Mistletoe. Prevalent, in Montane Coniferous Forest; montane, above 1000m: e.g. Pine Hills; Julian; Cuyamaca Lake; Cuyamaca Mountains; Garnet Peak; Laguna Resort. Aug.-Oct. Known hosts: *Pinus ponderosa*; *P. jeffreyi*; *P. coulteri*. *n*=14.

A. divaricatum Engelm. Pinyon Dwarf Mistletoe. Uncommon, in Pinyon Pines; desert, above 800m: desert slopes of Laguna Mountains. Aug.-Sep. Known host: *Pinus monophylla*. *n*=14.

Phoradendron Nutt.

Pistillate spikes with 2 flowers at each joint
Leaves reduced to connate scales
Plants canescent; pistillate spikes 3-4-jointed; berries usually red . . . . *P. californicum*
Plants glabrous; pistillate spikes 1-jointed; berries white . . . . *P. juniperinum*
Leaves green, oblanceolate
Plants 1-2dm high; leaves sessile. . . . . *P. bolleanum* ssp. *densum*
Plants 2-3dm high; leaves subpetiolate. . . . . *P. bolleanum* ssp. *pauciflorum*
Pistillate spikes with 6 or more flowers at each joint; leaves green
Internodes minutely canescent to canescent-tomentose; anthesis largely from October through March; fruit glabrous . . . . *P. tomentosum*
Internodes densely short-villous; anthesis largely July through September; fruit pubescent. . . . . *P. villosum*

P. bolleanum (Seem.)Eichler ssp. **densum** (Torr.)Wiens. [*P. d.* Torr.] Dense Mistletoe. Frequent, in open Chaparral; foothill and desert, 300-1100m: 16km west of Warner Springs; Banner; Bankhead Spring; Table Mountain; Jacumba; southeast corner of county; InKoPah County Park. Also Tecate Peak. May-Aug. Known hosts: *Juniperus californica*; *Cupressus guadalupensis* ssp. *forbesii*. *n*=14, 27.

P. bolleanum ssp. **pauciflorum** (Torr.)Wiens. [*P. p.* Torr.] White Fir Mistletoe. Rare, in Montane Coniferous Forest; montane, above 1500m: Cuyamaca Peak. July-Sep. Known host: *Abies concolor*. *n*=14.

P. californicum Nutt. Desert Mistletoe. Common, in Creosote Bush Scrub; desert, below 700m: e.g northeast corner of county; Fish Creek; Lone Palm; Borrego Springs; Borrego Palm Canyon; San Felipe Valley; Pinyon Mountain; Vallecito Stage Station; Mountain Palm Springs. Dec.-Mar. Known hosts: *Acacia greggii* var. *arizonica*; *Prosopis glandulosa* var. *torreyana*; *P. pubescens*; *Cercidium floridum*. *n*=14.

P. juniperinum Engelm. ex Gray ssp. **libocedri** (Engelm.)Wiens. Incense Cedar Mistletoe. Infrequent, in Montane Coniferus Forest; montane, above 1050m: Julian. July-Sep. Known host: *Calocedrus decurrens*. *n*=14.

P. tomentosum (DC.)Engelm. ex Gray ssp. **macrophyllum** (Engelm.)Wiens. [*P. longispicum* Trel.] Long-Spike Mistletoe.Frequent, in Southern Oak Woodland and Chaparral; foothill and montane, above 900m: e.g. Live Oak Park; Pala; San Luis Rey; Oak Grove; Witch Creek; San Felipe Valley; Volcan Indian Reservation; Inspiration Point; Mason Valley; Ballena; Pine Valley; Cottonwood Valley; Jacumba; San Vicente Reservoir; Alpine; Lakeside; Poway; Penasquitos Canyon. Dec.-May. Known hosts: *Platanus racemosa*; *Populus fremontii*; *Salix* spp. *n*=14.

P. villosum (Nutt. in T.& G.)Nutt. Hairy Mistletoe. Frequent, in Southern Oak Woodland and Chaparral; foothill and montane, above 700m: e.g. Lake Henshaw; Warner Hot Springs; Hot Springs Mountain; Morettis; San Felipe; Julian; Pine Hills; North Peak; Stonewall Mine; Laguna Resort; Cuyapaipe Indian Reservation; Barona Valley; Alpine; Mussey Grade; Buckman Spring; Campo; Jacumba. July-Sep. Known hosts: *Quercus* spp. *n*=14.

VITACEAE. Grape Family

Vitis L.

V. girdiana Munson. Desert Grape. Frequent, in canyons; coastal and foothill, below 1200m; e.g. San Luis Rey River at Pala; De Luz; Bear Valley; Borrego Palm Canyon; Julian; Vallecito Canyon; Jacumba; Dehesa; Jamacha; Sloane Canyon. May-June. Type from San Diego County.

ZYGOPHYLLACEAE. Caltrop Family

KEY TO GENERA

Stipules spiny; leaves 3-foliolate; flowers purplish . . . . *Fagonia*
Stipules not spiny; leaves 2-many-pinnate; flowers yellow or orange
  Leaflets 1 pair; stamens with scalelike appendages . . . . *Larrea*
  Leaflets 4 or more pairs; stamens not appendaged
    Fruit hemispheric or higher, not radiate, breaking up into 8-12 merely tubercled nutlets. . . . . *Kallstroemia*
    Fruit flat, radiate, breaking up into 5 very spiny nutlets. . . . . *Tribulus*

Fagonia L.

Flowers about 1cm in diameter; stipules 1-6mm long. . . . . *F. laevis*
Flowers about 1.5cm in diameter; stipules 3-16mm long. . . . . *F. pachyacantha*

F. laevis Standl. [*F. chilensis* var. *l.* (Standl.)Johnston] Frequent, in rocky areas; desert, below 700m: Coyote Creek; Borrego Palm Canyon; Borrego Springs; The Narrows; Fish Creek; Box Canyon; Vallecito Stage Station. Also upper Otay Dam; San Ysidro. Mar.-May; Nov.-Jan.

F. pachyacantha Rydb. in Vail & Rydb. [*F. chilensis* var. ***glutinosa*** (Vail)Jtn.] Scarce, in rocky places; desert, below 300m: northeast corner of county; Split Mountain; Mortero Wash. Feb.-May.

Kallstroemia Scop.

***K. parviflora** Nort. Native of central United States. Local, weed at Warner Hot Springs, 940m. Aug.-Oct.

Larrea Cav. Creosote Bush

L. divaricata Cav. ssp. **tridentata** (Sesse & Moc. ex DC.)Felger & Lowe. [*L. t.* (Sesse & Moc. ex DC.)Cov.] Prevalent, in sandy areas as an indicator of Creosote Bush Scrub; desert, below 1100m: e.g. Culp Valley; Borrego Valley; The Narrows; Vallecito and Carrizo Stage Stations; Dos Cabezas; Jacumba. Apr.-May. $2n$=52, 104.

Tribulus L. Puncture Vine

***T. terrestris** L. Native of Old World. Occasional, weed in sandy areas; cismontane and desert, below 1400m: e.g. Warner Hot Springs; Pine Valley; Jacumba; San Diego. Apr.-Oct. $2n$=24, 48.

INDEX TO FAMILIES AND GENERA IN THE CATALOG OF THE FLORA

Abies 40
Abronia 185
Abutilon 182
Acacia 154
Acalypha 150
Acamptopappus 93
ACANTHACEAE 79
Acanthomintha 175
Acer 79
ACERACEAE 79
Achillea 93
Achyrachaena 93
Achyranthes 80
Achyronychia 137
Acourtia 93
Acroptilon 93
Adenostegia 221, 222
Adenostoma 213
Adiantum 37
ADIANTACEAE 37, 38
Adolphia 211
ADOXACEAE 79
AGAVACEAE 42, 43
Agave 42
Ageratina 93
Agianthus 130
Agoseris 93
Agrimonia 214
Agropyron 58
Agrostis 58
Ailanthus 226
Aira 59
AIZOACEAE 79
Albizia 154
Alchemilla 214
Alisma 43
ALISMATACEAE 43
Allenrolfea 140
Allionia 185
ALLIACEAE 43
Allium 43
Allophyllum 195
Alnus 118
Alopecurus 59
Alternanthera 80
Alyssum 129
AMARANTHACEAE 80
Amaranthus 81
AMARYLLIDACEAE 45
Amblyopappus 94
Ambrosia 94
Amelanchier 214
Ammannia 181
Ammi 83
Ammophila 59
Ammoselinum 83
Amorpha 154
Amsinckia 119
ANACARDIACEAE 81
Anagallis 208
Andropogon 60, 70
Androsace 208
Anemone 209
Anemopsis 218
Anethum 83
Angelica 83
Anisocoma 94
Anthemis 94
Anthriscus 83
Antirrhinum 220
Aphanisma 140
APIACEAE 82
Apiastrum 83
Apium 83
APOCYNACEAE 86
Apocynum 86
Aponogeton 45
APONOGETONACEAE 45
Aptenia 79
Aquilegia 209
Arabis 124,*130*
ARACEAE 45
Aralia 87
ARALIACEAE 87
Araujia 87
Arbutus 148
Arceuthobium 233
Arctostaphylos 148,*149*
Arctotis 94
ARECACEAE 45
Arenaria 137
Argemone 192
Argythamnia 150
Aristida 59
Arnica 94
Aromia 94
Artemisia 95
Arthrocnemum 143
Arundo 60
ASCLEPIADACEAE 87
Asclepias 87
ASPARAGACEAE 53, 54
Asparagus 53
ASPHODELACEAE 53
Asphodelus 53
ASPIDIACEAE 38
Aspidotis 37
ASPLENIACEAE 37
Asplenium 37
Aster 95, *111, 117*
ASTERACEAE 88
Astragalus 154
Athysanus 125
Atriplex 140
Audibertia 177
Avena 60
Avicennia 117
AVICENNIACEAE 117
Ayenia 229
Azolla 39

Baccharis 96
Baeria 110
Baileya 96
Barbarea 125
BASELLACEAE 117
Bassia 141
BATACEAE 117
Batis 117
Bebbia 96
Bellis 96
Beloperone 79
BERBERIDACEAE 117
Berberis 118
Bergerocactus 131
Bernardia 150
Berula 84
Beta 141

BETULACEAE 118
Bidens 97
BIGNONIACEAE 118
BLECHNACEAE 39
Blennosperma 97
Bloomeria 44
Boerhavia 185
Bogenhardia 182
Boisduvalia 186
Bommeria 37
BORAGINACEAE 118
Borago 119
Bothriochloa 60
Boussingaultia 117
Bouteloua 60
Bowlesia 84
Boykinia 219
Brachypodium 60
Brahea 45
Brassica 125
BRASSICACEAE 123
Brickellia 97
Briza 60
Brodiaea 44
Bromus 60
Bupleurum 84
Burrielia 110
Bursera 131
BURSERACEAE 131

CACTACEAE 131
Caesalpinia 156
Cakile 126
Calais 112
Calamagrostis 62
Calandrinia 207
Calla 45
Calliandra 156
CALLITRICHACEAE 134
Callitriche 134
Calocedrus 39
Calochortus 53
Calycadenia 97
Calycoseris 97
Calyptridium 207
Calystegia 144
Camelina 126
Camissonia 187
CAMPANULACEAE 134
CANNABACEAE 135
Cannabis 135
CAPPARACEAE 136
CAPRIFOLIACEAE 136
Capsella 126, *128*
Caragana 156
Cardamine 126
Cardaria 126
Cardionema 137
Carduus 97
Carex 46
Carpobrotus 79
Carthamnus 97
CARYOPHYLLACEAE 137
Cassia 164
Castilleja 220
Catalpa 118
Caucalis 86
Caulanthus 126, *130*
Ceanothus 211
CELASTRACEAE 139
Celtis 230
Cenchrus 62
Centaurea 98, *93*
Centaurium 168
Centranthus 231
Centromadia 108
Centrostegia 202
Centunculus 208
Cerastium 137
CERATOPHYLLACEAE 139
Ceratophyllum 139
Cercidium 156
Cercis 156
Cercocarpus 214
Cereus 131, 132, 133
Chaenactis 98
Chaetopappa 99
Chamaebatia 214
Chamaenerion 189
Chamaesyce 150
Cheilanthes 37
CHENOPODIACEAE 140
Chenopodium 141
Chilopsis 118
Chimaphila 149
Chloris 62
Chlorogalum 54
Chorizanthe 201
Chrysanthemum 99
Chrysopsis 99
Chrysothamnus 99
Cichorium 100
Cicuta 84
Cirsium 100
CISTACEAE 143
Cistus 143
Clarkia 188
Claytonia 207
Clematis 209
Cleome 136
Cleomella 136
Cneoridium 217
Cnicus 100
Coldenia 123
Coleogyne 214
Collinsia 221
Collomia 196
Comarostaphylis 149
COMMELINACEAE 45
COMPOSITAE 88
Condalia 213
Condaliopsis 213
Conium 84
Consolida 210
CONVOLVULACEAE 143
Convolvulus 144
Conyza 100
Corallorhiza 55
Cordylanthus 221
Coreopsis 101
Corethrogyne 101
Coridandrum 84
CORNACEAE 145
Cornus 145
Coronopus 126
Cortaderia 62
Cotula 101
Crassula 145
CRASSULACEAE 145
Cressa 144
Crossosoma 147
CROSSOSOMATACEAE 147
Croton 151
CRUCIFERAE 123
Crunocallis 208
Cryophytum 80
Crypsis 62
Cryptantha 119
Cucurbita 147
CUCURBITACEAE 147
CUPRESSACEAE 39
Cupressus 39
Cuscuta 144

Cycloloma 142
Cynanchum 87
Cynara 102
Cynodon 62
Cynosurus 63
CYPERACEAE 45
Cyperus 47
Cypselea 80
Cyrtomium 38
Cystopteris 38
Cytisus 156

Dactylis 63
Dactyloctenium 63
Dalea 157, *162, 164*
Danthonia 63
Datisca 147
DATISCACEAE 147
Datura 227
Daucus 84
Delphinium 209
Dendromecon 192
DENNSTAEDTIACEAE 38
Dentaria 126
Deschampsia 63
Descurainia 127
Dianthus 137
Dicentra 168
Dichanthelium 63
Dichelostemma 44
Dichondra 145
Dicoria 102
Dicranostegia 222
Digitaria 63
Dimorphotheca 102
Diplachne 63
Diplacus 222
DIPSACACEAE 147
Dipsacus 148
Distichlis 63
Ditaxis 150
Dithyrea 127
Dodecatheon 208
Dolichos 157
Downingia 134
Draba 127
DRACAENACEAE 42
Drosanthemum 80
Dryopteris 38
Dudleya 146
Dyssodia 102

Echeveria 146
Echinocactus 131, *132*
Echinocereus 131
Echinochloa 64
Echinocystis 147
Echinodorus 43
Echinopsilon 141
Echium 121
Eclipta 102
Eichhornia 72
ELATINACEAE 148
Elatine 148
Eleocharis 48
Eleusine 64
Elymus 64
× Elysitanion 64
Emex 202
Emmenanthe 171
Encelia 102
Ephedra 41
EPHEDRACEAE 41
Epicampes 67
Epilobium 189
Epipactis 55
EQUISETACEAE 36
Equisetum 36
Eragrostis 64
Eremalche 182
Eremocarpus 151
Eriastrum 196
ERICACEAE 148
Ericameria 103
Erigeron *100*, 103
Eriodictyon 171, *174*
Eriogonum 202, *205*
Erioneuron 65
Eriophyllum 104
Erodium 169
Erophila 127
Eruca 127
Eryngium 84
Erysimum 127
Erythea 45
Eschscholzia 192
Eucalyptus 184
Eucnide 180
Eucrypta 171
Euonymus 139
Eupatorium 93
Euphorbia *151*, 152
EUPHORBIACEAE 150

Eustoma 169
Eutoca 173
Evax 104

FABACEAE 153
FAGACEAE 166
Fagonia 234
Fallopia 205
Fenzlia 199
Ferocactus 132
Festuca 65, *72*
Ficus 230
Filago 104
Fimbristylis 48
Foeniculum 84
Fouquieria 167
FOUQUIERIACEAE 167
Fragaria 214
Frankenia 167
FRANKENIACEAE 167
Franseria 94
Frasera 169
Fraxinus 186
Fremontia 230
Fremontodendron 229
Fritillaria 54
Fumaria 168
FUMARIACEAE 168
FUNKIACEAE 54

Galinsoga 105
Galium 216
Garrya 168
GARRYACEAE 168
Gasoul 80
Gastridium 65
Gaura 190
Gayophytum 190
GENTIANACEAE 168
Geraea 105
GERANIACEAE 169
Geranium 169
Gilia *195*, 196,
198, 199, 200
Githopsis 134
Glandularia 231
Glinus 80
Glycyrrhiza 157
Gnaphalium 105
Godetia 189
Gonolobus 88

GRAMINEAE 56
Grindelia 106
GROSSULARIACEAE 170
Guizotia 106
Gutierrezia 106
GUTTIFERAE 174
GYMNOGRAMMACEAE 37

Habenaria 55
HALORAGACEAE 170
Haplopappus 103, 106, 109, 111
Harpagonella 121
Hazardia 106
Hedera 87
Hedypnois 106
Helenium 107
Heleochloa 62
Helianthemum 143
Helianthus 107
Heliotropium 121
Hemicarpha 48
Hemizonella 111
Hemizonia *95*, 107
Herissantia 182
Herniaria 138
Hesperocallis 54
Hesperocnide 230
Hesperolinon 179
Heterocodon 134
Heterodraba 128
Heteromeles 214
Heteropogon 65
Heterotheca *99*, 108
Heuchera 219
Hibiscus 182
Hieracium 108
Hilaria 65
Hoffmannseggia *156*, 157
Hofmeisteria 113
Holcus 66, *70*
Holocarpha 108
Holostigma 188
Hookera 44
Hordeum 66
Horkelia 214
Horsfordia 182
Hosackia 158
Huegelia 196
Hulsea 108
Hutchinsia 128

HYACINTHACEAE 54
Hydrilla 49
HYDROCHARITACEAE 49
Hydrocotyle 84
HYDROPHYLLACEAE 171
Hylmenocyclus 80
Hylmenolobus 128
Hymenoclea 108
Hymenopappus 108, *109*
Hymenothrix 109
HYPERICACEAE 174
Hypericum 174
Hypochoeris 109
Hyptis 176

Ipomoea 145
Ipomopsis 198
IRIDACEAE 49
Iris 49
Isocoma 109
ISOETACEAE 35
Isoetes 35
Isomeris 136
Iva 109

Jaumea 109
Jepsonia 219
JUGLANDACEAE 175
Juglans 175
JUNCACEAE 49
JUNCAGINACEAE 51
Juncus 49
Juniperus 40
Jussiaea 190
Justicia 79

Kallstroemia 234
Keckiella 222
Kelloggia 217
Kickxia 223
Kochia 142
Koeleria 66
Krameria 175
KRAMERIACEAE 175
Kyllinga 48

LABIATAE 175
Lactuca 109
Lagophylla 110
Lamarckia 66
LAMIACEAE 175
Lamium 176
Lampranthus 80
Langloisia 198
Larrea 234
Lastarriaea 201
Lasthenia 110
Lathyrus 157
LAURACEAE 179
Lavatera 182
Layia 110
LEGUMINOSAE 153
Lemna 52
LEMNACEAE 51
LENNOACEAE 179
Leonotis 176
Lepechinia 176
Lepidium *126*, 128
Lepidospartum 110
Leptochloa 63
Leptodactylon 198
Lepturus 67
Lessingia 110
Leucanthemum 97
Lewisia 208
Leymus 64
Lilaea 51
LILIACEAE 52, 53, 54
Lilium 54
LIMNANTHACEAE 179
Limnanthes 179
Limnorchis 55
Limonium 195
Limosella 223
LINACEAE 179
Linanthus 198
Linaria 223
Linum 179
Lippia 231
Lithophragma 219
LOASACEAE 180
Lobelia 135
Lobularia 129
Loeflingia 138
Loeseliastrum 198
Logfia 105
Lolium 66
Lomatium 84
Lonicera 136
Lotus 157
Ludwigia 190
Lupinus 160

Luzula 51
Lycium 227
Lycopersicon 227
Lygodesmia 113
Lyrocarpa 129
LYTHRACEAE 181
Lythrum 181

Machaeranthera 111, *117*
Madia 111
Mahonia 117
Malacomeris 112
Malacothamnus 182
Malacothrix 111
Malephora 80
Malosma 81
Malperia 112
Malva 183
MALVACEAE 181
Malvastrum 182, 183
Malvella 183
Mammillaria 132
Marah 147
Marina 162
Marrubium 176
Marsilea 39
MARSILEACEAE 39
Martynia 184
MARTYNIACEAE 183
Matelea 88
Matricaria 112
Matthiola 129
Meconella 193
Medicago 162
MELANTHIACEAE 54
Melia 184
MELIACEAE 184
Melica 67
Melilotus 163
Mendora 186
Mentha 176
Mentzelia 180
Mesembryanthemum *79*, 80
Micrampelis 147
Micropus 112
Microseris 112
Microsteris 200
Mimetanthe 224
Mimulus *222*, 223
Mirabilis 185
Modiola 183
Mohavea 224
Mollugo 80
Monanthochloe 67
Monardella 176
Monerma 67
Monolepis 142
Monoptilon 112
Montia *207*, 208
Montiastrum 208
Muhlenbergia 67
Muilla 44
MYOPORACEAE 184
Myoporum 184
Myosurus 210
Myriophyllum 170
MYRTACEAE 184

NAJADACEAE 55
Najas 55
Nama 171
Narcissus 45
Nasturtium 129
Navarretia *196*, 200
Nelumbo 184
NELUMBONACEAE 184
Nemacaulis 205
Nemacladus 135
Nemophila 172
Nemoseris 114
Nicotiana 227
Nitrophila 142
Nolina 42
Notholaena 37
Nothoscordum 45
NYCTAGINACEAE 184

Obione 141
Oenanthe 85
Oenothera *187*, 190
OLEACEAE 186
Oligomeris 211
Olneya 163
ONAGRACEAE 186
OPHIOGLOSSACEAE 39
Ophioglossum 39
Opuntia 132
ORCHIDACEAE 55
Orcuttia 67
Ornithogalum 45
Ornithostaphylos 149
OROBANCHACEAE 191
Orobanche 191
Orthocarpus 224
Oryzopsis 67
Osmadenia 97
Osmorhiza 85
OXALIDACEAE 191
Oxalis 191
Oxybaphus 186
Oxytheca 205

Paeonia 192
PAEONIACEAE 192
Palafoxia 113
PALMAE 45
Palmerella 135
Panicum *62, 63*, 68
Papaver 193
PAPAVERACEAE 192
Parapholis 68
Parentucellia 225
Parietaria 230
Parkinsonia 163
Paspalum 68
Pastinaca 85
Pectis 113
Pectocarya 121
Pedicularis 225
Pelargonium 170
Pellaea 38
Pennisetum 68
Penstemon *222*, 225
Pentachaeta 99
Perezia 93
Perideridia 85
Perityle 113
Persicaria 205
Petalonyx 181
Petunia 228
Peucedanum 85
Peucephyllum 113
Phaca 155
Phacelia 172
Phalaris 68
Phleum 69
Phlox 200
Phoenix 45
Pholisma 179
Pholistoma 174
Phoradendron 233
Photinia 214
Phragmites 69

Phyla 231
Phyllospadix 73
Physalis 228
Phytolacca 193
PHYTOLACCACEAE 193
Pickeringia 163
Picris 113
Pilostyles 209
Pilularia 39
PINACEAE 40
Pinus 40
Piperia 55
PITTOSPORACEAE 194
Pittosporum 194
Pityrogramma 38
Plagiobothrys 122
PLANTAGINACEAE 194
Plantago 194
Platanthera 55
PLATANACEAE 195
Platanus 195
Platystemon 193
Plectritis 231
Pleurocoronis 113
Pluchea 113, *116*
PLUMBAGINACEAE 195
Poa 69
POACEAE 56
Pogogyne 177
POLEMONIACEAE 195
Polycarpon 138
Polygala 201
POLYGALACEAE 201
POLYGONACEAE 201
Polygonum 205
Polypappus 116
POLYPODIACEAE 36, 38
Polypodium 38
Polypogon *59*, 69
Polystichum 38
PONTEDERIACEAE 72
Populus 217
Porophyllum 113
Portulaca 208
PORTULACACEAE 206
Potamogeton 72
POTAMOGETONACEAE 72
Potentilla *214*, 215
Prenanthella 113
PRIMULACEAE 208
Proboscidea 183
Prosopis 163
Prunella 177
Prunus 215
Psathyrotes 113
Pseudotsuga 41
Psilocarphus 113
Psoralea 163
Psorothamnus 164
Pteridium 38
Pterostegia 206
Puccinellia 69
Purshia 215
Pycnanthemum 177
Pyrola 149

Quercus 166

RAFFLESIACEAE 209
Rafinesquia 114
RANUNCULACEAE 209
Ranunculus 210
Raphanus 129
Rapistrum 129
Reseda 211
RESEDACEAE 211
Rhagadiolus 106
RHAMNACEAE 211
Rhamnus 212
Rhododendron 149
Rhus *81*,82
Rhynchelytrum 69
Ribes 170
Ricinus 152
Robinia 164
Romneya 193
Rorippa 129
Rosa 215
ROSACEAE 213
Roubieva 142
RUBIACEAE 216
Rubus 216
Rumex 206
Ruppia 72
Ruta 217
RUTACEAE 217

Sagina 138
Salazaria 177
SALICACEAE 217
Salicornia 142
Salix 218
Salpichroa 228
Salsola 143
Salvia 177
SALVINIACEAE 39
Sambucus 79
Samolus 209
Sanguisorba 216
Sanicula 85
Saponaria 138, *139*
Sarcodes 149
Sarcostemma 88
Satureja 178
SAURURACEAE 218
Saviniona 182
Saxifraga 219
SAXIFRAGACEAE 219
Scabiosa 148
Scandix 86
Schinus 82
Schismus 70
Schizachyrium 70
Schmaltzia 82
Scirpus 49
Scleranthus 138
Scrophularia 226
SCROPHULARIACEAE 219
Scutellaria 178
Secale 70
Sedum *146*, 147
Selaginella 35
SELAGINELLACEAE 35
Senecio 114
Senna 164
Sesbania 164
Sesuvium 80
Setaria 70
Sherardia 217
Sibara130
Sida 183
Sidalcea 183
Silene 138
Silybum 114
SIMAROUBACEAE 226
Simmondsia 226
SIMMONDSIACEAE 226
Sisymbrium 130
Sisyrinchium49
Sitanion *64*, 70
Smilacina 54
SOLANACEAE 226
Solanum 228

Solidago 114
Soliva 115
Sonchus 115
Sorghum 70
SPARGANIACEAE 73
Sparganium 73
Spartina 70
Spartium 164
Specularia 135
Spergula 139
Spergularia 139
Spermolepis 86
Sphaeralcea 183
Spiranthes 55
Spirodela 52
Sporobolus *67*, 70
Stachys 178
Stanleya 130
Stellaria 139
Stemodia 226
Stenotaphrum 71
Stenotopsis 103
Stephanomeria 115
STERCULIACEAE 229
Stillingia 152
Stipa 71
Streptanthella 130
Streptanthus 130
Stylocline 116
Stylomecon 193
Stylophyllum 146
Styphonia 82
STYRACACEAE 230
Styrax 230
Suaeda 143
Swertia 169
Symphoricarpos 136

TAMARICACEAE 230
Tamarix 230
Taraxacum 116
Tauschia 86
Tessaria 116
Tetracoccus 152
Tetradymia 116
Tetragonia 80
Thalictrum 211
Thamnosma 217
Thelesperma 116
Thelypodium 126
THELYPTERIDACEAE 39
Thelypteris 39
Thermopsis 164
Thlaspi 130
Thymophylla 102
Thysanocarpus 131
Tidestromia *80*, 81
Tillaea 145
Tiquilia 123
Tithymalus 152
Torilis 86
Toxicodendron 82
Toxicoscordion 55
Tradescantia 45
Tragopogon 116
Tribulus 234
Tricholaena 69
Trichoptilium 116
Trichostema 179
Tridens 65
Trifolium 164
Triglochin 51
Triodanis 135
Triodia 65
Triticum 71
Trixis 116
TROPAEOLACEAE 230
Tropaeolum 230
Tropidocarpum 131
Turricula 174
Typha 73
TYPHACEAE 73
Ulex 166
ULMACEAE 230
UMBELLIFERAE 82
Umbellularia 179
Uropappus 112
Urtica 231
URTICACEAE 230

Vaccaria 139
Vaccinium
VALERIANACEAE 231
Velaea 86
Venegasia 117
Veratrum 54
Verbascum 226
Verbena 231
VERBENACEAE 231
Veronica 226
Vicia 166
Viguiera 117
Vinca 86
Viola 232
VIOLACEAE 232
VISCACEAE 233
VITACEAE 233
Vitis 233
Vulpia 71

Washingtonia 45
Wigandia 174
Wislizenia 136
Wolffia 52
Woodwardia 39
Wyethia 117

Xanthium 117
Xylococcus 149
Xylorhiza 117
Xylothermia 163

Yabea 86
Yucca 43

Zannichellia 73
ZANNICHELLIACEAE 73
Zantedeschia 45
Zauschneria 189
Zea 72
Zigadenus 54
Ziziphus 213
Zostera 73
ZOSTERACEAE 73
ZYGOPHYLLACEAE 234